T0321204

Variational Convergence and Stochastic Homogenization of Nonlinear Reaction-Diffusion Problems

Variational Convergence and Stochastic Homogenization of Nonlinear Reaction-Diffusion Problems

Omar Anza Hafsa
Université de Nîmes, France
Jean-Philippe Mandallena
Université de Nîmes, France
Gérard Michaille
Université de Montpellier, France & Université de Nîmes, France

NEW JERSEY • LONDON • SINGAPORE • BEIJING • SHANGHAI • HONG KONG • TAIPEI • CHENNAI • TOKYO

Published by

World Scientific Publishing Co. Pte. Ltd.
5 Toh Tuck Link, Singapore 596224
USA office: 27 Warren Street, Suite 401-402, Hackensack, NJ 07601
UK office: 57 Shelton Street, Covent Garden, London WC2H 9HE

Library of Congress Control Number: 2022026782

British Library Cataloguing-in-Publication Data
A catalogue record for this book is available from the British Library.

VARIATIONAL CONVERGENCE AND STOCHASTIC HOMOGENIZATION OF NONLINEAR REACTION-DIFFUSION PROBLEMS

ISBN 978-981-125-848-0 (hardcover)
ISBN 978-981-125-849-7 (ebook for institutions)
ISBN 978-981-125-850-3 (ebook for individuals)

For any available supplementary material, please visit
https://www.worldscientific.com/worldscibooks/10.1142/12896#t=suppl

Printed in Singapore

Preface

Most of the material in this book is derived from the research work developed from various seminars and work groups during the past four years, in collaboration with the laboratory MIPA at the University of Nîmes. The primary objective is to inform students and researchers about recent mathematical developments in the domain of variational convergence of sequences of reaction-diffusion equations. These theoretical results are mainly illustrated through the modeling of population dynamics, ecosystems or diseases, spreading in heterogeneous spatial environments. In particular, based on the stochastic homogenization framework specifically developed for this area, this book intend to develop an understanding of how interactions between various organisms, and their spatial environment, determine the distribution of their densities. A significant portion of the results of this book also apply in the field of heat conduction, biochemical systems or chemical physics, in order to describe temperature distribution or chemical substance concentration.

O. Anza Hafsa, J.P. Mandallena, G. Michaille

Contents

Chapter 1

Introduction

The chapters in Parts 1 and 2 are linked in pairs as follows: Chapters 2, 3, 4, 5, and 6 of Part 1 correspond with Chapters 7, 8, 9, 10, and 11 of Part 2, respectively.

A substantial number of problems in physics, chemical physics, and biology, are modeled through reaction-diffusion equations, to describe temperature distribution or chemical substance concentration. For problems arising from ecology, sociology, or more generally from population dynamics, they describe the density of some populations or species. In the first examples, heat and mass transfer are expressed by the diffusion term while the reaction term expresses the rate of heat or mass production. In the second examples, the diffusion term corresponds with a motion of individuals, dispersing from an area of high concentration to an area of low concentration, and the reaction term describes their rate reproduction. In this book $u(t, x)$ denotes a concentration, or a density according to the cases, at position x and time t. It is referred to as the state variable. The connection between the time derivative of the state variable, the flux $J(t, x)$ and the reaction term $F(t, u(t, x))$, is classically obtained through the conservation of mass (or the number of individuals). This principle states that the amount of u in any regular bounded domain $\mathcal{O} \subset \mathbb{R}^N$, i.e.

$$\int_{\mathcal{O}} u(t, x) dx,$$

changes over time either through the boundary $\partial\mathcal{O}$ given by

$$\int_{\partial\mathcal{O}} J(t, x) \cdot \mathbf{n}(x) \; d\mathcal{H}_{N-1}$$

or because of sources in $\mathcal{O}$ given by

$$\int_{\mathcal{O}} F(t, u(t, x)) dx.$$

In the surface integral, $\mathbf{n}$ is the unit outward normal to $\partial\mathcal{O}$. Recall that J is a vector field which points into the general direction of movement, and that $|J|$ is proportional to the number of particles which flow in that direction per unit time. The density $F(t, u(t, x))$ accounts for the nonlinear rate of inflow. Assume that the functions u, J and F are regular (these assumptions will be specified in the following chapters). Then, the change of the amount of u on $\mathcal{O}$ is given for all $t \in \mathbb{R}_+$, by

$$\begin{aligned}\frac{d}{dt}\int_{\mathcal{O}} u(t,x)dx &= \int_{\mathcal{O}} \frac{\partial u}{\partial t}(t,x)dx \\ &= -\int_{\partial\mathcal{O}} J(t,x)\cdot \mathbf{n}(x)\, d\mathcal{H}_{N-1} + \int_{\mathcal{O}} F(t,u(t,x))dx \\ &= -\int_{\mathcal{O}} \operatorname{div} J(t,x)dx + \int_{\mathcal{O}} F(t,u(t,x))dx,\end{aligned}$$

where $\mathbf{n}$ is the unit outward normal to $\partial\Omega$, and the surface flux integral has been converted to a volume integral according to the Green formula. Since $\mathcal{O}$ is arbitrary, the conservation law is then expressed by the so called *continuity equation for mass* (or second Fick's law)

$$\frac{\partial u}{\partial t}(t,x) + \operatorname{div} J(t,x) = F(t,u(t,x)). \tag{1.1}$$

In the following this equation is restricted to some bounded physical domain $\Omega \subset \mathbb{R}^N$. On the other hand, from the first Fick's law (Fourier's law in the context of thermal conduction), the flux has locally the direction of the negative spatial gradient of the state variable, and is given by

$$J(t,x) = -D(x)\nabla u(t,x). \tag{1.2}$$

This reflects the fact that the flux goes from regions of high concentration to regions of low concentration. The coefficient D (namely the diffusivity) accounts for the rate of movement. By combining (1.1) with (1.2) we obtain

$$\frac{\partial u}{\partial t}(t,x) - \operatorname{div}(D(x)\nabla u(t,x)) = F(t,u(t,x)), \quad x \in \Omega, \tag{1.3}$$

which is known as *a reaction-diffusion equation.* In some models, the flux has two components: one J_F, satisfies the first Fick's law, the other J_{NF}, takes into account a time memory. This last flux, namely the non Fickian flux, has locally the direction of the negative spatial gradient of the state variable at some past time $\tau > 0$, i.e. is given by

$$J_{NF}(t,x) = -D(x)\nabla u(t-\tau,x).$$

For example, in population dynamics, the non Fickian flux may models maturation period, resource regeneration time, mating processes, or incubation period; it is superimposed on the first flow at each time t. By using the first order approximation for small τ,

$$J_{NF}(t+\tau,x) \sim J_{NF}(t,x) + \tau\frac{\partial J_{NF}}{\partial t}(t,x),$$

and assuming that $J_{NF}(0,\cdot) = 0$, the non Fickian flux can be expressed as

$$J_{NF}(t,x) = -\frac{1}{\tau}\int_0^t \exp\left(\frac{s-t}{\tau}\right) D(x)\nabla u(s,x)ds,$$

and the continuity equation leads to the *integrodifferential reaction-diffusion equation*

$$\frac{\partial u}{\partial t}(t,x) - \operatorname{div}(D(x)\nabla u(t,x)) - \frac{1}{\tau}\int_0^t \exp\left(-\frac{t-s}{\tau}\right)\operatorname{div}(D(x)\nabla u(s,x))ds$$
$$= F(t,u(x)).$$

This case is treated in Chapters 5, 10. For diffusion models derived from random walk in ecological problems, we refer the reader to Okubo (1980); Turchin (2015).

We must take into account the boundary conditions. The state variable may be gained or lost by flowing through the boundary of the domain Ω. This implies that the flux J must be specified, equal to some flux J_0 on the domain boundary $\partial\Omega$. For the reaction-diffusion equation (1.3), this corresponds to the Neumann-type boundary condition $D(x)\nabla u(t,x)\cdot\mathbf{n}(x) = J_0(t,x)\cdot\mathbf{n}(x)$ for $x \in \partial\Omega$. Other types of boundary conditions can be considered in conservation laws. For example, u may be prescribed on $\partial\Omega$. This corresponds to a Dirichlet-type boundary condition. In this case the flux across the boundary is the flux necessary to maintain the Dirichlet boundary condition. More general examples involve mixed Dirichlet-Neumann boundary conditions. When at each time t, the flux across the boundary is proportional to the difference between the outside surrounding density and the density inside Ω, the boundary condition can be written as $D(x)\nabla u(t,x)\cdot\mathbf{n}(x) = c(x)(\phi(x) - u(t,x))$, $c(x) > 0$. This condition, called the Robin boundary condition, can be written as $a(x)u(t,x) + D(x)\nabla u(t,x)\cdot\mathbf{n}(x) = \phi(x)$.

In the standard presentations, the reaction functional F is assumed to be globally Lipschitz-continuous with respect to the state variable u. This hypothesis provides the existence of a solution through a fixed point procedure. Nevertheless, in many situations, this functional is only locally Lipschitz-continuous, furthermore, the expected solutions must be bounded, or at least positive in models involved in population dynamics or theoretical ecology. Chapter 2 originates in works of Anza Hafsa *et al.* (2019a), and introduces the notion of CP-structured reaction functionals (CP for Comparison Principle). These reaction functionals are locally Lipschitz-continuous regarding the state variable, and structured so that they generate sub and upper solutions of corresponding reaction-diffusion Cauchy problems. Consequently, by using a comparison principle, one can show that these problems admit a unique bounded solution, based on the boundedness of the initial function; it is positive in many examples derived from population dynamics, and admits a right derivative at all $t \in]0,T[$ under some additional regularity condition on F. Concerning the coefficient D, the dependence upon the spatial variable suggests that the dynamics may occur in heterogeneous media. It also may depend upon a small or large parameter, as well as the reaction term. From a purely formal perspective, these parameters are indexed by $n \in \mathbb{N}$ which tends to infinity. For instance, the diffusion and the reaction term may depend on a parameter ε_n which is intended to tend towards 0 when n goes to infinity. In this last situation, ε_n accounts for the size of small spatial discontinuities, or the size of a mosaic of small habitat in a patch dynamic approach to a nonhomogeneous ecosystem, such as those described in Chapters 7, 8 and 9. The reaction-diffusion equation then becomes

$$\frac{\partial u_n}{\partial t}(t,x) - \operatorname{div}(D_n(x)\nabla u_n(t,x)) = F_n(t, u_n(t,x)),$$

which, with suitable boundary and initial conditions, gives rise to a Cauchy problem $(\mathcal{P}_n)$. The primary result obtained in Chapter 2 is some stability at the limit for the class of these problems, when the class of energy functionals associated with the diffusion term $\operatorname{div}_x(D_n(x)\nabla u_n(t,x))$, is equipped with the *Mosco-convergence*, i.e. the Γ-convergence associated with both strong and weak topology of $L^2(\Omega)$. The whole book actually deals with more general diffusion terms, which are the subdifferentials of convex integral functionals Φ_n, whose domain contains the boundary conditions. Due to the variational nature of the Γ-convergence, the convergence of $(\mathcal{P}_n)$ to some Cauchy problem $(\mathcal{P})$ is qualified as variational. It is worth noticing that without additional hypotheses under which the variational limit of Φ_n admits a integral representation, we are not ensured that $(\mathcal{P})$ is of gradient flow type. For this reason, convergence theorems, Theorem 2.6, and Theorem 2.7, must rather be considered as relaxation results. Concerning the question of convergence, we have chosen a text structure that moves gradually towards an abstract framework, with the idea to first focus on applications. Therefore, in a last section of Chapter 2, we provide a weak abstract version of these two convergence results: the functionals Φ_n are no longer necessarily integral functionals and the reaction terms are no longer CP-structured reaction functionals.

The behavior of heterogeneous media in physics or mechanics has been thoroughly analyzed, from a mathematical perspective, through the framework of *homogenization*. In the particular context of random media, the variational convergence related to random energies has been fairly well analyzed in Dal Maso and Modica (1986); Messaoudi and Michaille (1994); Papanicolaou (1995); Duerinckx and Gloria (2016) and reference therein; for a similar analysis in the context of discrete random energies, we refer the reader to Iosifescu *et al.* (2001); Neukamm *et al.* (2017) and references therein. For quantitative estimates in stochastic homogenization, the reader may consult Gloria and Otto (2011); Gloria *et al.* (2015); Armstrong *et al.* (2019); Gloria *et al.* (2019); Duerinckx *et al.* (2020); Gloria and Otto (2021). In this book, the diffusion operator is the subdifferential of a random energy. Chapter 7 addresses a stochastic homogenization framework elaborated for the specific case of nonlinear reaction-diffusion problems, as for example, problems modeling biological invasion in the context of food-limited population dynamics or ecosystems. Indeed, empirical observations suggest that growth rates, or various thresholds which appear in the models are mostly influenced by the spatial scale and the statistical structure of the environment. From a mathematical perspective, both diffusion and reaction terms in problems modeling the propagation, present random coefficients and a small parameter ε which accounts for the size of heterogeneities. To identify the effective coefficients (effective growth rate, various effective thresholds etc.), Chapter 7 describes the equivalent homogenized problem when ε goes to zero, by suitably applying the results of Chapter 2 combined with probabilistic arguments. To illustrate the interplay between the environment and the evolution, the last section of Chapter 7 deals with the stochastic homogenization of the food-limited population model, whose reaction functional is that of the Fisher model with Allee effect. The growth rate, the critical density threshold, and the density associated with the random diffusion, correspond with two different values outside or inside the random spatial heterogeneities. In the homogenized problem, the density of the diffusion and the reaction terms are deterministic and homogeneous, and the critical density threshold of the Allee effect is now a function of the growth rate. In all examples treated in Part 2, the spatial environment is assumed to be statistically homogeneous. From a strictly mathematical point of view, this means that the probability dynamical system which models the spatial environment in $\mathbb{R}^N$, i.e. the probability space $(\Sigma, \mathcal{A}, \mathbb{P})$ equipped with a group $(T_z)_{z \in \mathbb{Z}^N}$ of measurable transformations $T_z : \Sigma \to \Sigma$, is such that the push-forward probability measure of $\mathbb{P}$ by T_z is invariant, equal to $\mathbb{P}$, for all z in $\mathbb{Z}^N$.

Chapter 3 extends the convergence process described in Chapter 2 to the more involved case of sequences of time delays reaction-diffusion problems, such as integroparabolic equations of the Volterra type. These problems are of the form

$$(\mathcal{P}_n) \begin{cases} \dfrac{du_n}{dt}(t) + \partial\Phi_n(u_n(t)) = F_n(t, u_n(t), v_n(t)) \text{ for a.e. } t \in (0,T) \\ \\ u_n(t) = \eta_n(t) \text{ for all } t \in (-\infty, 0], \end{cases}$$

where the function v_n is connected to the function $u_n \in L^2(0, T, L^2(\Omega))$ via a family $(\mathbf{m}_t^n)_{t \geq 0}$ of $L^\infty(\Omega)$-valued Borel vector measures $\mathbf{m}_t^n$, according to the formula

$$v_n(t) = \int_{-\infty}^{t} u_n(s) d\mathbf{m}_t^n(s).$$

The *history function* η_n, satisfies $\eta_n(0) \in \text{dom}(\Phi_n)$. These problems include single time delay where $\mathbf{m}_t^n = \delta_{t-\tau_n}$, $\tau_n > 0$, single time delay depending on the space variable where $\mathbf{m}_t^n = \frac{1}{\#(\tau_n(\mathbb{R}^N))} \delta_{t-\tau_n(\cdot)}$ (for the notation see Section 3.1.3), and potentially, a mixture of diffuse and discrete time delays. The reaction functional F_n is such that for fixed $v \in L^2(\Omega)$, $(t, u) \mapsto F_n(t, u, v)$ is a CP-structured reaction functional as defined in Chapter 2. Problems $(\mathcal{P}_n)$ model various situations involving a maturation period, resource regeneration time, mating processes, or the incubation period in vector disease models as illustrated in the four examples coming from ecology and biology models. As in Chapter 2, the stability of the class of problems $(\mathcal{P}_n)$ is first established when the class of functionals Φ_n, and the class of vector measures $\mathbf{m}_t^n$, are equipped with the *Mosco-convergence*, and the weak convergence of measures for each fixed t in $[0, T]$, respectively. When a global weak convergence is assumed for the product of the growth rate with the time delays term of the reaction functional, then the structure of the limit reaction functional is no longer preserved, and a mixing effect occurs in the limit reaction functional.

Chapter 8 delineates the stochastic homogenization of time delay reaction-diffusion problems defined in Chapter 3, within the framework of stochastic homogenization introduced in Chapter 7. The main results is stated in Theorem 8.1 and is first illustrated through the homogenization of a vector disease model with a single random delay depending on the space variable, or with multiple random delays. We show that for the homogenized problem, the growth rate among the uninfected population and the time delays coefficients are mixed. In the second example of a delay logistic equation with immigration, the homogenized problem may be interpreted as a diffuse delay logistic equation which models the evolution of a density population, spreading in an homogeneous environment; its carrying capacity is a function of the growth rate, time dependent, and is larger than the carrying capacity of the heterogeneous domain. From these examples, we illustrate some elementary concepts of percolation theory.

Chapter 4 is devoted to the convergence of sequences of two components nonlinear reaction-diffusion systems

$$(\mathcal{S}_n)\begin{cases}\dfrac{du_n}{dt}(t)+\partial\Phi_{1,n}(u_n(t))=F_{1,n}(t,u_n(t),v_n(t)) \text{ for a.e. } t\in(0,T)\\[2ex] \dfrac{dv_n}{dt}(t)+\partial\Phi_{2,n}(v_n(t))=F_{2,n}(t,u_n(t),v_n(t)) \text{ for a.e. } t\in(0,T)\\[2ex] \underline{\rho}_{1,n}\le u_n(0)\le\overline{\rho}_{1,n},\ \underline{\rho}_{2,n}\le v_n(0)\le\overline{\rho}_{2,n}\\[2ex] u_n(0)\in\overline{\mathrm{dom}(\partial\Phi_{1,n})},\ v_n(0)\in\overline{\mathrm{dom}(\partial\Phi_{2,n})},\end{cases}$$

in which both equations are of the type of reaction-diffusion equations considered in Chapter 2. The reaction functionals are such that for fixed $v\in L^2(\Omega)$, $(t,u)\mapsto F_{n,1}(t,u,v)$, and for fixed $u\in L^2(\Omega)$, $(t,v)\mapsto F_{n,2}(t,u,v)$ are CP-structured reaction functionals as defined in Chapter 2. Existence, uniqueness and boundedness or positivity in $C([0,T],L^2(\Omega))\times C([0,T],L^2(\Omega))$, are established through a suitable fixed point procedure combined with the result pertaining to existence from Chapter 2. Under the *Mosco-convergence* of functionals $\Phi_{i,n}$, and a suitable convergence of $F_{i,n}$ for $i=1,2$, the first primary result of the chapter, states the convergence of $(\mathcal{S}_n)$ toward a reaction-diffusion system $(\mathcal{S})$ of the same type, and can be interpreted as a stability result for the class of systems considered. Notably, this study includes systems (S_n) coupling a reaction-diffusion equation (r.d.e.) and a non-diffusive reaction equation (n.d.r.e.). This chapter is illustrated through various examples involving competition or symbiosis models, prey predator models in ecology, as well as thermo-chemical models, and models describing the evolution of the electrical potential across the axonal membrane as the FitzHugh-Nagumo system.

Chapter 9 discusses the themes of Chapter 7 in the framework of two components reaction-diffusion systems. The primary result is a stochastic homogenization theorem for two components nonlinear reaction-diffusion systems with random coefficients, obtained by applying the general convergence theorem established in Chapter 4, combined with probabilistic arguments invoked in Chapter 7. The last section addresses the stochastic homogenization of a prey-predator model with a saturation effect. The random environment is comprised of spherical heterogeneities with radius of size of order ε, whose centers are independently randomly distributed with a given frequency λ, based on a Poisson point process with intensity λ. The system models the evolution of two species with random density u_ε and v_ε of a prey and a predator respectively, with random birth growth rates. The prey population corresponds with a logistic growth, with some random and time-space-dependent maximum carrying capacity K_{car} (the carrying capacity of the prey when the density of the predator is equal to zero), perturbed by a "predator term" which saturates to a density proportional to $-v_\varepsilon$ when u_ε is large. This model reflects the limited capability of the predator when the prey are abundant. In the homogenized system, the effective birth growth rate of each two species is the mean value of corresponding random growth rate with respect to the product probability measure $\mathcal{L}_2\lfloor(0,1)^2 \otimes \mathbb{P}_\lambda$, where $\mathbb{P}_\lambda$ is the probability measure associated with the Poisson point process. The effective maximum carrying capacity is a deterministic and function of the random birth growth rates and the random carrying capacity K_{car}. This illustrates the interplay between the growth rate of the prey and the maximum carrying capacity of the environment when the size of the spatial heterogeneities is very small.

In the spirit of Chapter 2, Chapter 5 is devoted to the convergence of integro-differential reaction-diffusion problems of the type

$$(\mathcal{P}) \begin{cases} \dfrac{du}{dt}(t,\cdot) + \partial\Phi(u(t,\cdot)) + \displaystyle\int_0^t K(t-s)\partial\Psi(u(s))\, ds \\ \\ = F(t, u(t,\cdot)) \text{ for a.e. } t \in (0,T) \\ \\ u(0,\cdot) = u_0, \; u_0 \in \operatorname{dom}(\partial\Phi), \end{cases}$$

under suitable variational convergences on the classes of convex and lower semicontinuous, functionals Φ and Ψ from a Hilbert space X into $\mathbb{R} \cup \{+\infty\}$. As already said, when dealing with concrete functionals Φ and Ψ, such problems arise when the flux has two contributions: one depends only on the diffusion (the Fickian flux), and the second one (the non Fickian flux) takes into account a time memory effect. The convergence is obtained under the Mosco-convergence for the class of functionals Φ, and a suitable Γ-convergence for the class of functionals Ψ. This chapter is applied to proceed to the stochastic homogenization analysis of problems of the type $(\mathcal{P})$ in Chapter 10.

At first glance, the topic of Chapter 6, which originates among other papers in Tartar (1989, 1990); Mascharenas (1993); Toader (1999); Michaille and Valadier (2002); Anza Hafsa *et al.* (2020c), differs significantly from the previous themes, but is preparatory to Chapter 11 which is primarily concerned with the stochastic homogenization of non-diffusive reaction equations, with emergence of memory effects. Nevertheless, it seems to present a self-interest. Given $T > 0$, a polish subset

$\mathcal{X}$ of real valued functions from $[0,T] \times \mathbb{R}^N \times \mathbb{R}^2$, convex in $\mathbb{R}^2$, and a bounded domain Ω of $\mathbb{R}^N$, the chapter mostly concerns the continuity of a map $\mu \mapsto \Phi_\mu$ from the set of Young measures on $\Omega \times \mathcal{X}$ endowed with the narrow convergence, into a set of suitable functionals defined in $H^1(0,T,L^2(\Omega))$ weak, equipped with the Γ-convergence, and containing the class of convex functionals of the type

$$\int_{(0,T)\times\Omega} f_\varepsilon(x)(t,x,u(t,x),\dot{u}(t,x))\, dt \otimes dx + \frac{1}{2}\|u(T,x)\|^2_{L^2(\Omega)}$$

where $x \mapsto f_\varepsilon$ maps $\mathbb{R}^N$ into $\mathcal{X}$. Involving Young measures on $\Omega \times \mathcal{X}$ is quite natural since it is well known that Young measures $\mu_\varepsilon = dx \otimes \delta_{f_\varepsilon(x)}$ capture oscillations of the sequence of functions $(f_\varepsilon)_{\varepsilon>0}$. Therefore, a consequence of the abstract continuity result, Theorem 6.1, is the characterization of the homogenized functional of $u \mapsto \int_{(0,T)\times\Omega} f(\frac{x}{\varepsilon})(t,x,u(t,x),\dot{u}(t,x))\, dt \otimes dx + \frac{1}{2}\|u(T,x)\|^2_{L^2(\Omega)}$ in terms of the narrow limit of the Young measure $\mu_\varepsilon = \delta_{f(\frac{x}{\varepsilon})}$. One of the issues discussed in Chapter 11, obtained as a special case of the general principle of continuity, Theorem 6.1, and which brings us back to the general framework of this book, is the characterization of the weak limit in $H(0,T,L^2(\Omega))$ of the sequence $(u_\varepsilon(\omega,\cdot,\cdot))_{\varepsilon>0}$ in which $u_\varepsilon(\omega,\cdot,\cdot)$ solves the non-diffusive reaction differential equation, with a rapidly random oscillating reaction function,

$$\begin{cases} -\dfrac{\partial u_\varepsilon}{\partial t}(\omega,t,x) = \dfrac{\partial \psi}{\partial s}\left(\omega, \dfrac{x}{\varepsilon}, t, x, u_\varepsilon(\omega,t)\right), & \text{for a.e. } (t,x) \in (0,T) \times \Omega \\ \\ u_\varepsilon(\omega,0,x) = u_0(x), \quad u_0 \in L^2(\Omega). \end{cases}$$

The potential ψ is a stationary random process, and $\zeta \mapsto \psi(\omega,y,t,x,\zeta)$ is strictly convex. We show that almost surely, the solution $u_\varepsilon(\omega,\cdot,\cdot)$ weakly converges in $H^1(0,T,L^2(\Omega))$ toward the minimizer of a functional which is nonlocal in general. In the case of random checkerboard-like or Poisson point process spatial environments, we can specify the homogenized functional as the inf-convolution of two integral functionals. We provide a complete description of this minimizer, which is a convex combination of the solutions of two non-diffusive integro-differential equations. This last result illustrates, within the scope of stochastic homogenization, the memory effect induced by the homogenization of non-diffusive reaction differential equations.

Readers of this book should have some minimal background in functional analysis, Sobolev spaces of vector valued functions, ordinary and standard differential equations, and probability theory. We have endeavored to provide the most accessible writing possible with the help of the Appendix.

PART 1

Sequences of reaction-diffusion problems: Convergence

Chapter 2

Variational convergence of nonlinear reaction-diffusion equations

Let $\Omega \subset \mathbb{R}^N$ be a bounded regular domain and T a positive real number. This chapter aims to investigate the convergence of the sequence $(u_n)_{n\in\mathbb{N}}$, where $u_n \in L^2(0,T,L^2(\Omega))$ solves the Cauchy problem of the following type

$$(\mathcal{P}_n)\begin{cases} \dfrac{du_n}{dt}(t,\cdot) + \partial\Phi_n(u_n(t,\cdot)) = F_n(t,u_n(t,\cdot)) \text{ for a.e. } t \in (0,T) \\ u_n(0,\cdot) = u_n^0,\ u_n^0 \in \overline{\mathrm{dom}(\partial\Phi_n)}, u_n^0 \text{ suitably bounded according to } F_n. \end{cases}$$

Most of this study is devoted to the identification of the Cauchy evolution problem $(\mathcal{P})$ which governs the limit of the sequence $(u_n)_{n\in\mathbb{N}}$ in $C([0,T],L^2(\Omega))$. In summary $(\mathcal{P}_n)$ converges to $(\mathcal{P})$ when $(u_n)_{n\in\mathbb{N}}$ uniformly converges in $C([0,T],L^2(\Omega))$ to the unique solution of $(\mathcal{P})$. We specify this type of convergence below. The diffusion term $\partial\Phi_n(u_n(t,\cdot))$ of $(\mathcal{P}_n)$ is associated with a sequence $(\Phi_n)_{n\in\mathbb{N}}$ of convex proper and lower semicontinuous functions $\Phi_n : L^2(\Omega) \to \mathbb{R}\cup\{+\infty\}$, whose subdifferential $\partial\Phi_n$ is assumed to be single valued (see Remark 2.5). Apart from Sections 2.1 and 2.6, the mathematical analysis of this chapter is restricted to the case when Φ_n is a standard integral functional of the calculus of variations, with convex integrand $W_n(x,\xi)$. Therefore, the diffusion term is of the gradient flow type, expressed by $-\mathrm{div}_x\partial_\xi W_n(\cdot,\nabla u_n)$. The boundary conditions are implicitly included in the domain of $\partial\Phi_n$. Indeed, it is well known that for almost every t in $(0,T)$, the solution $u_n(t)$ belongs to $\mathrm{dom}(\partial\Phi_n)$, which clearly captures boundary conditions, even if the initial condition belongs to $\overline{\mathrm{dom}(\partial\Phi_n)}$. For these results, refer to (Brezis, 1973, Theorem 3.2), or (Attouch *et al.*, 2014, Section 17.2.3), and to Section 2.1.2.

The reaction functional F_n satisfies a comparison principle condition (CP) (see Definition 2.1) and is structured as follows:

$$\forall v \in L^2(\Omega), \quad F_n(t,v)(x) = r_n(t,x)\cdot g_n(v(t,x)) + q_n(t,x)$$

where $r_n \in L^\infty([0,+\infty)\times\mathbb{R}^N,\mathbb{R}^l)$, $q_n \in L^2([0,+\infty),L^2_{loc}(\mathbb{R}^N))$, and $g_n : \mathbb{R}\to\mathbb{R}^l$ is locally Lipschitz continuous. The map F_n is referred to as CP-structured reaction functional.

Under these conditions, $(\mathcal{P}_n)$ admits a unique solution, bounded according to the initial function u_n^0, positive in models specifically adapted to population dynamics.

Furthermore, when r_n and q_n are smooth enough with respect to the time variable, the solution admits a right time derivative at all $t \in]0,T[$ (see Section 2.2.5). In the quasilinear case, it satisfies an invasion property, i.e. is nondecreasing in time, whenever the initial function is sub solution of the corresponding elliptic problem (see Section 2.3). These reaction functionals encompass a broad variety of applications as illustrated in Section 2.2.2, Examples 2.1, 2.2, 2.3, 2.4. We show that these properties still hold when the reaction functional does not satisfy the structure condition but still satisfies (CP). We illustrate this fact in Examples 2.5 stemming from the dynamics of a phreatic water table.

Regarding the diffusion term, the convergence of $(\mathcal{P}_n)$ is established in Section 2.4 without resorting to the theory of semigroups of contraction and their approximation (for a knowledge of this theory, consult Brezis (1973); Brezis and Pazy (1972) and references therein). Our primary intended application concerns the stochastic homogenization of reaction-diffusion problems where Φ_n is a random convex potential. Therefore, the idea is to apply the sequential continuity of the map $\Phi \mapsto \partial\Phi$ when the class of convex functionals Φ is equipped with the Γ-convergence associated with both strong and weak topology of $L^2(\Omega)$, namely, the *Mosco-convergence*, and the class of subdifferential, with the graph convergence (see (Attouch *et al.*, 2014, Theorem 17.4.4) and references therein). In this manner, we take advantage of standard results involving the Γ-convergence of sequences of random functionals Φ_n. To overcome the difficulty due to the reaction term which is not in general a subdifferential operator, we apply an equivalent result which expresses the bicontinuity of the Fenchel duality transformation with respect to the Mosco-convergence (see Theorem B.2), combined with the Fenchel extremality condition (see the various references in the proofs of Theorem 2.6 and Theorem 2.7). For a similar strategy in homogenization we refer the reader to Visintin (2008); Nandakumaran and Visintin (2015). Because of the variational nature of the Γ-convergence (see Appendix B.1), the convergence of $(\mathcal{P}_n)$ towards $(\mathcal{P})$ is qualified as variational. Variational convergence of sequences of time delays reaction-diffusion equations, and coupled reaction-diffusion systems, are addressed in next chapters in which this chapter is applied as a first basic step. Since we have chosen to move gradually towards an abstract framework, we end the chapter with Section 2.6 where we provide an abstract version of Theorems 2.6, 2.7: the functionals Φ_n are no longer necessarily integral functionals and the reaction terms are no longer CP-structured reaction functionals. Nevertheless, the strong convergence hypothesis on the reaction term drastically limits the applications in homogenization. The proof of this abstract version is less complex than those of Theorems 2.6 and 2.7; therefore, for a first reading, we advise the reader to go directly to Section 2.6 after Section 2.1.

2.1 Existence and uniqueness for reaction-diffusion Cauchy problems in Hilbert spaces

In this section, X denotes a Hilbert space equipped with a scalar product denoted by $\langle \cdot, \cdot \rangle$, and the associated norm $\| \cdot \|_X$. In all along the book we use the same notation $| \cdot |$ to denote the norms of the euclidean spaces $\mathbb{R}^d$, $d \geq 1$, and by $\xi \cdot \xi'$ the standard scalar product of two elements ξ, ξ' in $\mathbb{R}^d$. Let T be a positive real number, we write $(0, T)$ to denote indifferently one of the two intervals $]0, T[$ or $[0, T]$.

In what follows, $\Phi : X \to \mathbb{R} \cup \{+\infty\}$ is a convex proper lower semicontinuous (lsc in short) functional, satisfying $\inf_X \Phi > -\infty$, and *Gâteaux-differentiable* so that its subdifferential is single valued (see Definition B.5). We make this choice in order to simplify the notation since in this section, we could use the subdifferential $\partial\Phi$ of Φ in place of its Gâteaux derivative denoted by $D\Phi$, without additional difficulties. We denote by $\text{dom}(\Phi)$, $\text{dom}(\partial\Phi)$ or $\text{dom}(D\Phi)$ the domain of Φ, $\partial\Phi$ or $D\Phi$ respectively.

Remark 2.1. One can establish that $\overline{\text{dom}(\Phi)} = \overline{\text{dom}(\partial\Phi)}$. The proof is a straightforward consequence of the Brønsted-Rockafellar Lemma (see (Attouch *et al.*, 2014, Lemma 17.4.1)). Another direct proof is to apply the convergence of the resolvent of index λ, $(I + \lambda\partial\Phi)^{-1}u \to u$ as $\lambda \to 0$ for all $u \in \overline{\text{dom}(\Phi)}$, and to note that $(I + \lambda\partial\Phi)^{-1}u$ belongs to $\text{dom}(\partial\Phi)$ (see (Attouch *et al.*, 2014, Remark 17.2.2)).

On the other hand, we consider a Borel measurable map $F : [0, +\infty) \times X \to X$ fulfilling the two following conditions:

(C_1) there exists $L \in L^2_{\text{loc}}(0, +\infty)$ such that

$$\|F(t, u) - F(t, v)\|_X \leq L(t)\|u - v\|_X$$

for all $(u, v) \in X^2$ and all $t > 0$;

(C_2) the map $t \mapsto \|F(t, 0)\|_X$ belongs to $L^2_{\text{loc}}(0, +\infty)$.

Let $T > 0$ and $u_0 \in \overline{\text{dom}(D\Phi)}$. The map F is referred to as *the reaction part*, and $D\Phi$ as *the diffusion part* of the Cauchy problem

$$(\mathcal{P}) \begin{cases} \dfrac{du}{dt}(t) + D\Phi(u(t)) = F(t, u(t)) \text{ for a.e. } t \in (0, T) \\ \\ u(0) = u_0, \ u_0 \in \overline{\text{dom}(D\Phi)}, \end{cases}$$

where $\frac{du}{dt}$ denotes the distributional derivative of u. We say that u is a solution of $(\mathcal{P})$ if u belongs to $L^2(0, T, X)$, is absolutely continuous, and satisfies $(\mathcal{P})$. In all the chapter, the space $C([0, T], X)$ is endowed with the sup-norm.

2.1.1 *Local existence and uniqueness*

The results stated in Theorem 2.1 are somewhat well known. For the sake of completeness we provide a complete proof based on Lemma 2.1 below in combination with a standard fixed point procedure. For a proof of Lemma 2.1 we refer the reader to (Attouch *et al.*, 2014, Theorems 17.2.5, 17.2.6), or to (Brezis, 1973, Theorem 3.7).

Lemma 2.1. *Let $T > 0$ and X be a Hilbert space. Let $\Phi : X \to \mathbb{R} \cup \{+\infty\}$ be a convex proper lsc and minorized functional, $f \in L^2(0,T,X)$, and $u_0 \in \overline{\mathrm{dom}(\partial\Phi)}$. Then there exists a unique solution u in $\mathcal{C}([0,T],X)$ of the Cauchy problem*

$$\begin{cases} \dfrac{du}{dt}(t) + D\Phi(u(t)) = f(t) \text{ for a.e. } t \in (0,T) \\ \\ u(0) = u_0,\ u_0 \in \overline{\mathrm{dom}(\partial\Phi)} \end{cases}$$

which satisfies

(S_1) *$u(t) \in \mathrm{dom}(D\Phi)$ for a.e. $t \in (0,T)$,*

(S_2) *u is almost everywhere differentiable in $(0,T)$ and $u'(t) = \dfrac{du}{dt}(t)$ for a.e. $t \in (0,T)$.*
If furthermore f belongs to $W^{1,1}(0,T,X)$, then

(S_3) *$u(t) \in \mathrm{dom}(D\Phi)$ for all $t \in]0,T]$, and u admits a right derivative $\dfrac{d^+u}{dt}(t)$ at every $t \in]0,T[$, which satisfies*

$$\frac{d^+u}{dt}(t) + D\Phi(u(t)) = f(t).$$

Remark 2.2. When $\partial\Phi$ is not single valued, then assertion (S_3) must be changed as follows:

$u(t) \in \mathrm{dom}(D\Phi)$ for all $t \in]0,T]$, u admits a right derivative $\dfrac{d^+u}{dt}(t)$ at every $t \in (0,T)$, which satisfies

$$\frac{d^+u}{dt}(t) + (\partial\Phi(u(t)) - f(t))^0 = 0,$$

where $(\partial\Phi(u(t)) - f(t))^0$ is the element of minimal norm of the set $\partial\Phi(u(t)) - f(t)$.

Theorem 2.1 (Local existence and uniqueness). *Assume that F satisfies (C_1) and (C_2). Then, there exists $T > 0$ small enough which depends only on L, such that ($\mathcal{P}$) admits a unique solution $u \in C([0,T],X)$ which satisfies (S_1) and (S_2). Assume furthermore that G defined by $G(t) = F(t,u(t))$ belongs to $W^{1,1}(0,T,L^2(\Omega))$, then u satisfies (S_3) with $f = G$.*

Proof. We claim that $G \in L^2(0,T,X)$. Indeed, from (C_1) we infer that

$$\|G(t)\|_X \le \|F(t,0)\|_X + L(t)\|u(t)\|_X,$$

so that for a.e. $t \in (0,T)$

$$\|G(t)\|_X^2 \le 2\|F(t,0)\|_X^2 + 2L^2(t)\|u\|^2_{C([0,T],X)}, \tag{2.1}$$

and the claim follows from the fact that L and $t \mapsto \|F(t,0)\|_X$ belong to $L^2_{\text{loc}}(0,+\infty)$.

For each $u \in C([0,T],X)$, we denote by Λu the solution in $C([0,T],X)$ of the Cauchy problem

$$(\mathcal{P}_u) \begin{cases} \dfrac{d\Lambda u}{dt}(t) + D\Phi(\Lambda u(t)) = F(t,u(t)) \text{ for a.e. } t \in (0,T) \\ \\ \Lambda u(0) = u_0, \ u_0 \in \overline{\text{dom}(\partial\Phi)} \end{cases} \tag{2.2}$$

whose existence is guaranteed by Lemma 2.1.

Step 1. We show that for $T > 0$ small enough, $\Lambda : C([0,T],X) \to C([0,T],X)$ is a contraction. Let $(u,v) \in C([0,T],X) \times C([0,T],X)$. Then, for a.e. $t \in (0,T)$ we have

$$\frac{d\Lambda u}{dt}(t) + D\Phi(\Lambda u(t)) = F(t,u(t)),$$
$$\frac{d\Lambda v}{dt}(t) + D\Phi(\Lambda v(t)) = F(t,v(t)).$$

By subtracting these two equalities and taking the scalar product in X with $\Lambda u(t) - \Lambda v(t)$ we obtain

$$\left\langle \frac{d}{dt}(\Lambda u - \Lambda v)(t), (\Lambda u - \Lambda v)(t) \right\rangle + \langle D\Phi(\Lambda u(t)) - D\Phi(\Lambda v(t)), \Lambda u(t) - \Lambda v(t)\rangle$$
$$= \langle F(t,u(t)) - F(t,v(t)), \Lambda u(t) - \Lambda v(t)\rangle .$$

Then, using the fact that $D\Phi$ is a monotone operator, we infer that for a.e. $t \in (0,T)$

$$\frac{1}{2}\frac{d}{dt}\|(\Lambda u - \Lambda v)(t)\|_X^2 \leq \langle F(t,u(t)) - F(t,v(t)), \Lambda u(t) - \Lambda v(t)\rangle .$$

Thus, for a.e. $t \in (0,T)$,

$$\begin{aligned} \frac{d}{dt}\|(\Lambda u(t) - \Lambda v(t))\|_X^2 &\leq 2\|F(t,u(t)) - F(t,v(t))\|_X \|\Lambda u(t) - \Lambda v(t)\|_X \\ &\leq \|F(t,u(t)) - F(t,v(t))\|_X^2 + \|\Lambda u(t) - \Lambda v(t)\|_X^2 \\ &\leq L^2(t)\|u(t) - v(t)\|_X^2 + \|\Lambda u(t) - \Lambda v(t)\|_X^2 \\ &\leq L^2(t)\|u - v\|_{C([0,T],X)}^2 + \|\Lambda u(t) - \Lambda v(t)\|_X^2. \end{aligned} \tag{2.3}$$

By integration over $(0,s)$ for $s \in [0,T]$, and taking into account that $\Lambda u(0) = \Lambda v(0) = u_0$, we obtain

$$\|\Lambda u(s) - \Lambda v(s)\|_X^2 \leq \|u - v\|_{C([0,T],X)}^2 \int_0^s L^2(t)dt + \int_0^s \|\Lambda u(t) - \Lambda v(t)\|_X^2 dt.$$

By using Grönwall's lemma (see Lemma A.1), we deduce that for all $s \in [0,T]$

$$\|\Lambda u(s) - \Lambda v(s)\|_X^2 \leq \|u - v\|_{C([0,T],X)}^2 \int_0^s L^2(t)dt \exp(T),$$

hence

$$\|\Lambda u-\Lambda v\|_{C([0,T],X)}\leq\|u-v\|_{C([0,T],X)}\left(\int_0^T L^2(t)dt\right)^{1/2}\exp(T/2).$$

Consequently the map Λ is a contraction provided that

$$\left(\int_0^T L^2(t)dt\right)^{1/2}\exp(T/2)<1$$

which is obtained for T small enough since $\lim_{T\to 0}\left(\int_0^T L^2(t)dt\right)^{1/2}\exp(T/2)=0.$

Step 2. According to the fact that $C([0,T],X)$ is a complete normed space, the map Λ admits a fixed point which is a solution of $(\mathcal{P})$ and inherits all the properties of the problem $(\mathcal{P}_u)$, in particular (S_1) and (S_2). Assertion (S_3) is a straightforward consequence of Lemma 2.1 if G belongs to $W^{1,1}(0,T,L^2(\Omega))$.

Step 3. For T small enough obtained in ***Step 1***, we establish the uniqueness of the solution of $(\mathcal{P})$ in $C([0,T],X)$. Let u and v be two solutions, then, from (2.3) (with $\Lambda u=u$ and $\Lambda v=v$), we infer that for a.e. $t\in(0,T)$

$$\frac{d}{dt}\|u(t)-v(t)\|_X^2\leq(L^2(t)+1)\,\|u(t)-v(t)\|_X^2\,.$$

By using Grönwall's Lemma after integrating over $(0,s)$, $s\in[0,T]$, we obtain that for all $s\in[0,T]$,

$$\|u(s)-v(s)\|_X^2\leq\|u(0)-v(0)\|_X^2\exp\left(\int_0^T L^2(t)+1\;dt\right)$$

and the claim follows from the fact that $u(0)=v(0)=0$. □

Assume now that F satisfies the following less restrictive conditions: there exists $T>0$ with

(C_1') there exists $L\in L^2(0,T)$ such that for all $(u,v)\in X^2$ and all $0\leq t\leq T$, $\|F(t,u)-F(t,v)\|_X\leq L(t)\|u-v\|_X$;

(C_2') the map $t\mapsto\|F(t,0)\|_X$ belongs to $L^2(0,T)$.

Then existence and uniqueness stated in Theorem 2.1 with T small enough, can be obtained with T given above. More precisely

Theorem 2.2. *Assume that there exists $T>0$ such that F satisfies* (C_1') *and* (C_2'). *Then, $(\mathcal{P})$ admits a unique solution $u\in C([0,T],X)$ which satisfies* (S_1) *and* (S_2).

Assume furthermore that G defined by $G(t)=F(t,u(t))$ belongs to $W^{1,1}(0,T,L^2(\Omega))$, then u satisfies (S_3).

Proof. With the notation of the proof of ***Step 1***, Theorem 2.1, we claim that the iterated map Λ^n is a strict contraction for n large enough. Indeed, from existence of a unique fixed point u for Λ^n we will deduce that Λu is a fixed point too. Thus, from uniqueness $\Lambda u = u$, so that u is a fixed point for Λ.

Let $(u, v) \in C([0,T], X) \times C([0,T], X)$ satisfying

$$\frac{d\Lambda u}{dt}(t) + D\Phi(\Lambda u(t)) = F(t, u(t)),$$
$$\frac{d\Lambda v}{dt}(t) + D\Phi(\Lambda v(t)) = F(t, v(t))$$

for a.e. $t \in (0,T)$. From the monotonicity of $\partial\Phi$, we infer that for a.e. $\sigma \in (0,T)$

$$\left\langle \frac{d\Lambda v}{dt}(\sigma) - \frac{d\Lambda u}{dt}(\sigma), \Lambda v(\sigma) - \Lambda u(\sigma) \right\rangle \leq \langle F(\sigma, u(\sigma)) - F(\sigma, v(\sigma)), \Lambda v(\sigma) - \Lambda u(\sigma) \rangle,$$

hence

$$\frac{1}{2}\frac{d}{dt}\|\Lambda v(\sigma) - \Lambda u(\sigma)\|_X^2 \leq \langle F(\sigma, u(\sigma)) - F(\sigma, v(\sigma)), \Lambda v(\sigma) - \Lambda u(\sigma) \rangle.$$

Integrating over $(0,t)$ where $0 \leq t \leq T$, we obtain for all $t \in [0,T]$

$$\begin{aligned} \frac{1}{2}\|\Lambda v(t) - \Lambda u(t)\|_X^2 &\leq \int_0^t \langle F(\sigma, u(\sigma)) - F(\sigma, v(\sigma)), \Lambda v(\sigma) - \Lambda u(\sigma) \rangle \, d\sigma \\ &\leq \int_0^t \|F(\sigma, u(\sigma)) - F(\sigma, v(\sigma))\|_X \|\Lambda v(\sigma) - \Lambda u(\sigma)\|_X d\sigma. \end{aligned}$$

Thus, according to the Grönwall type lemma, Lemma A.2 with $p = 2$, it follows that for all $t \in [0,T]$

$$\|\Lambda v(t) - \Lambda u(t)\|_X \leq \int_0^t \|F(\sigma, u(\sigma)) - F(\sigma, v(\sigma))\|_X d\sigma,$$

from which we infer that

$$\|\Lambda v - \Lambda u\|_{C([0,t],X)} \leq \int_0^t L(\sigma)\|u - v\|_{C([0,\sigma],X)} d\sigma$$

for all $t \in [0,T]$. By iterating above inequality, and according to the formula

$$\int_0^t L(\sigma_1) \int_0^{\sigma_1} L(\sigma_2) \dots \int_0^{\sigma_{n-1}} L(\sigma_n) d\sigma_n \dots d\sigma_1 = \frac{(\int_0^t L(\sigma)d\sigma)^n}{n!}$$

obtained by a standard calculus for multiple integrals, we obtain

$$\|\Lambda^n v - \Lambda^n u\|_{C([0,T],X)} \leq \frac{(\int_0^T L(\sigma)d\sigma)^n}{n!}\|u - v\|_{C([0,T],X)}.$$

The claim follows from

$$\lim_{n\to+\infty} \frac{(\int_0^t L(\sigma)d\sigma)^n}{n!} = 0.$$

The rest of the proof is similar to that of Theorem 2.1. □

Remark 2.3. When the second member is a map

$$G : C([0,T],X) \to C([0,T],X)$$

which satisfies

$$\|G(u) - G(v)\|_{C([0,T],X)} \le L(t)\|u - v\|_{C([0,T],X)}$$

for all $(u,v) \in C([0,T],X) \times C([0,T],X)$, where $L \in L^1(0,T)$, then clearly the same conclusion holds. Therefore the same conclusion holds too when the second member is of the form $G(u) + F(t,u(\cdot))$ where F fulfills the conditions (C'_1) and (C'_2).

2.1.2 *Global existence and uniqueness*

Denote by $T^* > 0$ a small enough real number so that $(\mathcal{P})$ admits a unique solution in $C([0,T^*],X)$. Existence of such T^* is asserted in Theorem 2.1. Under the initial condition $u_0 \in \overline{\mathrm{dom}(D\Phi)}$ we are not assured that the derivative $\frac{du}{dt}$ of the solution belongs to $L^2(0,T^*,X)$. Nevertheless we know that $\sqrt{t}\frac{du}{dt} \in L^2(0,T^*,X)$ (see (Attouch *et al.*, 2014, Theorem 17.2.5) or (Brezis, 1973, Theorem 3.6)). Hence, for $0 < \delta < T^*$, $\frac{du}{dt}$ belongs to $L^2(\delta,T^*,X)$. Fix $\delta \in (0,T^*)$, and set

$$E := \{T > \delta : \exists u \in C([0,T],X) \text{ solution of } (\mathcal{P})\}.$$

Since $T^* \in E$, we have $E \neq \emptyset$. We define the maximal time in $\overline{\mathbb{R}}_+$ by $T_{Max} := \sup E$ and denote by u the maximal solution of $(\mathcal{P})$ in $C([0,T_{Max}),X)$. We have the following alternative:

Theorem 2.3 (Global existence or blow-up in finite time). *Assume that F satisfies* (C_1), (C_2), *then we have the blow-up alternative*

(G_1) $T_{Max} = +\infty$ *(existence of a global solution);*

(G_2) $T_{Max} < +\infty$. *In this case* $\lim_{T \to T_{Max}} \|u\|_{C([0,T],X)} = +\infty$ *(blow-up in finite time).*

Moreover, for all $T \in (0,T_{Max})$, the restriction of u to $[0,T]$ satisfies assertions (S_1) *and* (S_2), *and satisfies* (S_3) *when $G : [0,T_{Max}) \to X$ defined by $G(t) = F(t,u(t))$, belongs to $W^{1,1}(0,T,X)$ for all T, $0 < T < T_{Max}$.*

Proof. The last assertion is clear: apply Lemma 2.1 with $f(t) = F(t,u(t))$. Assuming that $T_{Max} < +\infty$, we are reduced to show that $\lim_{T \to T_{Max}} \|u\|_{C([0,T],X)} = +\infty$. We argue by contradiction. Assume that u does not fulfills $\lim_{T \to T_{Max}} \|u\|_{C([0,T],X)} = +\infty$. Then there exist $\mathcal{G} > 0$ and a sequence $(T_n)_{n\in\mathbb{N}}$ in E such that $T_n \to T_{Max}$ and $\|u\|_{C([0,T_n],X)} \le \mathcal{G}$ for all $n \in \mathbb{N}$.

Step 1. We show that $\lim_{t \to T_{Max}} u(t)$ exists in X.

Let $n \in \mathbb{N}$. For a.e. $t \in (0,T_n)$ we have

$$\left\langle \frac{du}{dt}(t), \frac{du}{dt}(t) \right\rangle + \left\langle D\Phi(u(t)), \frac{du}{dt}(t) \right\rangle = \left\langle F(t,u(t)), \frac{du}{dt}(t) \right\rangle.$$

By integration over (δ, T_n) we obtain

$$\int_\delta^{T_n} \left\| \frac{du}{dt}(t) \right\|_X^2 dt + \Phi(u(T_n)) - \Phi(u(\delta))$$
$$\leq \left(\int_0^{T_n} \|F(t, u(t))\|_X^2 dt \right)^{\frac{1}{2}} \left(\int_\delta^{T_n} \left\| \frac{du}{dt}(t) \right\|_X^2 dt \right)^{\frac{1}{2}}. \qquad (2.4)$$

From (2.1) we have

$$\int_0^{T_n} \|F(t, u(t))\|_X^2 dt \leq 2 \int_0^{T_{Max}} \|F(t, 0)\|_X^2 dt + 2\|u\|^2_{C([0,T_n],X)} \int_0^{T_{Max}} L^2(t) dt$$
$$\leq 2 \int_0^{T_{Max}} \|F(t, 0)\|_X^2 dt + 2\mathcal{G}^2 \int_0^{T_{Max}} L^2(t) dt.$$

Therefore, since $\inf_X \Phi > -\infty$, (2.4) yields the estimate

$$\int_\delta^{T_n} \left\| \frac{du}{dt}(t) \right\|_X^2 dt \leq C(\Phi, \delta, T_{Max}, \mathcal{G}) \left(1 + \left(\int_\delta^{T_n} \left\| \frac{du}{dt}(t) \right\|_X^2 dt \right)^{\frac{1}{2}} \right)$$

where the constant $C(\Phi, \delta, T_{Max}, \mathcal{G})$ does not depend on T_n. We infer that

$$\sup_{n \in \mathbb{N}} \int_\delta^{T_n} \left\| \frac{du}{dt}(t) \right\|_X^2 dt < +\infty. \qquad (2.5)$$

From (2.5), we infer that $u : [\delta, T_{Max}) \to X$ is uniformly continuous. Indeed, let $s < t$ in $[\delta, T_{Max})$ and choose n large enough (depending on (s, t)) so that s and t belong to $[\delta, T_n]$. We have

$$\|u(t) - u(s)\|_X \leq \int_s^t \left\| \frac{du}{dt}(\tau) \right\|_X d\tau \leq (t - s)^{\frac{1}{2}} \left(\int_\delta^{T_n} \left\| \frac{du}{dt}(t) \right\|_X^2 dt \right)^{\frac{1}{2}}$$
$$\leq (t - s)^{\frac{1}{2}} \left(\sup_{n \in \mathbb{N}} \int_\delta^{T_n} \left\| \frac{du}{dt}(t) \right\|_X^2 dt \right)^{\frac{1}{2}}.$$

Hence u is more precisely $\frac{1}{2}$-Hölder continuous. According to the continuous extension principle in the complete normed space X, u possesses a unique continuous extension $\overline{u}$ in $[\delta, T_{Max}]$, i.e., $\lim_{t \to T_{Max}} u(t) = \overline{u}(T_{Max})$.

Step 2. (Contradiction) Consider the Cauchy problem

$$(\mathcal{P}') \begin{cases} \dfrac{dv}{dt}(t) + D\Phi(v(t)) = F(t, v(t)) \text{ for a.e. } t \in (0, T) \\ \\ v(0) = \overline{u}(T_{Max}). \end{cases}$$

Note that $\overline{u}(T_{Max}) \in \overline{\mathrm{dom}(D\Phi)}$. Indeed, $u(t) \in \mathrm{dom}(D\Phi)$ for a.e. $t \in (0, T)$ and $\overline{u}(T_{Max}) = \lim_{t \to T_{Max}} u(t)$ (choose $t_n \to T_{Max}$ with t_n outside the negligible set in which $u(t) \notin \mathrm{dom}(D\Phi)$). Then applying Theorem 2.1, there exists $T^{**} > 0$ small enough such that $(\mathcal{P}')$ admits a solution $v \in C([0, T^{**}], X)$. Set

$$\widetilde{u}(t) = \begin{cases} u(t) \text{ if } t \in [0, T_{Max}] \\ v(t - T_{Max}) \text{ if } t \in [T_{Max}, T_{Max} + T^{**}]. \end{cases}$$

Then $\widetilde{u} \in C([0, T_{Max} + T^{**}], X)$ is a solution of $(\mathcal{P})$. This leads to a contradiction with the maximality of T_{Max}. □

Proposition 2.1 below provides a condition on $D\Phi$ which ensures that $(\mathcal{P})$ satisfies (G_1). More precisely

Proposition 2.1. *Assume that* $\langle D\Phi(v), v\rangle \geq 0$ *for all* $v \in \mathrm{dom}(D\Phi)$. *Then* $(\mathcal{P})$ *admits a global solution.*

Proof. From Theorem 2.3, it suffices to prove that there is no blow-up in finite time. Assume that $T_{Max} < +\infty$ and let $T \in (0, T_{Max})$. Taking $u(t)$ as a test function, for a.e. $t \in (0, T)$ we have

$$\left\langle \frac{du}{dt}(t), u(t) \right\rangle + \langle D\Phi(u(t)), u(t)\rangle = \langle F(t, u(t)), u(t)\rangle .$$

Hence, using the fact that $\langle D\Phi(u(t)), u(t)\rangle \geq 0$ (recall that $u(t) \in \mathrm{dom}(D\Phi)$ for a.e. $t \in (0, T)$), we infer that

$$\begin{aligned} \frac{d}{dt}\|u(t)\|_X^2 &\leq 2\langle F(t, u(t)), u(t)\rangle \\ &\leq (\|F(t, 0)\|_X + L(t)\|u(t)\|_X)^2 + \|u(t)\|_X^2 \\ &\leq 2\|F(t, 0)\|_X^2 + (2L^2(t) + 1)\|u(t)\|_X^2. \end{aligned}$$

By integrating over $(0, s)$ for $s \in [0, T]$, we deduce

$$\|u(s)\|_X^2 \leq \|u_0\|_X^2 + 2\int_0^s \|F(t, 0)\|_X^2 dt + \int_0^s (2L^2(t) + 1)\|u(t)\|_X^2 dt$$

(note that from (C_2), $t \mapsto \|F(t, 0)\|_X^2$ belongs to $L^1(0, T)$, and $t \mapsto (2L^2(t) + 1)\|u(t)\|_X^2$ belongs to $L^1(0, T)$ since $\|u(t)\|_X \leq \|u\|_{C([0,T],X)}$ and $L \in L^2(0, T)$). By using Grönwall's lemma and the continuity in $[0, T]$ of

$$s \mapsto \|u_0\|_X^2 + 2\int_0^s \|F(t, 0)\|_X^2 dt + \int_0^s (2L^2(t) + 1)\|u(t)\|_X^2 dt - \|u(s)\|_X^2,$$

we obtain for all $t \in [0, T]$

$$\|u(t)\|_X^2 \leq \left(\|u_0\|_X^2 + 2\int_0^t \|F(s, 0)\|_X^2\, ds\right) \exp\left(\int_0^t (2L^2(s) + 1)\, ds\right).$$

Then, if $T_{Max} < +\infty$, we have

$$\begin{aligned} &\sup_{T < T_{Max}} \|u\|_{C([0,T],X)} \\ &\leq \left(\|u_0\|_X^2 + 2\int_0^{T_{Max}} \|F(s, 0)\|_X^2\, ds\right) \exp\left(\int_0^{T_{Max}} (2L^2(s) + 1)\, ds\right). \end{aligned}$$

This makes $\lim_{T \to +T_{Max}} \|u\|_{C([0,T],X)} = +\infty$ impossible. Thus $T_{Max} = +\infty$. □

From Proposition 2.1, when $\Phi = 0$ the condition (G_1) is automatically satisfied. Therefore we obtain the following global existence for non-diffusive problems (here $\mathrm{dom}(D\Phi) = X$).

Theorem 2.4 (Global existence for non-diffusive Cauchy problems). *Assume that F satisfies* (C_1), (C_2). *Then, there exists a unique global solution $u \in C([0, +\infty), X)$ of the non-diffusive Cauchy problem*

$$(\mathcal{P}) \begin{cases} \dfrac{du}{dt}(t) = F(t, u(t)) \text{ for a.e. } t \in (0, T) \\ \\ u(0) = u_0, \ u_0 \in X. \end{cases}$$

Moreover, for all $T < T_{Max}$ the restriction of u to $[0, T]$ satisfies assertion (S_2), *and furthermore* (S_3) *when $G : [0, T_{Max}) \to X$ defined by $G(t) = F(t, u(t))$ belongs to $W^{1,1}(0, T, X)$.*

2.2 Existence and uniqueness of bounded solution of reaction-diffusion problems associated with convex functionals of the calculus of variations and CP-structured reaction functionals

From now on, Ω is a domain of $\mathbb{R}^N$ of class C^1 and $\mathcal{L}_N$ denotes the Lebesgue measure on $\mathbb{R}^N$. We denote by $\partial\Omega$ its boundary and by Γ a subset of $\partial\Omega$ with $\mathcal{H}_{N-1}(\Gamma) > 0$, where $\mathcal{H}_{N-1}$ is the $N-1$-dimensional Hausdorff measure. To shorten the notation, we sometimes write X to denote the Hilbert space $L^2(\Omega)$ equipped with its standard scalar product and its associated norm, denoted by $\langle \cdot, \cdot \rangle$ and $\| \cdot \|_X$ respectively.

2.2.1 *The class of diffusion terms associated with convex integral functionals of the calculus of variations*

Throughout the rest of the chapter, we focus on the specific case of a standard convex functionals Φ of the calculus of variations, i.e. integral functionals $\Phi : L^2(\Omega) \to \mathbb{R} \cup \{+\infty\}$ of the form[1]

$$\Phi(u) = \begin{cases} \displaystyle\int_\Omega W(x, \nabla u(x))dx + \frac{1}{2}\int_{\partial\Omega} a_0 u^2 d\mathcal{H}_{N-1} - \int_{\partial\Omega} \phi u \, d\mathcal{H}_{N-1} & \text{if } u \in H^1(\Omega) \\ \\ +\infty & \text{otherwise} \end{cases} \tag{2.6}$$

where $\phi \in L^2_{\mathcal{H}_{N-1}}(\partial\Omega)$, and $a_0 \in L^\infty_{\mathcal{H}_{N-1}}(\partial\Omega)$ with

$$\begin{cases} a_0 \geq 0 \quad \mathcal{H}_{N-1}\text{-a.e. in } \partial\Omega \\ \\ \exists \sigma > 0 \quad a_0 \geq \sigma \quad \mathcal{H}_{N-1}\text{-a.e. in } \Gamma. \end{cases}$$

[1]In the integrals on $\partial\Omega$, we still denote by u the trace of u on $\partial\Omega$.

The density $W : \mathbb{R}^N \times \mathbb{R}^N \to \mathbb{R}$ is a Borel measurable function which satisfies condition (D) below:

(D) there exist $\alpha > 0$ and $\beta > 0$ such that for a.e. $x \in \mathbb{R}^N$ and all $\xi \in \mathbb{R}^N$

$$\alpha|\xi|^2 \leq W(x,\xi) \leq \beta(1+|\xi|^2);$$

for a.e. $x \in \mathbb{R}^N$, $W(x,\cdot)$ is a Gâteaux-differentiable convex function (we denote by $D_\xi W(x,\cdot)$ its Gâteaux derivative);

$$D_\xi W(x,0) = 0 \text{ for a.e. } x \in \mathbb{R}^N.$$

Remark 2.4. We introduce this type of functional in order to generate the Robin boundary condition $a_0 u + D_\xi W(\cdot,\nabla u)\cdot n = \phi$ on $\partial\Omega$, as stated in Lemma 2.3 below.

Remark 2.5. Recall that the Gâteaux-differentiability of $W(x,\cdot)$ is equivalent to the fact that the subdifferential $\partial W(x,\cdot)$ of $W(x,\cdot)$ is single valued. This hypothesis yields that the subdifferential $\partial\Phi$ of Φ is single valued too in its domain (see Lemma 2.3 below). We make this hypothesis to simplify the notation and proofs (see Remarks 2.2, 2.7).

Remark 2.6. Condition $D_\xi W(x,0) = 0$ does not come into play until Section 2.2.5. It is introduced so that some particular constant functions with respect to the space variable, are sub or upper-solutions of $(\mathcal{P})$.

By using the subdifferential inequality together with the upper growth condition of (D), it is easy to show that there exist nonnegative constants $L(\beta)$ and $C(\beta)$ such that, for all $(\xi,\xi') \in \mathbb{R}^N \times \mathbb{R}^N$, and for a.e. $x \in \mathbb{R}^N$,

$$\begin{cases} |W(x,\xi) - W(x,\xi')| \leq L(\beta)|\xi - \xi'|(1+|\xi|+|\xi'|), \\ \\ |D_\xi W(x,\xi)| \leq C(\beta)(1+|\xi|). \end{cases} \tag{2.7}$$

From the second estimate, we infer that if $u \in H^1(\Omega)$, then the function $D_\xi W(\cdot,\nabla u)$ belongs to $L^2(\Omega,\mathbb{R}^N)$.

By using standard arguments, it is easily seen that the functional (2.6) is proper convex. We show below that it is minorized and coercive. From the coercivity and the compact embedding $H^1(\Omega) \hookrightarrow L^2(\Omega)$, we can easily deduce that Φ is lower semicontinuous.

Lemma 2.2. *There exist two positive constants*

$$C_1 = C(\alpha,\sigma,C_p,C_{\text{trace}},\mathcal{L}_N(\Omega),\mathcal{H}_{N-1}(\Gamma)),$$
$$C_2 = C_2(\alpha,\sigma,C_p,C_{\text{trace}},\mathcal{L}_N(\Omega),\mathcal{H}_{N-1}(\Gamma))$$

such that

$$\Phi(v) \geq C_1\|v\|^2_{H^1(\Omega)} - C_2\|\phi\|^2_{L^2_{\mathcal{H}_{N-1}}(\partial\Omega)}$$

where C_p denotes the constant of Poincaré, and C_{trace} the constant of continuity of the trace operator from $H^1(\Omega)$ into $L^2(\partial\Omega)$.

Proof. We may assume that $v \in H^1(\Omega)$, thus

$$\begin{aligned}\Phi(v) &\geq \alpha \int_\Omega |\nabla v|^2 dx + \frac{\sigma}{2} \int_\Gamma v^2 d\mathcal{H}_{N-1} - \|\phi\|_{L^2_{\mathcal{H}_{N-1}}(\partial\Omega)} \|v\|_{L^2_{\mathcal{H}_{N-1}}(\partial\Omega)} \\ &\geq \alpha' \left(\int_\Omega |\nabla v|^2 dx + \int_\Gamma v^2 d\mathcal{H}_{N-1} \right) - \|\phi\|_{L^2_{\mathcal{H}_{N-1}}(\partial\Omega)} \|v\|_{L^2_{\mathcal{H}_{N-1}}(\partial\Omega)} \quad (2.8)\end{aligned}$$

where $\alpha' = \min(\alpha, \frac{\sigma}{2})$. On the other hand, according to the generalized Poincaré inequality (see (Attouch *et al.*, 2014, Theorem 5.4.3 and proof of Theorem 6.3.2)),

$$\left\| v - \fint_\Gamma v \, d\mathcal{H}_{N-1} \right\|_{L^2(\Omega)} \leq C_p \|\nabla v\|_{L^2(\Omega, \mathbb{R}^N)}.$$

We infer that

$$\|v\|^2_{L^2(\Omega)} \leq 2 \left\| \fint_\Gamma v \, d\mathcal{H}_{N-1} \right\|^2_{L^2(\Omega)} + 2C_p^2 \|\nabla v\|^2_{L^2(\Omega, \mathbb{R}^N)}.$$

Applying Jensen's inequality, we obtain

$$\|v\|^2_{L^2(\Omega)} \leq 2 \frac{\mathcal{L}_N(\Omega)}{\mathcal{H}_{N-1}(\Gamma)} \int_\Gamma v^2 \, d\mathcal{H}_{N-1} + 2C_p^2 \|\nabla v\|^2_{L^2(\Omega, \mathbb{R}^N)},$$

from which we deduce

$$\|v\|^2_{H^1(\Omega)} \leq \frac{\max\left(2 \frac{\mathcal{L}_N(\Omega)}{\mathcal{H}_{N-1}(\Gamma)}, (2C_p^2 + 1)\right)}{\alpha'} \alpha' \left(\int_\Omega |\nabla v|^2 \, dx + \int_\Gamma v^2 \, d\mathcal{H}_{N-1} \right),$$

hence

$$\alpha' \left(\int_\Omega |\nabla v|^2 \, dx + \int_\Gamma v^2 \, d\mathcal{H}_{N-1} \right) \geq \alpha'' \|v\|^2_{H^1(\Omega)} \quad (2.9)$$

where $\alpha'' = \alpha' \left(\max\left(2 \frac{\mathcal{L}_N(\Omega)}{\mathcal{H}_{N-1}(\Gamma)}, 2C_p^2 + 1\right)\right)^{-1}$. Combining (2.8), (2.9), applying the continuity of the trace operator from $H^1(\Omega)$ into $L^2(\partial\Omega)$, we infer that

$$\begin{aligned}\Phi(v) &\geq \alpha'' \|v\|^2_{H^1(\Omega)} - \|\phi\|_{L^2_{\mathcal{H}_{N-1}}(\partial\Omega)} \|v\|_{L^2_{\mathcal{H}_{N-1}}(\partial\Omega)} \\ &\geq \alpha'' \|v\|^2_{H^1(\Omega)} - \frac{C_{trace}\nu}{2} \|v\|^2_{H^1(\Omega)} - \frac{C_{trace}}{2\nu} \|\phi\|^2_{L^2_{\mathcal{H}_{N-1}}(\partial\Omega)}. \quad (2.10)\end{aligned}$$

Choosing $\nu = \frac{\alpha''}{C_{trace}}$, we obtain

$$\Phi(v) \geq \frac{\alpha''}{2} \|v\|^2_{H^1(\Omega)} - \frac{C_{trace}}{2\nu} \|\phi\|^2_{L^2_{\mathcal{H}_{N-1}}(\partial\Omega)},$$

which proves the thesis. □

We are going to express the subdifferential of the functional Φ (actually its Gâteaux derivative), whose domain contains Robin boundary conditions of Cauchy problems $(\mathcal{P})$. Consider the space $\mathbf{H}(\text{div}) := \{\sigma \in L^2(\Omega, \mathbb{R}^N) : \text{div}\sigma \in L^2(\Omega)\}$. It is well known that when Ω is an open domain of class C^1, with outer unit normal $\mathbf{n}$, the normal trace

$$\gamma_\mathbf{n} : \mathbf{H}(\text{div}) \cap C(\overline{\Omega}) \to H^{-\frac{1}{2}}(\partial\Omega) \cap C(\partial\Omega)$$

defined by $\gamma_{\mathbf{n}}(\sigma) = (\sigma \cdot \mathbf{n})\lfloor_{\partial\Omega}$, has a continuous extension from $\mathbf{H}(\mathrm{div})$ onto $H^{-\frac{1}{2}}(\partial\Omega)$, still denoted by $\gamma_{\mathbf{n}}$. Recall that the Green's formula holds: for every $\varphi \in H^1(\Omega)$ whose trace denoted by $\gamma_0(\varphi)$ belongs to $H^{\frac{1}{2}}(\partial\Omega)$, we have

$$\int_\Omega \mathrm{div}\sigma\varphi dx = -\int_\Omega \sigma \cdot \nabla\varphi \, dx + \langle \gamma_{\mathbf{n}}(\sigma), \gamma_0(\varphi)\rangle_{H^{-\frac{1}{2}}(\partial\Omega), H^{\frac{1}{2}}(\partial\Omega)}.$$

For any $\sigma \in \mathbf{H}(\mathrm{div})$ and any $\varphi \in H^1(\Omega)$, we (improperly) write $\int_{\partial\Omega} \sigma \cdot \mathbf{n}\varphi \, d\mathcal{H}_{N-1}$ instead of the last term $\langle \gamma_{\mathbf{n}}(\sigma), \gamma_0(\varphi)\rangle_{H^{-\frac{1}{2}}(\partial\Omega), H^{\frac{1}{2}}(\partial\Omega)}$, and, as for regular functions, we denote by $\sigma \cdot \mathbf{n}$ and φ the normal trace and the trace of σ and φ respectively.

Lemma 2.3. *The subdifferential of the functional Φ is the operator $A = \partial\Phi (= D\Phi)$ defined by*

$$\begin{cases} \mathrm{dom}(A) \\ = \Big\{ v \in H^1(\Omega) : \mathrm{div} D_\xi W(\cdot, \nabla v) \in L^2(\Omega), \ a_0 v + D_\xi W(\cdot, \nabla v) \cdot \mathbf{n} = \phi \text{ on } \partial\Omega \Big\} \\ \\ A(v) = -\mathrm{div} D_\xi W(\cdot, \nabla v) \text{ for all } v \in \mathrm{dom}(A) \end{cases}$$

where $a_0 v + D_\xi W(\cdot, \nabla v) \cdot \mathbf{n} = \phi$ must be taken in the trace sense for all $v \in H^1(\Omega)$.

Proof. The strategy of the proof consists in establishing that A is a maximal monotone operator included in the subdifferential $\partial\Phi$, which, in turn, is a maximal monotone operator (for this last point refer to (Attouch *et al.*, 2014, Theorem 17.4.1)). We then conclude that $A = \partial\Phi$.

Step 1 (Monotonicity of A). Let $(u, v) \in \mathrm{dom}(A)^2$. Note that $\mathrm{dom}(A) \subset \mathbf{H}(\mathrm{div})$. Then, according to the Green formula, to the convexity of $\xi \mapsto W(x, \xi)$ for fixed x, and to the boundary condition expressed in $\mathrm{dom}(A)$, we infer that

$$\begin{aligned}
&\langle A(u) - A(v), u - v\rangle \\
&= -\int_\Omega (\mathrm{div} D_\xi W(x, \nabla u(x)) - \mathrm{div} D_\xi W(x, \nabla v(x)))(u(x) - v(x))dx \\
&= \int_\Omega (D_\xi W(x, \nabla u(x)) - D_\xi W(x, \nabla v(x))).(\nabla u(x) - \nabla v(x))dx \\
&\quad - \int_{\partial\Omega} (D_\xi W(x, \nabla u(x)) - D_\xi W(x, \nabla v(x))) \cdot \mathbf{n}(x)(u(x) - v(x))d\mathcal{H}_{N-1} \\
&= \int_\Omega (D_\xi W(x, \nabla u(x)) - D_\xi W(x, \nabla v(x))).(\nabla u(x) - \nabla v(x))dx \\
&\quad - \int_{\partial\Omega} (\phi - a_0 u - \phi + a_0 v)(u - v)d\mathcal{H}_{N-1} \\
&= \int_\Omega (D_\xi W(x, \nabla u(x)) - D_\xi W(x, \nabla v(x))).(\nabla u(x) - \nabla v(x))dx \\
&\quad + \int_{\partial\Omega} a_0(u - v)^2 d\mathcal{H}_{N-1} \geq 0,
\end{aligned}$$

which proves the monotonicity of A.

Step 2 ($A \subset \partial\Phi$). Due to the definition of $\partial\Phi$ (see Definition B.5), it is enough to prove that for any $u \in \mathrm{dom}(A)$ and any $v \in \mathrm{dom}(\Phi) = H^1(\Omega)$, the following inequality holds: $\Phi(v) \geq \Phi(u) + \langle A(u), v-u\rangle$. From convexity of $\xi \mapsto W(x,\xi)$, for a.e. $x \in \mathbb{R}^N$ we have

$$W(x,\nabla v(x)) \geq W(x,\nabla u(x)) + D_\xi W(x,\nabla u(x))\cdot(\nabla v(x) - \nabla u(x)) \text{ a.e. } x \in \mathbb{R}^N.$$

By integrating over Ω and adding the surface energy

$$\frac{1}{2}\int_{\partial\Omega} a_0 v^2 d\mathcal{H}_{N-1} - \int_{\partial\Omega} \phi v \, d\mathcal{H}_{N-1},$$

we deduce

$$\begin{aligned}\Phi(v) \geq & \int_\Omega W(x,\nabla u(x))dx \\ & + \int_\Omega D_\xi W(x,\nabla u(x))\cdot(\nabla v(x) - \nabla u(x))dx \\ & + \frac{1}{2}\int_{\partial\Omega} a_0 v^2 d\mathcal{H}_{N-1} - \int_{\partial\Omega} \phi v \, d\mathcal{H}_{N-1}.\end{aligned}$$

By using Green's formula in the second integral, and taking into account that $D_\xi W(\cdot,\nabla u)\cdot\mathbf{n} = \phi - a_0 u$ on $\partial\Omega$ (recall that $u \in \mathrm{dom}(A)$), we infer that

$$\begin{aligned}\Phi(v) \geq & \int_\Omega W(x,\nabla u(x))dx + \int_\Omega (-\mathrm{div} D_\xi W(x,\nabla u(x))(v(x)-u(x)))dx \\ & + \int_{\partial\Omega} D_\xi W(x,\nabla u(x))\cdot\mathbf{n}(x)\,(v(x)-u(x))d\mathcal{H}_{N-1} \\ & + \frac{1}{2}\int_{\partial\Omega} a_0 v^2 d\mathcal{H}_{N-1} - \int_{\partial\Omega} \phi v \, d\mathcal{H}_{N-1} \\ = & \int_\Omega W(x,\nabla u(x))dx + \langle A(u), v-u\rangle \\ & + \int_{\partial\Omega} D_\xi W(x,\nabla u(x))\cdot\mathbf{n}(x)\,(v(x)-u(x))d\mathcal{H}_{N-1} \\ & + \frac{1}{2}\int_{\partial\Omega} a_0 v^2 d\mathcal{H}_{N-1} - \int_{\partial\Omega} \phi v \, d\mathcal{H}_{N-1} \\ = & \int_\Omega W(x,\nabla u(x))dx + \langle A(u), v-u\rangle \\ & + \int_{\partial\Omega} (\phi - a_0 u)(v-u)d\mathcal{H}_{N-1} + \frac{1}{2}\int_{\partial\Omega} a_0 v^2 d\mathcal{H}_{N-1} - \int_{\partial\Omega} \phi v \, d\mathcal{H}_{N-1} \\ = & \int_\Omega W(x,\nabla u(x))dx + \langle A(u), v-u\rangle \\ & + \int_{\partial\Omega} a_0 \left(u^2 + \frac{1}{2}v^2 - uv\right) d\mathcal{H}_{N-1} - \int_{\partial\Omega} \phi u \, d\mathcal{H}_{N-1} \\ \geq & \Phi(u) + \langle A(u), v-u\rangle\end{aligned}$$

where we have used inequality $u^2 + \frac{1}{2}v^2 - uv \geq \frac{1}{2}u^2$ in the second integral.

Step 3 (A is a maximal operator). According to Minty's theorem (see (Attouch *et al.*, 2014, Theorem 17.2.1) and references therein) it remains to prove that $R(I+A)=L^2(\Omega)$ where $R(I+A)$ denotes the range of the operator $(I+A)$. Equivalently, for any f in $L^2(\Omega)$, we have to establish the existence of a solution $u\in H^1(\Omega)$ of the homogeneous mixed Dirichlet-Neumann problem:

$$\begin{cases} u-\mathrm{div}D_\xi W(\cdot,\nabla u)=f \text{ in } \Omega \\ \\ a_0u+D_\xi W(\cdot,\nabla u)\cdot\mathbf{n}=\phi \text{ on } \partial\Omega. \end{cases} \tag{2.11}$$

This is a standard result; we provide a proof by using the direct method of the calculus of variations. Observe that the solutions of (2.11) are the minimizers of

$$\min_{v\in H^1(\Omega)}\left\{\Phi(v)+\frac{1}{2}\int_\Omega v^2dx-\int_\Omega fvdx\right\}. \tag{2.12}$$

Therefore the claim follows from the lover semicontinuity and the coercivity in $H^1(\Omega)$ of the map $\Psi: v\mapsto\Phi(v)+\frac{1}{2}\int_\Omega v^2dx-\int_\Omega fvdx$. The lower semicontinuity is easily seen, the coercivity follows directly from that of Φ obtained in Lemma 2.2. The end of the proof then follows by a standard argument: let $(u_n)_{n\in\mathbb{N}}$ be a $\frac{1}{n}$-minimizing sequence of (2.11); from coercivity, $(u_n)_{n\in\mathbb{N}}$ is bounded in $H^1(\Omega)$; hence there exists a subsequence which weakly converges to some function u in $H^1(\Omega)$; according to the lower semicontinuity, u is a minimizer of Ψ. □

Remark 2.7. When $W(x,\cdot)$ is not Gâteaux-differentiable, the subdifferential of Φ is the multivalued operator given by

$$\begin{cases} \mathrm{dom}(\partial\Phi) \\ =\left\{v\in H^1(\Omega):\mathrm{div}\partial_\xi W(\cdot,\nabla v)\in L^2(\Omega),\ a_0v+\partial_\xi W(\cdot,\nabla v)\cdot\mathbf{n}\ni\phi \text{ on } \partial\Omega\right\} \\ \\ \partial\Phi(v)=-\mathrm{div}\partial_\xi W(\cdot,\nabla v) \text{ for all } v\in\mathrm{dom}(\partial\Phi) \end{cases}$$

where $\mathrm{div}\partial_\xi W(\cdot,\nabla v)$ denotes the set $\{\mathrm{div}\xi^*(\cdot):\xi^*(\cdot)\in\partial_\xi W(\cdot,\nabla v)\}$. The proof is obtained following the same calculation with obvious adaptations.

Assume that $\phi=0$, and define $a_0:\partial\Omega\to[0,+\infty]$ by

$$a_0(x)=\begin{cases}0 & \text{if } x\in\partial\Omega\setminus\Gamma \\ +\infty & \text{if } x\in\Gamma.\end{cases}$$

Then, the integral $\int_{\partial\Omega}a_0u^2d\mathcal{H}_{N-1}$ may be view as a penalization which forces the function u to belong to $H^1_\Gamma(\Omega)=\{u\in H^1(\Omega):u=0 \text{ on } \Gamma\}$. By convention the functional Φ becomes

$$\Phi(u)=\begin{cases}\displaystyle\int_\Omega W(x,\nabla u(x))dx & \text{if } u\in H^1_\Gamma(\Omega) \\ \\ +\infty & \text{otherwise.}\end{cases} \tag{2.13}$$

We refer the reader to Corollary 2.6 for a justification of this convention. The subdifferential of Φ contains now the homogeneous Dirichlet-Neumann boundary conditions as stated in the following lemma, which can be proved by an easy adaptation of the proof of Lemma 2.3:

Lemma 2.4. *The subdifferential of the functional* (2.13) *is the operator* $\partial\Phi(= D\Phi)$ *defined by*

$$\begin{cases} \mathrm{dom}(D\Phi) \\ = \Big\{ v \in H^1_\Gamma(\Omega) : \mathrm{div} D_\xi W(\cdot, \nabla v) \in L^2(\Omega),\ D_\xi W(\cdot, \nabla v) \cdot \mathbf{n} = 0 \ \textit{on}\ \partial\Omega \setminus \Gamma \Big\} \\ D\Phi(v) = -\mathrm{div} D_\xi W(\cdot, \nabla v) \ \textit{for all}\ v \in \mathrm{dom}(D\Phi). \end{cases}$$

Remark 2.8. When $W(x, \cdot)$ is not Gâteaux-differentiable, the subdifferential of the functional (2.13) is the multivalued operator given by

$$\begin{cases} \mathrm{dom}(\partial\Phi) = \big\{ v \in H^1_\Gamma(\Omega) : \mathrm{div}\partial_\xi W(\cdot, \nabla v) \in L^2(\Omega),\ \partial_\xi W(\cdot, \nabla v) \cdot \mathbf{n} \ni 0 \text{ on } \partial\Omega \setminus \Gamma \big\} \\ \partial\Phi(v) = -\mathrm{div}\partial_\xi W(\cdot, \nabla v) \text{ for all } v \in \mathrm{dom}(\partial\Phi). \end{cases}$$

Remark 2.9. In some particular cases it is possible to concretely express $\mathrm{dom}(D\Phi)$ and its adherence. Take for example the functional (2.13) with $W(x, \xi) = A(x)\xi \cdot \xi$ where the measurable matrix valued function $A = (a_{i,j})_{i,j=1\ldots N} : \mathbb{R}^N \to \mathbb{M}_N$ satisfies for some $\alpha > 0$ and $\beta > 0$, the two bounds $\alpha|\xi|^2 \leq \sum_{i,j=1}^N a_{i,j}(x)\xi_i\xi_j \leq \beta|\xi|^2$ for a.e. $x \in \mathbb{R}^N$ and all $\xi \in \mathbb{R}^N$. Assume that $a_{i,j} \in C^1(\mathbb{R}^N)$. According to Lemma 2.4, it is easy to show that

$$\mathrm{dom}(D\Phi) = \Big\{ v \in H^1_\Gamma(\Omega) \cap H^2(\Omega) : A(\cdot)\nabla v \cdot \mathbf{n} = 0 \text{ on } \partial\Omega \setminus \Gamma \Big\}.$$

Hence, since $\mathcal{D}(\Omega)$ is dense in $L^2(\Omega)$, we infer that $\overline{\mathrm{dom}(D\Phi)} = L^2(\Omega)$, i.e. $\mathrm{dom}(D\Phi)$ is dense in $L^2(\Omega)$.

2.2.2 *The class of CP-structured reaction functionals*

The reaction-diffusion problems that model a wide variety of applications and amenable to analytical manipulation in homogenization, involve the use of a special class of functionals that we define below.

Definition 2.1. A map $F : [0, +\infty) \times L^2(\Omega) \to \mathbb{R}^\Omega$ is called a *CP-structured reaction functional*, if there exists a Borel measurable function $f : [0, +\infty) \times \mathbb{R}^N \times \mathbb{R} \to \mathbb{R}$ such that for all $t \in [0, +\infty)$ and all $v \in L^2(\Omega)$, $F(t, v)(x) = f(t, x, v(x))$, and fulfilling the following structure conditions:

- $f(t, x, \zeta) = r(t, x) \cdot g(\zeta) + q(t, x)$;
- $g : \mathbb{R} \to \mathbb{R}^l$ is a locally Lipschitz continuous function;
- for all $T > 0$, r belongs to $L^\infty([0, T] \times \mathbb{R}^N, \mathbb{R}^l)$;
- for all $T > 0$, q belongs to $L^2(0, T, L^2_{\mathrm{loc}}(\mathbb{R}^N))$.

Furthermore f must satisfy the following condition:

(CP) there exist a pair $(\underline{f}, \overline{f})$ of functions $\underline{f}, \overline{f} : [0, +\infty) \times \mathbb{R} \to \mathbb{R}$ with $\underline{f} \leq 0 \leq \overline{f}$ and a pair $(\underline{\rho}, \overline{\rho})$ in $\mathbb{R}^2$ with $\underline{\rho} \leq \overline{\rho}$, such that each of the two following ordinary differential equations

$$\underline{\text{ODE}} \begin{cases} \underline{y}'(t) = \underline{f}(t, \underline{y}(t)) \text{ for a.e. } t \in [0, +\infty) \\ \underline{y}(0) = \underline{\rho} \end{cases}$$

$$\overline{\text{ODE}} \begin{cases} \overline{y}'(t) = \overline{f}(t, \overline{y}(t)) \text{ for a.e. } t \in [0, +\infty) \\ \overline{y}(0) = \overline{\rho} \end{cases}$$

admits at least a solution satisfying for a.e. $(t, x) \in (0, +\infty) \times \mathbb{R}$

$$\underline{f}(t, \underline{y}(t)) \leq f(t, x, \underline{y}(t)) \quad \text{and} \quad f(t, x, \overline{y}(t)) \leq \overline{f}(t, \overline{y}(t)).$$

The map F is referred to as a *CP-structured reaction functional associated with* (r, g, q), and f as a *CP-structured reaction function associated with* (r, g, q). If furthermore, for all $T > 0$, r belongs to $W^{1,1}(0, T, L^2_{\text{loc}}(\mathbb{R}^N, \mathbb{R}^l))$ and q belongs to $W^{1,1}(0, T, L^2_{\text{loc}}(\mathbb{R}^N))$, then the map F is referred to as a *regular CP-structured reaction functional and* f *as a* regular CP-structured reaction function.

Remark 2.10. 1) The reason why we introduce condition (CP) may be summarized as follows: in the proof of Corollary 2.1, we show that condition (CP) generates sub and upper solutions of $(\mathcal{P})$; then, according to a comparison principle established in Propositions 2.2 and 2.3, we can prove that reaction-diffusion problems associated with a CP-structured reaction functional admits a unique solution which satisfies $\underline{y}(T) \leq u \leq \overline{y}(T)$ whenever the initial condition satisfies $\underline{\rho} \leq u_0 \leq \overline{\rho}$. This justifies the terminology (CP) (for "Comparison Principle").

2) Since $\underline{y}$ and $\overline{y}$ are nonincreasing and nondecreasing respectively, for any $T > 0$ we have

$$\underline{y}(T) \leq \underline{y}(0) = \underline{\rho} \leq \overline{\rho} = \overline{y}(0) \leq \overline{y}(T).$$

3) We introduce the spaces $L^2_{\text{loc}}(\mathbb{R}^N)$ and $L^2_{\text{loc}}(\mathbb{R}^N, \mathbb{R}^l)$ because of the specific form of sequences of CP-structured reaction functionals F_ε in the framework of homogenization where the scaling $x \mapsto \frac{x}{\varepsilon}$ appears. Nevertheless, in this chapter, we can replace these two spaces by $L^2(\Omega)$ and $L^2(\Omega, \mathbb{R}^l)$ respectively. Note that when X is a reflexive space, $W^{1,1}(0, T, X)$ is exactly the space of absolutely continuous functions from $[0, T]$ into X (see (Brezis, 1973, Corollary A4)).

2.2.3 *Examples of CP-structured reaction functionals*

We give four classes of CP-structured reaction functionals, and the way of finding $\underline{\rho}, \overline{\rho}, \underline{y}, \overline{y}$ in condition (CP).

Examples 2.1. Let us examine a first class of examples of CP-structured reaction functionals for which condition (CP) is readily checked. Assume that for a.e. $(t, x) \in (0, +\infty) \times \mathbb{R}$, $f(t, x, 0) \geq 0$ and that there exists $\rho > 0$ such that $f(t, x, \rho) \leq 0$. Then (CP) is satisfied. Indeed, take $\underline{f} = \overline{f} = 0$ and $\underline{\rho} = 0$, $\overline{\rho} = \rho$. Then $\underline{y} = 0$ and $\overline{y} = \rho$ are solution of $\underline{\text{ODE}}$ and $\overline{\text{ODE}}$ respectively, and

$$\begin{aligned} \underline{f}(t, \underline{y}(t)) &= 0 \leq f(t, x, 0) = f(t, x, \underline{y}(t)) \\ f(t, x, \rho) &= f(t, x, \overline{y}(t)) \leq 0 = \overline{f}(t, \overline{y}(t)). \end{aligned}$$

For various discussions and references about examples c), d), e) and f) below, we refer the reader to Pao (1992).

a) Example derived from food-limited population models.

The Fisher logistic growth model. The reaction function is given by

$$f(t, x, \zeta) = r(t, x)\zeta\left(1 - \frac{\zeta}{K_{car}}\right)$$

where $r \in L^\infty([0, T] \times \mathbb{R}^N, \mathbb{R})$ for all $T > 0$, $r \geq 0$, and $K_{car} > 0$. The function g defined by $g(\zeta) = \zeta\left(1 - \frac{\zeta}{K_{car}}\right)$ is locally Lipschitz continuous. Moreover, $f(t, x, 0) = 0$ and $f(t, x, \rho) \leq 0$[2] for all $\rho \geq K_{car}$. Therefore the functional F is a CP-structured reaction functional associated with $(r, g, 0)$, and $l = 1$.

The interpretation of the model (with this reaction function) is the following:

- $u(t, x)$ is the population density of some specie at time t located at x,
- $r(t, x)$ is the growth rate of the population at time t, located at x,
- K_{car} is the carrying capacity, i.e., the capacity of the environment to sustain the population,
- $\dfrac{1}{u}\dfrac{du}{dt}$ is the per-capita growth rate.

The same conclusion holds for the following extension of the previous logistic function proposed by ***Turner-Bradley-Kirk***

$$f(t, x, \zeta) = r(t, x)\zeta^{1+\beta(1-\gamma)}\left(1 - \left(\frac{\zeta}{K_{car}}\right)^\beta\right)^\gamma$$

where $\beta > 0$, $\gamma > 0$ and $\gamma < 1 + \frac{1}{\beta}$ (this last condition ensures that the maximal growth is obtained for $\zeta > 0$). For the analysis of this function and various logistic growth models, we refer the reader to Tsoularis (2001).

The logistic growth model with immigration (or stocking). The reaction function is given by

$$f(t, x, \zeta) = r(t, x)\zeta\left(1 - \frac{\zeta}{K_{car}}\right) + q(t, x)$$

[2] From now on, inequalities relating to L^p-functions, $p \in [1, +\infty]$, must be taken in the sense of a.e.

where we assume that

$$M := \operatorname*{ess\,sup}_{(t,x)\in\mathbb{R}_+\times\mathbb{R}^N} \frac{q}{r}(t,x) < +\infty.$$

The interpretation is that of the logistic growth model, perturbed by $q \in L^2(0,T,L^2_{\text{loc}}(\mathbb{R}^N))$ for all $T > 0$, $q \geq 0$. This additional term accounts for the immigration rate. We have $f(t,x,0) \geq 0$. We see that for a.e. $(t,x) \in (0,+\infty) \times \mathbb{R}$, $f(t,x,\rho) \leq 0$ for

$$\rho \geq K_{car} \frac{1+\sqrt{1+\frac{4M}{K_{car}}}}{2}.$$

The functional F is then a CP-structured reaction functional associated with (r,g,q), and $l = 1$. We will consider the logistic growth model with emigration (or harvesting) in Example 2.4 because it does not fall into this category.

The Fisher logistic growth model with Allee effect. The reaction function is given by

$$f(t,x,\zeta) = r(t,x)\zeta\left(1-\frac{\zeta}{K_{car}}\right)\left(\frac{\zeta-a(t,x)}{K_{car}}\right)$$

where r and a belong to $L^\infty([0,T]\times\mathbb{R}^N,\mathbb{R})$ for all $T > 0$, $0 < a \leq K_{car}$ and $r \geq 0$. We have $f(t,x,0) = 0$ and $f(t,x,\rho) \leq 0$ for all $\rho \geq K_{car}$ and a.e. $(t,x) \in [0,T]\times\mathbb{R}^N$. The function f may be written

$$f(t,x,\zeta) = r(t,x)\frac{\zeta^2}{K_{car}}\left(1-\frac{\zeta}{K_{car}}\right) - r(t,x)a(t,x)\frac{\zeta}{K_{car}}\left(1-\frac{\zeta}{K_{car}}\right).$$

Therefore, the functional F is a CP-structured reaction functional associated with $((r_i)_{i=1,2},(g_i)_{i=1,2},0)$ where $g_1(\zeta) = \frac{\zeta^2}{K_{car}}\left(1-\frac{\zeta}{K_{car}}\right)$, $g_2(\zeta) = \frac{\zeta}{K_{car}}\left(1-\frac{\zeta}{K_{car}}\right)$ and $r_1 = r$, $r_2 = -ra$.

The interpretation of the model is that of Fisher model with the additional critical density a below which the per-capita growth rate turns negative. We can also consider the logistic growth model with Allee effect and immigration by setting

$$f(t,x,\zeta) = r(t,x)\zeta\left(1-\frac{\zeta}{K_{car}}\right)\left(\frac{\zeta-a(t,x)}{K_{car}}\right) + q(t,x)$$

with the stocking rate $q \in L^2(0,T,L^2_{\text{loc}}(\mathbb{R}^N))$ for all $T > 0$, $q(t,x) \geq 0$. We have $f(t,x,0) \geq 0$ and, as previously, $f(t,x,\rho) \leq 0$ for ρ large enough depending on $\operatorname{ess\,sup}_{(t,x)\in\mathbb{R}_+\times\mathbb{R}^N} \frac{q}{r}(t,x)$.

b) Example derived from haematopoiesis (Wazewska-Czyziewska & Lasota model). The reaction function is given by

$$f(t,x,\zeta) = -\mu(t,x)\zeta + P(t,x)\exp(-\theta\zeta)$$

where μ and P belong to $L^\infty([0,T]\times\mathbb{R}^N,\mathbb{R})$ for all $T > 0$, $\mu > 0$, $P > 0$, $\theta > 0$. We assume that

$$M := \operatorname*{ess\,sup}_{(t,x)\in\mathbb{R}_+\times\mathbb{R}^N} \frac{P(t,x)}{\mu(t,x)} < +\infty.$$

We have $f(t,x,0) = P(t,x) > 0$ and $f(t,x,\rho) \leq 0$ for $\rho \geq M$. Therefore, by setting $g_1(\zeta) = \zeta$, $g_2(\zeta) = \exp(-\theta\zeta)$, $r_1 = -\mu$ and $r_2 = P$, we can conclude that the functional F is a CP-structured reaction functional associated with $((r_i)_{i=1,2}, (g_i)_{i=1,2}, 0)$.

The interpretation of the model with this reaction function is the following:

- $u(t,x)$ is the number of red-blood cell at time t located at x,
- $\mu(t,x)$ is the probability of death of red-blood cells, P and θ are two coefficients related to the production of red-blood cells per unit time.

For a generalization of the reaction function f in the context of delay ordinary differential equations, we refer the reader to Ruan (2006), and to Chapter 3 for time delays reaction-diffusion equations.

c) Example derived from nuclear reactor dynamics and heat conduction. The reaction function is given by

$$f(t,x,\zeta) = r(t,x)\zeta(a - b\zeta) + q(x)$$

where a and b are two positive constants, $r \in L^\infty([0,T] \times \mathbb{R}^N, \mathbb{R})$ for all $T > 0$, $r \geq 0$, and $q \in L^2(0,T,L^2_{\mathrm{loc}}(\mathbb{R}^N))$ for all $T > 0$, $q \geq 0$. We assume that

$$M := \operatorname*{ess\,sup}_{(t,x)\in\mathbb{R}_+\times\mathbb{R}^N} \frac{q(t,x)}{r(t,x)} < +\infty.$$

We have $f(t,x,0) = q(t,x) \geq 0$ and $f(t,x,\rho) \leq 0$ for

$$\rho \geq \frac{a + \sqrt{a^2 + 4bM}}{2b} \text{ for a.e. } (t,x) \in [0,T] \times \mathbb{R}^N.$$

Therefore, the functional F is a CP-structured reaction functional associated with (r,g,q) with $g(\zeta) = \zeta(a - b\zeta)$.

The interpretation of the model with this reaction function is the following:

- $u(t,x)$ is the one velocity neutron flux at time t located at x, i.e. the total path length covered by all neutrons in one cubic centimeter during one second, of the beam of neutrons traveling in a single direction. Mathematically, $u(t,x) = m(t,x)v(t,x)$ where $m(t,x)$ is the neutron density $(neutrons/cm^3)$ and $v(t,x)$ the neutron velocity (cm/sec).
- $f_0(\zeta) := a - b\zeta$ is called the multiplication factor. It represents the feedback effect in the reactor.
- $r(t,x)$ is the total cross fission.
- $q(t,x)$ is an additional source.

Note that the reaction function f is similar to that of the logistic growth model with immigration above, but with another interpretation. In a second model, where the multiplication factor is of the form u^p, $p \geq 1$, the reaction function is given by $f(t,x,\zeta) = r(t,x)u^p$. We have $f(t,x,0) = q(t,x) \geq 0$ but there is no $\rho > 0$ such that $f(t,x,\rho) \leq 0$ for a.e. $(t,x) \in [0,T] \times \mathbb{R}^N$. Therefore this function is not within the framework of this first class of examples. It will be treated in the class of Example 2.3 below.

d) Example derived from heat transfer: the Stefan-Boltzmann fourth-power law in heat transfer. The reaction function is given by

$$f(t,x,\zeta) = r(t,x)(a^4 - \zeta^4)$$

with $r \in L^\infty([0,T] \times \mathbb{R}^N, \mathbb{R})$ for all $T > 0$, $r \geq 0$, and $a > 0$. We have $f(t,x,0) = r(t,x)a^4 \geq 0$ and $f(t,x,\rho) \leq 0$ for all $\rho \geq a$ and for a.e. $(t,x) \in [0,T] \times \mathbb{R}^N$. The functional F is then a CP-structured reaction functional associated with $(r,g,0)$ where $g(\zeta) = a^4 - \zeta^4$.

The interpretation of the model with this reaction function is the following:

- $u(t,x)$ is temperature radiated by a black body at time t located at x,
- a is the temperature of surroundings,
- r is related to the radiating area and the emissivity of the radiator.

e) Example derived from chemical reactor and combustion models. The reaction function is given by

$$f(t,x,\zeta) = -r(t,x)\zeta^p$$

with $p \geq 1$ and $r \in L^\infty([0,T] \times \mathbb{R}^N, \mathbb{R})$ for all $T > 0$, $r \geq 0$, or its generalization

$$f(t,x,\zeta) = -(r_1(t,x)\zeta^{p_1} + r_2(t,x)\zeta^{p_2})$$

with $p_i \geq 1$, and $r_i \in L^\infty([0,T] \times \mathbb{R}^N, \mathbb{R})$ for all $T > 0$, $r_i \geq 0$, $i = 1,2$. We have $f(t,x,0) = 0$, and $f(t,x,\rho) \leq 0$ for all $\rho > 0$ and a.e. $(t,x) \in [0,T] \times \mathbb{R}^N$. Consequently, F is a CP-structured reaction functional associated with $(-(r_i)_{i=1,2}, (g_i)_{i=1,2}, 0)$ where $g_i(\zeta) = \zeta^{p_i}$, $i = 1,2$.

The interpretation is the following for $i = 1$:

- $u(t,x)$ is the mass concentration of the combustible material at time t located at x in a nonisothermal reaction,
- r is given according to Arrhenius kinetics by $r(t,x) = \exp\left(\gamma - \frac{\gamma}{v(t,x)}\right)$ where $v(t,x)$ is the temperature, γ the Arrhenius number, and p is the order of the reaction.

f) Example derived from enzyme kinetics models in biochemical system. The reaction function is given by

$$f(t,x,\zeta) = -r(t,x)\frac{\zeta}{1+a\zeta}, \quad \text{or} \quad f(t,x,\zeta) = -r(t,x)\frac{\zeta}{1+a\zeta+b\zeta^2},$$

with $a > 0$, $b > 0$, and $r \in L^\infty([0,T] \times \mathbb{R}^N, \mathbb{R})$ for all $T > 0$, $r \geq 0$. We have $f(t,x,0) = 0$ and $f(t,x,\rho) \leq 0$ for any $\rho > 0$. Therefore, F is a CP-structured reaction functional associated with $(-r, g, 0)$ where $g(\zeta) = \frac{\zeta}{1+a\zeta}$ or $\frac{\zeta}{1+a\zeta+b\zeta^2}$, extended by 0 for $\zeta < 0$.

The interpretation is the following:

- $u(t,x)$ is the substrate concentration,
- r depends on the total amount of enzyme and various rates of the reaction,
- a, b depend on various rates of the reaction.

Examples 2.2. We examine now a second class of examples. We assume, as in the previous examples, that for a.e. $(t,x) \in (0,+\infty) \times \mathbb{R}$, $f(t,x,0) \geq 0$, but the second condition is no longer satisfied. Nevertheless we assume that there exists a constant $M \geq 0$ such that $f \leq M$. Then (CP) is satisfied. Indeed, take $\underline{f} = 0$, $\underline{\rho} = 0$ as for the previous class of examples, and $\overline{f} = M$ and $\overline{\rho}$ any positive ρ. Then $\underline{y} = 0$ and $\overline{y} = Mt + \rho$ are solution of $\underline{\text{O}\small{\text{DE}}}$ and $\overline{\text{O}\small{\text{DE}}}$ respectively, with

$$\underline{f}(t, \underline{y}(t)) = 0 \leq f(t,x,0) = f(t,x,\underline{y}(t))$$
$$f(t,x,Mt+\rho) = f(t,x,\overline{y}(t)) \leq M = \overline{f}(t,\overline{y}(t)).$$

Example derived from thermal explosions in the theory of combustion. The reaction function is given by

$$f(t,x,\zeta) = \begin{cases} r(t,x)\exp\left(\gamma\left(1 - \dfrac{1}{\zeta}\right)\right) & \text{if } \zeta > 0 \\ 0 & \text{if } \zeta \leq 0, \end{cases}$$

where γ is a positive constant. We assume that $r \in L^\infty([0,T] \times \mathbb{R}^N, \mathbb{R})$ for all $T > 0$, $r \geq 0$, and that

$$\overline{r} := \operatorname*{ess\,sup}_{(t,x) \in \mathbb{R}_+ \times \mathbb{R}^N} r(t,x) < +\infty.$$

We have $f(t,x,0) = 0$ and $f \leq \exp(\gamma)\overline{r}$. Consequently, F is a CP-structured reaction functional associated with $(r, g, 0)$ where $g = f/r$.

The interpretation of the model with this reaction function is the following:

- $u(t,x)$ is temperature at time t located at x in thermal explosion,
- γ and r are physical coefficients (see Pao (1992) and references therein).

Examples 2.3. We deal with a third class of examples where we still assume that for a.e. $(t,x) \in (0,+\infty) \times \mathbb{R}$, $f(t,x,0) \geq 0$, but f does not satisfies the second condition fulfilled by the two previous class of examples, but satisfies $f(t,x,\zeta) \leq a\zeta^p$ for some $a > 0$ and $p \geq 1$. Then (CP) is satisfied. Indeed, take $\underline{f} = 0$, $\underline{\rho} = 0$ as for the previous class of examples, $\overline{f}(t,\zeta) = a\zeta^p$ and $\overline{\rho}$ any positive ρ. Then $\underline{y} = 0$ is solution of $\underline{\text{O}\small{\text{DE}}}$ and $\overline{y}$ defined by

$$\overline{y}(t) = \begin{cases} \rho\exp(at) & \text{when } p = 1 \\ ((1-\rho)at + \rho^{1-p})^{\frac{1}{1-p}} & \text{when } p > 1 \end{cases}$$

is solution of $\overline{\text{O}_{\text{DE}}}$ (when $p > 1$, $\overline{\text{O}_{\text{DE}}}$ is the Bernouilli o.d.e $y' = ay^p$) with

$$\begin{aligned} \underline{f}(t, \underline{y}(t)) = 0 \leq f(t, x, 0) = f(t, x, \underline{y}(t)) \\ f(t, x, \overline{y}(t)) \leq a\overline{y}(t)^p = \overline{f}(t, \overline{y}(t)). \end{aligned}$$

Example derived from nuclear reactor dynamics and heat conduction or from chemical reactor. The reaction function is given by

$$f(t, x, \zeta) = r(t, x)\zeta^p$$

where $p \geq 1$. We assume that $r \in L^\infty([0, T] \times \mathbb{R}^N, \mathbb{R})$ for all $T > 0$, $r \geq 0$, and that

$$\overline{r} := \operatorname*{ess\,sup}_{(t,x) \in \mathbb{R}_+ \times \mathbb{R}^N} r(t, x) < +\infty.$$

We have $f(t, x, 0) = 0$ and $f \leq \overline{r}\zeta^p$. The functional F is then a CP-structured reaction functional associated with $(r, g, 0)$, where $g(\zeta) = \zeta^p$.

The interpretation of the model with this reaction function is the following:

- $u(t, x)$ represents the one velocity neutron flux at time t located at x in case there is a positive temperature feedback. A second interpretation occurs in the scope of chemical reactor, where $u(t, x)$, this time, is the concentration of a chemical labile specie (see Pao (1992) and references therein).
- g is the multiplication factor. It represents the feedback effect in the reactor (see example c) of Examples 2.1 as a first model).

For the specific case $p = 2$, we refer to Pohožaev (1960).

Examples 2.4. We finally complete our examples by examining a last class for which the first condition $f(t, x, 0) \geq 0$ is not satisfied. We assume that there exists $\rho > 0$ such that for all $(t, x) \in [0, +\infty) \times \mathbb{R}^N$, $f(t, x, \rho) \leq 0$ and $f(t, x, \zeta) \geq \zeta(1 - \zeta) - a$ for some $a > \frac{1}{4}$. Since $\zeta(1 - \zeta) - a < 0$, it is not assured that $f(t, x, 0) \geq 0$. Nevertheless we claim that condition (CP) is satisfied. Indeed take $\overline{\rho} = \rho$, $\overline{f} = 0$ as in Examples 2.1. On the other hand, take $\underline{f}(t, \zeta) = \zeta(1 - \zeta) - a$ and $\underline{\rho}$ any negative number. Then $\underline{y}$ is the solution to the ordinary differential equation

$$\underline{\text{O}_{\text{DE}}} \begin{cases} \underline{y}' = \underline{y}(1 - \underline{y}) - a \\ \underline{y}(0) = \underline{\rho}. \end{cases}$$

We let the reader to check that the solution is

$$\underline{y}(t) = \frac{\underline{\rho} - \frac{1 - 2\underline{\rho} - \lambda^2}{2\lambda} \tan(\frac{\lambda t}{2})}{1 - \frac{1 - 2\underline{\rho}}{\lambda} \tan(\frac{\lambda t}{2})},$$

where $\lambda = \sqrt{4a - 1}$.

Example derived from food-limited population models with emigration or (harvesting). The reaction function is given by

$$f(t, x, \zeta) = r(t, x)\zeta \left(1 - \frac{\zeta}{K_{car}}\right) - q(t, x).$$

The interpretation is that of the logistic growth model, perturbed by in the emigration rate q, where $q \in L^2(0,T,L^2_{\rm loc}(\mathbb{R}^N))$ for all $T > 0$, $q \geq 0$. We further assume that

$$\underline{r} := \operatorname*{ess\,inf}_{(t,x)\in\mathbb{R}_+\times\mathbb{R}^N} r(t,x) > 0,$$

$$\overline{q} := \operatorname*{ess\,sup}_{(t,x)\in\mathbb{R}_+\times\mathbb{R}^N} q(t,x) < +\infty.$$

The change of variable $\frac{\zeta}{K_{car}} = s$, and the change of function $\widetilde{f}(t,x,s) = \frac{1}{\underline{r}\ K_{car}} f(t,x,K_{car}s)$ leads to $\widetilde{f}(t,x,s) \geq s(1-s) - \overline{q}/\underline{r}K_{car}$. We are in the general situation described above provided that we assume $a := \overline{q} > \frac{\underline{r}\ K_{car}}{4}$. For further examples and discussions on logistic growth models with migration in the context of ordinary differential equations, we refer the reader to (Banks, 1994, Section 2.7).

2.2.4 *The comparison principle*

The comparison result stated in the two propositions below will be used for proving existence of bounded solutions of reaction-diffusion problems associated with CP-structured reaction functionals (see Remark 2.10). For similar notion and applications of sub and upper solution related to elliptic boundary valued problems we refer the reader to Berestycki and Lions (1980a,b) and for parabolic problems, to Pao (1992).

We are given two functionals $F_1, F_2 : [0,+\infty) \times L^2(\Omega) \to L^2(\Omega)$ defined by

$$F_1(t,u)(x) = f_1(t,x,u(x)), \quad F_2(t,u)(x) = f_2(t,x,u(x))$$

where $f_1, f_2 : [0,+\infty) \times \mathbb{R}^N \times \mathbb{R} \to \mathbb{R}$ are two measurable functions. The function f_2 is assumed to be Lipschitz continuous, uniformly with respect to (t,x), i.e. there exists $L > 0$ such that for every $(t,x) \in (0,+\infty) \times \mathbb{R}^N$ and every $(\zeta,\zeta') \in \mathbb{R}^2$,

$$|f_2(t,x,\zeta) - f_2(t,x,\zeta')| \leq L|\zeta - \zeta'|.$$

Let $W : \mathbb{R}^N \times \mathbb{R}^N \to \mathbb{R}$ be a Borel measurable function. We assume that for a.e. $x \in \mathbb{R}^N$, $W(x,\cdot)$ is a Gâteaux-differentiable convex function and we set $V := \{v \in H^1(\Omega) : \mathrm{div} D_\xi W(\cdot,\nabla v) \in L^2(\Omega)\}$.

Proposition 2.2. *Let $T > 0$, u_0, v_0 in $L^2(\Omega)$, ϕ_1, ϕ_2 in $L^2(0,T,L^2_{\mathcal{H}_{N-1}}(\partial\Omega))$, and $u \in C([0,T],L^2(\Omega))$, $v \in C([0,T],L^2(\Omega))$ be a sub-solution and a upper-solution of the following reaction-diffusion problems with respect to the data (u_0,ϕ_1,F_1) and (v_0,ϕ_2,F_2) respectively:*

$$\underline{\mathcal{P}(u_0,\phi_1,F_1)}\begin{cases} u(t) \in V,\ \dfrac{du}{dt}(t) \in L^2(\Omega) \text{ for a.e. } t \in (0,T), \\[2ex] \dfrac{du}{dt}(t) - \mathrm{div} D_\xi W(\cdot,\nabla u(t)) \leq F_1(t,u(t)) \text{ for a.e. } t \in (0,T), \\[2ex] u(0) = u_0 \in L^2(\Omega), \\[2ex] a_0 u(t) + D_\xi W(\cdot,\nabla u(t)) \cdot \mathbf{n} = \phi_1(t) \text{ on } \partial\Omega \text{ for a.e. } t \in (0,T), \end{cases}$$

$$\overline{\mathcal{P}(v_0,\phi_2,F_2)}\begin{cases} v(t)\in V,\ \dfrac{dv}{dt}(t)\in L^2(\Omega) \text{ for a.e. } t\in(0,T),\\[2mm] \dfrac{dv}{dt}(t)-\mathrm{div}D_\xi W(\cdot,\nabla v(t))\geq F_2(t,v(t)) \text{ for a.e. } t\in(0,T),\\[2mm] v(0)=v_0\in L^2(\Omega),\\[2mm] a_0v(t)+D_\xi W(\cdot,\nabla v(t))\cdot\mathbf{n}=\phi_2(t) \text{ on } \partial\Omega \text{ for a.e. } t\in(0,T). \end{cases}$$

Then the following comparison principle holds:

$$\left.\begin{array}{l} u_0\leq v_0 \text{ in } L^2(\Omega),\\ \phi_1(t)\leq\phi_2(t) \text{ on } \partial\Omega, \text{ for a.e. } t\in(0,T),\\ F_1\leq F_2 \end{array}\right\}\Longrightarrow u(t)\leq v(t) \text{ for all } t\in[0,T].$$

Proof. Set $w=v-u$. We are going to prove that $w(t)^-=0$ for a.e. $t\in(0,T)$. Indeed, for a.e. $t\in(0,T)$ we have

$$\frac{dw}{dt}(t)-[\mathrm{div}D_\xi W(\cdot,\nabla v(t))-\mathrm{div}D_\xi W(\cdot,\nabla u(t))]\geq F_2(t,v(t))-F_1(t,u(t)). \tag{2.14}$$

Fix t satisfying (2.14). Take $w(t)^-$ as a test function. By integrating over Ω, and using Green's formula we obtain

$$\begin{aligned} \int_\Omega \frac{dw}{dt}(t)w(t)^-dx &+\int_\Omega (D_\xi W(x,\nabla v(t))-D_\xi W(x,\nabla u(t)))\cdot\nabla w(t)^-dx\\ &-\int_{\partial\Omega}(D_\xi W(x,\nabla v(t))-D_\xi W(x,\nabla u(t)))\cdot\mathbf{n}\ w(t)^-d\mathcal{H}_{N-1}\\ &\geq\int_\Omega (f_2(t,x,v(t))-f_1(t,x,u(t)))w(t)^-dx. \end{aligned}$$

Noticing that $D_\xi W(x,\nabla u(t))\cdot\mathbf{n}=\phi_1(t)-a_0u(t)$, and $D_\xi W(x,\nabla v(t))\cdot\mathbf{n}=\phi_2(t)-a_0v(t)$ on $\partial\Omega$, we infer that

$$\begin{aligned} \int_\Omega \frac{dw}{dt}(t)w(t)^-dx &+\int_\Omega (D_\xi W(x,\nabla v(t))-D_\xi W(x,\nabla u(t)))\cdot\nabla w(t)^-dx\\ &+\int_{\partial\Omega}(\phi_1(t)-\phi_2(t))w(t)^-d\mathcal{H}_{N-1}+\int_{\partial\Omega}a_0(v-u)w^-d\mathcal{H}_{N-1}\\ &\geq\int_\Omega (f_2(t,x,v(t))-f_1(t,x,u(t)))w(t)^-dx, \end{aligned}$$

from which we deduce

$$\begin{aligned}
&-\int_\Omega \frac{dw^-}{dt}(t)w(t)^- dx\\
&-\int_{[w(t)\le 0]} (D_\xi W(x,\nabla v(t)) - D_\xi W(x,\nabla u(t)))\cdot(\nabla v(t)-\nabla u(t))dx\\
&-\int_{\partial\Omega}(\phi_2(t)-\phi_1(t))w(t)^- d\mathcal{H}_{N-1} - \int_{[w(t)\le 0]\cap\partial\Omega} a_0(v-u)^2 d\mathcal{H}_{N-1}\\
&\ge \int_\Omega (f_2(t,x,v(t)) - f_1(t,x,u(t)))w(t)^- dx,
\end{aligned}$$

where we have used the relations $\left(\frac{dw}{dt}(t)\right)^+ = \frac{dw^+}{dt}(t)$, $\left(\frac{dw}{dt}(t)\right)^- = \frac{dw^-}{dt}(t)$, $\frac{dw}{dt}(t) = \frac{dw^+}{dt}(t) - \frac{dw^-}{dt}(t)$ and $\frac{dw^+}{dt}(t)w(t)^- = 0$ in the distributional sense. Noticing that the three last integrands of the first member are nonnegative, and $f_1 \le f_2$, we obtain

$$\frac{1}{2}\frac{d}{dt}\int_\Omega |w(t)^-|^2 dx \le \int_\Omega (f_2(t,x,u(t)) - f_2(t,x,v(t)))w(t)^- dx. \tag{2.15}$$

From (2.15) and the Lipschitz continuity of the function f_2, we deduce that

$$\frac{1}{2}\frac{d}{dt}\int_\Omega |w(t)^-|^2 dx \le L\int_\Omega |w(t)|w(t)^- dx = L\int_\Omega |w(t)^-|^2 dx.$$

Integrating this inequality over $(0,s)$ for $s\in[0,T]$, we obtain

$$\int_\Omega |w(s)^-|^2\,dx - \int_\Omega |w(0)^-|^2 dx \le 2L\int_0^s \left(\int_\Omega |w(t)^-|^2\,dx\right)dt.$$

Note that since $w^- \in C([0,T],L^2(\Omega))$, $t\mapsto \int_\Omega |w(t)^-|^2\,dx$ is continuous. Then, according to Grönwall's lemma we finally obtain that for all $s\in[0,T]$

$$\int_\Omega |w(s)^-|^2\,dx \le \int_\Omega |w(0)^-|^2 dx \exp(2Ls),$$

from which we deduce, since $w(0)^- = (v_0-u_0)^- = 0$, that $w^-(s)=0$ in $L^2(\Omega)$ for all $s\in[0,T]$, i.e., $u(s)\le v(s)$ for all $s\in[0,T]$. □

Let us consider the case

$$a_0(x) = \begin{cases} 0 & \text{if } x\in\partial\Omega\setminus\Gamma\\ +\infty & \text{if } x\in\Gamma,\end{cases}$$

with $\phi_i = 0$, $i=1,2$, and set

$$\widetilde{V} = \{v\in H^1(\Omega) : \mathrm{div} D_\xi W(\cdot,\nabla v)\in L^2(\Omega),\ D_\xi W(\cdot,\nabla v)\cdot\mathbf{n} = 0 \text{ on } \partial\Omega\setminus\Gamma\}.$$

By an easy adaptation of the previous proof, we obtain the following comparison principle.

Proposition 2.3. *Let $T > 0$, $u \in C([0,T], L^2(\Omega))$ and $v \in C([0,T], L^2(\Omega))$ be a sub solution and a upper solution of the following reaction-diffusion problems with respect to the data (u_0, F_1) and (v_0, F_2) respectively:*

$$\underline{\mathcal{P}(u_0,F_1)}\begin{cases} u(t) \in \widetilde{V}, \ \dfrac{du}{dt}(t) \in L^2(\Omega) \text{ for a.e. } t \in (0,T), \\ \dfrac{du}{dt}(t) - \mathrm{div} D_\xi W(\cdot, \nabla u(t)) \leq F_1(t, u(t)) \text{ for a.e. } t \in (0,T), \\ u(0) = u_0 \in L^2(\Omega), \end{cases}$$

$$\overline{\mathcal{P}(v_0,F_2)}\begin{cases} v(t) \in \widetilde{V}, \ \dfrac{dv}{dt}(t) \in L^2(\Omega) \text{ for a.e. } t \in (0,T), \\ \dfrac{dv}{dt}(t) - \mathrm{div} D_\xi W(\cdot, \nabla v(t)) \geq F_2(t, v(t)) \text{ for a.e. } t \in (0,T), \\ v(0) = v_0 \in L^2(\Omega). \end{cases}$$

Then the following comparison principle holds:

$$\left.\begin{array}{l} u_0 \leq v_0 \text{ in } L^2(\Omega), \\ u(t) \leq v(t) \text{ on } \Gamma, \text{ for a.e. } t \in (0,T), \\ F_1 \leq F_2 \end{array}\right\} \Longrightarrow u(t) \leq v(t) \text{ for all } t \in [0,T].$$

In the case of non-diffusive problems, the same proof leads to the following comparison principle.

Proposition 2.4. *Let $T > 0$, $u \in C([0,T], L^2(\Omega))$ and $v \in C([0,T], L^2(\Omega))$ be a sub-solution and a upper-solution of the following reaction problems with respect to the data (u_0, F_1) and (v_0, F_2) respectively:*

$$\underline{\mathcal{P}(u_0,F_1)}\begin{cases} u(t) \in L^2(\Omega), \ \dfrac{du}{dt}(t) \in L^2(\Omega) \text{ for a.e. } t \in (0,T), \\ \dfrac{du}{dt}(t) \leq F_1(t, u(t)) \text{ for a.e. } t \in (0,T), \\ u(0) = u_0 \in L^2(\Omega), \end{cases}$$

$$\overline{\mathcal{P}(v_0,F_2)}\begin{cases} v(t) \in L^2(\Omega), \ \dfrac{dv}{dt}(t) \in L^2(\Omega) \text{ for a.e. } t \in (0,T), \\ \dfrac{dv}{dt}(t) \geq F_2(t, v(t)) \text{ for a.e. } t \in (0,T), \\ v(0) = v_0 \in L^2(\Omega). \end{cases}$$

Then the following comparison principle holds:

$$\left.\begin{array}{l} u_0 \leq v_0 \text{ in } L^2(\Omega), \\ F_1 \leq F_2 \end{array}\right\} \Longrightarrow u(t) \leq v(t) \text{ for all } t \in [0,T].$$

2.2.5 *Existence and uniqueness of bounded solutions*

Combining Theorem 2.1 and Theorem 2.3 with the comparison principle, we are now able to prove the existence of a bounded global solution of the Cauchy problem with CP-structured reaction functionals.

Corollary 2.1. *Let F be a CP-structured reaction functional, with $\underline{\rho}$, $\overline{\rho}$ and $\underline{y}$, $\overline{y}$ given by* (CP)*, and let Φ be a standard functional of the calculus of variations* (2.6)*. Assume that $a_0\underline{\rho} \leq \phi \leq a_0\overline{\rho}$. Then for any $T > 0$, the Cauchy problem*

$$(\mathcal{P}) \begin{cases} \dfrac{du}{dt}(t) + D\Phi(u(t)) = F(t, u(t)) \text{ for a.e. } t \in (0,T) \\ \\ u(0) = u_0, \quad \underline{\rho} \leq u_0 \leq \overline{\rho}, \quad u_0 \in \overline{\mathrm{dom}(D\Phi)} \end{cases}$$

admits a unique solution $u \in C([0,T], L^2(\Omega))$ satisfying assertions (S_1), (S_2) *and the following bounds for all t in $[0,T]$: $\underline{y}(T) \leq \underline{y}(t) \leq u(t) \leq \overline{y}(t) \leq \overline{y}(T)$. If furthermore F is a regular CP-structured reaction functional, then u satisfies* (S_3).

Proof. ***Step 1.*** We prove existence of a solution u of $(\mathcal{P})$ for $T = T^* > 0$ small enough, which satisfies (S_1), (S_2) and the bounds $\underline{y}(T^*) \leq \underline{y}(t) \leq u(t) \leq \overline{y}(t) \leq \overline{y}(T^*)$.

By definition of CP-structured reaction functionals, $F : [0, +\infty) \times L^2(\Omega) \to \mathbb{R}^\Omega$ is defined for all $t \in [0, +\infty)$, all $v \in L^2(\Omega)$, and for a.e. $x \in \Omega$ by $F(t,v)(x) = f(t,x,v(x))$, where for all $\zeta \in \mathbb{R}$

$$f(t,x,\zeta) = r(t,x) \cdot g(\zeta) + q(t,x),$$

and $g : \mathbb{R} \to \mathbb{R}^l$ is locally Lipschitz continuous. Fix arbitrary $T' > 0$. The restriction of g to the interval $[\underline{y}(T'), \overline{y}(T')]$ is Lipschitz continuous with some Lipschitz constant L_g.[3] Consequently $\zeta \mapsto f(t,x,\zeta)$ is Lipschitz continuous with respect to ζ, uniformly with respect to (t,x) in $[\underline{y}(T'), \overline{y}(T')]$, with

$$|f(t,x,\zeta) - f(t,x,\zeta')| \leq L|\zeta - \zeta'|,$$

where $L = L_g \|r\|_{L^\infty([0,T]\times\mathbb{R}^N, \mathbb{R}^l)}$. According to the Mac Shane extension lemma (see (Dacorogna, 1989, Lemma 3.2)), g can be extended into a Lipschitz continuous function $\widetilde{g}$ in $\mathbb{R}$. Hence the extension $\widetilde{f}$ of f defined by $\widetilde{f}(t,x,\zeta) = r(t,x) \cdot \widetilde{g}(\zeta) + q(t,x)$ is Lipschitz continuous with respect to ζ in $\mathbb{R}$, uniformly with respect to (t,x), with the same Lipschitz constant L. Consequently, the functional $\widetilde{F} : [0,+\infty) \times L^2(\Omega) \to L^2(\Omega)$ defined by $\widetilde{F}(t,v)(x) = \widetilde{f}(t,x,v(x))$ fulfills the two conditions (C_1) and (C_2) with $L(t) = L$.

[3]To simplify the notation, we do not indicate the dependence on T'.

Therefore, according to Theorem 2.1, for $T^* > 0$ small enough, that we can choose such that $T^* \le T'$, the problem

$$(\widetilde{\mathcal{P}}) \begin{cases} \dfrac{d\widetilde{u}}{dt}(t) + D\Phi(\widetilde{u}(t)) = \widetilde{F}(t, \widetilde{u}(t)) \text{ for a.e. } t \in (0, T^*) \\ \\ \widetilde{u}(0) = u_0 \end{cases}$$

admits a unique solution in $C([0, T^*], X)$ which satisfies (S_1) and (S_2). From Lemma 2.3, $\widetilde{u}$ satisfies the boundary condition

$$a_0 \widetilde{u}(t) + D_\xi W(x, \nabla \widetilde{u}(t)) \cdot \mathbf{n} = \phi \text{ on } \partial\Omega$$

for a.e. $t \in (0, T^*)$.

By applying the comparison principle of Proposition 2.2, we are going to establish that for all $t \in [0, T^*]$, $\widetilde{u}(t) \in \lfloor \underline{y}(t), \overline{y}(t) \rfloor \subset \lfloor \underline{y}(T^*), \overline{y}(T^*) \rfloor$. From condition (CP), the function $\underline{y}$, which does not depend on x, is a sub solution of the reaction-diffusion problem $\overline{\widetilde{\mathcal{P}}(\underline{\rho}, a_0 \underline{y}(t), \widetilde{F})}$ in the sense of Proposition 2.2. Indeed since $\underline{y}(t) \in [\underline{y}(T'), \overline{y}(T')]$, $\overline{D_\xi W(x, 0) = 0}$, and $\nabla \underline{y} = 0$, we have for all $t \in [0, T^*]$

$$\begin{cases} \widetilde{F}(t, \underline{y}(t)) = F(t, \underline{y}(t)) = f(t, \cdot, \underline{y}(t)) \ge \underline{f}(t, \underline{y}(t)) = \underline{y}'(t) \\ \qquad\qquad\qquad\qquad\qquad\qquad = \dfrac{d\underline{y}}{dt}(t) - \operatorname{div} D_\xi W(\cdot, \nabla \underline{y}(t)), \\ \\ \text{initial condition } \underline{y}(0) = \underline{\rho}, \\ \\ \text{boundary condition } a_0 \underline{y}(t) + D_\xi W(x, \nabla \underline{y}(t)) \cdot \mathbf{n} = a_0 \underline{y}(t) \text{ on } \partial\Omega. \end{cases}$$

On the other hand $\widetilde{u}$ is a solution of $(\widetilde{\mathcal{P}})$, thus a upper solution of $\overline{\widetilde{\mathcal{P}}(u_0, \phi, \widetilde{F})}$. Since, from hypothesis,

$$\begin{aligned} &\underline{\rho} \le u_0, \\ &a_0 \underline{y}(t) \le a_0 \underline{y}(0) = a_0 \underline{\rho} \le \phi, \end{aligned}$$

according to the comparison principle, Proposition 2.2, we infer that $\underline{y}(t) \le \widetilde{u}(t)$ for a.e. $t \in (0, T^*)$. Actually, inequality $\underline{y}(t) \le \widetilde{u}(t)$ holds for all $t \in [0, T^*]$ (invoke the continuity of $t \mapsto \|(\widetilde{u}(t) - \underline{y}(t))^- \|_X$). Reasoning similarly with $\overline{y}$ which is a upper solution of $\overline{\widetilde{\mathcal{P}}(\overline{\rho}, a_0 \overline{y}(t), \widetilde{F})}$, we obtain that $\overline{y}(t) \ge \widetilde{u}(t)$ for all $t \in [0, T^*]$. To sum up we have $\widetilde{u}(t) \in \lfloor \underline{y}(t), \overline{y}(t) \rfloor \subset \lfloor \underline{y}(T^*), \overline{y}(T^*) \rfloor$ for all $t \in [0, T^*]$.

We claim that $\widetilde{u}$ is actually solution of $(\mathcal{P})$ in $C([0, T^*], X)$. For all $t \in [0, T^*]$ we have $\widetilde{u}(t) \in \lfloor \underline{y}(t), \overline{y}(t) \rfloor \subset \lfloor \underline{y}(T^*), \overline{y}(T^*) \rfloor$ which in its turn is included in $\lfloor \underline{y}(T'), \overline{y}(T') \rfloor$. Therefore $\widetilde{F}(t, \widetilde{u}(t)) = F(t, \widetilde{u}(t))$ so that $\widetilde{u}$ is solution of $(\mathcal{P})$. From now on we write u for $\widetilde{u}$.

Step 2. We prove that there exists a global solution of $(\mathcal{P})$ which satisfies (S_1), (S_2) and the bounds $\underline{y}(T) \le \underline{y}(t) \le u(t) \le \overline{y}(t) \le \overline{y}(T)$ for all $T > 0$. We use the notation of Section 2.1.2, and still denote by $u \in C([0, T_{Max}), X)$ the maximal solution of $(\mathcal{P})$. By applying Theorem 2.3 it suffices to establish that there is no blow-up in finite time. Assume that $T_{Max} < +\infty$. Reproducing ***Step 1*** with T substitute for T^*, we infer that for all $T < T_{Max}$, and all $t \in [0,T]$ we have $u(t) \in [\underline{y}(T_{Max}), \overline{y}(T_{Max})]$. Hence

$$\|u\|_{C([0,T],X)} \leq \mathcal{L}_N(\Omega)^{\frac{1}{2}} \max(\left|\underline{y}(T_{Max})\right|, |\overline{y}(T_{Max})|),$$

which makes $\lim_{T\to T_{Max}} \|u\|_{C([0,T],X)} = +\infty$ impossible.

Step 3. We finally establish that if F is a regular CP-structured reaction functional, then $G : [0,+\infty) \to L^2(\Omega)$, defined by $G(t) = F(t, u(t))$, belongs to $W^{1,1}(0,T,L^2(\Omega))$ for all $T > 0$. Indeed, according to Theorem 2.3, we will infer that u satisfies (S_3).

For all $s < t$ in $[0,T]$, and from the fact that q, r and u are absolutely continuous, we have

$$\begin{aligned}
&\|F(t,u(t)) - F(s,u(s))\|_X \\
&\leq \|F(t,u(t)) - F(s,u(t))\|_X + \|F(s,u(t)) - F(s,u(s))\|_X \\
&\leq \|q(t) - q(s)\|_X + \sup_{\zeta\in[\underline{y}(T),\overline{y}(T)]} |g(\zeta)| \|r(t) - r(s)\|_{L^2(\Omega,\mathbb{R}^l)} \\
&\qquad + \|r\|_{L^\infty([0,T]\times\mathbb{R}^N,\mathbb{R}^l)} L_g \|u(t) - u(s)\|_X \\
&\leq \int_s^t \varphi_u(\tau)d\tau \qquad (2.16)
\end{aligned}$$

where[4] the function $\varphi_u : [0,T] \to \mathbb{R}_+$, given by

$$\begin{aligned}
&\varphi_u(\tau) \qquad (2.17)\\
&= \left\|\frac{dq}{dt}(\tau)\right\|_X + \sup_{\zeta\in[\underline{y}(T),\overline{y}(T)]} |g(\zeta)| \left\|\frac{dr}{dt}(\tau)\right\|_{L^2(\Omega,\mathbb{R}^l)} + \|r\|_{L^\infty([0,T]\times\mathbb{R}^N,\mathbb{R}^l)} L_g \left\|\frac{du}{dt}(\tau)\right\|_X
\end{aligned}$$

belongs to $L^1(0,T)$. From (2.16), we easily deduce that G is absolutely continuous, then belongs to $W^{1,1}(0,T,X)$. This completes the proof. □

By an easy adjustment of the proof above, applying this time Proposition 2.3, we obtain the following result.

Corollary 2.2. *Let F be a CP-structured reaction functional, with $\underline{\rho}$, $\overline{\rho}$ and $\underline{y}$, $\overline{y}$ given by* (CP)*, and let Φ be the functional of the calculus of variations* (2.13)*. Assume that $\underline{\rho} \leq 0 \leq \overline{\rho}$. Then for any $T > 0$, the Cauchy problem*

$$(\mathcal{P}) \begin{cases} \dfrac{du}{dt}(t) + D\Phi(u(t)) = F(t,u(t)) \text{ for a.e. } t \in (0,T) \\[2ex] u(0) = u_0, \quad \underline{\rho} \leq u_0 \leq \overline{\rho}, \quad u_0 \in \overline{\operatorname{dom}(D\Phi)} \end{cases}$$

admits a unique solution $u \in C([0,T], L^2(\Omega))$ satisfying assertions (S_1), (S_2) *and the following bounds in $[0,T]$: $\underline{y}(T) \leq \underline{y}(t) \leq u(t) \leq \overline{y}(t) \leq \overline{y}(T)$. If furthermore F is a regular CP-structured reaction functional, then u satisfies* (S_3).

[4] We still denote by L_g the Lipchitz constant of the restriction of g to $[\underline{y}(T), \overline{y}(T)]$.

Remark 2.11. The set of functions $u_0 \in \overline{\mathrm{dom}(D\Phi)}$ satisfying $\underline{\rho} \leq u_0 \leq \overline{\rho}$ is non empty. For the functional (2.6) of Corollary 2.1, any constant in $[\underline{\rho}, \overline{\rho}]$ is suitable since $H^1(\Omega) = \mathrm{dom}(\Phi) \subset \overline{\mathrm{dom}(\Phi)} = \overline{\mathrm{dom}(D\Phi)}$ (see Remark 2.1). For the functional (2.13) of Corollary 2.2, under condition $\underline{\rho} \leq 0 \leq \overline{\rho}$, $u_0 = 0$ is suitable.

Remark 2.12. Regarding Corollary 2.1, when F is a regular CP-structured reaction functional, $(\mathcal{P})$ may be written as

$$(\mathcal{P}) \begin{cases} \dfrac{du}{dt}(t) - \mathrm{div} D_\xi W(\cdot, \nabla u(t)) = F(t, u(t)) \text{ for a.e. } t \in (0,T), \\ u(0) = u_0, \quad \underline{\rho} \leq u_0 \leq \overline{\rho}, \quad u_0 \in \overline{\mathrm{dom}(D\Phi)}, \\ u(t) \in H^1(\Omega), \mathrm{div} D_\xi W(\cdot, \nabla u(t)) \in L^2(\Omega) \text{ for all } t \in]0,T], \\ a_0 u(t) + D_\xi W(\cdot, \nabla u(t)) \cdot \mathbf{n} = \phi \text{ on } \partial\Omega \text{ for all } t \in]0,T]. \end{cases}$$

Regarding Corollary 2.2, the same remark holds, i.e. problem $(\mathcal{P})$ may be written as

$$(\mathcal{P}) \begin{cases} \dfrac{du}{dt}(t) - \mathrm{div} D_\xi W(\cdot, \nabla u(t)) = F(t, u(t)) \text{ for a.e. } t \in (0,T), \\ u(0) = u_0, \quad \underline{\rho} \leq u_0 \leq \overline{\rho}, \quad u_0 \in \overline{\mathrm{dom}(D\Phi)}, \\ u(t) \in H^1_\Gamma(\Omega), \mathrm{div} D_\xi W(\cdot, \nabla u(t)) \in L^2(\Omega) \text{ for all } t \in]0,T], \\ D_\xi W(\cdot, \nabla u(t)) \cdot \mathbf{n} = 0 \text{ on } \partial\Omega \setminus \Gamma \text{ for all } t \in]0,T]. \end{cases}$$

Corollary 2.2 holds for non-diffusive reaction equations whose reaction functional is a CP-structured reaction functional. Note however that the functional Φ defined by

$$\Phi(u) = \begin{cases} 0 & \text{if } u \in H^1(\Omega), \\ +\infty & \text{if } u \in L^2(\Omega) \setminus H^1(\Omega) \end{cases}$$

is not lower semicontinuous because of the lack of coercivity. For this reason, and in order to take advantage of the compactness of the embedding $H^1(\Omega) \hookrightarrow L^2(\Omega)$ in the convergence processes for problems coupling reaction-diffusion equations with non-diffusive reaction equations in Chapter 4 (cf. Theorem 4.4), we consider the non-diffusive reaction Cauchy problem $(\mathcal{P})$ in the space $L^2(0,T,X)$ with the Hilbert space $X = H^1(\Omega)$. More precisely we have

Corollary 2.3. *Let F be a CP-structured reaction functional, with $\underline{\rho}$, $\overline{\rho}$ and $\underline{y}$, $\overline{y}$ given by* (CP)*. Then for any $T > 0$, the Cauchy problem in $L^2(0,T,H^1(\Omega))$*

$$(\mathcal{P}) \begin{cases} \dfrac{du}{dt}(t) = F(t, u(t)) \text{ for a.e. } t \in (0,T), \\ u(0) = u_0, \quad \underline{\rho} \leq u_0 \leq \overline{\rho}, \quad u_0 \in H^1(\Omega), \end{cases}$$

admits a unique solution $u \in C([0,T], H^1(\Omega))$ satisfying assertion (S$_2$) *and the following bounds in $[0,T]$: $\underline{y}(T) \leq \underline{y}(t) \leq u(t) \leq \overline{y}(t) \leq \overline{y}(T)$. If furthermore F is a regular CP-structured reaction functional, then u satisfies* (S$_3$).

Proof. Reproduce the proof of Corollary 2.1 by using the comparison principle stated in Proposition 2.4 and Theorem 2.4 with $X = H^1(\Omega)$. □

In the definition of CP-structured reaction functionals, in addition to condition (CP), we impose the structure $f(t,x,\zeta) = r(t,x) \cdot g(\zeta) + q(t,x)$ to proceed to the variational convergence of reaction-diffusion problems in Section 2.4, and because most of problems resulting from modelings, involve these functionals as we have underlined in examples of Section 2.2.3. Nevertheless, this condition is not essential in the proofs of Corollaries 2.1, 2.2 and 2.3. The key hypothesis is the (CP)-condition. The proofs can be reproduced with minor adaptations for reactions functions f fulfilling the two following conditions:

(i) $\zeta \mapsto f(t,x,\zeta)$ is locally Lipschitz continuous, uniformly with respect to $(t,x) \in \mathbb{R} \times \mathbb{R}^N$;
(ii) f satisfies (CP).

We leave to the reader to establish the following result by mimic the proof of Corollary 2.1.

Corollary 2.4 (Generalization of Corollaries 2.1, 2.2). *Assume that the function f satisfies (i) and (ii), and let $\underline{\rho}$, $\overline{\rho}$ and $\underline{y}$, $\overline{y}$ given by* (CP). *Assume that $a_0\underline{\rho} \leq \phi \leq a_0\overline{\rho}$ when Φ is the standard functional* (2.6), *or $\underline{\rho} \leq 0 \leq \overline{\rho}$ when Φ is the standard functional* (2.13). *Then for any $T > 0$, the Cauchy problem $(\mathcal{P})$ admits a unique solution $u \in C([0,T], L^2(\Omega))$ satisfying assertions* (S$_1$), (S$_2$), *and the following bounds for all t in $[0,T]$: $\underline{y}(T) \leq \underline{y}(t) \leq u(t) \leq \overline{y}(t) \leq \overline{y}(T)$. If furthermore G defined by $G(t) = F(t,u(t))$, belongs to $W^{1,1}(0,T,L^2(\Omega))$, then u satisfies* (S$_3$).

We give below an example taken from the modeling of the dynamics of a phreatic water table.

Example 2.5. Consider the reaction-diffusion Cauchy problem

$$(\mathcal{P}) \begin{cases} \dfrac{du}{dt}(t) - \mathrm{div} D_\xi W(\cdot, \nabla u(t)) = F(t, u(t)) \text{ for a.e. } t \in (0,T), \\[2ex] u(0) = u_0, \quad 0 \leq u_0 \leq \overline{\rho}, \quad u_0 \in \overline{\mathrm{dom}(D\Phi)}, \\[2ex] u(t) \in H^1_\Gamma(\Omega), \mathrm{div} D_\xi W(\cdot, \nabla u(t)) \in L^2(\Omega) \text{ for a.e. } t \in (0,T), \\[2ex] D_\xi W(\cdot, \nabla u(t)) \cdot \mathbf{n} = 0 \text{ on } \partial\Omega \setminus \Gamma \text{ for a.e. } t \in (0,T), \end{cases}$$

where $\overline{\rho} > 0$, and, for every $v \in L^2(\Omega)$, $F(t,v)(x) = S\left(t - g(\frac{L-v(x)}{K}), x\right)$. Due to the entanglement between the time variable and the state variable, in general, the reaction function given by $f(t,x,\zeta) = S\left(t - g(\frac{L-\zeta}{K}), x\right)$, does not fulfills the structure condition $r(t,x) \cdot g(\zeta) + q(t,x)$.

We assume that $S(t,x) = a(x)R(t)$, and that R, g and a fulfills the following conditions: $L > 0$ and $K > 0$ are two constants, and

- $R : \mathbb{R} \to \mathbb{R}_+$ is Lipschitz continuous: there exists $L_R > 0$ such that
$$|R(t) - R(t')| \leq L_R|t - t'|$$
for all $(t,t') \in \mathbb{R}^2$;
- $g : \mathbb{R} \to \mathbb{R}$ is locally Lipschitz continuous: for every interval $I \subset \mathbb{R}$, there exists $L_I > 0$ such that $|g(\zeta) - g(\zeta')| \leq L_I|\zeta - \zeta'|$ for all $(\zeta,\zeta') \in I \times I$;
- $a \in L^\infty(\mathbb{R}^N)$, and $a \geq 0$.

We infer that $\zeta \mapsto f(t,x,\zeta) = a(x)R\left(t - g(\frac{L-\zeta}{K})\right)$ is locally Lipschitz continuous, uniformly with respect to $(t,x) \in \mathbb{R} \times \mathbb{R}^N$: this results from

$$|f(t,x,\zeta) - f(t,x,\zeta')| \leq \|a\|_{L^\infty(\mathbb{R}^N)} \frac{L_R L_I}{K}|\zeta - \zeta'|$$

for all $(\zeta,\zeta') \in I \times I$ and all $(t,x) \in \mathbb{R} \times \mathbb{R}^N$.

To show that f satisfies (CP), take

$\underline{f} = 0,\ \underline{y} = 0, \underline{\rho} = 0,$

$\overline{f}(t,\zeta) = \|a\|_{L^\infty(\mathbb{R}^N)} R\left(t - g\left(\frac{\zeta}{k}\right)\right),$

$\overline{y}$ solution of o.d.e. $\overline{y}' = \|a\|_{L^\infty(\mathbb{R}^N)} R\left(t - g\left(\frac{L-\overline{y}}{K}\right)\right)$, with $\overline{y}(0) = \overline{\rho}$.

Then, from Corollary 2.4, $(\mathcal{P})$ admits a unique solution which satisfies (S_1), (S_2), and $0 \leq u(t) \leq \overline{y}(t)$ for all $t \in [0,T]$.

One of the possible origins of problem $(\mathcal{P})$ is as follows. If we disregard the coefficients of physical origin of concrete models, within the framework of Dupuit's 2d-approximation in hydrogeology (cf. Dauvargne (2006); Dupuit (1863)), state variable $z := u(t,x)$ may denotes the piezometric load, or the height at time t and position $x \in \Omega \subset \mathbb{R}^2$, of a water table. The reaction function S denotes the source, or the recharge rate, due to the precipitation which recharges the water table with a delay time which is function of $\frac{1}{K}(L - u(t,x))$. The coefficient $K > 0$ denotes the speed infiltration of the water, from the soil to the groundwater. The density W takes into account the stored capacity of the porous media, and the transmissivity, i.e. the rate at which groundwater flows horizontally.

2.2.6 *Estimate of the $L^2(\Omega)$-norm of the right derivative*

From above we know that when the CP-structured reaction functional is regular, the solution of $(\mathcal{P})$ admits a right derivative at each $t \in]0,T[$. The next estimate below is central for proving the weak convergence of the right derivatives in (***Step 6***) of the proof of Theorem 2.6.

Proposition 2.5. *Under hypotheses of Corollary 2.1 or 2.2, when F is a regular CP-structured reaction functional, we have for all $t \in]0,T[$*

$$\left\|\frac{d^+u}{dt}(t)\right\|_X \leq C + \left(C + \frac{1}{t}\right)\int_0^t \left\|\frac{du}{dt}(\tau)\right\|_X d\tau$$

where

$$C = \max(L_g \|r\|_{L^\infty([0,T]\times\mathbb{R}^N,\mathbb{R}^l)},\ V(q,r)) \tag{2.18}$$

where

$$V(q,r) := \int_0^T \left\|\frac{dq}{dt}(\tau)\right\|_X d\tau + \sup_{\zeta\in[\underline{y}(T),\overline{y}(T)]} |g(\zeta)| \int_0^T \left\|\frac{dr}{dt}(\tau)\right\|_{L^2(\Omega,\mathbb{R}^l)} d\tau,$$

and L_g denotes the Lipschitz constant of the restriction of g to $[\underline{\rho},\overline{\rho}]$.

Proof. ***Step 1.*** We establish the following lemma.

Lemma 2.5. *Let X be a Hilbert space, $T > 0$, $G \in W^{1,1}(0,T,X)$ and $\Phi : X \to \mathbb{R}\cup\{+\infty\}$ be a convex proper lower semicontinuous functional. Let u satisfy*

$$\begin{cases} \dfrac{du}{dt}(t) + \partial\Phi(u(t)) \ni G(t) \text{ for a.e. } t \in (0,T), \\[2ex] u(0) \in \overline{\mathrm{dom}(\partial\Phi)}. \end{cases} \tag{2.19}$$

Then the right derivative of u satisfies for all $t \in]0,T[$ the following estimate

$$\left\|\frac{d^+u}{dt}(t)\right\|_X \leq \frac{1}{t}\int_0^t \left\|\frac{du}{dt}(s)\right\|_X ds + \int_0^t \left\|\frac{dG}{dt}(s)\right\|_X ds.$$

Proof. For $h > 0$ intended to tend to 0, set $H := G(\cdot + h)$ and let v be the solution of

$$\begin{cases} \dfrac{dv}{dt}(t) + \partial\Phi(v(t)) \ni H(t) \text{ for a.e. } t \in (0,T), \\[2ex] v(0) = u(h). \end{cases} \tag{2.20}$$

Clearly $v(t) = u(t+h)$ (recall that u which solves (2.19) is the restriction to $[0,T]$ of a unique global solution $u \in C([0,+\infty),X)$ of (2.19) in $(0,+\infty)$, and that $u(t) \in \mathrm{dom}(\Phi)$ for all $t \in (0,T)$). From (2.19), (2.20), and the monotonicity of $\partial\Phi$, we infer that for a.e. $\sigma \in (0,T)$

$$\left\langle \frac{dv}{dt}(\sigma) - \frac{du}{dt}(\sigma), v(\sigma) - u(\sigma)\right\rangle \leq \langle H(\sigma) - G(\sigma), v(\sigma) - u(\sigma)\rangle,$$

hence

$$\frac{1}{2}\frac{d}{dt}\|v(\sigma)-u(\sigma)\|_X^2 \le \langle H(\sigma)-G(\sigma), v(\sigma)-u(\sigma)\rangle .$$

Integrating over (s,t) where $0 \le s \le t \le T$, we obtain for all $t \in [0,T]$

$$\begin{aligned}\frac{1}{2}\|v(t)-u(t)\|_X^2 &\le \frac{1}{2}\|v(s)-u(s)\|_X^2 + \int_s^t \langle H(\sigma)-G(\sigma), v(\sigma)-u(\sigma)\rangle\, d\sigma \\ &\le \frac{1}{2}\|v(s)-u(s)\|_X^2 + \int_0^t \|H(\sigma)-G(\sigma)\|_X \|v(\sigma)-u(\sigma)\|_X d\sigma.\end{aligned}$$

Thus, according to the Grönwall type lemma, Lemma A.2 with $p = 2$, it follows that for all $t \in [0,T]$ and all $s \in [0,t]$

$$\|v(t)-u(t)\|_X \le \|v(s)-u(s)\|_X + \int_0^t \|H(\sigma)-G(\sigma)\|_X d\sigma,$$

that is

$$\|u(t+h)-u(t)\|_X \le \|u(s+h)-u(s)\|_X + \int_0^t \|G(\sigma+h)-G(\sigma)\|_X d\sigma.$$

Dividing by h and letting $h \to 0$, we infer that for all $t \in]0,T[$ and all $s \in]0,t]$

$$\left\|\frac{d^+u}{dt}(t)\right\|_X \le \left\|\frac{d^+u}{dt}(s)\right\|_X + \int_0^t \left\|\frac{dG}{dt}(\sigma)\right\|_X d\sigma$$

(for a justification in order to obtain the last integral, we refer the reader to (Brezis, 1973, Proposition A2)). By integration over $(0,t)$, we obtain for all $t \in]0,T[$

$$\begin{aligned} t\left\|\frac{d^+u}{dt}(t)\right\|_X &\le \int_0^t \left\|\frac{d^+u}{dt}(s)\right\|_X + t\int_0^t \left\|\frac{dG}{dt}(s)\right\|_X ds \\ &= \int_0^t \left\|\frac{du}{dt}(s)\right\|_X + t\int_0^t \left\|\frac{dG}{dt}(s)\right\|_X ds.\end{aligned}$$

This ends the proof of Lemma 2.5. □

Last step. The thesis of Proposition 2.5 follows by combining Lemma 2.5 and the expression of the total variation of G, given by (2.16) and (2.17) where $G(t) = F(t, u(t))$. □

2.3 Invasion property

This section discusses the following problem: providing a sufficient condition on the initial value u_0, that implies the growth in time of the solution of the Cauchy problem $(\mathcal{P})$. This question is important when it concerns for example the evolution of a density in population dynamics. We restrict the analysis to the case of functional (2.13), when the diffusion part is quadratic, i.e. when the functional Φ is of the form

$$\Phi(u) = \begin{cases} \int_\Omega A(x)\nabla u(x)\cdot\nabla u(x)dx & \text{if } u \in H^1_\Gamma(\Omega) \\ +\infty & \text{otherwise.}\end{cases} \tag{2.21}$$

We assume that the matrix valued function $A = (a_{i,j})_{i,j=1...N} : \mathbb{R}^N \to \mathbb{M}_N$ satisfies for some $\alpha > 0$ and $\beta > 0$, $\alpha|\xi|^2 \leq \sum_{i,j=1}^N a_{i,j}(x)\xi_i\xi_j \leq \beta|\xi|^2$ for all $x \in \mathbb{R}^N$, all $\xi \in \mathbb{R}^N$, and that $a_{i,j} \in C^1(\mathbb{R}^N)$. Clearly Φ satisfies (D). Then, from Lemma 2.4, the subdifferential $\partial\Phi$ $(= D\Phi)$ of the functional Φ is given by

$$\begin{cases} \text{dom}(D\Phi) = \left\{ v \in H^1_\Gamma(\Omega) : \text{div}(A(\cdot)\nabla v) \in L^2(\Omega),\ A(\cdot)\nabla v(t) \cdot \mathbf{n} = 0 \text{ on } \partial\Omega \backslash \Gamma \right\} \\ \\ D\Phi(v) = -\text{div}(A(\cdot)\nabla v) \text{ for } v \in \text{dom}(D\Phi). \end{cases}$$

Recall that from Remark 2.9, $\text{dom}(D\Phi)$ is dense in $L^2(\Omega)$, and more explicitly given by

$$\text{dom}(D\Phi) = \left\{ v \in H^1_\Gamma(\Omega) \cap H^2(\Omega) : A(\cdot)\nabla v(t) \cdot \mathbf{n} = 0 \text{ on } \partial\Omega \setminus \Gamma \right\}.$$

We assume that the CP-structured reaction functional F does not depend on the time variable. We are concerned with the semilinear evolution problems of the form

$$(\mathcal{P}) \begin{cases} \dfrac{du}{dt}(t) - \text{div}(A(\cdot)\nabla u(t)) = r(\cdot)g(u(t)) + q(\cdot) \text{ for a.e. } t \in (0,T), \\ \\ u(0) = u_0, \quad \underline{\rho} \leq u_0 \leq \overline{\rho},\ u_0 \in \text{dom}(\partial\Phi), \\ \\ u(t) \in H^1_\Gamma(\Omega) \cap H^2(\Omega) \text{ for all } t \in]0,T], \\ \\ A(\cdot)\nabla u(t) \cdot \mathbf{n} = 0 \text{ on } \partial\Omega \setminus \Gamma \text{ for all } t \in]0,T], \end{cases}$$

where $\underline{\rho}$ and $\overline{\rho}$, $\underline{\rho} \leq 0 \leq \overline{\rho}$, are given by (CP). Existence and uniqueness are given by Corollary 2.1, which states that u satisfies (S_2) and (S_3). Theorem below shows that, when g is regular and the initial function is a sub solution of the elliptic semilinear problem associated with $(\mathcal{P})$, then the solution u is nondecreasing in time.

Theorem 2.5 (Invasion Property). *Assume that $g \in C^1([\underline{\rho}, \overline{\rho}])$. If the initial function u_0 satisfies*

$$(\mathcal{P}_{ell}) \begin{cases} -\text{div}(A(\cdot)\nabla u_0) \leq F(u_0) \ (\textit{inequality in } L^2(\Omega)) \\ u_0 \in \text{dom}(D\Phi), \end{cases}$$

then $\frac{du}{dt}(t) \geq 0$ for all $t \in [0,T]$.

Proof. First observe that in this particular case of semilinear evolution problem, we can assert that the solution belongs to $C^1([0,T], L^2(\Omega))$ and solves $(\mathcal{P})$ for all $t \in [0,T]$. This follows from (Cazenave and Haraux, 1998, Corollary 4.1.2, Proposition 4.1.6) applied to the linear evolution problem 2.2 of the fixed point procedure in the proof of Theorem 2.1. Set $v = \frac{du}{dt}$. By taking the time derivative of each equation of $(\mathcal{P})$ we infer that v solves the problem

$$(\mathcal{P}') \begin{cases} \dfrac{dv}{dt}(t) - \text{div}(A(\cdot)\nabla v(t)) = r(\cdot)g'(u(t))v(t) \text{ for all } t \in [0,T], \\ \\ v(0) = \frac{du}{dt}(0),\ A(\cdot)\nabla v(t) \cdot \mathbf{n} = 0 \text{ on } \partial\Omega \text{ for all } t \in [0,T]. \end{cases}$$

Take v^- as a test function. According to Green's formula, we obtain

$$\int_\Omega \frac{dv}{dt}(t)v^-(t)dx + \int_\Omega A(x)\nabla v(t).\nabla v^-(t)dx - \int_{\partial\Omega} A(x)\nabla v(t)\cdot \mathbf{n}\ v^-(t)\ d\mathcal{H}_{N-1}$$
$$= \int_\Omega r(x)g'(u(t))(v^-)^2(t)dx.$$

Hence, since $v^+(t)v^-(t) = 0$ a.e., $\frac{dv^+}{dt}(t)v^-(t) = 0$ a.e., $A(\cdot)\nabla v^+(t)\cdot\nabla v^- = 0$ a.e., and $A(\cdot)\nabla v(t)\cdot\mathbf{n} = -a_0 v(t)$, we infer that

$$\begin{aligned}\frac{1}{2}\frac{d}{dt}\int_\Omega (v^-)^2(t)dx &\le -\int_{\partial\Omega} a_0(v^-)^2(t)d\mathcal{H}_{N-1} + \int_\Omega r(x)g'(v(t))(v^-)^2(t)dx\\ &\le \int_\Omega r(x)g'(u(t))(v^-)^2(t)dx\\ &\le \sup_{(x,\zeta)\in\mathbb{R}^N\times[\underline{y}(T),\overline{y}(T)]} |r(x)g'(\zeta)| \int_\Omega (v^-)^2(t)dx\end{aligned}$$

(recall that from Corollary 2.1, $u(t)\in[\underline{y}(T),\overline{y}(T)]$ for all $t\in[0,T]$). From Grönwall's lemma we deduce that for all $t\in[0,T]$

$$\left\|v^-(t)\right\|^2_{L^2(\Omega)} \le \left\|v^-(0)\right\|^2_{L^2(\Omega)}\exp(Ct) \tag{2.22}$$

where $C := \sup_{(x,\zeta)\in\mathbb{R}^N\times[\underline{y}(T),\overline{y}(T)]}|r(x)g'(\zeta)|$. But

$$v(0) = \frac{du}{dt}(0) = \mathrm{div}(A(\cdot)\nabla u_0) + F(u_0) \ge 0$$

(the first equality holds because $v\in C([0,T],L^2(\Omega))$, the second because u solves $(\mathcal{P})$ for all $t\in[0,T]$, and the last inequality holds because u_0 solves $(\mathcal{P}_{ell})$ by hypothesis). Hence $v^-(0) = 0$ and, from (2.22), we infer that $v^-(t) = 0$ for all $t\in[0,T]$. □

2.4 Variational convergence of reaction-diffusion problems with CP-structured reaction functionals

We are interested in the convergence of problems $(\mathcal{P}_n)$ associated with the sequence $(\Phi_n)_{n\in\mathbb{N}}$ of functionals of the calculus of variations$\Phi_n : L^2(\Omega)\to\mathbb{R}\cup\{+\infty\}$ defined by

$$\Phi_n(u) = \begin{cases} \displaystyle\int_\Omega W_n(x,\nabla u(x))dx + \frac{1}{2}\int_{\partial\Omega} a_{0,n}u^2 d\mathcal{H}_{N-1} - \int_{\partial\Omega}\phi_n u\ d\mathcal{H}_{N-1} & \text{if } u\in H^1(\Omega)\\ +\infty & \text{otherwise}\end{cases} \tag{2.23}$$

where $\phi_n\in L^2_{\mathcal{H}_{N-1}}(\partial\Omega)$, and $a_{0,n}\in L^\infty_{\mathcal{H}_{N-1}}(\partial\Omega)$ with

$$\begin{cases} a_{0,n}\ge 0 \quad \mathcal{H}_{N-1}\text{-a.e. in } \partial\Omega\\ \exists\sigma>0 \quad a_{0,n}\ge\sigma \quad \mathcal{H}_{N-1}\text{-a.e. in } \Gamma\subset\partial\Omega,\end{cases}$$

or by

$$\Phi_n(u) = \begin{cases} \displaystyle\int_\Omega W_n(x, \nabla u(x))dx & \text{if } u \in H^1_\Gamma(\Omega) \\ +\infty & \text{otherwise.} \end{cases} \tag{2.24}$$

The density $W_n : \mathbb{R}^N \times \mathbb{R}^N \to \mathbb{R}$ is a Borel measurable function which fulfills the condition

(D_n) there exist $\alpha > 0$ and $\{\beta_n\}_{n\in\mathbb{N}} \subset \mathbb{R}^*_+$ such that for a.e. $x \in \mathbb{R}^N$, all $\xi \in \mathbb{R}^N$, and all $n \in \mathbb{N}$

$$\alpha|\xi|^2 \leq W_n(x,\xi) \leq \beta_n(1 + |\xi|^2);$$

for a.e. $x \in \mathbb{R}^N$ and all $n \in \mathbb{N}$, $W_n(x, \cdot)$ is a Gâteaux-differentiable convex function;

$$D_\xi W_n(x, 0) = 0 \text{ for a.e. } x \in \mathbb{R}^N.$$

In the following, we fix $T > 0$ and consider a sequence $(F_n)_{n\in\mathbb{N}}$ of CP-structured reaction functionals, each of them being associated with (r_n, g_n, q_n), i.e. $F_n(t, v)(x) = f_n(t, x, v(x))$ for a.e. $x \in \Omega$ for all $t \in [0, T]$, and all $v \in L^2(\Omega)$, where

$$f_n(t, x, \zeta) = r_n(t, x) \cdot g_n(\zeta) + q_n(t, x) \text{ for all } (t, x, \zeta) \in [0, +\infty) \times \mathbb{R}^N \times \mathbb{R}. \tag{2.25}$$

We assume that for all $n \in \mathbb{N}$, the function g_n is a locally Lipschitz continuous, uniformly with respect to n, i.e. for every interval $I \subset \mathbb{R}$ there exists $L_I \geq 0$ such that for every $(\zeta, \zeta') \in I \times I$,

$$\sup_{n\in\mathbb{N}} |g_n(\zeta) - g_n(\zeta')| \leq L_I|\zeta - \zeta'|. \tag{2.26}$$

For example, this condition is fulfilled by $g_n = (g_{n,i})_{i=1,\dots,l}$ where the scalar functions $g_{n,i}$ are convex and satisfy for all $\zeta \in \mathbb{R}$, $0 \leq g_{n,i}(\zeta) \leq \beta_i(1 + |\zeta|^{p_i})$ for some $\beta_i > 0$ and $p_i \geq 1$. This is the case of Example 2.1 ***b)*** with $\theta_n > 0$ substitute for $\theta > 0$.

We also assume that

$$-\infty < \inf_{n\in\mathbb{N}} \underline{y}_n(T) \text{ and } \sup_{n\in\mathbb{N}} \overline{y}_n(T) < +\infty, \tag{2.27}$$

and, for all $n \in \mathbb{N}$,

$$a_{0,n}\underline{\rho}_n \leq \phi_n \leq a_{0,n}\overline{\rho}_n \ \mathcal{H}_{N-1} \text{ a.e. in } \partial\Omega \tag{2.28}$$

where $\underline{y}_n$, $\overline{y}_n$, $\underline{\rho}_n$ and $\overline{\rho}_n$ are given by condition (CP) fulfilled by each F_n. Recall that $\underline{y}_n$ and $\overline{y}_n$ are solution of suitable o.d.e. with initial condition $\underline{\rho}_n$ and $\overline{\rho}_n$ respectively. Note that when the functional Φ_n is of the form (2.24), condition (2.28) becomes

$$\underline{\rho}_n \leq 0 \leq \overline{\rho}_n \tag{2.29}$$

for all $n \in \mathbb{N}$.

In order to establish the variational convergence of reaction-diffusion problems $(\mathcal{P}_n)$ with diffusion part $D\Phi_n$, we take advantage of standard results involving the Γ-convergence of the functionals Φ_n to Φ, particularly in homogenization framework (see (Attouch *et al.*, 2014, Subsection 12.4)). More precisely, we establish the convergence of the sequence of problems $(\mathcal{P}_n)$ under the hypothesis of the Mosco-convergence of the sequence $(\Phi_n)_{n\in\mathbb{N}}$, i.e. the Γ-convergence of the functionals Φ_n when $L^2(\Omega)$ is equipped both with the strong and its weak topology. This notion of convergence introduced in Mosco (1969, 1971), occurs quite naturally since, under a normalization condition, this convergence is equivalent to the graph-convergence of the subdifferentials (see Theorem B.4). Note that although from Lemma 2.3, Φ_n is Gâteaux-differentiable, we are not ensured that its Mosco-limit is. For the definition and variational properties of this notion we refer the reader to Appendix B or (Attouch *et al.*, 2014, Section 17.4.2). For the connection with Moreau-Yosida approximations we refer to Attouch (1984), Dal Maso (1993). We denote by $\Phi_n \overset{M}{\to} \Phi$ the Mosco-convergence of the sequence $(\Phi_n)_{n\in\mathbb{N}}$ to Φ. A first important lemma concerns the Mosco-convergence of functionals defined in $L^2(0,T,X)$.

Lemma 2.6. *Let $(\psi_n)_{n\in\mathbb{N}}$, ψ, be a sequence of nonnegative convex lower semicontinuous proper functions from a Hilbert space X into $[0,+\infty]$, such that $\psi_n \overset{M}{\to} \psi$, and consider $(\Psi_n)_{n\in\mathbb{N}}$, $\Psi : L^2(0,T,X) \to \mathbb{R}_+ \cup \{+\infty\}$ defined by*

$$\Psi_n(v) := \int_0^T \psi_n(v(t))dt; \ \Psi(v) := \int_0^T \psi(v(t))dt.$$

Then $\Psi_n \overset{M}{\to} \Psi$.

Proof. A first proof is given in Attouch (1977). We propose a proof using the Moreau-Yosida approximations (see Definition B.3). We are going to apply Proposition B.4, which states the equivalence between the Mosco-convergence and the piecewise convergence of the Moreau-Yosida approximations.

Step 1. Denote by ψ_n^λ, ψ^λ, Ψ_n^λ, and Ψ^λ the Moreau-Yosida approximation of index $\lambda > 0$ of ψ_n, ψ, Ψ_n, and Ψ respectively. We establish that for every $u \in L^2(0,T,X)$:

$$\Psi_n^\lambda(u) = \int_0^T \psi_n^\lambda(u(t))dt \quad \text{and} \quad \Psi^\lambda(u) = \int_0^T \psi^\lambda(u(t))dt.$$

This result is an elementary case of interchange of infimum and integral (see for instance Anza Hafsa and Mandallena (2003)). In order to make the reading self-contained we give a direct proof of the second equality, the proof of the first one is similar. By definition, we have

$$\begin{aligned}\Psi^\lambda(u) &= \inf_{v\in L^2(0,T,X)}\left\{\int_0^T\left(\psi(v(t))+\frac{1}{2\lambda}\|v(t)-u(t)\|_X^2\right)dt\right\}\\ &= \int_0^T\left(\psi(J_\lambda u(t))+\frac{1}{2\lambda}\|J_\lambda u(t)-u(t)\|_X^2\right)dt\\ &\geq \int_0^T\inf_{w\in X}\left\{\psi(w)+\frac{1}{2\lambda}\|w-u(t)\|_X^2\right\}dt\\ &= \int_0^T\psi^\lambda(u(t))dt.\end{aligned}$$

We have used the fact that the infimum in the Moreau-Yosida approximation $\Psi^\lambda(u)$ is attained at a unique point denoted by $J_\lambda u$ (see (Attouch *et al.*, 2014, Proposition 17.2.1), or (Attouch, 1984, Theorem 3.24)). Conversely

$$\begin{aligned}\int_0^T\psi^\lambda(u(t))dt &= \int_0^T\inf_{w\in X}\left\{\psi(w)+\frac{1}{2\lambda}\|w-u(t)\|_X^2\right\}dt\\ &= \int_0^T\left(\psi(J_\lambda(u(t)))+\frac{1}{2\lambda}\|J_\lambda(u(t))-u(t)\|_X^2\right)dt \end{aligned}\tag{2.30}$$

where we have used the fact that the infimum in the definition of $\psi^\lambda(u(t))$ is achieved at $J_\lambda(u(t))$ for the same reason. Assume for the moment that the map $t\mapsto J_\lambda(u(t))$ belongs to $L^2(0,T,X)$. Then (2.30) yields

$$\int_0^T\psi^\lambda(u(t))dt \geq \inf_{v\in L^2(0,T,X)}\left\{\int_0^T\left(\psi(v(t))+\frac{1}{2\lambda}\|v(t)-u(t)\|_X^2\right)dt\right\} := \Psi^\lambda(u),$$

which proves the claim. It remains to prove that $t\mapsto J_\lambda(u(t))$ belongs to $L^2(0,T,X)$. According to (Attouch *et al.*, 2014, Proposition 17.2.1), $J_\lambda : X\to X$ is non expansive, then continuous. Therefore the measurability of $J_\lambda\circ u$ from $[0,T]$ into X follows from the measurability of u from $[0,T]$ into X. On the other hand, since ψ is proper, there exists w_0 in X such that $\psi(w_0)<+\infty$. Hence

$$\frac{1}{2\lambda}\|J_\lambda(u(t))-u(t)\|_X^2 \leq \psi^\lambda(J_\lambda(u(t))) \leq \psi(w_0)+\frac{1}{2\lambda}\|w_0-u(t)\|_X^2,$$

from which we deduce that $\int_0^T\|J_\lambda(u(t))\|^2\,dt<+\infty$ since u belongs to $L^2(0,T,X)$.

Step 2. We claim that $\psi_n \overset{M}{\to}\psi \Longrightarrow \Psi_n\overset{M}{\to}\Psi$. From Proposition B.4 we have

$$\psi_n \overset{M}{\to}\psi \iff \forall u\in X,\quad \forall\lambda>0,\quad \psi_n^\lambda(u)\to\psi^\lambda(u).$$

Let $u\in L^2(0,T,X)$, then from above, for a.e. $t\in(0,T)$ and for all $\lambda>0$, we have

$$\psi_n \overset{M}{\to}\psi \Longrightarrow \psi_n^\lambda(u(t))\to\psi^\lambda(u(t)). \tag{2.31}$$

Let v be arbitrarily chosen in $\mathrm{dom}(\psi)$. Since $\psi_n \overset{M}{\to} \psi$, there exists a sequence $(v_n)_{n\in\mathbb{N}}$ in X such that $v_n \to v$ and $\psi_n(v_n) \to \psi(v)$. Then there exists $N \in \mathbb{N}$ which depends only on $\psi(v)$ and $\|v\|_X$ such that for all $n \geq N$

$$\begin{aligned} 0 \leq \psi_n^\lambda(u(t)) &\leq \psi_n(v_n) + \frac{1}{2\lambda}\|v_n - u(t)\|_X^2 \\ &\leq \psi(v) + 1 + \frac{1}{2\lambda}\|v_n - u(t)\|_X^2 \\ &\leq \psi(v) + 1 + \frac{1}{\lambda}\|v_n\|_X^2 + \frac{1}{\lambda}\|u(t)\|_X^2 \\ &\leq \psi(v) + 2 + \frac{1}{\lambda}\|v\|_X^2 + \frac{1}{\lambda}\|u(t)\|_X^2, \end{aligned} \tag{2.32}$$

where $\psi(v) + 2 + \frac{1}{\lambda}\|v\|_X^2 + \frac{1}{\lambda}\|u(\cdot)\|_X^2$ belongs to $L^1(0,T)$. Then, from (2.31) and the Lebesgue dominated convergence theorem, we infer that for all $\lambda > 0$

$$\int_0^T \psi_n^\lambda(u(t))dt \to \int_0^T \psi^\lambda(u(t))dt,$$

that is, from ***Step 1***, $\Psi_n^\lambda(u) \to \Psi^\lambda(u)$. According to Proposition B.4, we deduce that $\Psi_n \overset{M}{\to} \Psi$. □

The main result below states some stability with respect to the convergence of problems $(\mathcal{P}_n)$. Note, however, that the limit problem $(\mathcal{P})$ is a differential inclusion. Furthermore, without additional hypotheses under which the variational limit Φ admits a integral representation (we make no assumption on the sequence $\{\beta_n\}_{n\in\mathbb{N}}$), we are not ensured that $(\mathcal{P})$ is of gradient flow type. For integral representation of Γ-limits we refer the reader to (Dal Maso, 1993, Chapter 20). For this reason, Theorem 2.6 can be considered rather as a relaxation result.

Theorem 2.6 (General convergence theorem). *Assume that the sequence of densities $(W_n)_{n\in\mathbb{N}}$ of functionals* (2.23), *or* (2.24), *satisfies* (D_n), *and that the sequence of CP-structured reaction functionals $(F_n)_{n\in\mathbb{N}}$, of the form* (2.25), *satisfies* (2.26), (2.27), (2.28) *or* (2.29). *Let u_n be the unique solution of the Cauchy problem*

$$(\mathcal{P}_n)\begin{cases} \dfrac{du_n}{dt}(t) + D\Phi_n(u_n(t)) = F_n(t, u_n(t)) \text{ for a.e. } t \in (0,T) \\ u_n(0) = u_n^0, \quad \underline{\rho}_n \leq u_n^0 \leq \overline{\rho}_n, \quad u_n^0 \in \mathrm{dom}(\Phi_n). \end{cases}$$

Assume that

(H_1) $\Phi_n \overset{M}{\to} \Phi$, *and* $\sup_{n\in\mathbb{N}} \|\phi_n\|_{L^2_{\mathcal{H}_{N-1}}(\partial\Omega)} < +\infty$;

(H_2) $\sup_{n\in\mathbb{N}} \Phi_n(u_n^0) < +\infty$;

(H_3) *there exists u^0 such that $u_n^0 \to u^0$ strongly in $L^2(\Omega)$;*

(H_4) *there exists g such that g_n pointwise converge to g;*

(H$_5$) $\sup_{n\in\mathbb{N}} \|r_n\|_{L^\infty([0,T]\times\mathbb{R}^N,\mathbb{R}^l)} < +\infty$, *and there exists* r *in* $L^\infty([0,T]\times\mathbb{R}^N,\mathbb{R}^l)$ *such that* $r_n \rightharpoonup r$ *weakly in* $L^2(0,T,L^2(\Omega,\mathbb{R}^l))$;

(H$_6$) *there exists* q *in* $L^2(0,T,L^2(\Omega))$ *such that* $q_n \rightharpoonup q$ *weakly in* $L^2(0,T,L^2(\Omega))$.

Then $(u_n)_{n\in\mathbb{N}}$ *uniformly converges in* $C([0,T],L^2(\Omega))$ *to the unique solution of the problem*

$$(\mathcal{P})\begin{cases} \dfrac{du}{dt}(t) + \partial\Phi(u(t)) \ni F(t,u(t)) \text{ for a.e. } t\in(0,T) \\ u(0)=u^0, \quad \inf_n \underline{\rho}_n \le u^0 \le \sup_n \overline{\rho}_n, \quad u_0 \in \mathrm{dom}(\Phi). \end{cases}$$

The reaction functional $F:[0,+\infty)\times L^2(\Omega)\to\mathbb{R}^\Omega$ *is defined, for all* $t\in[0,T]$, *all* $v\in L^2(\Omega)$ *and for a.e.* $x\in\Omega$, *by*

$$F(t,v)(x) = f(t,x,v(x)) \quad \text{and} \quad f(t,x,\zeta) = r(t,x)\cdot g(\zeta) + q(t,x).$$

Moreover, $\inf_n \underline{y}_n(t) \le u(t) \le \sup_n \overline{y}_n(t)$ *for all* $t\in[0,T]$, *and*

$$\frac{du_n}{dt} \rightharpoonup \frac{du}{dt} \text{ weakly in } L^2(0,T,L^2(\Omega)).$$

If furthermore $\Phi_n(u_n^0)\to\Phi(u^0)$, $r_n\to r$ *strongly in* $L^2(0,T,L^2(\Omega,\mathbb{R}^l))$, *and* $q_n\to q$ *strongly in* $L^2(0,T,L^2(\Omega))$, *then*

$$\frac{du_n}{dt} \to \frac{du}{dt} \text{ strongly in } L^2(0,T,L^2(\Omega)),$$

and $\Phi_n(u_n(t))$ *converges to* $\Phi(u(t))$ *for all* $t\in[0,T]$.

Assume that $(F_n)_{n\in\mathbb{N}}$ *is a sequence of regular CP-structured reaction functionals and that* $\frac{dq_n}{dt} \rightharpoonup \frac{dq}{dt}$ *weakly in* $L^1(0,T,L^2(\Omega))$, *and* $\frac{dr_n}{dt} \rightharpoonup \frac{dr}{dt}$ *weakly in* $L^1(0,T,L^2(\Omega,\mathbb{R}^l))$. *Then the functional* G *defined by* $G(t)=F(t,u(t))$ *belongs to* $W^{1,1}(0,T,L^2(\Omega))$, *and* $u(t)\in\mathrm{dom}(\partial\Phi)$ *for all* $t\in]0,T]$. *Assume furthermore that for each* $t\in]0,T[$, $q_n(t,\cdot)\rightharpoonup q(t,\cdot)$ *weakly in* $L^2(\Omega)$, $r_n(t,\cdot)\rightharpoonup r(t,\cdot)$ *weakly in* $L^2(\Omega,\mathbb{R}^l)$, *and that* $\partial\Phi$ *is single valued. Then for each* $t\in]0,T[$

$$\frac{d^+u_n}{dt}(t) \rightharpoonup \frac{d^+u}{dt}(t) \text{ weakly in } L^2(\Omega).$$

Proof. Note that in the statement of Theorem 2.6, we assume that $u_n^0\in H^1(\Omega)=\mathrm{dom}(\Phi_n)$. Since $\mathrm{dom}(\Phi_n)\subset\overline{\mathrm{dom}(\Phi_n)}=\overline{\mathrm{dom}(D\Phi_n)}$ (see Remark 2.1), we infer that $u_n^0\in\overline{\mathrm{dom}(D\Phi_n)}$. Therefore, according to Corollary 2.1, $(\mathcal{P}_n)$ admits a unique solution u_n which satisfies (S$_1$), (S$_2$), the bounds $\underline{y}_n(t)\le u_n(t)\le\overline{y}_n(t)$ for all $t\in[0,T]$. Furthermore u_n fulfills (S$_3$) when F_n is a regular CP-structured reaction functional. Finally, from $u_n^0\in\mathrm{dom}(\Phi_n)$, we infer that $\frac{du_n}{dt}$ belongs to $L^2(0,T,L^2(\Omega))$ (see (Attouch *et al.*, 2014, Theorem 17.2.5) or (Brezis, 1973, Theorem 3.6)).

Step 1. We establish

$$\sup_{n\in\mathbb{N}} \|u_n\|_{C(0,T,X)} < +\infty, \tag{2.33}$$

$$\sup_{n\in\mathbb{N}} \left\| \frac{du_n}{dt} \right\|_{L^2(0,T,X)} < +\infty. \tag{2.34}$$

Estimate (2.33) follows directly from $\inf_n \underline{y}_n(T) \leq \underline{y}_n(T) \leq u_n \leq \overline{y}_n(T) \leq \sup_n \overline{y}_n(T)$. Let us establish (2.34). In what follows the letter C denotes a constant which can vary from line to line. Fix $n \in \mathbb{N}$. From $(\mathcal{P}_n)$ we deduce that for a.e. $t \in (0,T)$,

$$\left\| \frac{du_n}{dt}(t) \right\|_X^2 + \left\langle D\Phi_n(u_n(t)), \frac{du_n}{dt}(t) \right\rangle = \left\langle F_n(t, u_n(t)), \frac{du_n}{dt}(t) \right\rangle.$$

We have used the fact that $F_n(t, u_n(t))$ belongs to $L^2(\Omega)$ which is easy to establish. By integrating this equality over $(0,T)$, we obtain

$$\int_0^T \left\| \frac{du_n}{dt}(t) \right\|_X^2 dt + \int_0^T \left\langle D\Phi_n(u_n(t)), \frac{du_n}{dt}(t) \right\rangle dt$$
$$= \int_0^T \left\langle F_n(t, u_n(t)), \frac{du_n}{dt}(t) \right\rangle dt. \tag{2.35}$$

Since $\frac{du_n}{dt}$ belongs to $L^2(0,T,X)$ and $t \mapsto \Phi_n(u_n(t))$ is absolutely continuous (see (Brezis, 1973, Theorem 3.6)), we infer that for a.e. $t \in (0,T)$, $\frac{d}{dt}\Phi(u_n(t)) = \langle D\Phi u_n(t), \frac{du_n}{dt}(t) \rangle$ (for a proof refer to (Attouch *et al.*, 2014, Proposition 17.2.5)). Therefore from (2.35), we obtain

$$\int_0^T \left\| \frac{du_n}{dt}(t) \right\|_X^2 dt$$
$$= -\Phi_n(u_n(T)) + \Phi_n(u_n^0) + \int_0^T \left\langle F_n(t, u_n(t)), \frac{du_n}{dt}(t) \right\rangle dt \tag{2.36}$$
$$\leq - \inf_{n\in\mathbb{N}, v\in L^2(\Omega)} \Phi_n(v) + \sup_{n\in\mathbb{N}} \Phi_n(u_n^0)$$
$$+ \left(\int_0^T \|F_n(t, u_n(t))\|_X^2 dt \right)^{\frac{1}{2}} \left(\int_0^T \left\| \frac{du_n}{dt}(t) \right\|_X^2 dt \right)^{\frac{1}{2}}. \tag{2.37}$$

On one hand, from Lemma 2.2 and (H_1) we have

$$\inf_{n\in\mathbb{N}, v\in L^2(\Omega)} \Phi_n(v) \geq -C_2 \sup_{n\in\mathbb{N}} \|\phi_n\|_{L^2_{\mathcal{H}_{N-1}(\partial\Omega)}} > -\infty, \tag{2.38}$$

and, from (H_2), $\sup_{n\in\mathbb{N}} \Phi_n(u_n^0) < +\infty$. On the other hand, according to the structure of the CP-structured reaction functional F_n, we have

$$\|F_n(t, u_n(t))\|_X^2 = \left\| \widetilde{F}_n(t, u_n(t)) \right\|_X^2 \leq 2\|F_n(t,0)\|_X^2 + 2L^2 \|u_n(t)\|_{C([0,T],X)}^2 \tag{2.39}$$

where $L = L_I \sup_{n\in\mathbb{N}} \|r_n\|_{L^\infty([0,T]\times\mathbb{R}^N, \mathbb{R}^l)}$, and L_I is the Lipschitz constant of $(g_n)_{n\in\mathbb{N}}$ in the interval $I = \left[\inf_n \underline{y}_n(T), \sup_n \overline{y}_n(T)\right]$. It is easily seen that

$$\|F_n(t,0)\|_X^2 \leq C(1 + \|q_n(t,\cdot)\|_X^2) \tag{2.40}$$

where, from hypothesis (H_4) and (H_5), C is a nonnegative constant which does not depend on n. Hence

$$\int_0^T \|F_n(t,0)\|_X^2 dt \leq C\left(1 + \|q_n\|^2_{L^2(0,T,X)}\right)$$

so that, from (H_6)

$$\sup_{n\in\mathbb{N}} \int_0^T \|F_n(t,0)\|_X^2 dt < +\infty. \tag{2.41}$$

Combining (2.33) and (2.41), (2.39) yields

$$\sup_{n\in\mathbb{N}} \int_0^T \|F_n(t,u_n(t))\|_X^2 dt < +\infty. \tag{2.42}$$

From (2.37), and (2.38), (2.42), we infer that there exists a nonnegative constant C, which does not depend on n, such that

$$\int_0^T \left\|\frac{du_n}{dt}(t)\right\|_X^2 dt \leq C\left(1 + \left(\int_0^T \left\|\frac{du_n}{dt}(t)\right\|_X^2 dt\right)^{\frac{1}{2}}\right),$$

from which we deduce (2.34).

Step 2. (Compactness) We prove that there exist $u \in C([0,T],X)$, and a subsequence of $(u_n)_{n\in\mathbb{N}}$, not relabeled, satisfying $u_n \to u$ in $C([0,T],X)$. As usual, we apply the Ascoli-Arzela compactness theorem in the space $C([0,T],X)$. From (2.33), $(u_n)_{n\in\mathbb{N}}$ is bounded in $C([0,T],X)$. Moreover, for all $(s,t) \in [0,T]^2$ with $s < t$, we have

$$\begin{aligned}\|u_n(t) - u_n(s)\|_X &\leq \int_s^t \left\|\frac{du_n}{dt}(\tau)\right\|_X d\tau \\ &\leq (t-s)^{\frac{1}{2}} \left\|\frac{du_n}{dt}\right\|_{L^2(0,T,X)} \leq (t-s)^{\frac{1}{2}} \sup_{n\in\mathbb{N}} \left\|\frac{du_n}{dt}\right\|_{L^2(0,T,X)}\end{aligned}$$

which, from (2.34), yields the equicontinuity of the sequence $(u_n)_n \in \mathbb{N}$. It remains to establish the relative compactness in X of the set $E_t := \{u_n(t) : n \in \mathbb{N}\}$ for each $t \in [0,T]$. For $t = 0$ there is nothing to prove because of hypothesis (H_3) on the initial condition. Let $t \in]0,T]$. Substituting t for T in (2.36) gives

$$\int_0^t \left\|\frac{du_n}{ds}(s)\right\|_X^2 ds = -\Phi_n(u_n(t)) + \Phi_n(u_n^0) + \int_0^t \left\langle F_n(s,u_n(s)), \frac{du_n}{ds}(s)\right\rangle ds.$$

From (H_2), (2.34) and (2.42), we infer that $\limsup_{n\to+\infty} \Phi_n(u_n(t)) < +\infty$. Hence, from the equi-coerciveness property deduced from Lemma 2.2 (equi-coercivity of functional (2.24) is obvious), we infer that $\limsup_{n\to+\infty} \|\nabla u_n(t)\|_X < +\infty$. The compactness of the set E_t then follows from (2.33) and the compact embedding $H^1(\Omega) \hookrightarrow L^2(\Omega)$.

Step 3. We assert that $\frac{du_n}{dt} \rightharpoonup \frac{du}{dt}$ weakly in $L^2(0,T,X)$ for a non relabeled subsequence, and that $\inf_n \underline{y}_n(t) \leq u(t) \leq \sup_n \overline{y}_n(t)$ for all $t \in [0,T]$. The first claim is a straightforward consequence of (2.34) and ***Step 2***. The second one follows from inequality $\inf_n \underline{y}_n(t) \leq u_n(t) \leq \sup_n \overline{y}_n(t)$ for all $t \in [0,T]$, and the convergence $u_n \to u$ in $C([0,T],X)$.

Step 4. We prove that u is the unique solution of $(\mathcal{P})$. We denoted by Φ_n^* the Legendre-Fenchel conjugate of Φ_n. To simplify the notation, we still write $G_n(t) = F_n(t,u_n(t))$, $\Phi_n(u_n(t))$, and $\Phi_n^*(G_n(t) - \frac{du_n}{dt}(t))$ for their subsequences obtained from that of $(u_n)_{n\in\mathbb{N}}$, whose existence has been established in ***Step 3***. According to the Fenchel extremality condition (see (Attouch *et al.*, 2014, Proposition 9.5.1)) $(\mathcal{P}_n)$ is equivalent to

$$\Phi_n(u_n(t)) + \Phi_n^*\left(G_n(t) - \frac{du_n}{dt}(t)\right) + \left\langle \frac{du_n}{dt}(t) - G_n(t), u_n(t)\right\rangle = 0$$

for a.e. $t \in (0,T)$ (together with the initial condition that we do not write), which is also equivalent to

$$\int_0^T \left[\Phi_n(u_n(t)) + \Phi_n^*\left(G_n(t) - \frac{du_n}{dt}(t)\right) + \left\langle \frac{du_n}{dt}(t) - G_n(t), u_n(t)\right\rangle\right] dt = 0.$$

The last equivalence above is due to the Legendre-Fenchel inequality which asserts that inequality $\Phi_n(u_n(t)) + \Phi_n^*(G_n(t) - \frac{du_n}{dt}(t)) + \langle \frac{du_n}{dt}(t) - G_n(t), u_n(t)\rangle \geq 0$ for a.e. $t \in (0,T)$, is always true (see (Attouch *et al.*, 2014, Remark 9.5.1)). Therefore, $(\mathcal{P}_n)$ is equivalent to

$$\int_0^T \left[\Phi_n(u_n(t)) + \Phi_n^*\left(G_n(t) - \frac{du_n}{dt}(t)\right)\right.$$
$$\left. + \frac{d}{dt}\left(\frac{1}{2}\|u_n(\cdot)\|^2\right)(t) - \langle G_n(t), u_n(t)\rangle\right] dt = 0,$$

hence

$$\int_0^T \left[\Phi_n(u_n(t)) + \Phi_n^*\left(G_n(t) - \frac{du_n}{dt}(t)\right)\right] dt$$
$$+ \frac{1}{2}\left(\|u_n(T)\|^2 - \|u_n^0\|^2\right) - \int_0^T \langle G_n(t), u_n(t)\rangle \, dt = 0. \quad (2.43)$$

We are going to pass to the limit in (2.43). On the one hand we have from (H_3)

$$\|u_n^0\|_X \to \|u^0\|_X. \quad (2.44)$$

Combining $u_n(T) = u_n^0 + \int_0^T \frac{du_n}{dt}(t)dt$ with $\frac{du_n}{dt} \rightharpoonup \frac{du}{dt}$ in $L^2(0,T,X)$, we infer that

$$\liminf_{n\to+\infty} \|u_n(T)\|^2 \geq \|u(T)\|^2. \quad (2.45)$$

On the other hand we have the following convergence, established after the end of the proof:

Lemma 2.7. *The functional $G_n = F_n(\cdot, u_n)$ weakly converges in $L^2(0,T,X)$ to G defined by $G(t) = F(t,u(t))$ where $F(t,u(t))(x) = r(t,x)\cdot g(u(t,x)) + q(t,x)$.*

Observe now that $w \mapsto \int_0^T \Phi_n(w(t))dt$ Mosco-converges to $w \mapsto \int_0^T \Phi(w(t))dt$. Indeed, apply Lemma 2.6 to $\Phi_n + C \geq 0$ and $\Phi + C \geq 0$[5] where $C = C_2 \sup_{n\in\mathbb{N}} \|\phi_n\|_{L^2_{\mathcal{H}_{N-1}(\partial\Omega)}}$. Hence, from ***Step 2***, ***Step 3***, Lemma 2.7, and Theorem B.2, the following estimates hold

$$\liminf_{n\to+\infty} \int_0^T \Phi_n(u_n(t))dt \geq \int_0^T \Phi(u(t))dt,$$
$$\liminf_{n\to+\infty} \int_0^T \Phi_n^*\left(G_n(t) - \frac{du_n}{dt}(t)\right)dt \geq \int_0^T \Phi^*\left(G(t) - \frac{du}{dt}(t)\right)dt.$$

Therefore, passing to the limit in (2.43), from (2.44), (2.45), Lemma 2.7, we obtain

$$\int_0^T \left[\Phi(u(t)) + \Phi^*\left(G(t) - \frac{du}{dt}(t)\right)\right] dt$$
$$+ \frac{1}{2}(\|u(T)\|^2 - \|u^0\|^2) - \int_0^T \langle G(t), u(t)\rangle \, dt \leq 0$$

or equivalently,

$$\int_0^T \left[\Phi(u(t)) + \Phi^*\left(G(t) - \frac{du}{dt}(t)\right) + \left\langle \frac{du}{dt}(t) - G(t), u(t)\right\rangle\right] dt \leq 0. \tag{2.46}$$

From the Legendre-Fenchel inequality, we have $\Phi(u(t)) + \Phi^*(G(t) - \frac{du}{dt}(t)) + \langle \frac{du}{dt}(t) - G(t), u(t)\rangle \geq 0$, so that (2.46) yields that for a.e. $t \in (0,T)$, $\Phi(u(t)) + \Phi^*(G(t) - \frac{du}{dt}(t)) + \langle \frac{du}{dt}(t) - G(t), u(t)\rangle = 0$ which is, according to (Attouch *et al.*, 2014, Proposition 9.5.1), equivalent to

$$\frac{du}{dt}(t) + \partial\Phi(u(t)) \ni G(t) \text{ for a.e. } t \in (0,T).$$

We have already proved that $\inf_n \underline{\rho}_n \leq u_0 \leq \sup_n \overline{\rho}_n$ in ***Step 3***. It remains to establish that $u_0 \in \mathrm{dom}(\Phi)$. From ($H_1$) and ($H_2$), we infer that

$$\Phi(u_0) \leq \liminf_{n\to+\infty} \Phi_n(u_n^0) \leq \sup_{n\in\mathbb{N}} \Phi_n(u_n^0) < +\infty,$$

which proves the claim. For the proof of uniqueness of

$$(\mathcal{P}) \begin{cases} \dfrac{du}{dt}(t) + \partial\Phi(u(t)) \ni F(t,u(t)) \text{ for a.e. } t \in (0,T) \\[2ex] u(0) = u^0, \quad \inf_n \underline{\rho}_n \leq u_0 \leq \sup_n \overline{\rho}_n, \quad u^0 \in H^1(\Omega), \end{cases}$$

it is enough to reproduce the proof of uniqueness in ***Step 3*** of Theorem 2.1, with a Lipschitz constant given by $L(t) = \|r\|_{L^\infty([0,T]\times\mathbb{R}^N,\mathbb{R}^l)} L_I$, where $I =$

[5]Since $\Phi_n \xrightarrow{M} \Phi$, for any $u \in \mathrm{dom}(\Phi)$, there exists $v_n \to v$ in $L^2(0,T,X)$ such that $\Phi(u) = \lim_{n\to+\infty} \Phi_n(v_n) \geq -C$.

$\Big[\inf_n \underline{y}_n(T), \sup_n \overline{y}_n(T)\Big]$. Since every subsequence of the subsequence of $(u_n)_{n\in\mathbb{N}}$ obtained above, converges to the same limit u in $C([0,T],X)$, the sequence $(u_n)_{n\in\mathbb{N}}$ converges to u in $C([0,T],X)$. For the same reason, the sequence $(\frac{du_n}{dt})_{n\in\mathbb{N}}$ converges to $\frac{du}{dt}$, weakly in $L^2(0,T,X)$.

Step 5. We show that if $\Phi_n(u_n^0) \to \Phi(u^0)$, $r_n \to r$ strongly in $L^2(0,T,L^2(\Omega,\mathbb{R}^l))$ and $q_n \to q$ strongly in $L^2(0,T,X)$, then $\frac{du_n}{dt} \to \frac{du}{dt}$ strongly in $L^2(0,T,X)$. From ***Step 3*** and ***Step 4***, we have $\frac{du_n}{dt} \rightharpoonup \frac{du}{dt}$ weakly in $L^2(0,T,X)$. Therefore, it suffices to establish that $\left\|\frac{du_n}{dt}\right\|^2_{L^2(0,T,X)} \to \left\|\frac{du}{dt}\right\|^2_{L^2(0,T,X)}$ to prove the claim. By repeating the proof of Lemma 2.7 under the hypotheses of strong convergence of r_n and q_n to r and q respectively, it is easily seen that G_n strongly converges to G in $L^2(0,T,X)$. Consequently, passing to the limit on (2.36), i.e. on

$$\int_0^T \left\|\frac{du_n}{dt}(t)\right\|_X^2 dt = -\Phi_n(u_n(T)) + \Phi_n(u_n^0) + \int_0^T \left\langle F_n(t,u_n(t)), \frac{du_n}{dt}(t)\right\rangle dt,$$

and since $\Phi_n \overset{M}{\to} \Phi$, we deduce that

$$\begin{aligned}
&\limsup_{n\to+\infty} \int_0^T \left\|\frac{du_n}{dt}(t)\right\|_X^2 dt \\
&= -\liminf_{n\to+\infty} \Phi_n(u_n(T)) + \Phi(u^0) + \int_0^T \left\langle F(t,u(t)), \frac{du}{dt}(t)\right\rangle dt \\
&\le -\Phi(u(T)) + \Phi(u^0) + \int_0^T \left\langle F(t,u(t)), \frac{du}{dt}(t)\right\rangle dt \\
&= \int_0^T \left\|\frac{du}{dt}(t)\right\|_X^2 dt.
\end{aligned}$$

The conclusion follows from the lower semicontinuity of the convex function $\zeta \mapsto \|\zeta\|^2_{L^2(0,T,X)}$ in $L^2(0,T,X)$.

For the last claim, observe that in (2.36), T can be replaced by an arbitrary $t \in [0,T]$, which, using Lemma 2.7 and the strong convergence of the time-derivatives, implies that $\lim_{n\to+\infty} \Phi_n(u_n(t)) = \Phi(u(t))$.

Step 6. We establish the last assertion of Theorem 2.6. We start by establishing the following estimate on the total variation Var $(G,[0,T])$ of G in $[0,T]$, from which we deduce that $G \in W^{1,1}(0,T,L^2(\Omega))$: set $\overline{g} = \sup_{\zeta\in I} |g(\zeta)|$ where $I = \Big[\inf_n \underline{y}_n(T), \sup_n \overline{y}_n(T)\Big]$, and denote by L_I the Lipchitz constant of g in I, then

$$\mathrm{Var}(G,[0,T]) = \int_0^T \left\|\frac{G}{dt}(t)\right\|_{L^2(\Omega)} dt \le C'\left(1 + \int_0^T \left\|\frac{du}{dt}(t)\right\|_{L^2(\Omega)} dt\right) \tag{2.47}$$

where $C' = C\left(\int_0^T \left\|\frac{dr}{dt}(\tau,\cdot)\right\|_{L^2(\Omega,\mathbb{R}^l)}, \int_0^T \left\|\frac{dq}{dt}(\tau,\cdot)\right\|_{L^2(\Omega,\mathbb{R})}, \overline{g}, L_I\right)$ is a nonnegative constant.

Indeed, according to the structure of F, we have for all (t,s), $0 \le t \le s \le T$

$$
\begin{aligned}
&\|G(t) - G(s)\|_X \\
&\le \bar{g}\|r(t) - r(s)\|_{L^2(\Omega,\mathbb{R}^l)} + \|q(t) - q(s)\|_{L^2(\Omega)} + \|r\|_\infty \|g(u(t)) - g(u(s))\|_{L^2(\Omega)} \\
&\le \bar{g}\int_s^t \left\|\frac{dr}{dt}(\tau,\cdot)\right\|_{L^2(\Omega,\mathbb{R}^l)} d\tau + \int_s^t \left\|\frac{dq}{dt}(\tau,\cdot)\right\|_{L^2(\Omega)} d\tau + L_I \int_s^t \left\|\frac{du}{dt}(\tau,\cdot)\right\|_{L^2(\Omega,\mathbb{R})} d\tau \\
&\le \bar{g}\int_0^T \left\|\frac{dr}{dt}(\tau,\cdot)\right\|_{L^2(\Omega,\mathbb{R}^l)} d\tau + \int_0^T \left\|\frac{dq}{dt}(\tau,\cdot)\right\|_{L^2(\Omega)} d\tau + L_I \int_0^T \left\|\frac{du}{dt}(\tau,\cdot)\right\|_{L^2(\Omega,\mathbb{R})} d\tau,
\end{aligned}
$$

and estimate (2.47) follows, since $\int_0^T \|\frac{dr}{dt}(\tau,\cdot)\|_{L^2(\Omega,\mathbb{R}^l)} d\tau < +\infty$ and $\int_0^T \left\|\frac{dq}{dt}(\tau,\cdot)\right\|_{L^2(\Omega)} d\tau < +\infty$.

From Lemma 2.1 and Remark 2.2, we infer that $u(t) \in \operatorname{dom}(\partial\Phi)$ for all $t \in]0,T]$. On the other hand, since F_n is a regular CP-structured reaction functionals, by using the same calculation, we can easily show that $G_n \in W^{1,1}(0,T,L^2(\Omega))$. Hence, according to Corollary 2.1, u_n satisfies (S_3), then possesses a right derivative at each $t \in]0,T[$. Moreover,

$$\frac{d^+u_n}{dt}(t) + D\Phi_n(u_n(t)) = F_n(t,u_n(t)) \text{ for all } t \in]0,T[. \tag{2.48}$$

Fix $t \in]0,T[$. We first establish successively

$$F_n(t,u_n(t)) \rightharpoonup F(t,u(t)) \text{ weakly in } L^2(\Omega); \tag{2.49}$$

$$\sup_{n\in\mathbb{N}} \left\|\frac{d^+u_n}{dt}(t)\right\|_X < +\infty; \tag{2.50}$$

$$\sup_{n\in\mathbb{N}} \|D\Phi_n(u_n(t))\|_X < +\infty. \tag{2.51}$$

Proof of (2.49). This convergence follows straightforwardly from $q_n(t,\cdot) \rightharpoonup q(t,\cdot)$ weakly in $L^2(\Omega)$, $r_n(t,\cdot) \rightharpoonup r(t,\cdot)$ weakly in $L^2(\Omega,\mathbb{R}^l)$, and $g_n(u_n(t)) \to g(u(t))$ strongly in $L^2(\Omega,\mathbb{R}^l)$ (see the proof of Lemma 2.7 below).

Proof of (2.50). By applying Proposition 2.5, we deduce that there exists $C_n \ge 0$ given by (2.18), such that

$$\left\|\frac{d^+u_n}{dt}(t)\right\|_X \le C_n + \left(C_n + \frac{1}{t}\right)\int_0^T \left\|\frac{du_n}{dt}(t)\right\|_X dt.$$

From the uniform Lipschitz condition (2.26), the weak convergences of $\frac{dr_n}{dt}$ in $L^1(0,T,L^2(\Omega,\mathbb{R}^l))$, $\frac{dq_n}{dt}$ in $L^1(0,T,L^2(\Omega))$, and (H_5), we infer that $\sup_{n\in\mathbb{N}} C_n < +\infty$. Hence (2.50) follows from (2.34). We deduce (2.51) from these two estimates and (2.48).

The idea is now to identify the weak limit of $D\Phi_n(u_n(t))$ by using a suitable graph-convergence and to pass to the weak limit in (2.48). From the uniform convergence of $(u_n)_{n\in\mathbb{N}}$ to u in $C([0,T],L^2(\Omega))$, (2.50) and (2.51), there exists a subsequence of $(u_n(t))_{n\in\mathbb{N}}$ (non relabeled), A, and θ in $L^2(\Omega)$, such that

$$u_n(t) \to u(t) \text{ strongly in } L^2(\Omega),$$

$$D\Phi_n(u_n(t)) \rightharpoonup A(t) \text{ weakly in } L^2(\Omega),$$

$$\frac{d^+u_n}{dt}(t) \rightharpoonup \theta(t) \text{ weakly in } L^2(\Omega).$$

Recall that from Theorem B.4,

$$\Phi_n \overset{\Gamma_{w-X}}{\to} \Phi \text{ and } (\Phi_n)_{n\in\mathbb{N}} \text{ equi-coercive} \Longrightarrow \partial\Phi_n \overset{G_{s,w}}{\to} \partial\Phi.$$

Hence from the two first assertions above, we infer that $A(t) \in \partial\Phi(u(t)) = \{D\Phi(u(t))\}$. Consequently, by going to the limit in (2.48), from (2.49), we obtain

$$\theta(t) + D\Phi(u(t)) = F(t,u(t)).$$

Since $G \in W^{1,1}(0,T,L^2(\Omega))$ we know that $\frac{d^+u}{dt}(t) + D\Phi(u(t)) = F(t,u(t))$. Hence $\theta(t) = \frac{d^+u}{dt}(t)$. Therefore $\frac{d^+u_n}{dt}(t) \rightharpoonup \frac{d^+u}{dt}(t)$ weakly in $L^2(\Omega)$. □

It remains to establish the proof of Lemma 2.7 invoked in ***Step 4***.

Proof of Lemma 2.7. Recall that $G_n(t) = H_n(t) + q_n(t)$ where

$$H_n(t)(x) = r_n(t,x) \cdot g_n(u_n(t,x)).$$

Hence, since $q_n \rightharpoonup q$ in $L^2(0,T,X)$, it remains to prove that $H_n \rightharpoonup H$ in $L^2(0,T,X)$ where $H(t)(x) = r(t,x) \cdot g(u(t,x))$. According to (2.26) in the interval $I = [\underline{\rho}, \overline{\rho}]$, we have[6]

$$\|g_n(u_n(t)) - g(u(t))\|_{L^2(\Omega,\mathbb{R}^l)} \le L_I\|u_n(t) - u(t)\|_X + \|g_n(u(t)) - g(u(t))\|_{L^2(\Omega,\mathbb{R}^l)}.$$

Hence

$$\int_0^T \|g_n(u_n(t)) - g(u(t))\|^2_{L^2(\Omega,\mathbb{R}^l)}\, dt$$
$$\le 2L_I^2 \int_0^T \|u_n(t) - u(t)\|_X^2\, dt + \int_0^T \|g_n(u(t)) - g(u(t))\|^2_{L^2(\Omega,\mathbb{R}^l)}\, dt. \tag{2.52}$$

On the other hand, from (2.26), and hypothesis (H_4), we clearly deduce that $|g_n(\zeta)| \le C(1+|\zeta|)$ for all $\zeta \in \mathbb{R}$, where C is a nonnegative constant depending only on L_I and $g(0)$. Consequently, applying the Lebesgue dominated convergence theorem and (H_4), we infer that

$$\lim_{n\to+\infty} \int_0^T \|g_n(u(t)) - g(u(t))\|^2_{L^2(\Omega,\mathbb{R}^l)}\, dt = 0.$$

Passing to the limit in (2.52) we deduce that $g_n(u_n(.)) \to g(u(.))$ strongly in $L^2(0,T,L^2(\Omega,\mathbb{R}^l))$. The conclusion of Lemma 2.7 follows from the fact that $r_n \rightharpoonup r$ weakly in $L^2(0,T,L^2(\Omega,\mathbb{R}^l))$. □

[6]To simplify the notation we write $g_n(u_n(t))$ for the function $x \mapsto g_n(u_n(t,x))$.

Remark 2.13. In Anza Hafsa *et al.* (2019a) one can find another proof of the compactness ***Step 2***, under the equi-ellipticity condition: there exists $\gamma > 0$ such that for a.e. $x \in \mathbb{R}^N$,

$$\inf_n \langle D_\xi W_n(x,\xi), \xi\rangle \geq \gamma|\xi|^2 \text{ for all } \xi \in \mathbb{R}^N.$$

The proof is more involved and based on the $L^2(\Omega)$-norm estimate of the right derivative, Proposition 2.5. However, for nonnegative functions W_n, equi-ellipticity yields the lower bound of (D_n). Indeed, from the subdifferential inequality, we infer that

$$\begin{aligned} W_n(x,\xi) &\geq W_n\left(x, \frac{1}{2}\xi\right) + \left\langle D_\xi W_n\left(x, \frac{1}{2}\xi\right), \frac{1}{2}\xi\right\rangle \\ &\geq \frac{\gamma}{4}|\xi|^2. \end{aligned}$$

Remark 2.14. When Φ admits an integral representation, then, under a suitable condition on the Legendre-Fenchel conjugate of its density, one can assert that $\partial\Phi$ is single valued (see condition (D_3^*) in Part 2).

Some conclusions of Theorem 2.6 are still valid for more general sequences of reaction functionals F_n. Indeed, assume that associated reactions functions f_n fulfill the two conditions:

(i) $\zeta \mapsto f_n(t,x,\zeta)$ is locally Lipschitz continuous, uniformly with respect to $(t,x,n) \in \mathbb{R} \times \mathbb{R}^N \times \mathbb{N}$, i.e. for all interval $I \subset \mathbb{R}$, there exists $L_I \geq 0$ such that, for all $(\zeta,\zeta') \in I \times I$, and all $(t,x) \in \mathbb{R} \times \mathbb{R}^N$,
$$\sup_{n\in\mathbb{N}} |f_n(t,x,\zeta) - f_n(t,x,\zeta')| \leq L_I|\zeta - \zeta'|;$$
(ii) f_n satisfies (CP).

Then, reproducing ***Step 1–Step 4*** of the proof of Theorem 2.6, and from Corollary 2.4, we easily establish the following convergence result:

Theorem 2.7. *Assume that the sequence of densities* $(W_n)_{n\in\mathbb{N}}$ *of functionals* (2.23), *or* (2.24), *satisfies* (D_n), *and that the sequence of reaction functionals* $(F_n)_{n\in\mathbb{N}}$, *satisfies (i), (ii),* (2.27), (2.28) *or* (2.29). *Let* u_n *be the unique solution of the Cauchy problem*

$$(\mathcal{P}_n) \begin{cases} \dfrac{du_n}{dt}(t) + D\Phi_n(u_n(t)) = F_n(t,u_n(t)) \text{ for a.e. } t \in (0,T) \\ \\ u_n(0) = u_n^0, \quad \underline{\rho}_n \leq u_n^0 \leq \overline{\rho}_n, \quad u_n^0 \in \mathrm{dom}(\Phi_n). \end{cases}$$

Assume that (H_1), (H_2), (H_3) *hold, and that* G_n *defined by* $G_n(t) = F_n(t,u_n(t))$ *weakly converges to some functional* G *in* $L^2(0,T,L^2(\Omega))$. *Then* $(u_n)_{n\in\mathbb{N}}$ *uniformly converges in* $C([0,T],L^2(\Omega))$ *to the unique solution of the problem*

$$(\mathcal{P}) \begin{cases} \dfrac{du}{dt}(t) + \partial\Phi(u(t)) \ni G(t) \text{ for a.e. } t \in (0,T) \\ \\ u(0) = u^0, \quad \inf_n \underline{\rho}_n \leq u^0 \leq \sup_n \overline{\rho}_n, \quad u_0 \in \mathrm{dom}(\Phi). \end{cases}$$

Moreover, $\inf_{n\in\mathbb{N}} \underline{y}_n(t) \leq u(t) \leq \sup_{n\in\mathbb{N}} \overline{y}_n(t)$ *for all* $t \in [0,T]$, *and* $\frac{du_n}{dt} \rightharpoonup \frac{du}{dt}$ *weakly in* $L^2(0,T,L^2(\Omega))$.

Remark 2.15. If G belongs to $W^{1,1}(0,T,L^2(\Omega))$, then, from Lemma 2.1 and Remark 2.2, we can assert that $u(t) \in \mathrm{dom}(\partial\Phi)$ for all $t \in]0,T]$.

Example 2.6. Consider the sequence of reaction-diffusion Cauchy problems corresponding to Example 2.5:

$$(\mathcal{P}_n)\begin{cases} \dfrac{du_n}{dt}(t) - \mathrm{div}D_\xi W_n(\cdot,\nabla u(t)) = F_n(t,u_n(t)) \text{ for a.e. } t\in(0,T),\\ u(0)=u_0,\quad 0\leq u_n^0\leq\overline{\rho},\quad u_0\in\mathrm{dom}(\Phi_n),\\ u_n(t)\in H^1_\Gamma(\Omega), \mathrm{div}D_\xi W_n(\cdot,\nabla u(t))\in L^2(\Omega) \text{ for a.e. } t\in(0,T),\\ D_\xi W_n(\cdot,\nabla u_n(t))\cdot\mathbf{n}=0 \text{ on } \partial\Omega\setminus\Gamma \text{ for a.e. } t\in(0,T), \end{cases}$$

with $F_n(t,u_n(t))(x) = a_n(x)R\left(t - g\left(\frac{L-u_n(t,x)}{K}\right)\right)$, where

- $R:\mathbb{R}\to\mathbb{R}_+$ is Lipschitz continuous: there exists $L_R>0$ such that
$$|R(t)-R(t')|\leq L_R|t-t'|$$
for all $(t,t')\in\mathbb{R}^2$;
- $g:\mathbb{R}\to\mathbb{R}$ is locally Lipschitz continuous: for all interval $I\subset\mathbb{R}$, there exists $L_I>0$ such that $|g(\zeta)-g(\zeta')|\leq L_I|\zeta-\zeta'|$ for all $(\zeta,\zeta')\in I\times I$;
- $a_n\in L^\infty(\mathbb{R}^N)$, and $a_n\geq 0$.

Assume that Φ_n of the form (2.24) satisfies: $\Phi_n \overset{M}{\to} \Phi$, and $\sup_{n\in\mathbb{N}}\Phi_n(u_n^0) < +\infty$. Assume furthermore that $u_n^0 \to u^0$ strongly in $L^2(\Omega)$ and $a_n \rightharpoonup a$ for the $\sigma(L^\infty,L^1)$-weak topology. Then, from Theorem 2.7, $(u_n)_{n\in\mathbb{N}}$ uniformly converges in $C([0,T],L^2(\Omega))$ to the unique solution of the problem

$$(\mathcal{P})\begin{cases} \dfrac{du}{dt}(t)+\partial\Phi(u(t))\ni G(t) \text{ for a.e. } t\in(0,T)\\ u(0)=u^0,\quad 0\leq u^0\leq\overline{\rho},\quad u^0\in\mathrm{dom}(\Phi), \end{cases}$$

with

$$G(t)(x) = a(x)S\left(t-g\left(\frac{L-u(t,x)}{K}\right)\right).$$

Moreover $\frac{du_n}{dt} \rightharpoonup \frac{du}{dt}$ weakly in $L^2(0,T,L^2(\Omega))$, and, for all $t\in[0,T]$, $0\leq u(t)\leq \overline{y}(t)$, where $\overline{y}$ is the solution of o.d.e. $\overline{y}' = \sup_n \|a_n\|_{L^\infty(\mathbb{R}^N)}R\left(t-g\left(\frac{L-\overline{y}}{K}\right)\right)$, with initial condition $\overline{y}(0)=\overline{\rho}$ (we take $\overline{y}_n=\overline{y}$ for all $n\in\mathbb{N}$). This example is treated within the framework of stochastic homogenization in Chapter 7.

In some cases, we can specify the domain of the limit functional Φ as stated in the following Proposition.

Proposition 2.6. *i) Assume* (H_1). *Then* $\mathrm{dom}(\Phi) \subset H^1(\Omega)$.

ii) Under (H_1), *assume that* $\liminf_{n\to+\infty} \beta_n < +\infty$, $a_{0,n} \rightharpoonup a_0$ *for the* $\sigma\left(L^\infty_{\mathcal{H}_{N-1}}(\partial\Omega), L^1_{\mathcal{H}_{N-1}}(\partial\Omega)\right)$ *topology, and* $\phi_n \rightharpoonup \phi$ *weakly in* $L^2_{\mathcal{H}_{N-1}}(\partial\Omega)$. *Then* $\mathrm{dom}(\Phi) = H^1(\Omega)$.

Proof. Proof of i). Let $v \in \mathrm{dom}(\Phi)$, then from (H_1) and Proposition B.3, there exists $v_n \to v$ strongly in $L^2(\Omega)$ such that $\lim_{n\to+\infty} \Phi_n(v_n) = \Phi(v) < +\infty$. From Lemma 2.2 we see that Φ is equi-coercive so that

$$\sup_{n\in\mathbb{N}} \|v_n\|_{H^1(\Omega)} < +\infty.$$

Therefore, there exists a subsequence, that we do not relabel, and $w \in H^1(\Omega)$, such that $v_n \rightharpoonup w$ weakly in $H^1(\Omega)$ and strongly in $L^2(\Omega)$. Hence $v = w \in H^1(\Omega)$.

Proof of ii). It remains to prove that $H^1(\Omega) \subset \mathrm{dom}(\Phi)$. Observe that from ($\mathrm{H}_1$), Proposition B.3, and the upper growth condition in (D_n), we have for all $v \in H^1(\Omega)$,

$$\begin{aligned}\Phi(v) &\le \liminf_{n\to+\infty} \Phi_n(v)\\ &\le \liminf_{n\to+\infty} \beta_n \left(1 + \int_\Omega |\nabla v|^2 dx\right) + \frac{1}{2}\int_{\partial\Omega} a_0 v^2 d\mathcal{H}_{N-1} - \int_{\partial\Omega} \phi v d\mathcal{H}_{N-1} < +\infty.\end{aligned}$$

This completes the proof. □

For each $n \in \mathbb{N}$, let us write the functional (2.23) as follows:

$$\Phi_n(u) = \begin{cases} \widetilde{\Phi}_n(u) + \dfrac{1}{2}\displaystyle\int_{\partial\Omega} a_{0,n} u^2 \, d\mathcal{H}_{N-1} - \int_{\partial\Omega} \phi_n u \, d\mathcal{H}_{N-1} & \text{if } u \in H^1(\Omega)\\ +\infty & \text{otherwise,}\end{cases}$$

where $\widetilde{\Phi}_n : H^1(\Omega) \to \mathbb{R}_+$ is defined by $\widetilde{\Phi}_n(u) = \displaystyle\int_\Omega W_n(x, \nabla u(x))\, dx$. The proposition below gives sufficient conditions which imply the Mosco-convergence of $(\Phi_n)_{n\in\mathbb{N}}$ when we assume that $\widetilde{\Phi}_n$ Γ-converges to $\widetilde{\Phi}$ with respect to the $L^2(\Omega)$ topology.

Proposition 2.7. *Assume that the sequence* $(W_n)_{n\in\mathbb{N}}$ *satisfies* (D_n) *with* $\beta_n = \beta$ *for all* $n \in \mathbb{N}$, *and that*

(H'_1)
- $\widetilde{\Phi}_n$ *Γ-converges to* $\widetilde{\Phi}$ *when* $H^1(\Omega)$ *is equipped with the strong topology of* $L^2(\Omega)$;
- $a_{0,n} \to a_0$ *strongly in* $L^\infty_{\mathcal{H}_{N-1}}(\partial\Omega)$;
- $\phi_n \to \phi$ *strongly in* $L^2_{\mathcal{H}_{N-1}}(\partial\Omega)$.

Then $\Phi_n \stackrel{M}{\to} \Phi$ *where* $\Phi : L^2(\Omega) \to \mathbb{R} \cup \{+\infty\}$ *is given by*

$$\Phi(u) = \begin{cases} \widetilde{\Phi}(u) + \dfrac{1}{2}\displaystyle\int_{\partial\Omega} a_0 u^2 \, d\mathcal{H}_{N-1} - \int_{\partial\Omega} \phi u \, d\mathcal{H}_{N-1} & \text{if } u \in H^1(\Omega) \\ +\infty & \text{otherwise.} \end{cases}$$

Proof. The proof falls into two steps. For the definition of Γ-convergence, we refer the reader to Appendix B.1, Definition B.1.

Step 1. Let $v_n \rightharpoonup v$ weakly in $L^2(\Omega)$, we establish that $\Phi(v) \le \liminf_{n\to+\infty} \Phi_n(v_n)$.

We assume that $\liminf_{n\to+\infty} \Phi_n(v_n) < +\infty$ and reason with various subsequences that we do not relabel. The letter C denotes various positive constants. From (H_1') and Lemma 2.2, we have

$$\sup_{n\in\mathbb{N}} \|v_n\|_{H^1(\Omega)} < +\infty.$$

Consequently, there exist a subsequence and $w \in H^1(\Omega)$ such that $v_n \rightharpoonup w$ weakly in $H^1(\Omega)$ and strongly in $L^2(\Omega)$. Thus $w = v$ so that $v \in H^1(\Omega)$ and $v_n \to v$ strongly in $L^2(\Omega)$. According to (H_1'), we infer that

$$\widetilde{\Phi}(v) \le \liminf_{n\to+\infty} \widetilde{\Phi}_n(v_n). \tag{2.53}$$

On the other hand

$$\begin{aligned}\int_{\partial\Omega} a_{0,n} v_n^2 d\mathcal{H}_{N-1} &= \int_{\partial\Omega} (a_{0,n} - a_0) v_n^2 d\mathcal{H}_{N-1} + \int_{\partial\Omega} a_0 v_n^2 d\mathcal{H}_{N-1} \\ &\ge -\|a_{0,n} - a_0\|_{L^\infty_{\mathcal{H}_{N-1}}(\partial\Omega)} \sup_{n\in\mathbb{N}} \int_{\partial\Omega} v_n^2 d\mathcal{H}_{N-1} + \int_{\partial\Omega} a_0 v_n^2 d\mathcal{H}_{N-1}.\end{aligned}$$

According to the weak continuity of the trace operator from $H^1(\Omega)$ into $L^2_{\mathcal{H}_{N-1}}(\partial\Omega)$ and to the lower semicontinuity of the map $w \mapsto \int_{\partial\Omega} a_0 w^2 \, d\mathcal{H}_{N-1}$ defined in $L^2_{\mathcal{H}_{N-1}}(\partial\Omega)$ equipped with its weak convergence, we infer that

$$\int_{\partial\Omega} a_0 v^2 \, d\mathcal{H}_{N-1} \le \liminf_{n\to+\infty} \int_{\partial\Omega} a_{0,n} v_n^2 \, d\mathcal{H}_{N-1}. \tag{2.54}$$

Finally, since $\phi_n \to \phi$ strongly in $L^2_{\mathcal{H}_{N-1}}(\partial\Omega)$, and $v_n \rightharpoonup v$ weakly in $L^2_{\mathcal{H}_{N-1}}(\partial\Omega)$, we have

$$\lim_{n\to+\infty} \int_{\partial\Omega} \phi_n v_n \, d\mathcal{H}_{N-1} = \int_{\partial\Omega} \phi v \, d\mathcal{H}_{N-1}. \tag{2.55}$$

The proof of the claim is obtained by collecting (2.53), (2.54), and (2.55).

Step 2. Assume that $\Phi(v) < +\infty$. We prove that there exists a sequence $(v_n)_{n\in\mathbb{N}}$ strongly converging to v in $L^2(\Omega)$ such that $\limsup_{n\to+\infty} \Phi_n(v_n) \le \Phi(v)$.

Since $\Phi(v) < +\infty$, we have $v_n \in H^1(\Omega)$, and, according to hypothesis (H_1'), there exists a sequence $(w_n)_{n\in\mathbb{N}}$ in $H^1(\Omega)$ strongly converging to v in $L^2(\Omega)$, such that

$$\lim_{n\to+\infty} \widetilde{\Phi}_n(w_n) = \Phi(v).$$

By using the well known De Giorgi slicing method, (see (Attouch *et al.*, 2014, proof of Corollary 11.2.1), it is precisely at this point that we use the uniform growth condition), we can modify w_n into a function v_n in $H^1(\Omega)$ satisfying $v_n = v$ on $\partial\Omega$ and

$$\limsup_{n\to+\infty} \widetilde{\Phi}_n(v_n) \leq \widetilde{\Phi}(v).$$

Then clearly $\limsup_{n\to+\infty} \Phi_n(v_n) \leq \Phi(v)$, which proof the claim. □

Proposition 2.7 leads straight to the following corollary of Theorem 2.6 which is applied in the context of stochastic homogenization in Part 2.

Corollary 2.5. *Assume that the sequence $(W_n)_{n\in\mathbb{N}}$ satisfies (D_n) with $\beta_n = \beta$ for all $n \in \mathbb{N}$. Under hypotheses of Theorem 4.1 where (H_1') is substituted for (H_1), the same conclusions hold.*

As stated in the next corollary, we can, in some sense, justify our convention which consists to see the functional

$$\widetilde{\Phi}(u) = \begin{cases} \displaystyle\int_\Omega W(x, \nabla u(x))dx & \text{if } u \in H^1_\Gamma(\Omega) \\ +\infty & \text{otherwise} \end{cases}$$

as a particular case of

$$\Phi(u) = \begin{cases} \displaystyle\int_\Omega W(x, \nabla u(x))dx + \frac{1}{2}\int_{\partial\Omega} a_0 u^2 d\mathcal{H}_{N-1} - \int_{\partial\Omega} \phi u \; d\mathcal{H}_{N-1} & \text{if } u \in H^1(\Omega) \\ +\infty & \text{otherwise} \end{cases}$$

with $\phi = 0$ and $a_0(x) = \begin{cases} 0 & \text{if } x \in \partial\Omega \setminus \Gamma \\ +\infty & \text{if } x \in \Gamma. \end{cases}$

For this we are going to suitably apply Theorem 2.6.

Corollary 2.6. *The problem with Robin boundary conditions*

$$(\mathcal{P}_n) \begin{cases} \dfrac{du}{dt}(t) - \mathrm{div} D_\xi W(\cdot, \nabla u(t)) = F(t, u(t)) \textit{ for a.e. } t \in (0, T), \\ u(0) = u_0, \quad \underline{\rho} \leq u_0 \leq \overline{\rho}, \quad u_0 \in \mathrm{dom}(D\widetilde{\Phi}), \\ u(t) \in H^1(\Omega), \ \mathrm{div} D_\xi W(\cdot, \nabla u(t)) \in L^2(\Omega) \textit{ for a.e. } t \in (0, T) \\ n\ u(t) + D_\xi W(\cdot, \nabla u(t)) \cdot \mathbf{n} = 0 \textit{ on } \partial\Omega \textit{ for a.e. } t \in (0, T), \end{cases}$$

converges in the sense of Theorem 2.6, to the problem with homogeneous Dirichlet-Neumann boundary conditions

$$(\mathcal{P})\begin{cases}\dfrac{du}{dt}(t)-\operatorname{div}D_\xi W(\cdot,\nabla u(t))=F(t,u(t)) \textit{ for a.e. } t\in(0,T),\\[2ex] u(0)=u_0,\quad \underline{\rho}\le u_0\le\overline{\rho},\quad u_0\in\operatorname{dom}(D\widetilde{\Phi})\\[2ex] u(t)=0 \textit{ on } \Gamma \textit{ for a.e. } t\in(0,T),\\[2ex] u(t)\in H^1(\Omega),\ \operatorname{div}D_\xi W(\cdot,\nabla u(t))\in L^2(\Omega) \textit{ for a.e. } t\in(0,T),\\[2ex] u=0 \textit{ on } \Gamma,\ D_\xi W(\cdot,\nabla u(t))\cdot\mathbf{n}=0 \textit{ on } \partial\Omega\setminus\Gamma \textit{ for a.e. } t\in(0,T).\end{cases}$$

Proof. Set $a_{0,n}(x)=\begin{cases}0 & \text{if } x\in\partial\Omega\setminus\Gamma\\ n & \text{if } x\in\Gamma\end{cases}$. We have

$$\Phi_n(u)=\begin{cases}\displaystyle\int_\Omega W(x,\nabla u(x))dx+\frac{n}{2}\int_\Gamma u^2 d\mathcal{H}_{N-1} & \text{if } u\in H^1(\Omega)\\[2ex] +\infty & \text{otherwise.}\end{cases}$$

On the other hand, set $F_n=F$ and $u_0^n=u_0$, $\underline{\rho}\le u_0\le\overline{\rho}$ (note that $\operatorname{dom}(D\widetilde{\Phi})=H^1_\Gamma(\Omega)\subset\operatorname{dom}(D\Phi_n)=H^1(\Omega)$). Problem $(\mathcal{P}_n)$ is clearly associated with Φ_n. In order to apply Theorem 2.6, it remains to establish that Φ_n Mosco converges to $\widetilde{\Phi}$. Indeed all other conditions of Theorem 2.6 are clearly fulfilled.

Let $v_n\to v$ strongly in $L^2(\Omega)$ and assume that $\liminf_{n\to+\infty}\Phi_n(v_n)<+\infty$. In what follows, we reason with various subsequences that we do not relabel. From

$$\sup_{n\in\mathbb{N}}\frac{n^2}{2}\int_\Gamma v_n^2 d\mathcal{H}_{N-1}<+\infty$$

we infer that

$$v_n\to 0 \text{ strongly in } L^2_{\mathcal{H}_{N-1}}(\Gamma). \tag{2.56}$$

On the other hand, from

$$\sup_{n\in\mathbb{N}}\int_\Omega W(x,\nabla v_n(x))dx<+\infty$$

and the lower bound condition fulfilled by W, we deduce that the sequence $(v_n)_{n\in\mathbb{N}}$ is bounded in $H^1(\Omega)$ (recall that $v_n\to v$ in $L^2(\Omega)$). Therefore $v_n\rightharpoonup v$ weakly in $H^1(\Omega)$, and, according to the continuity of the trace operator from $H^1(\Omega)$ into $L^2_{\mathcal{H}_{N-1}}(\partial\Omega)$, $v_n\rightharpoonup v$ weakly in $L^2_{\mathcal{H}_{N-1}}(\Gamma)$. From (2.56) we infer that $v=0$ in Γ, hence $v\in H^1_\Gamma(\Omega)$ and $\widetilde{\Phi}(v)=\displaystyle\int_\Omega W(x,\nabla v(x))\,dx$. Since for all $n\in\mathbb{N}$,

$$\int_\Omega W(x,\nabla v(x))dx\le\Phi_n(v_n)$$

we deduce that $\widetilde{\Phi}(v)\le\liminf_{n\to+\infty}\Phi_n(v_n)$.

Take now $v \in H^1_\Gamma(\Omega)$ (otherwise we have nothing to prove), and set $v_n = v$ for all $n \in \mathbb{N}$. Since $\Phi_n(v) = \widetilde{\Phi}(v)$ for all $n \in \mathbb{N}$, we have $\lim_{n\to+\infty} \Phi_n(v_n) = \widetilde{\Phi}(v)$, which proves the claim. □

2.5 Stability of the invasion property

This section deals with the stability in terms of variational convergence, of the invasion property discussed in Section 2.3. The density W_n is quadratic and the functional (2.23) is of the form (2.21), i.e.

$$\Phi_n(u) = \begin{cases} \displaystyle\int_\Omega A_n(x)\nabla u(x) \cdot \nabla u(x)dx & \text{if } u \in H^1_\Gamma(\Omega) \\ +\infty & \text{otherwise,} \end{cases} \tag{2.57}$$

where $A_n = (a_{i,j,n})_{i,j=1\dots N} : \mathbb{R}^N \to \mathbb{M}_N$ with $a_{i,j,n} \in C^1(\mathbb{R}^N)$. We assume that there exists $\alpha > 0$ and $\{\beta_n\}_{n\in\mathbb{N}} \subset \mathbb{R}^*_+$, such that

$$\alpha|\xi|^2 \leq \sum_{i,j=1}^{N} a_{i,j,n}(x)\xi_i\xi_j \leq \beta_n|\xi|^2$$

for all $x \in \mathbb{R}^N$ and all $\xi \in \mathbb{R}^N$.

The theorem below, which is a straightforward consequence of Theorem 2.6, states that the limit problem $(\mathcal{P})$ of $(\mathcal{P}_n)$, defined in the previous section, satisfies the invasion property even if the function g entering the structure of the limit reaction functional F is not in general C^1-regular (recall that g is assigned to be C^1-regular in Theorem 2.5). Nevertheless, we obtain a weak form of the invasion principle in the following sense: we cannot claim that the limit solution is nondecreasing for all $t \in [0,T]$, but only for a.e. t in $[0,T]$.

Theorem 2.8 (Stability of the invasion principle). *Assume* (H_1)–(H_4), *that* F_n *does not depend on* t, *and*

$\sup_{n\in\mathbb{N}} \|r_n\|_{L^\infty(\mathbb{R}^N,\mathbb{R}^l)} < +\infty$ *and* $r_n \to r$ *strongly in* $L^2(\Omega, \mathbb{R}^l)$;
$q_n \to q$ *strongly in* $L^2(\Omega)$;
$g_n \in C^1\left(\left[-\underline{\rho}_n, \overline{\rho}_n\right]\right)$ *where* $\underline{\rho}_n$, $\overline{\rho}_n$ *are given by* (2.29);
$\Phi_n(u_n^0) \to \Phi(u^0)$.

Assume furthermore that u_n^0 *is a sub solution of the elliptic problem associated with* $(\mathcal{P}_n)$, *i.e.*

$$(\mathcal{P}_{n,ell}) \begin{cases} -\mathrm{div}(A_n(\cdot)\nabla u_n^0) \leq F_n(u_n^0) \ (\textit{inequality in } L^2(\Omega)), \\ u_n^0 \in \mathrm{dom}(D\Phi_n). \end{cases}$$

Then the solution u *of* $(\mathcal{P})$ *satisfies for a.e.* $t \in [0,T]$, $\frac{du}{dt}(t) \geq 0$ *a.e. in* Ω.

Proof. From Theorem 2.5, for each $n \in \mathbb{N}$, the solution u_n of $(\mathcal{P}_n)$ satisfies $\frac{du_n}{dt}(t) \geq 0$ a.e. in Ω, for all $t \in [0,T]$. On the other hand, according to Theorem 2.6 we have $\frac{du_n}{dt} \to \frac{du}{dt}$ strongly in $L^2(0,T,L^2(\Omega))$. The conclusion then follows directly. Note that the strong convergence $\frac{du_n}{dt} \to \frac{du}{dt}$ in $L^2(0,T,L^2(\Omega))$ does not allow to claim that $\frac{du}{dt}(t) \geq 0$ a.e. in Ω for all $t \in [0,T]$. □

2.6 Variational convergence of reaction-diffusion problems: abstract version

In this section $(X, \|\cdot\|)$ is a general Hilbert space and we want to get rid of the particular structure of functions Φ_n, and F_n. Precisely, $(\Phi_n)_{n\in\mathbb{N}}$ is a sequence of lower semicontinuous and uniformly proper convex functions from X into $\mathbb{R} \cup \{+\infty\}$. Recall that this last assumption means that there exists a sequence $(v_n^0)_{n\in\mathbb{N}}$ such that $\sup_{n\in\mathbb{N}} \Phi_n(v_n^0) < +\infty$. Reaction functionals F_n are those described in Section 2.1, fulfilling conditions (C_1') and (C_2'). Let us start by generalizing Lemma 2.6.

Lemma 2.8. *Let $(\psi_n)_{n\in\mathbb{N}}$, ψ, be a sequence of lower semicontinuous, uniformly proper convex functions from X into $\mathbb{R} \cup \{+\infty\}$ such that $\psi_n \overset{M}{\to} \psi$, and consider $(\Psi_n)_{n\in\mathbb{N}}$, $\Psi : L^2(0,T,X) \to \mathbb{R}_+ \cup \{+\infty\}$ defined by*

$$\Psi_n(v) := \int_0^T \psi_n(v(t))dt; \ \Psi(v) := \int_0^T \psi(v(t))dt.$$

Then $\Psi_n \overset{M}{\to} \Psi$.

Proof. The proof is that of Lemma 2.6; we only need to adapt the proof of the domination property (2.32). For this, we take advantage of the convergence $\psi_n \overset{M}{\to} \psi$ to invoke Lemma B.3 which states that the sequence $(\psi_n)_{n\in\mathbb{N}}$ is uniformly bounded from below, i.e. there exists $r > 0$ such that

$$\forall n \in \mathbb{N}, \ \forall v \in X, \quad \|\psi_n(v)\| \geq -r(\|v\| + 1). \tag{2.58}$$

Then, with the notation of the proof of Lemma 2.6, ***Step 2***, we infer that

$$\begin{aligned}\psi_n^\lambda(u(t)) &\geq \inf_{v\in X}\left(-r\|v\| + \frac{1}{2\lambda}\|v - u(t)\|^2\right) - r\\ &= -r\|K_\lambda u(t)\| + \frac{1}{2\lambda}\|v - K_\lambda u(t)\|^2 - r =: m(t)\end{aligned}$$

where $K_\lambda u(t)$ is a minimizer of $\inf_{v\in X}(-r\|v\| + \frac{1}{2\lambda}\|v - u(t)\|^2)$. Proceeding as in the proof of Lemma 2.6, it is easily seen that $K_\lambda u(\cdot)$ belongs to $L^2(0,T,X)$, so that m belongs to $L^1(0,T)$. Therefore, from (2.32), we infer that

$$|\psi_n^\lambda(u(t))| \leq \max(|m(t)|, |M(t)|)$$

where, for v arbitrary chosen in $\mathrm{dom}(\psi)$, $M(t) := \psi(v) + 2 + \frac{1}{\lambda}\|v\|_X^2 + \frac{1}{\lambda}\|u(t)\|_X^2$. □

Theorem below is an abstract version of Theorems 2.6, 2.7. The diffusions terms together with the reaction terms do not require a special structure. Nevertheless this theorem requires the strong convergence of the reaction terms, which reduces its application, in particular to homogenization.

Theorem 2.9. *Let $(\Phi_n)_{n\in\mathbb{N}}$, Φ be a sequence of convex, lower semicontinuous, uniformly proper functionals from an Hilbert space $(X, \|\cdot\|)$ into $\mathbb{R}\cup\{+\infty\}$, satisfying* (H_1),[7] (H_2), *and* (H_3). *Let $(F_n)_{n\in\mathbb{N}}$ be a sequence of reaction functionals fulfilling* (C'_1), (C'_2), *and u_n be the unique solution of the Cauchy problem*

$$(\mathcal{P}_n)\begin{cases}\dfrac{du_n}{dt}(t) + D\Phi_n(u_n(t)) = F_n(t, u_n(t)) \text{ for a.e. } t \in (0,T)\\[2ex] u_n(0) = u_n^0, \quad u_n^0 \in \mathrm{dom}(\Phi_n).\end{cases}$$

Assume that G_n defined by $G_n(t) = F_n(t, u_n(t))$ strongly converges to some functional G in $L^2(0,T,L^2(\Omega))$. Then $(u_n)_{n\in\mathbb{N}}$ uniformly converges in $C([0,T],L^2(\Omega))$ to the unique solution of the problem

$$(\mathcal{P})\begin{cases}\dfrac{du}{dt}(t) + \partial\Phi(u(t)) \ni G(t) \text{ for a.e. } t \in (0,T)\\[2ex] u(0) = u^0, \quad u_0 \in \mathrm{dom}(\Phi).\end{cases}$$

Moreover, $\frac{du_n}{dt} \rightharpoonup \frac{du}{dt}$ weakly in $L^2(0,T,L^2(\Omega))$.

Proof. We follow the scheme of the proof of Theorem 2.6, with some minor adaptations.

Step 1. We establish

$$\sup_{n\in\mathbb{N}}\left\|\frac{du_n}{dt}\right\|_{L^2(0,T,X)} < +\infty; \tag{2.59}$$

$$\sup_{n\in\mathbb{N}}\|u_n\|_{C(0,T,X)} < +\infty. \tag{2.60}$$

From $(\mathcal{P}_n)$ we deduce that for a.e. $t \in (0,T)$,

$$\left\|\frac{du_n}{dt}(t)\right\|^2 + \left\langle D\Phi_n(u_n(t)), \frac{du_n}{dt}(t)\right\rangle = \left\langle G_n, \frac{du_n}{dt}(t)\right\rangle.$$

By integrating this equality over $(0,T)$, we obtain

$$\int_0^T \left\|\frac{du_n}{dt}(t)\right\|^2 dt + \int_0^T \left\langle D\Phi_n(u_n(t)), \frac{du_n}{dt}(t)\right\rangle dt = \int_0^T \left\langle f_n(t), \frac{du_n}{dt}(t)\right\rangle dt$$

from which we deduce

$$\int_0^T \left\|\frac{du_n}{dt}(t)\right\|^2 dt = -\Phi_n(u_n(T)) + \Phi_n(u_n^0) + \int_0^T \left\langle f_n(t), \frac{du_n}{dt}(t)\right\rangle dt. \tag{2.61}$$

[7]Here ϕ_n is not involved.

Since $\Phi_n \overset{M}{\to} \Phi$, from Lemma 2.8, there exists $r > 0$ such that $\forall n \in \mathbb{N}$, $\Phi_n \geq -r(\|\cdot\| + 1)$. Therefore (2.61) yields

$$\int_0^T \left\| \frac{du_n}{dt}(t) \right\|^2 dt$$
$$\leq \Phi_n(u_n^0) + r(\|u_n(T)\| + 1) + \|G_n\|_{L^2(0,T,X)} \left(\int_0^T \left\| \frac{du_n}{dt}(t) \right\|^2 dt \right)^{1/2}. \tag{2.62}$$

From

$$u_n(T) = u_n^0 + \int_0^T \frac{du_n}{dt}(t)\, dt$$

we infer that

$$\|u_n(T)\| \leq \|u_n^0\| + T^{1/2} \left(\int_0^T \left\| \frac{du_n}{dt}(t) \right\|^2 dt \right)^{1/2}. \tag{2.63}$$

Combining (2.62) and (2.63) we finally obtain

$$\int_0^T \left\| \frac{du_n}{dt}(t) \right\|^2 dt$$
$$\leq \Phi_n(u_n^0) + r\left(\|u_n^0\| + 1\right) + \left(rT^{1/2} + \|G_n\|_{L^2(0,T,X)}\right) \left(\int_0^T \left\| \frac{du_n}{dt}(t) \right\|^2 dt \right)^{1/2}$$

and estimate (2.59) follows from the equiboundedness of $(u_n^0)_{n\in\mathbb{N}}$ in X, that of $(G_n)_{n\in\mathbb{N}}$ in $L^2(0,T,X)$ and from the equiboundedness of $(\Phi_n(u_n^0))_{n\in\mathbb{N}}$. Estimate (2.60) follows easily from (2.59).

Step 2. (Compactness) Unlike ***Step 2*** of the proof of Theorem 2.6, the compactness takes place in $L^2(0,T,X)$ equipped with its weak topology. Precisely, there exists $u \in L^2(0,T,X)$ and a subsequence of $(u_n)_{n\in\mathbb{N}}$ (not relabeled) such that

$$\frac{du_n}{dt} \rightharpoonup \frac{du}{dt} \text{ in } L^2(0,T,X); \tag{2.64}$$
$$u_n \rightharpoonup u \text{ in } L^2(0,T,X). \tag{2.65}$$

These convergences are easy to established. For a complete proof we refer the reader to (Attouch *et al.*, 2014, Theorem 17.4.7).

Step 3. We prove that u is the solution of $(\mathcal{P})$. The proof is that of ***Step 3*** of the proof of Theorem 2.6, where Lemma 2.8 is substituted for Lemma 2.6.

Step 4. We establish

$$u_n \to u \text{ in } C(0,T,X); \tag{2.66}$$
$$\frac{du_n}{dt} \to \frac{du}{dt} \text{ in } L^2(0,T,X) \text{ under hypothesis } \Phi_n(u_n^0) \to \Phi(u^0). \tag{2.67}$$

For proving (2.66), we are going to apply to the sequence $(u_n)_{n\in\mathbb{N}}$, the Ascoli compactness theorem in $C(0,T,X)$ equipped with the uniform convergence. From

(2.60) we already have $\sup_{n\in\mathbb{N}} \|u_n\|_{C(0,T,X)} < +\infty$. The uniform equicontinuity of the sequence $(u_n)_{n\in\mathbb{N}}$ is a straightforward consequence of (2.59). It remains to establish that $u_n(t) \to u(t)$ in X for every $t \in [0,T]$. Noticing that (2.64) holds in $L^2(0,t,X)$ for all $0 \le t \le T$, from

$$u_n(t) = u_n^0 + \int_0^t \frac{du_n}{ds}(s)\, ds$$

we infer that $u_n(t) \rightharpoonup u(t)$ in X for all $t \in [0,T]$. To prove the strong convergence $u_n(t) \to u(t)$, it remains to establish that $\|u_n(t)\| \to \|u(t)\|$. Observe that from the previous steps, and the hypothesis, we have

$$a := \int_0^t \Phi(u(s))\, ds \le \liminf_{n\to+\infty} \int_0^t \Phi(u_n(s))\, ds;$$

$$b := \int_0^t \Phi^*\left(G(s) - \frac{du}{ds}(s)\right)\, ds \le \liminf_{n\to+\infty} \int_0^t \Phi^*\left(G_n(s) - \frac{du_n}{ds}(s)\right)\, ds;$$

$$c := \frac{1}{2}\|u(t)\|^2 - \frac{1}{2}\|u^0\|^2 \le \liminf_{n\to+\infty} \frac{1}{2}\|u_n(t)\|^2 - \frac{1}{2}\|u^0\|^2;$$

$$d := -\int_0^t \langle G(s), u(s)\rangle = \lim_{n\to+\infty} -\int_0^t \langle G_n(s), u_n(s)\rangle$$

with, according to ***Step 3***, $a+b+c+d=0$. Denoting by a_n, b_n, c_n and d_n each of the four sequences of right members terms, we have $a_n+b_n+c_n+d_n = 0$. Therefore, from above, we have

$$\begin{aligned}
&a \le \liminf_{n\to+\infty} a_n;\\
&b \le \liminf_{n\to+\infty} b_n;\\
&c \le \liminf_{n\to+\infty} c_n;\\
&d = \lim_{n\to+\infty} d_n;\\
&a+b+c+d = a_n + b_n + c_n + d_n = 0.
\end{aligned}$$

We easily infer that $a = \lim_{n\to+\infty} a_n$, $b = \lim_{n\to+\infty} b_n$, and $c = \lim_{n\to+\infty} c_n$. In particular $\|u_n(t)\| \to \|u(t)\|$. The proof of (2.67) is exactly that of ***Step 5*** in the proof of Theorem 2.6. □

Chapter 3

Variational convergence of nonlinear distributed time delays reaction-diffusion equations

This chapter extends the convergence process discussed in Chapter 2, to the case when the reaction term includes single or distributed time delays. More precisely, it addresses the variational convergence of sequences $(\mathcal{P}_n)_{n\in\mathbb{N}}$ of reaction-diffusion equations with unknown u_n, i.e.

$$(\mathcal{P}_n)\begin{cases}\dfrac{du_n}{dt}(t)+D\Phi_n(u_n(t))=F_n(t,u_n(t),v_n(t)) \text{ for a.e. } t\in(0,T)\\ \\ u_n(t)=\eta_n(t) \text{ for all } t\in(-\infty,0].\end{cases}$$

The function v_n is connected to the function $u_n\in L^2(0,T,L^2(\Omega))$ via a family $(\mathbf{m}_t^n)_{t\geq 0}$ of vector measures $\mathbf{m}_t^n$ from the Borel field $\mathcal{B}(\mathbb{R})$ into $L^\infty(\Omega)$, according to the formula

$$v_n(t)=\int_{-\infty}^{t}u_n(s)d\mathbf{m}_t^n(s),$$

where the integral is defined in Section 3.1.1. The function η_n, referred to as history function, is nonnegative, bounded, absolutely continuous in the space $C((-\infty,0],L^2(\Omega))$, and satisfies $\eta_n(0)\in\mathrm{dom}(\Phi_n)$. When $\mathbf{m}_t^n=\delta_{t-\tau}$, where $\tau>0$ does not depend on the space variable, we recover the single time delay reaction-diffusion problems, i.e.

$$(\mathcal{P}_n)\begin{cases}\dfrac{du_n}{dt}(t)+D\Phi_n(u_n(t))=F_n(t,u_n(t),u_n(t-\tau)) \text{ for a.e. } t\in(0,T)\\ \\ u_n(t)=\eta_n(t) \text{ for all } t\in[-\tau,0].\end{cases}$$

When $\mathbf{m}_t^n=\sum_{k\in\mathbb{N}}\mathbf{d}_k^n\delta_{t-\tau_k}$, with $\mathbf{d}_k^n\in L^\infty(\Omega)$ and $\sum_{k\in\mathbb{N}}\|\mathbf{d}_k^n\|_{L^\infty(\Omega)}=1$, $(\mathcal{P}_n)$ is a multiple time delays reaction-diffusion problem whose time delays may depend of the space variable through the functions $\mathbf{d}_k^n$. For $\mathbf{m}_t^n=\frac{1}{\#(\tau_n(\mathbb{R}^N))}\delta_{t-\tau_n(\cdot)}$ where $\tau_n:\mathbb{R}^N\to[0,+\infty)$, $(\mathcal{P}_n)$ is a single time delay reaction-diffusion problem whose time delay may vary with respect to the space variable. When $\mathbf{m}_t^n=\mathcal{K}_n(\cdot,t-\tau)\,d\tau$, with for all $\sigma\in\mathbb{R}$, $\mathcal{K}_n(\cdot,\sigma)\in L^\infty(\Omega)$, $\mathcal{K}_n(\cdot,\sigma)\geq 0$ and $\mathcal{K}_n(\cdot,\sigma)=0$ for $\sigma<0$, $(\mathcal{P}_n)$ is a diffuse time delays reaction-diffusion problem with a distributed delays associated with the kernel $\mathcal{K}_n$. For a general Borel vector measure $\mathbf{m}_t^n$, $(\mathcal{P}_n)$ is associated with potentially a mixture of diffuse and discrete time delays (see Section 3.1).

As in Chapter 2, the diffusion term is the subdifferential of a standard integral functional of the calculus of variations $\Phi_n : L^2(\Omega) \to \mathbb{R} \cup \{+\infty\}$, whose domain contains the boundary conditions. The reaction functional F_n belongs to the class of DCP-structured reaction functionals which is defined in Section 3.2.1. Roughly speaking, for fixed $v \in L^2(\Omega)$, $(t, u) \mapsto F_n(t, u, v)$ is a CP-structured reaction functional as defined in Chapter 2. Problems $(\mathcal{P}_n)$ model various situations involving for example reaction time, maturation period, resource regeneration time, time required for substance production, mating processes, or incubation period in vector disease models (see Examples 3.4, 3.5, 3.6, and 3.7).

Section 3.2 is devoted to the existence and uniqueness of bounded nonnegative solutions of problems $(\mathcal{P}_n)$. The proof is based on Corollary 2.1 combined with a fixed point procedure.

The first main result of the chapter, Theorem 3.3, states the stability at the limit, i.e. the convergence of $(\mathcal{P}_n)$ toward a distributed time delays reaction-diffusion problem of the same type. The proof is established under the hypotheses of the Mosco-convergence of functionals Φ_n, the weak convergence of the vector measure $\mathbf{m}_t^n$ for each fixed t in $[0, T]$, and a suitable convergence of F_n to F. In the case of single time delay, we also propose a constructive alternative proof, by applying suitably a finite number of times the convergence result established in Theorem 2.6 to each $(\mathcal{P}_{n,i})$, with $i = 1, \ldots, \left[\frac{T}{\tau}\right] + 1$, defined by

$$(\mathcal{P}_{n,i}) \begin{cases} \dfrac{du_n^i}{dt}(t) + D\Phi_n(u_n^i(t)) = F_n^i(t, u_n^i(t)) \text{ for a.e. } t \in ((i-1)\tau, i\tau) \\ \\ u_n^i((i-1)\tau) = u_n^{i-1}((i-1)\tau) \end{cases}$$

with $u_n^0 := \eta_n$ in $[-\tau, 0]$, and where F_n^i is the CP-structured reaction functional (see Chapter 2) defined by $F_n^i(t, u_n^i(t)) := F_n(t, u_n^i(t), u_n^{i-1}(t))$.

When a global weak convergence of the product of the growth rate of F_n with the time delays term is substituted for the weak convergence of the vector measure $\mathbf{m}_t^n$, a second theorem (Theorem 3.4) states the convergence toward a delay reaction-diffusion problem where a mixing effect appears in the limit reaction functional.

Notation

We use the notation from the previous chapter. Recall that to shorten the notation, we sometimes write X to denote the Hilbert space $L^2(\Omega)$ equipped with its standard scalar product. We also denote by $\xi \odot \xi'$ the Hadamard (or Schur) product of two elements ξ and ξ' in $\mathbb{R}^d$, $d \geq 1$, i.e. $\xi \odot \xi'$ is the vector of $\mathbb{R}^d$ with components $\xi_i \xi_i'$, $i = 1, \ldots, d$. For any topological space $\mathbb{T}$, we denote by $\mathcal{B}(\mathbb{T})$ its Borel σ-algebra, and by $C_c(\mathbb{T}, X)$ the space of continuous function from $\mathbb{T}$ into X, with compact support.

3.1 The time-delays operator

3.1.1 *Integration with respect to vector measures*

We apply Appendix E.1 with $\mathbb{T} = \mathbb{R}$, $\mathbb{X} = L^2(\Omega)$, and $\mathbb{Y} = L^\infty(\Omega)$. Let $\mathbf{B} : L^\infty(\Omega) \times L^2(\Omega) \to L^2(\Omega)$ be the bilinear mapping defined by $\mathbf{B}(v, u) := vu$. According to the abstract scheme described in Appendix E.1, we can define the Borel vector measures $\mathbf{m} : \mathcal{B}(\mathbb{R}) \to L^\infty(\Omega)$ of finite variation associated with $\mathbf{B}$. Recall that we can integrate with respect to $\mathbf{m}$ the step functions of the form $S = \sum_{i \in I} \mathbb{1}_{B_i} S_i$, where I is any finite set, $B_i \in \mathcal{B}(\mathbb{R})$, and $S_i \in L^2(\Omega)$, according to the formula

$$\int S d\mathbf{m} := \sum_{i \in I} \mathbf{m}(B_i) S_i,$$

which defines an element of $L^2(\Omega)$. This integral can be extended to the space $\mathcal{E}^\infty(\mathbb{R}, L^2(\Omega))$ of functions $u : \mathbb{R} \to L^2(\Omega)$ which are uniform limit of step functions, thus defining an element of $L^2(\Omega)$ (see Proposition E.11). Moreover we have

$$\left\| \int u d\mathbf{m} \right\|_{L^2(\Omega)} \leq \int \|u\|_{L^2(\Omega)} d\|\mathbf{m}\|,$$

or, if we want to highlight the time variable,

$$\left\| \int u(t) d\mathbf{m}(t) \right\|_{L^2(\Omega)} \leq \int \|u(t)\|_{L^2(\Omega)} d\|\mathbf{m}\|(t).$$

Recall that $C_c(\mathbb{R}, L^2(\Omega)) \subset \mathcal{E}^\infty(\mathbb{R}, L^2(\Omega))$.

3.1.2 *Time-delays operator associated with vector measures*

The distributed delays considered in this chapter, are modeled by a family $(\mathbf{m}_t)_{t \geq 0}$ of Borel vector measures $\mathbf{m}_t : \mathcal{B}(\mathbb{R}) \to L^\infty(\Omega)$ whose total variation satisfies

(M_1) $\sup_{t \geq 0} \|\mathbf{m}_t\|(\mathbb{R}) \leq 1$.

Moreover, for all $t \geq 0$, we assume that $\mathbf{m}_t$ is positive in the following sense:

(M_2) $\forall t \geq 0, \quad \forall u \in \mathcal{E}^\infty(\mathbb{R}, L^2(\Omega)), \quad u \geq 0 \Longrightarrow \int u d\mathbf{m}_t \geq 0.$

From (M_1) and (M_2), we infer that for all $t \geq 0$, $\mathbf{m}_t$ takes its values in the subset of $L^\infty(\Omega)$ made up of functions v satisfying $0 \leq v \leq 1$. Indeed, the claim follows from

$$0 \leq \mathbf{m}_t(B) = |\mathbf{m}_t(B)| \leq \|\mathbf{m}_t(B)\|_{L^\infty(\Omega)} \leq \|\mathbf{m}_t\|(B) \leq \|\mathbf{m}_t\|(\mathbb{R}) \leq 1$$

for all $B \in \mathcal{B}(\mathbb{R})$.

Let denote by $\mathbf{M}_1^+(\mathbb{R}, L^\infty(\Omega))$ the set of positive vector Borel measures $\mathbf{m} : \mathcal{B}(\mathbb{R}) \to L^\infty(\Omega)$ satisfying $\|\mathbf{m}\|(\mathbb{R}) \leq 1$. Concerning the family $(\mathbf{m}_t)_{t \geq 0}$, we assume furthermore that:

(M$_3$) the map

$$\begin{array}{rcl}[0,+\infty) & \longrightarrow & \mathbf{M}_1^+(\mathbb{R}, L^\infty(\Omega)) \\ t & \longmapsto & \mathbb{1}_{(-\infty,t]}\mathbf{m}_t\end{array}$$

is measurable when $\mathbf{M}_1^+(\mathbb{R}, L^\infty(\Omega))$ is equipped with the σ-algebra generated by the family of evaluation maps $(\mathscr{E}_u)_{u\in\mathcal{E}^\infty(\mathbb{R},L^2(\Omega))}$, $\mathscr{E}_u$: $\mathbf{M}_1^+(\mathbb{R}, L^\infty(\Omega)) \to L^2(\Omega)$ defined by

$$\mathscr{E}_u(\mathbf{m}) := \int u d\mathbf{m}.$$

On other word, for all $u \in \mathcal{E}^\infty(\mathbb{R}, L^2(\Omega))$, the maps $t \mapsto \int_{-\infty}^t u d\mathbf{m}_t$[1] are measurable from $[0, +\infty)$ into $L^2(\Omega)$.

Definition 3.1. Let $T > 0$ be fixed. Let $(\mathbf{m}_t)_{t\geq 0}$ be a family of vector measures in $\mathbf{M}_1^+(\mathbb{R}, L^\infty(\Omega))$ satisfying (M$_1$), (M$_2$) and (M$_3$). We call *time-delays operator*, the linear continuous mapping $\mathcal{T}$ defined on $\mathcal{C}_c((-\infty, T], L^2(\Omega))$ by:

$$\begin{array}{rcl}\mathcal{T} : C_c((-\infty,T], L^2(\Omega)) & \longrightarrow & L^2(0,T,L^2(\Omega)) \\ u & \longmapsto & \mathcal{T}u,\end{array}$$

$$\text{where for all t in } [0,T],\ \mathcal{T}u(t) = \int_{-\infty}^t u(s)d\mathbf{m}_t(s).$$

Note that for each $t \in [0, T]$, $\mathcal{T}u(t)$ is well defined. Indeed, let us extend the restriction to $(-\infty, t]$ of $u \in C_c((-\infty, T], L^2(\Omega))$ by $\widetilde{u}_t \in C_c(\mathbb{R}, L^2(\Omega))$ defined as follows

$$\widetilde{u}_t(s) = \begin{cases} u(s) & \text{for } s \in (-\infty, t], \\ \dfrac{t-s}{\delta}u(t) + u(t) & \text{for } s \in]t, t+\delta], \\ 0 & \text{for } s \in]t+\delta, +\infty). \end{cases} \tag{3.1}$$

Then $\mathbb{1}_{(-\infty,t]}u = \mathbb{1}_{(-\infty,t]}\widetilde{u}_t$, $\mathbb{1}_{(-\infty,t]}\widetilde{u}_t \in \mathcal{E}^\infty(\mathbb{R}, L^2(\Omega))$, and

$$\mathcal{T}u(t) = \int \mathbb{1}_{(-\infty,t]}\widetilde{u}_t(s)\ d\mathbf{m}_t(s).$$

Furthermore, the operator $\mathcal{T}$ takes its values in $L^2(0, T, L^2(\Omega))$ and is continuous. Indeed $t \mapsto \mathcal{T}u(t)$ is clearly measurable, $\|\widetilde{u}_t\|_{C_c(\mathbb{R},L^2(\Omega))} \leq 2\|u\|_{C_c((-\infty,T],L^2(\Omega))}$, and

$$\|\mathcal{T}u(t)\|_{L^2(\Omega)} \leq \int \|\widetilde{u}_t(s)\|_{L^2(\Omega)} d\|\mathbf{m}_t\|(s) \leq 2\|u\|_{C_c((-\infty,T],L^2(\Omega))}, \tag{3.2}$$

$$\|\mathcal{T}u\|_{L^2(0,T,L^2(\Omega))} \leq 2T^{\frac{1}{2}}\|u\|_{C_c((-\infty,T],L^2(\Omega))}. \tag{3.3}$$

For $T' > T$, let denote by $\mathcal{T}'$ the time-delays operator defined on $C_c((-\infty, T'], L^2(\Omega))$. Then, the restriction of $\mathcal{T}'$ to $C_c((-\infty, T], L^2(\Omega))$ is clearly the time-delays operator $\mathcal{T}$. According to this remark, we do not indicate the dependance of $\mathcal{T}$ with respect to T.

---

[1] In a standard way, we write $\int_{-\infty}^t u\ d\mathbf{m}_t$ for $\int \mathbb{1}_{(-\infty,t]}\ u\ d\mathbf{m}_t$.

Lemma 3.1. *Let $u \in C_c((-\infty, T], L^2(\Omega))$, $u \geq 0$, and assume that there exists $\overline{u} \in \mathbb{R}_+$ such that $0 \leq u \leq \overline{u}$. Then $0 \leq \mathcal{T}u \leq \overline{u}$.*

Proof. From (M_2), for all $t \in [0, T]$, we have $\mathcal{T}u(t) = \displaystyle\int \mathbb{1}_{(-\infty,t]}\widetilde{u}_t(s)d\mathbf{m}_t(t) \geq 0$. On the other hand, for all $t \in [0, T]$, since $\mathbf{m}_t((-\infty, t]) \leq 1$ in $L^\infty(\Omega)$, we infer that

$$\overline{u} - \mathcal{T}u(t) \geq \mathbf{m}_t((-\infty, t])\overline{u} - \mathcal{T}u(t) = \int_{-\infty}^{t} (\overline{u} - u)d\mathbf{m}_t = \int \mathbb{1}_{(-\infty,t]}(\overline{u} - \widetilde{u}_t)d\mathbf{m}_t$$

with $\mathbb{1}_{(-\infty,t]}(\overline{u} - \widetilde{u}_t) \geq 0$, so that $\mathcal{T}u \leq \overline{u}$. □

In order that the solutions to problems of the type ($\mathcal{P}_n$) admit a right derivative at each $t \in [0, T]$, we assume that $(\mathbf{m}_t)_{t\geq 0}$ satisfies some regularity hypothesis through the operator $\mathcal{T}$, made precise in (M_4) below. Let $\eta \in C_c((-\infty, 0], L^2(\Omega))$ satisfying

$$\int_{-\infty}^{0} \left\| \frac{d\eta}{d\sigma} \right\|_{L^2(\Omega)} d\sigma < +\infty, \tag{3.4}$$

we denote by $\mathcal{C}_\eta((-\infty, T], L^2(\Omega))$ the subset of $C_c((-\infty, T], L^2(\Omega))$ made up of functions u which are *absolutely continuous on* $[0, T]$, and whose restriction to $(-\infty, 0]$ is equal to η. Then condition (M_4) is expressed as follows:

(M_4) For all $T > 0$, and all function u in $\mathcal{C}_\eta((-\infty, T], L^2(\Omega))$, there exists a locally integrable function $\varphi_u : [0, T] \to \mathbb{R}_+$, such that

$$\forall (s,t) \in [0, T]^2 \quad s < t \implies \|\mathcal{T}u(t) - \mathcal{T}u(s)\|_{L^2(\Omega)} \leq \int_s^t \varphi_u(\sigma)d\sigma.$$

Let us examine three basic examples of vector measures in $\mathbf{M}_1^+(\mathbb{R}, L^\infty(\Omega))$ and time-delays operators.

3.1.3 *Examples of time-delays operators*

Example 3.1 (**Multi-delays case**). Consider $(\mathbf{m}_t)_{t\geq 0}$ given by $\mathbf{m}_t = \sum_{k\in\mathbb{N}} \mathbf{d}_k \delta_{t-\tau_k}$ where, for all $k \in \mathbb{N}$, $\tau_k > 0$, $\mathbf{d}_k \in L^\infty(\Omega)$, $\mathbf{d}_k \geq 0$, $\sum_{k\in\mathbb{N}} \|\mathbf{d}_k\|_{L^\infty(\Omega)} = 1$. Clearly $\mathbf{m}_t$ belongs to $\mathbf{M}_1^+(\mathbb{R}, L^\infty(\Omega))$. For any $u \in \mathcal{E}^\infty(\mathbb{R}, L^2(\Omega))$, the map $t \mapsto \int_{-\infty}^{t} u(s)d\mathbf{m}_t$ is nothing but $t \mapsto \sum_{k\in\mathbb{N}} \mathbf{d}_k u(t - \tau_k)$ so that (M_1), (M_2) and (M_3) are fulfilled. The operator $\mathcal{T}$ is defined for every $u \in C_c((-\infty, T], L^2(\Omega))$ by $\mathcal{T}u(t) = \displaystyle\sum_{k\in\mathbb{N}} \mathbf{d}_k u(t - \tau_k)$.

Let $(s,t) \in [0,T]^2$ be such that $s < t$. Using Fubini-Tonelli theorem, for all $u \in C_\eta((-\infty,T],L^2(\Omega))$ we have

$$\begin{aligned}
\|\mathcal{T}u(t) - \mathcal{T}u(s)\|_{L^2(\Omega)} &\leq \sum_{k\in\mathbb{N}} \|\mathbf{d}_k\|_{L^\infty(\Omega)} \|u(t-\tau_k) - u(s-\tau_k)\|_{L^2(\Omega)} \\
&\leq \sum_{k\in\mathbb{N}} \|\mathbf{d}_k\|_{L^\infty(\Omega)} \int_{s-\tau_k}^{t-\tau_k} \left\|\frac{du}{d\sigma}(\sigma)\right\|_{L^2(\Omega)} d\sigma \\
&= \int_s^t \left(\sum_{k\in\mathbb{N}} \|\mathbf{d}_k\|_{L^\infty(\Omega)} \left\|\frac{du}{d\sigma}(\sigma-\tau_k)\right\|_{L^2(\Omega)}\right) d\sigma
\end{aligned}$$

so that, $\varphi_u : [0,T] \to \mathbb{R}_+$ defined by

$$\varphi_u(\sigma) = \sum_{k\in\mathbb{N}} \|\mathbf{d}_k\|_{L^\infty(\Omega)} \left\|\frac{du}{d\sigma}(\sigma-\tau_k)\right\|_{L^2(\Omega)},$$

is suitable for satisfying (M_4). Note that φ_u is nothing but the discrete convolution

$$\left\|\frac{du}{d\sigma}(\cdot)\right\|_{L^2(\Omega)} \star \sum_{k\in\mathbb{N}} \|\mathbf{d}_k\|_{L^\infty(\Omega)} \delta_{\tau_k}.$$

Example 3.2 (Single delay depending on the space variable). Given a measurable map $\tau : \mathbb{R}^N \to [0,+\infty)$ with $\#(\tau(\mathbb{R}^N)) < +\infty$ (i.e. with finite range), we set $\mathbf{m}_t = \frac{1}{\#(\tau(\mathbb{R}^N))}\delta_{t-\tau(\cdot)}$. It is easily seen that $\mathbf{m}_t$ is a vector measure from $\mathcal{B}(\mathbb{R})$ into $L^\infty(\Omega)$ satisfying (M_2) and (M_3). Let us show that $\mathbf{m}_t$ satisfies (M_1). Indeed, we have

$$\begin{aligned}
&\|\mathbf{m}_t\|(\mathbb{R}^N) \\
&= \sup_I \left\{\sum_{i\in I} \left\|\frac{1}{\#(\tau(\mathbb{R}^N))}\delta_{t-\tau(\cdot)}(B_i)\right\|_{L^\infty(\Omega)} : B_i \in \mathcal{B}(\mathbb{R}), (B_i)_{i\in I} \text{finite partition of } \mathbb{R}\right\} \\
&= \sup_I \left\{\frac{\#(A_I)}{\#(\tau(\mathbb{R}))} : B_i \in \mathcal{B}(\mathbb{R}), (B_i)_{i\in I} \text{ finite partition of } \mathbb{R}\right\}
\end{aligned}$$

where $A_I := \{i \in I : \tau^{-1}(t - B_i) \neq \emptyset\}$. For each fixed I, and for each $i \in A_I$, choose x_i in $\mathbb{R}^N$ such that $\tau(x_i) \in t - B_i$, and consider the map $\Gamma_I : A_I \to \tau(\mathbb{R}^N)$ defined by $\Gamma_I(i) = \tau(x_i)$. From the fact that $t - B_i$ and $t - B_j$ are disjoint sets for $i \neq j$, we infer that Γ_I is an injection, thus $\#(A_I) \leq \#(\tau(\mathbb{R}^N))$ which proves the claim.

The operator $\mathcal{T}$ is then defined for all $t \in [0,T]$ and all $u \in C_c((-\infty,T],L^2(\Omega))$ by $\mathcal{T}u(t) = \frac{1}{\#(\tau(\mathbb{R}^N))}u(t-\tau(\cdot))$, i.e. for a.e. $x \in \Omega$, $\mathcal{T}u(t)(x) = \frac{1}{\#(\tau(\mathbb{R}^N))}u(t - \tau(x), x)$.[2] Let us show that ($\mathrm{M}_4$) is satisfied. For all $u \in C_\eta((-\infty,T],L^2(\Omega))$ we

[2]To shorten the notation, we sometimes write $v(t,x)$ for $v(t)(x)$ for the functions $v \in L^2(0,T,L^2(\Omega))$.

have

$$\begin{aligned}
&\left\| \frac{1}{\#(\tau(\mathbb{R}^N))}(u(t-\tau(\cdot)) - u(s-\tau(\cdot))) \right\|_{L^2(\Omega)} \\
&\leq \frac{1}{\#(\tau(\mathbb{R}^N))} \int_s^t \left\| \frac{du}{d\sigma}(\sigma - \tau(\cdot)) \right\|_{L^2(\Omega)} d\sigma \\
&\leq \frac{1}{\#(\tau(\mathbb{R}^N))} \int_s^t \left(\sum_{\tau_i \in \tau(\mathbb{R}^N)} \int_{\tau^{-1}(\tau_i)\cap\Omega} \left| \frac{du}{d\sigma}(\sigma - \tau_i, x) \right|^2 dx \right)^{\frac{1}{2}} d\sigma \\
&\leq \frac{1}{\#(\tau(\mathbb{R}^N))} \int_s^t \sum_{\tau_i \in \tau(\mathbb{R}^N)} \left\| \frac{du}{d\sigma}(\sigma - \tau_i) \right\|_{L^2(\Omega)} d\sigma
\end{aligned}$$

so that φ_u given by

$$\varphi_u(\sigma) = \frac{1}{\#(\tau(\mathbb{R}^N))} \sum_{\tau_i \in \tau(\mathbb{R}^N)} \left\| \frac{du}{d\sigma}(\sigma - \tau_i) \right\|_{L^2(\Omega)}$$

is suitable to check (M_4). Note that φ_u is the discrete convolution

$$\left\| \frac{du}{d\sigma}(\cdot) \right\|_{L^2(\Omega)} \star \frac{1}{\#(\tau(\mathbb{R}^N))} \sum_{\tau_i \in \tau(\mathbb{R}^N)} \delta_{\tau_i}.$$

Example 3.3 (Distributed delays by a diffuse delays kernel). Consider $(\mathbf{m}_t)_{t\geq 0}$ given by $\mathbf{m}_t = \mathcal{K}_t(\cdot,\tau)d\tau$ where

i) $\forall \tau \in \mathbb{R}_+ \quad \forall t \geq 0 \qquad \mathcal{K}_t(\cdot,\tau) = \mathcal{K}(\cdot, t-\tau)$;

ii) $\forall \sigma \in \mathbb{R} \quad \mathcal{K}(\cdot,\sigma) \in L^\infty(\Omega)$;

iii) $\displaystyle\int_{-\infty}^{+\infty} \|\mathcal{K}(\cdot,\sigma)\|_{L^\infty(\Omega)} d\sigma = 1$;

iv) $\forall \sigma \in \mathbb{R} \qquad \mathcal{K}(x,\sigma) \geq 0$ a.e. in Ω;

v) $\forall \sigma \in (-\infty, 0) \quad \mathcal{K}(\cdot,\sigma) = 0$.

It is easy to show that $\mathbf{m}_t$ satisfies (M_1) and (M_2). An example of kernel $\mathcal{K}$ is given by $\mathcal{K}(\cdot,\sigma) = \mathbf{d}(\cdot)K(\sigma)$ where $\mathbf{d} \in L^\infty(\Omega)$, $\|\mathbf{d}\|_{L^\infty(\Omega)} = 1$, and K is the standard Γ_ℓ-distribution delays defined by

$$K(\sigma) = \begin{cases} \dfrac{\gamma\sigma^{\ell-1}\exp(-\gamma\sigma)}{(\ell-1)!} & \text{if } \sigma \geq 0 \\ \\ 0 & \text{if } \sigma < 0 \end{cases}$$

with $\ell \in \mathbb{N}^*$ and $\gamma > 0$.

For any $u \in \mathcal{E}^\infty(\mathbb{R}, L^2(\Omega))$, the map $t \mapsto \int_{-\infty}^t u(s)\mathcal{K}(\cdot, t-s)ds = \int_0^{+\infty} u(t-s)\mathcal{K}(\cdot,s)ds$ is clearly measurable so that (M_3) is satisfied. For all $u \in C_c$

$((-\infty,T],L^2(\Omega))$ we have

$$\forall t\in[0,T],\ \mathcal{T}u(t)=\int_{-\infty}^{t}u(s)\mathcal{K}(\cdot,t-s)ds$$
$$=\int_0^{+\infty}u(t-s)\mathcal{K}(\cdot,s)\,ds=u\star\mathcal{K}(\cdot,\cdot)(t).$$

An elementary calculation shows that $\varphi_u d=\left\|\frac{du}{d\sigma}(\cdot)\right\|_{L^2(\Omega)}\star\|\mathcal{K}(\cdot,\tau)\|_{L^\infty(\Omega)}d\tau$ is suitable to check (M$_4$). Indeed, for all $u\in C_c((-\infty,T],L^2(\Omega))$ we have

$$\left\|\int_0^{+\infty}\mathcal{K}(\cdot,\tau)u(t-\tau)d\tau-\int_0^{+\infty}\mathcal{K}(\cdot,\tau)u(s-\tau)d\tau\right\|_{L^2(\Omega)}$$
$$\leq\int_0^{+\infty}\|\mathcal{K}(\cdot,\tau)\|_{L^\infty(\Omega)}\|u(t-\tau)-u(s-\tau)\|_{L^2(\Omega)}d\tau$$
$$\leq\int_0^{+\infty}\|\mathcal{K}(\cdot,\tau)\|_{L^\infty(\Omega)}\left(\int_{s-\tau}^{t-\tau}\left\|\frac{du}{d\sigma}(\sigma)\right\|_{L^2(\Omega)}d\sigma\right)d\tau$$
$$=\int_s^t\left(\int_0^{+\infty}\|\mathcal{K}(\cdot,\tau)\|_{L^\infty(\Omega)}\left\|\frac{du}{d\sigma}(\sigma-\tau)\right\|_{L^2(\Omega)}d\tau\right)d\sigma,$$

and the claim follows by setting

$$\varphi_u(\sigma)=\int_0^{+\infty}\left\|\frac{du}{d\sigma}(\sigma-\tau)\right\|_{L^2(\Omega)}\|\mathcal{K}(\cdot,\tau)\|_{L^\infty(\Omega)}d\tau.$$

Remark 3.1. Denote by $\mathcal{P}_+(\mathbb{R})$ the set of all Borel probability measures on $\mathbb{R}$ concentrated on $\mathbb{R}_+$. One can express the two functions φ_u obtained in Examples 3.1, 3.2 and 3.3 in an unified way: there exists $\mu\in\mathcal{P}_+(\mathbb{R})$ such that

$$\varphi_u=\left\|\frac{du}{d\sigma}(\cdot)\right\|_{L^2(\Omega)}\star\mu,$$

i.e. such that for all $\sigma\in[0,T]$, $\varphi_u(\sigma)=\int_{\mathbb{R}}\left\|\frac{du}{d\sigma}(\sigma-\tau)\right\|_{L^2(\Omega)}d\mu(\tau)$.

In Example 3.1, $\mu=\sum_{k\in\mathbb{N}}\|\mathbf{d}_k\|_{L^\infty(\Omega)}\delta_{\tau_k}$, in Example 3.2, $\mu=\frac{1}{\#(\tau(\mathbb{R}^N))}\sum_{\tau_i\in\tau(\mathbb{R}^N)}\delta_{\tau_i}$, and in Example 3.3, $\mu=\|\mathcal{K}(\cdot,\tau)\|_{L^\infty(\Omega)}d\tau$.

In Section 3.3, we strengthen condition (M$_4$) by the more explicit condition (M$'_4$) below:

(M$'_4$) There exists $\mu\in\mathcal{P}_+(\mathbb{R})$ such for all $T>0$, all $u\in C_\eta((-\infty,T],L^2(\Omega))$, and all $(s,t)\in[0,T]^2$ with $s<t$,

$$\|\mathcal{T}u(t)-\mathcal{T}u(s)\|_{L^2(\Omega)}\leq\int_s^t\left(\left\|\frac{du}{d\sigma}(\cdot)\right\|_{L^2(\Omega)}\star\mu\right)(\sigma)d\sigma.$$

The Lemma below states that $\sigma\mapsto\left\|\frac{du}{d\sigma}(\cdot)\right\|_{L^2(\Omega)}\star\mu(\sigma)$ is σ-a.e. well defined and that under (M$'_4$) $\mathcal{T}u$ is absolutely continuous.

Lemma 3.2. *Let $\mu \in \mathcal{P}_+(\mathbb{R})$ and u in $C_\eta((-\infty,T],L^2(\Omega))$. Then for a.e. $\sigma \in [0,T]$, $\|\frac{du}{d\sigma}(\cdot)\|_{L^2(\Omega)} \star \mu(\sigma) < +\infty$. Assume (M_4'), then for all $u \in C_\eta((-\infty,T],L^2(\Omega))$, $\mathcal{T}u$ is absolutely continuous and*

$$\int_0^T \left\| \frac{d\mathcal{T}u}{d\sigma}(\sigma) \right\|_{L^2(\Omega)} d\sigma \leq \int_0^T \left(\left\| \frac{du}{d\sigma}(\cdot) \right\|_{L^2(\Omega)} \star \mu \right)(\sigma) d\sigma$$
$$\leq \int_{-\infty}^0 \left\| \frac{d\eta}{d\sigma}(\sigma) \right\|_X d\sigma + \int_0^T \left\| \frac{du}{d\sigma}(\sigma) \right\|_X d\sigma.$$

Proof. Fix $u \in C_\eta((-\infty,T],L^2(\Omega))$. From Fubini-Tonelli theorem, we have

$$\begin{aligned}
&\int_0^T \left(\left\| \frac{du}{d\sigma}(\cdot) \right\|_X \star \mu \right)(\sigma) d\sigma \\
&= \int_{\mathbb{R}_+} \left(\int_0^T \left\| \frac{du}{d\sigma}(\sigma-\tau) \right\|_X d\sigma \right) d\mu(\tau) \\
&= \int_{\mathbb{R}_+} \left(\int_{-\tau}^{T-\tau} \left\| \frac{du}{d\sigma}(\sigma) \right\|_X d\sigma \right) d\mu(\tau) \\
&\leq \int_{\mathbb{R}_+} \left(\int_{-\infty}^0 \left\| \frac{d\eta}{d\sigma}(\sigma) \right\|_X d\sigma + \int_0^T \left\| \frac{du}{d\sigma}(\sigma) \right\|_X d\sigma \right) d\mu(\tau) \\
&= \int_{-\infty}^0 \left\| \frac{d\eta}{d\sigma}(\sigma) \right\|_X d\sigma + \int_0^T \left\| \frac{du}{d\sigma}(\sigma) \right\|_X d\sigma,
\end{aligned}$$

which is finite since $u \in C_\eta((-\infty,T],L^2(\Omega))$ and because η satisfies (3.4). Therefore for a.e. $\sigma \in [0,T]$, $\|\frac{du}{d\sigma}(\cdot)\|_{L^2(\Omega)} \star \mu(\sigma) < +\infty$. On the other hand, for any finite partition $(s_i,t_i)_{i\in I}$ of $[0,T]$, from (M_4') and the previous calculation, we have

$$\begin{aligned}
\sum_{i\in I} \|\mathcal{T}u(t_i) - \mathcal{T}u(s_i)\|_{L^2(\Omega)} &\leq \sum_{i\in I} \int_{s_i}^{t_i} \left(\left\| \frac{du}{d\sigma}(\cdot) \right\|_X \star \mu \right)(\sigma) d\sigma \\
&\leq \int_0^T \left(\left\| \frac{du}{d\sigma}(\cdot) \right\|_X \star \mu \right)(\sigma) d\sigma \\
&\leq \int_{-\infty}^0 \left\| \frac{d\eta}{d\sigma}(\sigma) \right\|_X d\sigma + \int_0^T \left\| \frac{du}{d\sigma}(\sigma) \right\|_X d\sigma.
\end{aligned}$$

The claim follows by taking the supremum on all the finite partitions of $[0,T]$. □

3.2 Reaction-diffusion problems associated with convex functionals of the calculus of variations and DCP-structured reaction functionals

We are mainly concerned with sequences of delays reaction-diffusion problems of the form

$$(\mathcal{P}_{\mathcal{T},\eta}) \begin{cases} \dfrac{du}{dt}(t) + D\Phi(u(t)) = F(t,u(t),\mathcal{T}u(t)) \text{ for a.e. } t \in (0,T) \\ u(t) = \eta(t) \text{ for all } t \in (-\infty,0] \end{cases}$$

where $\mathcal{T}$ is a time-delays operator as defined in Definition 3.1 from a family $(\mathbf{m}_t)_{t\geq 0}$ of Borel vector measures satisfying (M_1), (M_2), (M_3). We sometimes assume that $(\mathbf{m}_t)_{t\geq 0}$ fulfills one of the additional conditions (M_4) or (M'_4). The function η belongs to $C_c((-\infty,0],L^2(\Omega))$ and $\eta(0)\in\overline{\mathrm{dom}(\partial\Phi)}$, where $\partial\Phi$ is the subdifferential of a standard convex functional Φ of the calculus of variations as discussed in the previous chapter. We recall that $\Phi: L^2(\Omega)\to\mathbb{R}\cup\{+\infty\}$ is defined by

$$\Phi(u)=\begin{cases}\displaystyle\int_\Omega W(x,\nabla u(x))dx+\frac{1}{2}\int_{\partial\Omega}a_0u^2d\mathcal{H}_{N-1}-\int_{\partial\Omega}\phi u\,d\mathcal{H}_{N-1} & \text{if } u\in H^1(\Omega)\\ +\infty & \text{otherwise,}\end{cases}$$

where $\phi\in L^2_{\mathcal{H}_{N-1}}(\partial\Omega)$ and $a_0\in L^\infty_{\mathcal{H}_{N-1}}(\partial\Omega)$ with

$$\begin{cases}a_0\geq 0 \quad \mathcal{H}_{N-1}\text{-a.e. in } \partial\Omega\\ \exists\sigma>0 \quad a_0\geq\sigma \quad \mathcal{H}_{N-1}\text{-a.e. in } \Gamma\subset\partial\Omega.\end{cases}$$

The density $W:\mathbb{R}^N\times\mathbb{R}^N\to\mathbb{R}$ is a measurable function which satisfies (D). We briefly recall some properties stated in Chapter 2. By using the subdifferential inequality together with the growth conditions in (D), it is easy to show that W fulfills (2.7). From the second estimate of (2.7), we infer that if $u\in H^1(\Omega)$, then the function $D_\xi W(\cdot,\nabla u)$ belongs to $L^2(\Omega,\mathbb{R}^N)$. The subdifferential $\partial\Phi$ of Φ (actually its Gâteaux derivative $D\Phi$), whose domain captures the boundary condition, is given by:

$$\begin{cases}\mathrm{dom}(\partial\Phi)\\ =\{v\in H^1(\Omega):\mathrm{div}D_\xi W(\cdot,\nabla v)\in L^2(\Omega),\ a_0v+D_\xi W(\cdot,\nabla v)\cdot\mathbf{n}=\phi \text{ on } \partial\Omega\}\\ \partial\Phi(v)=-\mathrm{div}D_\xi W(\cdot,\nabla v) \text{ for } v\in\mathrm{dom}(\partial\Phi)\end{cases}$$

where $a_0v+D_\xi W(\cdot,\nabla v)\cdot\mathbf{n}$ must be taken in the trace sense (see Lemma 2.3). Finally, recall that our analysis includes the case of functionals defined by

$$\Phi(u)=\begin{cases}\displaystyle\int_\Omega W(x,\nabla u(x))dx & \text{if } u\in H^1_\Gamma(\Omega)\\ +\infty & \text{otherwise.}\end{cases}$$

Indeed, take $\phi=0$ and set

$$a_0(x)=\begin{cases}0 & \text{if } x\in\partial\Omega\setminus\Gamma,\\ +\infty & \text{if } x\in\Gamma.\end{cases}$$

Then, as stated in Corollary 2.6, the integral $\int_{\partial\Omega}a_0u^2d\mathcal{H}_{N-1}$ is a penalization which forces the function u to belongs to $H^1_\Gamma(\Omega)=\{u\in H^1(\Omega):u=0 \text{ on } \Gamma\}$. The subdifferential $\partial\Phi(=D\Phi)$ of Φ contains now the homogeneous Dirichlet-Neumann boundary conditions and is given by:

$$\begin{cases}\mathrm{dom}(\partial\Phi)\\ =\{v\in H^1_\Gamma(\Omega):\mathrm{div}D_\xi W(\cdot,\nabla v)\in L^2(\Omega),D_\xi W(\cdot,\nabla v)\cdot\mathbf{n}=0 \text{ on } \partial\Omega\setminus\Gamma\}\\ \partial\Phi(v)=-\mathrm{div}D_\xi W(\cdot,\nabla v) \text{ for } v\in\mathrm{dom}(\partial\Phi).\end{cases}$$

3.2.1 *The class of DCP-structured reaction functionals*

The delays reaction-diffusion problems modeling a wide class of applications with a potential mathematical tractability in the homogenization framework (periodic or stochastic), involve a special class of reaction functionals that we define below.

Definition 3.2. A functional $F : [0, +\infty) \times L^2(\Omega) \times L^2(\Omega) \to \mathbb{R}^\Omega$ is called a *DCP-structured reaction functional*, if there exists a measurable function $f : [0, +\infty) \times \mathbb{R}^N \times \mathbb{R} \times \mathbb{R} \to \mathbb{R}$ such that for all $t \in [0, +\infty)$ and all $(u, v) \in L^2(\Omega) \times L^2(\Omega)$, $F(t, u, v)(x) = f(t, x, u(x), v(x))$, and fulfilling the following structure conditions:

- $f(t, x, \zeta, \zeta') = r(t, x) \odot h(\zeta') \cdot g(\zeta) + q(t, x)$;[3]
- $g, h : \mathbb{R} \to \mathbb{R}^l$ are locally Lipschitz continuous functions;
- $r \in L^\infty([0, T] \times \mathbb{R}^N, \mathbb{R}^l)$ for all $T > 0$;
- $q \in L^2(0, T, L^2_{loc}(\mathbb{R}^N))$ for all $T > 0$.

Furthermore f must satisfy the condition (DCP) below. (DCP) stands for Delays Comparison Principle):

(DCP) there exist $\overline{f} : [0, +\infty) \times \mathbb{R} \to \mathbb{R}_+$ and $\overline{\rho} \in \mathbb{R}^*_+$ such that the ordinary differential equation

$$\overline{\mathrm{O_{DE}}} \begin{cases} \overline{y}'(t) = \overline{f}(t, \overline{y}(t)) \text{ for a.e. } t \in [0, +\infty) \\ \overline{y}(0) = \overline{\rho} \end{cases}$$

admits at least a solution $\overline{y}$ such that for all $T > 0$, for a.e. $(t, x) \in (0, T) \times \mathbb{R}^N$ and for all $\zeta' \in [0, \overline{y}(T)]$

$$0 \le f(t, x, 0, \zeta') \quad \text{and} \quad f(t, x, \overline{y}(t), \zeta') \le \overline{f}(t, \overline{y}(t)).$$

The functional F is referred to as a *DCP-structured reaction functional associated with* (r, g, h, q) and f as a *DCP-structured reaction function associated with* (r, g, h, q). If furthermore, for all $T > 0$, $r \in W^{1,1}(0, T, L^2_{\mathrm{loc}}(\mathbb{R}^N, \mathbb{R}^l))$ and $q \in W^{1,1}(0, T, L^2_{\mathrm{loc}}(\mathbb{R}^N))$, the map F is referred to as a regular DCP-structured reaction functional and f as a regular DCP-structured reaction function.

Note that, since $\overline{y}$ is non decreasing, for any $T > 0$ we have $0 < \overline{\rho} = \overline{y}(0) \le \overline{y}(T)$.

Remark 3.2. 1) The regularity of DCP-structured reaction functional is unnecessary to assert existence of a bounded solution to our problems. It will be invoked for proving the following additional regularity: when furthermore η satisfies (3.4) and $(\mathbf{m}_t)_{t \ge 0}$ satisfies (M$_4$), the solution u of $(\mathcal{P}_{T,\eta})$ satisfies $u(t) \in \mathrm{dom}(D\Phi)$ for all $t \in]0, T]$ and possesses a right derivative at each $t \in]0, T[$.

[3]Using the coordinates of r, g and h we have $f(t, x, \zeta, \zeta') = \sum_{i=1}^{l} r_i(t, x) g_i(\zeta) h_i(\zeta') + q(t, x)$.

2) It is worth noting that for each fixed ζ' in $[0, \overline{y}(T)]$, the function $\zeta \mapsto f(t, x, \zeta, \zeta')$ is a CP-structured reaction functional associated with $(r \odot h(\zeta'), g, q)$ in the sense of Definition 2.1, where $\underline{\rho} = 0$ and $\underline{f} = 0$.
3) According to (DCP), we show in Theorem 3.1, that, if the history function η satisfies $0 \leq \eta \leq \overline{\rho}$, then $(\mathcal{P}_{\tau,\eta})$ admits a unique solution u satisfying $0 \leq u(t) \leq \overline{y}(T)$ for all $t \in [0, T]$.

3.2.2 *Some examples of DCP-structured reaction functions coming from ecology and biology models*

Example 3.4. ***Example derived from vector disease model: the Cooke model.*** Let $f : \mathbb{R}_+ \times \mathbb{R}^N \times \mathbb{R}^2 \to \mathbb{R}$ be defined by

$$f(t, x, \zeta, \zeta') = a(t, x)\zeta'(1 - \zeta) - b(t, x)\zeta$$

where $a > 0$, $b > 0$ belong to $L^\infty([0, T] \times \mathbb{R}^N)$ for all $T > 0$. Clearly the function f satisfies the structure condition of DCP-structured reaction function with $l = 2$ and

$$r(t, x) = (a(t, x), -b(t, x)), \ h(\zeta') = (\zeta', 1), \ g(\zeta) = (1 - \zeta, \zeta).$$

Let us show that (DCP) is fulfilled. Take $\overline{f} = 0$, and for $\overline{\rho}$, any real number greater or equal to 1. We have $\overline{y} = \overline{\rho}$ and, for all $\zeta' \geq 0$,

$$0 \leq a(t, x)\zeta' = f(t, x, 0, \zeta') \text{ and } f(t, x, \overline{y}(t), \zeta') = f(t, x, \overline{\rho}, \zeta') \leq 0 = \overline{f}(t, \overline{y}(t)).$$

This example of reaction function corresponds to the diffusive vector disease model of Cooke (see Pujo-Menjouet (2015), Ruan (2006) in the case of single delay):

$$\begin{cases} \dfrac{du}{dt}(t) + D\Phi(u(t)) = a(t, x)(1 - u(t)) \displaystyle\int_{-\infty}^{t} u(s) d\mathbf{m}_t(s) - b(t, \cdot)u(t) \\ \text{for a.e. } t \in (0, T) \\ \\ u(t) = \eta(t) \text{ for all } t \in (-\infty, 0] \end{cases}$$

where

- $u(t, x)$ denotes the density of the host (i.e. infected) population,
- $1 - u(t, x)$ the density of uninfected population,
- when $\mathbf{m}_t = \delta_{t-\tau}$, then τ denotes the incubation period before the disease agent can infect the host.

The choice of more general measures $\mathbf{m}_t$ corresponds to more complex incubation processes. Then if the history function η satisfies $0 \leq \eta \leq \overline{\rho}$ for some positive real number $\overline{\rho} \geq 1$, we prove in Theorem 3.1, that the problem admits a unique solution in $C((-\infty, T], L^2(\Omega))$ satisfying $0 \leq u \leq \overline{\rho}$. Moreover, if f is regular, η satisfies (3.4) and $(\mathbf{m}_t)_{t \geq 0}$ satisfies (M_4), then u is regular in the sense of Remark 3.2 1).

Example 3.5. ***Example derived from food limited models: the delay logistic equation with immigration (or stocking).*** This example corresponds to Example 2.1, a) wherein some time-delays distribution is incorporated.

$$f(t,x,\zeta,\zeta') = a(t,x)\zeta\left(1 - \frac{\zeta'}{K_{car}}\right) + q(t,x)$$

where $a \in L^\infty([0,T] \times \mathbb{R}^N)$ for all $T > 0$, $0 < a(t,x) \leq \overline{a}$, $K_{car} > 0$, and $q \in L^2(0,T,L^2_{loc}(\mathbb{R}^N))$ for all $T > 0$, $0 \leq q(t,x) \leq \overline{q}$. The function f verifies either of the two structures of DCP-structured reaction functions with $l = 1$ or $l = 2$. Indeed, with $l = 1$ take

$$r(t,x) = a(t,x),\ g(\zeta) = \zeta,\ h(\zeta') = \left(1 - \frac{\zeta'}{K_{car}}\right).$$

We can also write $f(t,x,\zeta,\zeta') = a(t,x)\zeta - a(t,x)\frac{\zeta\zeta'}{K_{car}} + q(t,x)$, so that f satisfies also the structure condition with $l = 2$ and

$$r(t,x) = \left(a(t,x), -\frac{a(t,x)}{K_{car}}\right),\ h(\zeta') = (1,\zeta'),\ g(\zeta) = (\zeta,\zeta).$$

Let us show that f fulfills (DCP). Take $\overline{f}(t,\zeta) = \overline{a}\zeta + \overline{q}$, and $\overline{\rho}$, any positive real number. Then $\overline{y}(t) = \left(\overline{\rho} + \frac{\overline{q}}{\overline{a}}\right)\exp(\overline{a}t) - \frac{\overline{q}}{\overline{a}}$ for all $t \in [0,+\infty)$. Thus for all $\zeta' \in \left[0, \left(\overline{\rho} + \frac{\overline{q}}{\overline{a}}\right)\exp(\overline{a}T) - \frac{\overline{q}}{\overline{a}}\right]$ we have

$0 \leq f(t,x,0,\zeta') = q(t,x)$ and

$$f(t,x,\overline{y}(t),\zeta') = a(t,x)\overline{y}(t) - a(t,x)\frac{\overline{y}(t)\zeta'}{K_{car}} + q(t,x) \leq \overline{a}\overline{y}(t) + \overline{q} = \overline{f}(t,\overline{y}(t)).$$

This reaction function corresponds to the diffusive delay logistic equation with immigration, which models the evolution of a population density with resource regeneration time, growth rate a, and carrying capacity K_{car}:

$$\begin{cases} \dfrac{du}{dt}(t) + D\Phi(u(t)) = a(t,\cdot)u(t)\left(1 - \dfrac{1}{K_{car}}\displaystyle\int_{-\infty}^{t} u(\tau)d\mathbf{m}_t(\tau)\right) + q(t,\cdot) \\ \text{for a.e. } t \in (0,T) \\ u(t) = \eta(t) \text{ for all } t \in (-\infty,0]. \end{cases}$$

We prove in Theorem 3.1, that if the history function η satisfies $0 \leq \eta \leq \overline{\rho}$ for some $\overline{\rho} > 0$, then this problem admits a unique solution u in $C((-\infty,T],L^2(\Omega))$ with $0 \leq u \leq (\overline{\rho} + \frac{\overline{q}}{\overline{a}})\exp(\overline{a}T) - \frac{\overline{q}}{\overline{a}}$. Moreover, if f is regular, η satisfies (3.4), and $(\mathbf{m}_t)_{t\geq 0}$ satisfies (M_4), then u is regular in the sense of Remark 3.2 1).

Example 3.6. ***Example derived from haematopoiesis: the Wazewska-Czyziewska and Lasota model.*** This example corresponds to Example 2.1, b) wherein some time-delays distribution is incorporated for modelling the regeneration time.

$$f(t,x,\zeta,\zeta') = -\mu(t,x)\zeta + P(t,x)\exp(-\gamma\zeta')$$

with μ and P belong to $L^\infty([0,T]\times\mathbb{R}^N)$ for all $T>0$, $\mu>0$, $P>0$, and $\gamma>0$. The function f clearly satisfies the structure condition of DCP-structured reaction functions with $l=2$ and

$$r(t,x)=(-\mu(t,x),P(t,x)),\ h(\zeta')=(1,\exp(-\gamma\zeta')),\ g(\zeta)=(\zeta,1).$$

We assume moreover that

$$\operatorname*{ess\,sup}_{(t,x)\in\mathbb{R}_+\times\mathbb{R}^N}\frac{P(t,x)}{\mu(t,x)}<+\infty.$$

Then, noticing that $f(t,x,0,\zeta')\geq 0$ and that $\forall\zeta'\geq 0,\ f(t,x,\rho,\zeta')\leq 0$ for $\rho\geq \operatorname{ess\,sup}_{(t,x)\in\mathbb{R}_+\times\mathbb{R}^N}\frac{P(t,x)}{\mu(t,x)}$, we can easily show that f satisfies (DCP). Indeed, take $\overline{f}=0$ and, for $\overline{\rho}$, any positive real number satisfying

$$\overline{\rho}\geq\operatorname*{ess\,sup}_{(t,x)\in\mathbb{R}_+\times\mathbb{R}^N}\frac{P(t,x)}{\mu(t,x)}.\tag{3.5}$$

This reaction function corresponds to the diffusive Wazewska-Czyziewska and Lasota model (see Pujo-Menjouet (2015), Ruan (2006) in the case of single delay)

$$\begin{cases}\dfrac{du}{dt}(t)+D\Phi(u(t))=-\mu(t,x)u(t)+P(t,\cdot)\exp\left(-\gamma\displaystyle\int_{-\infty}^{t}u(s)d\mathbf{m}_t(s)\right)\\ \text{for a.e. } t\in(0,T)\\ \\ u(t)=\eta(t)\text{ for all } t\in(-\infty,0]\end{cases}$$

where

- $u(t,x)$ is the number of red-blood cell at time t located at x,
- $\mu(t,x)$ is the probability of death of red-blood cells,
- P and γ are two coefficients related to the production of red-blood cells per unit time.

When τ is the time required to produce a red-blood, we take $\mathbf{m}_t=\delta_{t-\tau}$. The choice of more general measures $\mathbf{m}_t$ corresponds to more complex regeneration time. Then if the history function η satisfies $0\leq\eta\leq\overline{\rho}$ for some positive real number $\overline{\rho}$ satisfying (3.5), we prove in Theorem 3.1, that the problem admits a unique solution u in $C((-\infty,T],L^2(\Omega))$ satisfying $0\leq u\leq\overline{\rho}$. If f is regular, η satisfies (3.4) and $(\mathbf{m}_t)_{t\geq 0}$ satisfies (M_4), then u is regular in the sense of Remark 3.2 1).

Example 3.7. ***Example derived from haematopoiesis: the Mackey-Glass model.*** Let $f:\mathbb{R}_+\times\mathbb{R}^N\times\mathbb{R}^2\to\mathbb{R}$ be defined by

$$f(t,x,\zeta,\zeta')=-a(t,x)\zeta+b(t,x)\frac{\zeta'}{1+\zeta'^m}$$

where $a>0$, $b>0$, and $m>1$. We assume that for all $T>0$, a and b belong to $L^\infty([0,T]\times\mathbb{R}^N)$. The function f satisfies the structure condition of DCP-structured reaction functions with $l=2$ and

$$r(t,x)=(-a(t,x),b(t,x)),\ h(\zeta')=\left(1,\frac{\zeta'}{1+\zeta'^m}\right),\ g(\zeta)=(\zeta,1).$$

We assume moreover that

$$\operatorname*{ess\,sup}_{(t,x)\in\mathbb{R}_+\times\mathbb{R}^N}\frac{b(t,x)}{a(t,x)}<+\infty.$$

Let us check that (DCP) is satisfied. We have

$$0\leq f(t,x,0,\zeta')=b(t,x)\frac{\zeta'}{1+\zeta'^m}.$$

On the other hand, set $\overline{f}=0$ and take

$$\overline{\rho}\geq\left[\operatorname*{ess\,sup}_{(t,x)\in\mathbb{R}_+\times\mathbb{R}^N}\frac{b(t,x)}{a(t,x)}\right]h_2\left(\left(\frac{1}{m-1}\right)^{\frac{1}{m}}\right) \tag{3.6}$$

where $h_2(\zeta')=\frac{\zeta'}{1+\zeta'^m}$. An elementary calculation shows that $\sup_{\zeta'\geq 0}h_2(\zeta')=h_2\left(\left(\frac{1}{m-1}\right)^{\frac{1}{m}}\right)$. Then for all $\zeta'\geq 0$, $\overline{y}=\overline{\rho}$ satisfies

$$f(t,x,\overline{y}(t),\zeta')=f(t,x,\overline{\rho},\zeta')\leq 0=\overline{f}(t,\overline{y}(t)).$$

This example corresponds to the diffusive Mackey-Glass model whose reaction term is proposed to model the production of white blood cells (see Berezansky and Braverman (2006), Pujo-Menjouet (2015), Ruan (2006) in the case of a single delay):

$$\begin{cases}\dfrac{du}{dt}(t)+D\Phi(u(t))=-a(t,x)u(t)+b(t,\cdot)\dfrac{\displaystyle\int_{-\infty}^{t}u(s)d\mathbf{m}_t(s)}{1+\left(\displaystyle\int_{-\infty}^{t}u(s)d\mathbf{m}_t(s)\right)^m}\\ \text{for a.e. } t\in(0,T)\\ u(t)=\eta(t)\text{ for all } t\in(-\infty,0],\end{cases}$$

where

- $u(t,x)$ denotes the concentration of white-blood cells at time t located at x in circulating blood.
- The term $-a(t,x)u(t,x)$ represents the reduction of white-blood cells at time t, located at x.
- When $\mathbf{m}_t=\delta_{t-\tau}$, the term $b(t,\cdot)\frac{u(t-\tau)}{1+u(t-\tau)'^m}$ models the production of white-blood cells after a delay τ before the marrow releases further cells to replenish the deficiency.

As in the previous example, if f is regular, the history function η satisfies $0\leq\eta\leq\overline{\rho}$ for some positive real number $\overline{\rho}$ satisfying (3.6), then the problem admits a unique solution in $C((-\infty,T],L^2(\Omega))$ satisfying $0\leq u\leq\overline{\rho}$. If f is regular, η satisfies (3.4) and $(\mathbf{m}_t)_{t\geq 0}$ satisfies (M_4), then u is regular in the sense of Remark 3.2 1).

3.2.3 *Existence and uniqueness of bounded nonnegative solution*

Combining Corollary 2.1 with a suitable fixed point procedure, we establish the existence of a nonnegative bounded strong solution to the Cauchy problem associated with DCP-structured reaction functionals. Recall that u is called strong solution if it is absolutely continuous in times and satisfies a.e. the equation of $(\mathcal{P}_{\mathcal{T},\eta})$.

Theorem 3.1. *Let F be a DCP-structured reaction functional associated with (r, g, h, q), with $\overline{\rho}$ and $\overline{y}$ given by* (DCP), *Φ a standard functional* (2.6), *and assume that $0 \le \phi \le a_0\overline{\rho}$. Let $(\mathbf{m}_t)_{t\ge 0}$ be a family of vector measure in $\mathbf{M}_1^+(\mathbb{R}, L^\infty(\Omega))$ satisfying* (M_1), (M_2), (M_3), *and η in $C_c((-\infty, 0], L^2(\Omega))$ satisfying $\eta(0) \in \overline{\mathrm{dom}(D\Phi)}$, and $0 \le \eta \le \overline{\rho}$. Then, for any given $T > 0$, the time delays reaction-diffusion problem*

$$(\mathcal{P}_{\mathcal{T},\eta}) \begin{cases} \dfrac{du}{dt}(t) + D\Phi(u(t)) = F(t, u(t), \mathcal{T}u(t)) \text{ for a.e. } t \in (0, T) \\ \\ u(t) = \eta(t) \text{ for all } t \in (-\infty, 0] \end{cases}$$

admits a unique solution $u \in C((-\infty, T], L^2(\Omega))$ satisfying:

(SD_1) $u(t) \in \mathrm{dom}(D\Phi)$ *for a.e. $t \in (0, T)$,*

(SD_2) *u is almost everywhere derivable in $(0, T)$ and $u'(t) = \dfrac{du}{dt}(t)$ for a.e. $t \in (0, T)$,*

(SD_3) $u(t) \in [0, \overline{y}(T)]$ *for all $t \in [0, T]$.*

If moreover F is a regular DCP-structured reaction functional, η satisfies (3.4) *and $(\mathbf{m}_t)_{t>0}$ fulfills condition* (M_4), *then u satisfies*

(SD_4) $u(t) \in \mathrm{dom}(D\Phi)$ *for all $t \in]0, T]$, u admits a right derivative $\dfrac{d^+u}{dt}(t)$ at every $t \in]0, T[$ and*

$$\frac{d^+u}{dt}(t) + D\Phi(u(t)) = F(t, u(t), \mathcal{T}u(t)).$$

Proof. ***Step 1 (local existence).*** We establish existence of a unique solution of $(\mathcal{P}_{\mathcal{T},\eta})$ satisfying (SD_1)–(SD_4) for T small enough. For all $T > 0$, set

$$X_T := \left\{ v \in C((-\infty, T], X) : \ 0 \le v \le \overline{y}(T) \text{ and } v_{\lfloor(-\infty,0]} = \eta \right\}.$$

Choose M large enough so that Support$(\eta) \subset [-M, 0]$. Clearly X_T is a closed subset of the Banach space $C([-M, T], X)$ equipped with the uniform norm. Therefore, X_T equipped with the distance associated with the uniform norm, is a complete metric space. For each fixed u in X_T, consider the reaction-diffusion problem with unknown Λu defined by

$$(\mathcal{P}_\Lambda) \begin{cases} \dfrac{d\Lambda u}{dt}(t) + D\Phi(\Lambda u(t)) = F(t, \Lambda u(t), \mathcal{T}u(t)) \text{ for a.e. } t \in (0, T) \\ \\ \Lambda u(t) = \eta(t) \text{ for all } t \in (-\infty, 0], \end{cases}$$

and first prove that $(\mathcal{P}_\Lambda)$ admits a unique solution Λu satisfying (SD_1), (SD_2) and (SD_3). Note that it suffices to prove existence of Λu in $C([0,T],L^2(\Omega))$ which solves

$$\begin{cases} \dfrac{d\Lambda u}{dt}(t) + D\Phi(\Lambda u(t)) = F(t, \Lambda u(t), \mathcal{T}u(t)) \text{ for a.e. } t \in (0,T) \\ \\ \Lambda u(0) = \eta(0), \end{cases}$$

and to extend Λu by η on $(-\infty, 0]$. Since F is a DCP-structured reaction functional, F is associated with f given by $f(t,x,\zeta,\zeta') = r(t,x) \odot h(\zeta') \cdot g(\zeta) + q(t,x)$. Set

$$r_{\mathcal{T}}(t,x) := r(t,x) \odot h(\mathcal{T}u(t)(x)), \qquad f_{\mathcal{T}}(t,x,\zeta) := r_{\mathcal{T}}(t,x) \cdot g(\zeta) + q(t,x),$$

and, for $u \in L^2(\Omega)$, $F_{\mathcal{T}}(t,u)(x) := f_{\mathcal{T}}(t,x,u(x))$. We have

$$\begin{aligned} F(t, \Lambda u(t), \mathcal{T}u(t))(x) &= r(t,x) \odot h(\mathcal{T}u(t)(x)) \cdot g(\Lambda u(t,x)) + q(t,x) \\ &= r_{\mathcal{T}}(t,x) \cdot g(\Lambda u(t,x)) + q(t,x) \\ &= f_{\mathcal{T}}(t,x,\Lambda u(t,x)) := F_{\mathcal{T}}(t, \Lambda u(t))(x). \end{aligned}$$

The functional $F_{\mathcal{T}}$ is a CP-structured reaction functional associated with $(r_{\mathcal{T}}, g, q)$ as defined in Chapter 2, Definition 2.1. Condition (CP) is fulfilled because, from Lemma 3.1, $0 \le \mathcal{T}u \le \overline{y}(T)$, and f satisfies (DCP). Therefore, existence and uniqueness of Λu, which fulfills (SD_1), (SD_2), (SD_3) is ensured by Corollary 2.1.

Assume that F is regular, i.e. $r \in W^{1,1}(0,T,L^2(\Omega,\mathbb{R}^l))$, $q \in W^{1,1}(0,T,L^2(\Omega))$ for all $T > 0$, that η satisfies (3.4), and $(\mathbf{m}_t)_{t>0}$ fulfills condition (M_4). To show that Λu satisfies (SD_4), according to Corollary 2.1, it remains to prove that $r_{\mathcal{T}} : t \mapsto r_{\mathcal{T}}(t,\cdot)$ is absolutely continuous from $[0,T]$ into $L^2(\Omega)$. The claim follows from the absolute continuity of r, condition (M_4) (note that $u \in C_\eta((-\infty,T],L^2(\Omega)))$, and the following estimate for s and t in $[0,T]$:

$$\begin{aligned} \|r_{\mathcal{T}}(t,\cdot) - r_{\mathcal{T}}(s,\cdot)\|_{L^2(\Omega,\mathbb{R}^l)} =& \|r(t,\cdot) \odot h(\mathcal{T}u(t)) - r(s,\cdot) \odot h(\mathcal{T}u(s))\|_{L^2(\Omega,\mathbb{R}^l)} \\ \le& \|r(t,\cdot) \odot h(\mathcal{T}u(t)) - r(s,\cdot) \odot h(\mathcal{T}u(t))\|_{L^2(\Omega,\mathbb{R}^l)} \\ &+ \|r(s,\cdot) \odot h(\mathcal{T}u(t)) - r(s,\cdot) \odot h(\mathcal{T}u(s))\|_{L^2(\Omega,\mathbb{R}^l)} \\ \le& \|h\|_{L^\infty([0,\overline{y}(T)],\mathbb{R}^l)} \|r(t,\cdot) - r(s,\cdot)\|_{L^2(\Omega,\mathbb{R}^l)} \\ &+ \|r\|_{L^\infty([0,T]\times\mathbb{R}^N,\mathbb{R}^l)} L_{h,T} \|\mathcal{T}u(t) - \mathcal{T}u(s)\|_{L^2(\Omega)} \\ \le& \|h\|_{L^\infty([0,\overline{y}(T)],\mathbb{R}^l)} \|r(t,\cdot) - r(s,\cdot)\|_{L^2(\Omega,\mathbb{R}^l)} \\ &+ \|r\|_{L^\infty([0,T]\times\mathbb{R}^N,\mathbb{R}^l)} L_{h,T} \int_s^t \varphi_u(\tau)d\tau, \end{aligned}$$

where $L_{h,T}$ denotes the Lipschitz constant of the restriction of h to $[0,\overline{y}(T)]$.

We are going to apply the Banach fixed point theorem to $\Lambda : X_T \to C([-M,T],X)$. Clearly $\Lambda(X_T) \subset X_T$. Indeed estimate $0 \le \Lambda u \le \overline{y}(T)$ for all $t \in (-\infty,T]$ is due to hypothesis $0 \le \eta \le \overline{\rho} \le \overline{y}(T)$, to $\Lambda u(t) = \eta(t)$ for $t \in (-\infty,0]$, and to the fact that Λu satisfies (SD_3). To complete the proof, it remains to establish that $\Lambda : X_T \to X_T$ is a contraction for T small enough.

Let $(u,v) \in X_T \times X_T$. Then, for a.e. $t \in (0,T)$ we have

$$\frac{d\Lambda u}{dt}(t) + D\Phi(\Lambda u(t)) = F(t, \Lambda u(t), \mathcal{T}u(t)),$$
$$\frac{d\Lambda v}{dt}(t) + D\Phi(\Lambda v(t)) = F(t, \Lambda v(t), \mathcal{T}v(t)).$$

Subtract these two equalities and take the scalar product with $\Lambda u(t) - \Lambda v(t)$ in X. This yields

$$\left\langle \frac{d}{dt}(\Lambda u - \Lambda v)(t), (\Lambda u - \Lambda v)(t) \right\rangle + \langle D\Phi(\Lambda u(t)) - D\Phi(\Lambda v(t)), \Lambda u(t) - \Lambda v(t)\rangle$$
$$= \langle F(t, \Lambda u(t), \mathcal{T}u(t)) - F(t, \Lambda v(t), \mathcal{T}v(t)), \Lambda u(t) - \Lambda v(t)\rangle .$$

Then, using the fact that $D\Phi$ is a monotone operator, we infer that for a.e. $t \in (0,T)$

$$\frac{1}{2}\frac{d}{dt}\|(\Lambda u - \Lambda v)(t)\|_X^2 \leq \langle F(t, \Lambda u(t), \mathcal{T}u(t)) - F(t, \Lambda v(t), \mathcal{T}v(t)), \Lambda u(t) - \Lambda v(t)\rangle .$$

Thus, for a.e. $t \in (0,T)$,

$$\begin{aligned}
&\frac{d}{dt}\|\Lambda u(t) - \Lambda v(t)\|_X^2 \\
&\leq 2\|F(t, \Lambda u(t), \mathcal{T}u(t)) - F(t, \Lambda v(t), \mathcal{T}v(t))\|_X \|\Lambda u(t) - \Lambda v(t)\|_X \\
&\leq \|F(t, \Lambda u(t), \mathcal{T}u(t)) - F(t, \Lambda v(t), \mathcal{T}v(t))\|_X^2 + \|\Lambda u(t) - \Lambda v(t)\|_X^2. \qquad (3.7)
\end{aligned}$$

According to the structure of the DCP-structured reaction functional F, from (3.3) and the fact that $u = v$ in $(-\infty, 0]$, we have for a.e. $t \in (0,T)$

$$\begin{aligned}
&\|F(t, \Lambda u(t), \mathcal{T}u(t)) - F(t, \Lambda v(t), \mathcal{T}v(t))\|_X^2 \\
&\leq C(T,g,h)\|\mathcal{T}u(t) - \mathcal{T}v(t)\|_X^2 + C'(T,g,h)\|\Lambda u(t) - \Lambda v(t)\|_X^2 \\
&\leq 4TC(T,g,h)\|u(t) - v(t)\|_X + C'(T,g,h)\|\Lambda u(t) - \Lambda v(t)\|_X^2 \qquad (3.8)
\end{aligned}$$

where

$$C(T,g,h) = 2 \sup_{\zeta \in [0,\overline{y}(T)]} |g(\zeta)|^2 \|r\|^2_{L^\infty(\mathbb{R}^N,\mathbb{R}^l)} L_{h,T}(T)^2,$$
$$C'(T,g,h) = 2 \sup_{\zeta' \in [0,\overline{y}(T)]} |h(\zeta')|^2 \|r\|^2_{L^\infty(\mathbb{R}^N,\mathbb{R}^l)} L_{g,T}(T)^2$$

(recall that $L_{g,T}$ and $L_{h,T}$ denote the Lipschitz constants of the restrictions of g and h on $[0, \overline{y}(T)]$ respectively). Combining (3.7) and (3.8) we infer that for a.e. $t \in (0,T)$

$$\begin{aligned}
&\frac{d}{dt}\|\Lambda u(t) - \Lambda v(t)\|_X^2 \\
&\leq 4TC(T,g,h)\|u(t) - v(t)\|_X^2 + (1 + C'(T,g,h))\|\Lambda u(t) - \Lambda v(t)\|_X^2. \qquad (3.9)
\end{aligned}$$

By integrating this inequality over $(0,s)$ for $s \in (0,T)$ and noticing that $\Lambda u(0) = \Lambda v(0) = \eta(0)$, we obtain

$$\begin{aligned}
&\|\Lambda u(s) - \Lambda v(s)\|_X^2 \\
&\leq 4T^2\, C(T,g,h)\|u - v\|^2_{C((-\infty,T],X)} + (1 + C'(T,g,h)) \int_0^s \|\Lambda u(t) - \Lambda v(t)\|_X^2 dt
\end{aligned}$$

from which, according to Grönwall's lemma, we deduce that for all $s \in (0,T)$,

$$\|\Lambda u(s) - \Lambda v(s)\|_X^2 \leq 4T^2.C(T,g,h)\|u-v\|^2_{C((-\infty,T],X)} \exp((1+C'(T,g,h))T).$$

Since $\Lambda u = \Lambda v = \eta$ on $(-\infty, 0]$, this inequality holds for all $s \in (-\infty, T]$ so that

$$\|\Lambda u - \Lambda v\|_{C((-\infty,T],X)} \leq K(T)\|u-v\|_{C((-\infty,T],X)}$$

where $K(T) := 2T \ (C(T,g,h))^{\frac{1}{2}} \exp((1 + C'(T,g,h))\frac{T}{2})$. We claim that $\lim_{T\to 0} K(T) = 0$. The proof will be completed by choosing $T^* > 0$ small enough so that $K(T^*) < 1$. For this, note that the maps $T \mapsto \overline{y}(T)$, $T \mapsto L_{g,T}$, and $T \mapsto L_{h,T}$ are nondecreasing so that $C(T,g,h)$ and $C(T,g,h)$ decrease to finite limits when T decreases to 0.

Step 2 (uniqueness). For T^* obtained above, we establish the uniqueness of the solution of $(\mathcal{P}_{T,\eta})$. Let u and v be two solutions, then, by taking u for Λu and v for Λv in (3.9), we infer that for a.e. $t \in (0, T^*)$

$$\frac{d}{dt}\|u(t)-v(t)\|_X^2 \leq (1 + (C(T^*,g,h) + C'(T^*,g,h)))\|u(t)-v(t)\|_X^2.$$

Applying Grönwall's Lemma after integrating, we obtain for a.e. $t \in (0,T^*)$

$$\|u(t)-v(t)\|_X^2 \leq \|u(0)-v(0)\|_X^2 \exp((1+(C(T^*,g,h)+C'(T^*,g,h)))T^*)$$

and the claim follows since $u(0) = v(0) = \eta(0)$.

Step 3 (existence of a global solution). We use the notation of Section 2.1.2. Let denote by $T^* > 0$ a small enough number obtained in ***Step 1*** so that $(\mathcal{P}_{T,\eta})$ admits a unique solution in $C([0,T^*],X)$. We know that for $0 < \delta < T^*$, $\frac{du}{dt}$ belongs to $L^2(\delta, T^*, X)$ (see Chapter 2, proof of Theorem 2.3 and references therein). Set

$$E := \{T > \delta : \exists u \in C((-\infty,T],X),\ u \text{ solution of } (\mathcal{P}_{T,\eta})\}.$$

Since $T^* \in E$, we have $E \neq \emptyset$. Set $T_{Max} := \sup E$ in $\overline{\mathbb{R}}_+$. In the following we denote by u the maximal solution of $(\mathcal{P}_{T,\eta})$ in $C([0,T_{Max}),X)$ and argue by contradiction by assuming that there is blow-up in finite time, i.e. $T_{Max} < +\infty$.

a) We first show that $\lim_{t\to T_{Max}} u(t)$ exists in X. Let $T \in E$, then for a.e. $t \in (0,T)$ we have

$$\left\langle \frac{du}{dt}(t), \frac{du}{dt}(t)\right\rangle + \left\langle D\Phi u(t), \frac{du}{dt}(t)\right\rangle = \left\langle F(t,u(t),\mathcal{T}u(t)), \frac{du}{dt}(t)\right\rangle.$$

We know that for a.e. $t \in (\delta,T)$, $\frac{d}{dt}\Phi(u(t)) = \langle D\Phi(u(t)), \frac{du}{dt}(t)\rangle$ (see (Attouch *et al.*, 2014, Proposition 17.2.5)) so that by integration over (δ,T) we obtain

$$\int_\delta^T \left\|\frac{du}{dt}(t)\right\|_X^2 dt + \Phi(u(t)) - \Phi(u(\delta))$$
$$\leq \left(\int_0^T \|F(t,u(t),\mathcal{T}u(t))\|_X^2 dt\right)^{\frac{1}{2}} \left(\int_\delta^T \left\|\frac{du}{dt}(t)\right\|_X^2 dt\right)^{\frac{1}{2}}. \tag{3.10}$$

Note that for all $T \in E$, we have $[0, \overline{y}(T)] \subset [0, \overline{y}(T_{Max})]$. Thus, from the estimates $0 \le u \le \overline{y}(T)$ and $0 \le \mathcal{T}u \le \overline{y}(T)$, according to the structure of the DCP-structured reaction functional F, there exists a constant

$$C(\|r\|_{L^\infty(\mathbb{R}^N,\mathbb{R}^l)}, \|g\|_{L^\infty([0,\overline{y}(T_{Max})],\mathbb{R}^l)}, \|h\|_{L^\infty([0,\overline{y}(T_{Max})],\mathbb{R}^l)})$$

(that we write C in short) such that

$$\|F(t,u(t),\mathcal{T}u(t))\|_X^2 \le 2C^2\mathcal{L}_N(\Omega) + 2\|q(t,\cdot)\|_X^2.$$

Therefore, since $\inf \Phi > -\infty$, and $q \in L^2(0, T_{Max}, L^2(\Omega))$, (3.10) yields

$$\int_\delta^T \left\|\frac{du}{dt}(t)\right\|_X^2 dt \le C\left(1 + \left(\int_\delta^T \left\|\frac{du}{dt}(t)\right\|_X^2 dt\right)^{\frac{1}{2}}\right)$$

where the new constant C does not depend on T. We infer that

$$\int_\delta^{T_{Max}} \left\|\frac{du}{dt}(t)\right\|_X^2 dt = \sup_{T\in E}\int_\delta^T \left\|\frac{du}{dt}(t)\right\|_X^2 dt < +\infty. \tag{3.11}$$

From (3.11) we deduce that $u : [\delta, T_{Max}) \to X$ is uniformly continuous. Indeed, for $s < t$ in $[\delta, T_{Max})$ we have

$$\|u(t) - u(s)\|_X \le \int_s^t \left\|\frac{du}{d\tau}(\tau)\right\|_X^2 d\tau \le (t-s)^{\frac{1}{2}}\left(\int_\delta^{T_{Max}} \left\|\frac{du}{d\tau}(t)\right\|_X^2 dt\right)^{\frac{1}{2}}.$$

Since X is a Banach space, according to the continuous extension principle, u possesses a unique continuous extension $\overline{u}$ in $[\delta, T_{Max}]$ i.e. $\lim_{t\to T_{Max}} u(t) = \overline{u}(T_{Max})$, which proves the claim.

b) Contradiction: consider the distributed delays reaction-diffusion problem

$$(\mathcal{P}_{\mathcal{T},\eta'})\begin{cases} \dfrac{dv}{dt}(t) + D\Phi(v(t)) = F(t, v(t), \mathcal{T}v(t)) \text{ for a.e. } t \in (0,T) \\ \\ v(t) = \eta'(t) \text{ for } t \in (-\infty, 0] \end{cases}$$

where η' is defined by $\eta'(t) = \overline{u}(t + T_{Max})$. It is easily seen that $\eta' \in C_c((-\infty, 0], L^2(\Omega))$ and satisfies (3.4). Note that $\eta'(0) = \overline{u}(T_{Max}) \in \overline{\mathrm{dom}(D\Phi)}$. Indeed, choose $t_n \to T_{Max}$ with t_n outside the negligible set in which $u(t) \notin \mathrm{dom}(D\Phi)$ and use the fact that $\overline{u}(T_{Max}) = \lim_{t\to T_{Max}} u(t)$. Moreover $0 \le \eta' \le \overline{\rho}'$ where $\overline{\rho}' = \overline{y}(T_{Max})$. Then according to step 1, there exists $T^{**} > 0$ small enough such that $(\mathcal{P}'_{\mathcal{T},\eta})$ possesses a solution $v \in C([0, T^{**}], X)$. Set

$$\widetilde{u}(t) = \begin{cases} u(t) \text{ if } t \in (-\infty, T_{Max}] \\ v(t - T_{Max}) \text{ if } t \in [T_{Max}, T_{Max} + T^{**}]. \end{cases}$$

Then $\widetilde{u} \in C([0, T_{Max} + T^{**}], X)$ is a solution of $(\mathcal{P}_{\mathcal{T},\eta})$, a contradiction with the maximality of T_{Max}. $\square$

By an easy adaptation of the proof above, and applying Corollary 2.2, we establish the same result in the context of the convention above, i.e. when $a_0 = 0$ on $\Omega \setminus \Gamma$, is extended by $+\infty$ on Γ, and $\phi = 0$. More precisely,

Theorem 3.2. *Let F be a DCP-structured reaction functional associated with (r, g, h, q), with $\overline{\rho}$ and $\overline{y}$ given by (DCP), and Φ given by (2.13). Let T be any positive numbers, and η in $C_c((-\infty, 0], L^2(\Omega))$ satisfying $\eta(0) \in \overline{\mathrm{dom}(D\Phi)}$, and $0 \leq \eta \leq \overline{\rho}$. Then the time delays reaction-diffusion problem*

$$(\mathcal{P}_{T,\eta}) \begin{cases} \dfrac{du}{dt}(t) + D\Phi(u(t)) = F(t, u(t), \mathcal{T}u(t)) \text{ for a.e. } t \in (0, T) \\ \\ u(t) = \eta(t) \text{ for all } t \in (-\infty, 0] \end{cases}$$

admits a unique solution $u \in C((-\infty, T], L^2(\Omega))$ satisfying (SD_1), (SD_2) *and* (SD_3). *If moreover r belongs to $W^{1,1}(0, T, L^2(\Omega, \mathbb{R}^l))$, q belongs to $W^{1,1}(0, T, L^2(\Omega))$, η satisfies* (3.4) *and $(\mathbf{m}_t)_{t>0}$ fulfills condition* (M_4), *then u satisfies* (SD_4).

Remark 3.3. In the case of single time delay reaction-diffusion problems, for establishing existence and uniqueness for $(\mathcal{P}_{T,\eta})$ we could develop the following alternative constructive proof: $(\mathcal{P}_{T,\eta})$ is equivalent to the finite recursive sequence $(\mathcal{P}_i)$, $i = 1 \dots [\frac{T}{\tau}] + 1$ of reaction-diffusion Cauchy problem of the form

$$(\mathcal{P}_i) \begin{cases} \dfrac{du^i}{dt}(t) + D\Phi(u^i(t)) = F_i(t, u^i(t)) \text{ for a.e. } t \in ((i-1)\tau, i\tau) \\ \\ u^i((i-1)\tau) = u^{i-1}((i-1)\tau) \end{cases}$$

where $u^0 := \eta$, and $F_i(t, u^i(t)) := F(t, u^i(t), u^{i-1}(t))$ is clearly a CP-structured reaction functional. By applying Corollary 2.1, we see that there exists a unique solution u^i of $(\mathcal{P}_i)$. The function u defined by $u(t) = u^i(t)$ for $t \in [(i-1)\tau, i\tau]$ is a solution of $(\mathcal{P}_{T,\eta})$. Uniqueness is a consequence of the following remark: if u solves $(\mathcal{P}_{\tau,\eta})$, then the restriction of u in $[(i-1)\tau, i\tau]$ is the unique solution of $(\mathcal{P}_i)$.

3.3 Convergence theorems

As in Section 2.4, we are given a sequence $(\Phi_n)_{n\in\mathbb{N}}$ of functionals of the calculus of variations $\Phi_n : L^2(\Omega) \to \mathbb{R} \cup \{+\infty\}$ defined by

$$\Phi_n(u) = \begin{cases} \displaystyle\int_\Omega W_n(x, \nabla u(x))dx + \frac{1}{2}\int_{\partial\Omega} a_{0,n} u^2 d\mathcal{H}_{N-1} - \int_{\partial\Omega} \phi_n u \, d\mathcal{H}_{N-1} & \text{if } u \in H^1(\Omega) \\ \\ +\infty & \text{otherwise,} \end{cases} \tag{3.12}$$

where $\phi_n \in L^2_{\mathcal{H}_{N-1}}(\partial\Omega)$, and $a_{0,n} \in L^\infty_{\mathcal{H}_{N-1}}(\partial\Omega)$ with

$$\begin{cases} a_{0,n} \geq 0 \quad \mathcal{H}_{N-1}\text{-a.e. in } \partial\Omega \\ \\ \exists \sigma > 0 \quad a_{0,n} \geq \sigma \quad \mathcal{H}_{N-1}\text{-a.e. in } \Gamma \subset \partial\Omega \text{ with } \mathcal{H}_{N-1}(\Gamma) > 0. \end{cases}$$

We also include the case of functionals

$$\Phi_n(u) = \begin{cases} \displaystyle\int_\Omega W_n(x, \nabla u(x))dx & \text{if } u \in H^1_\Gamma(\Omega) \\ +\infty & \text{otherwise.} \end{cases} \tag{3.13}$$

Each $W_n : \mathbb{R}^N \times \mathbb{R}^N \to \mathbb{R}$ is a measurable function which fulfills condition (D_n).

From now on, we fix $T > 0$, and we assume the following structural conditions.

(SC1) We are given a sequence $(F_n)_{n\in\mathbb{N}}$ of DCP-structured reaction functionals, each F_n is associated with (r_n, g_n, h_n, q_n), i.e. $F_n(t, u, v)(x) = f_n(t, x, u(x), v(x))$ for all $t \in [0, T]$, a.e. $x \in \Omega$, and all $(u, v) \in L^2(\Omega)^2$, where $f_n : \mathbb{R}_+ \times \mathbb{R}^N \times \mathbb{R} \times \mathbb{R} \to \mathbb{R}$ is defined by

$$f_n(t, x, \zeta, \zeta') = r_n(t, x) \odot h_n(\zeta') \cdot g_n(\zeta) + q_n(t, x). \tag{3.14}$$

For all $n \in \mathbb{N}$, h_n and g_n are assumed to be locally Lipschitz functions, uniformly with respect to n, i.e., for all interval $I \subset \mathbb{R}$, there exist $L_I \geq 0$ and $L'_I \geq 0$ such that for all $(\zeta, \zeta') \in I \times I$,

$$\begin{aligned} &\sup_{n\in\mathbb{N}} |g_n(\zeta) - g_n(\zeta')| \leq L_I|\zeta - \zeta'| \text{ and} \\ &\sup_{n\in\mathbb{N}} |h_n(\zeta) - h_n(\zeta')| \leq L'_I|\zeta - \zeta'|. \end{aligned} \tag{3.15}$$

(SC2) We are given a sequence $(\eta_n)_{n\in\mathbb{N}}$ of history functions in $C_c((-\infty, 0], L^2(\Omega))$ whose support is included in a fixed compact set, satisfying $\eta_n(0) \in \mathrm{dom}(\Phi_n)$, $0 \leq \eta_n \leq \overline{\rho}_n$, and

$$\sup_{n\in\mathbb{N}} \int_{-\infty}^0 \left\| \frac{d\eta_n}{dt}(t, \cdot) \right\|_{L^2(\Omega)} dt < +\infty, \tag{3.16}$$

where $\overline{\rho}_n$ is given by the condition (DCP) fulfilled by each F_n for all $n \in \mathbb{N}$.

(SC3) We are given a sequence $(\mathbf{m}^n_t)_{t\geq 0, n\in\mathbb{N}}$ of vector measures in $\mathbf{M}^+_1(\mathbb{R}, L^\infty(\Omega))$ such that each $(\mathbf{m}^n_t)_{t\geq 0}$ satisfies (M_1), (M_2) and (M_3). The sequence of time-delays operator $(\mathcal{T}_n)_{n\in\mathbb{N}}$ associated with $(\mathbf{m}^n_t)_{t\geq 0, n\in\mathbb{N}}$, is defined for each v in $C_c((-\infty, T], L^2(\Omega))$, by

$$\mathcal{T}_n v(t) = \int_{-\infty}^t v(s) d\mathbf{m}^n_t(s)$$

(see Definition 3.1).

When F_n is regular, we assume furthermore that each family of vector measure $(\mathbf{m}^n_t)_{t\geq 0}$ fulfills condition (M'_4). More precisely, there exists a sequence of Borel probability measures $(\mu_n)_{n\in\mathbb{N}}$ in $\mathcal{P}_+(\mathbb{R})$, such that for all v in $C_{\eta_n}((-\infty, T], L^2(\Omega))$, and all $(s, t) \in [0, T]^2$ with $s < t$,

$$\|\mathcal{T}_n v(t) - \mathcal{T}_n v(s)\|_{L^2(\Omega)} \leq \int_s^t \left(\left\| \frac{dv}{d\sigma}(\cdot) \right\|_{L^2(\Omega)} \star \mu_n \right)(\sigma) d\sigma. \tag{3.17}$$

(SC4) Finally, we assume that

$$\sup_n \overline{y}_n(T) < +\infty, \tag{3.18}$$

and, for functionals (3.12)

$$\forall n \in \mathbb{N} \qquad 0 \leq \phi_n \leq a_{0,n}\overline{\rho}_n \text{ on } \partial\Omega, \tag{3.19}$$

where $\overline{y}_n$ is given by the condition (DCP) fulfilled by each F_n. For functionals (3.13), no additional condition is necessary.

3.3.1 *Stability at the limit*

According to the structure of the reaction functional F_n, the time delays term, and a suitable notion of weak convergence relating to $(\mathbf{m}_t^n)_{t\geq 0}$, we show below that the limit of the time delays problems $(\mathcal{P}_n)$ associated with above datas, keeps the same structure. An other proof can be found in Anza Hafsa *et al.* (2020b) under an equi-ellipticity condition.

Theorem 3.3 (No mixing effect). *Assume that the sequence $(W_n)_{n\in\mathbb{N}}$ of functionals* (3.12) *or* (3.13) *satisfies conditions* (D_n), *and that the sequences of DCP-structured reaction functionals $(F_n)_{n\in\mathbb{N}}$, history functions $(\eta_n)_{n\in\mathbb{N}}$, and vector measures $(\mathbf{m}_t^n)_{t\geq 0, n\in\mathbb{N}}$ satisfy* (SC1), (SC2), (SC3) *and* (SC4). *Let u_n be the unique solution of the time delays problem*

$$(\mathcal{P}_n)\begin{cases} \dfrac{du_n}{dt}(t) + D\Phi_n(u_n(t)) = F_n(t, u_n(t), \mathcal{T}_n u_n(t)) \text{ for a.e. } t \in (0,T), \\ \\ u_n(t) = \eta_n(t) \text{ for all } t \in (-\infty, 0]. \end{cases}$$

Assume that

(Hd_1) $\Phi_n \overset{M}{\to} \Phi$ *and* $\sup_{n\in\mathbb{N}} \|\phi_n\|_{L^2_{\mathcal{H}_{N-1}}(\partial\Omega)} < +\infty$;

(Hd_2) $\sup_{n\in\mathbb{N}} \Phi_n(\eta_n(0)) < +\infty$;

(Hd_3) *there exists η in $C_c((-\infty,0], L^2(\Omega))$ such that $\eta_n \to \eta$ in $C((-\infty,0], L^2(\Omega))$;*

(Hd_4) *there exists a family of $L^\infty(\Omega)$-valued measures $(\mathbf{m}_t)_{t\geq 0}$ with finite variation, such that for all $t \in [0,T]$, $\mathbf{m}_t^n \lfloor_{(-\infty,t]} \rightharpoonup \mathbf{m}_t \lfloor_{(-\infty,t]}$*[4] *for the weak convergence of measures, in $L^2(\Omega)$ strong, i.e. for all $\psi \in C_c(\mathbb{R}, L^2(\Omega))$,* $\lim_{n\to\infty} \int_{\mathbb{R}} \mathbb{1}_{(-\infty,t]}\psi \, d\mathbf{m}_t^n = \int_{\mathbb{R}} \mathbb{1}_{(-\infty,t]}\psi \, d\mathbf{m}_t$ *strongly in $L^2(\Omega)$;*

(Hd_5) *g_n and h_n pointwise converge to g and h respectively;*

[4]For any Borel vector measure μ, and Borel set B, we denote by $\mu\lfloor_B$ the restriction of μ to the σ-algebra restriction of $\mathcal{B}(\mathbb{R})$ to B.

(Hd$_6$) $\overline{r} := \sup_{n\in\mathbb{N}} \|r_n\|_{L^\infty([0,T]\times\mathbb{R}^N,\mathbb{R}^l)} < +\infty$ *and there exists* r *in* $L^\infty([0,T]\times\mathbb{R}^N,\mathbb{R}^l)$ *such that* $r_n \rightharpoonup r$ *in* $L^2(0,T,L^2(\Omega,\mathbb{R}^l))$;

(Hd$_7$) *there exists* q *in* $L^2(0,T,L^2(\Omega))$ *such that* $q_n \rightharpoonup q$ *in* $L^2(0,T,L^2(\Omega))$.

Then $(u_n)_{n\in\mathbb{N}}$ *uniformly converges in* $C([0,T],L^2(\Omega))$ *to a function* u *whose extension by* η *in* $(-\infty,0]$, *is the unique solution of the time delays problem*

$$(\mathcal{P})\begin{cases}\dfrac{du}{dt}(t)+\partial\Phi(u(t))\ni F(t,u(t),\mathcal{T}u(t)) \text{ for a.e. } t\in(0,T)\\[2ex] u(t)=\eta(t),\ 0\le\eta\le\sup\limits_n\overline{\rho}_n \text{ for all } t\in(-\infty,0],\ \eta(0)\in\mathrm{dom}(\Phi).\end{cases}$$

The reaction functional $F:[0,+\infty)\times L^2(\Omega)\times L^2(\Omega)\to\mathbb{R}^\Omega$ *is defined for all* $t\in[0,T]$, *all* $(u,v)\in L^2(\Omega)^2$ *and for a.e.* $x\in\Omega$, *by*

$$F(t,u,v)(x)=f(t,x,u(x),v(x))$$

where, for every $(t,x,\zeta,\zeta')\in\mathbb{R}_+\times\mathbb{R}^N\times\mathbb{R}\times\mathbb{R}$,

$$f(t,x,\zeta,\zeta')=r(t,x)\odot h(\zeta')\cdot g(\zeta)+q(t,x).$$

The operator $\mathcal{T}:C_c((-\infty,T],L^2(\Omega))\to L^2(0,T,L^2(\Omega))$ *is defined by*

$$\forall t\in[0,T]\quad \mathcal{T}u(t)=\int_{-\infty}^t u(s)d\mathbf{m}_t(s).$$

Moreover

$$\frac{du_n}{dt}\rightharpoonup\frac{du}{dt} \text{ weakly in } L^2(0,T,L^2(\Omega)),$$

and $0\le u(t)\le\sup_n\overline{y}_n(t)$ *for all* $t\in[0,T]$.

If furthermore $\Phi_n(u_n^0)\to\Phi(u^0)$, $r_n\to r$ *strongly in* $L^2(0,T,L^2(\Omega,\mathbb{R}^l))$, *and* $q_n\to q$ *strongly in* $L^2(0,T,L^2(\Omega))$, *then*

$$\frac{du_n}{dt}\to\frac{du}{dt} \text{ strongly in } L^2(0,T,L^2(\Omega)).$$

Assume that $\partial\Phi$ *is single valued and that* $(F_n)_{n\in\mathbb{N}}$ *is a sequence of regular DCP-structured reaction functionals. Assume furthermore that* $\frac{dq_n}{dt}\rightharpoonup\frac{dq}{dt}$ *weakly in* $L^1(0,T,L^2(\Omega))$, $\dfrac{dr_n}{dt}\rightharpoonup\dfrac{dr}{dt}$ *weakly in* $L^1(0,T,L^2(\Omega,\mathbb{R}^l))$, *and for each* $t\in]0,T[$, *that* $q_n(t,\cdot)\rightharpoonup q(t,\cdot)$ *weakly in* $L^2(\Omega)$ *and* $r_n(t,\cdot)\rightharpoonup r(t,\cdot)$ *weakly in* $L^2(\Omega,\mathbb{R}^l)$. *Then, for each* $t\in]0,T[$, $\dfrac{d^+u_n}{dt}(t)\rightharpoonup F(t,u(t),\mathcal{T}u)-D\Phi(u(t))$ *weakly in* $L^2(\Omega)$.

Proof. We follow the outlines of the proof of Theorem 2.6. The only additional difficulty is the presence of the delay term in the reaction functional. In the statement of Theorem 3.3, from (SC2), we assume that $u_n^0=\eta_n(0)\in H^1(\Omega)=\mathrm{dom}(\Phi_n)$. But $\mathrm{dom}(\Phi_n)\subset\overline{\mathrm{dom}(\Phi_n)}=\overline{\mathrm{dom}(D\Phi_n)}$ (see Remark 2.1), hence $u_n^0\in\overline{\mathrm{dom}(D\Phi_n)}$. Therefore, according to Theorem 3.1, or 3.2, $(\mathcal{P}_n)$ possesses a unique solution

u_n which satisfies (SD_1)–(SD_3), and (SD_4) when F_n is regular. Note that, from Lemma 3.1, we have $0 \leq \mathcal{T}_n u_n \leq \overline{y}_n(T)$.

Step 1. We establish

$$0 \leq \eta \leq \sup_n \overline{\rho}_n < +\infty; \tag{3.20}$$

$$\sup_{n\in\mathbb{N}} \|u_n\|_{C(0,T,X)} < +\infty; \tag{3.21}$$

$$\overline{g} := \sup_{(\zeta,n)\in[0,\sup_n \overline{y}_n(T)]\times\mathbb{N}} |g_n(\zeta)| < +\infty, \ \overline{h} := \sup_{(\zeta,n)\in[0,\sup_n \overline{y}_n(T)]\times\mathbb{N}} |h_n(\zeta)| < +\infty; \tag{3.22}$$

$$\sup_{n\in\mathbb{N}} \left\| \frac{du_n}{dt} \right\|_{L^2(0,T,X)} < +\infty. \tag{3.23}$$

Let $n \in \mathbb{N}$. Inequality (3.20) follows directly from $0 \leq \eta_n \leq \overline{\rho}_n = \overline{y}_n(0) \leq \overline{y}_n(T)$, and ($Hd_3$). Estimate (3.21) follows from $0 \leq u_n \leq \overline{y}_n(T) \leq \overline{\rho}$. We easily deduce (3.22) from (3.15), hypothesis (Hd_5) and estimate $|g_n(\zeta)| \leq |g_n(0)| + L_{[0,\sup_n \overline{y}_n(T)]}|\zeta|$ for all $\zeta \in [0, \sup_n \overline{y}_n(T)]$. Idem for h_n.

Let us establish (3.23). In what follows the letter C denotes various constants which can vary from line to line. By using the structure of the DCP-structured reaction functional F_n, and from (3.22) and hypothesis (Hd_6), we infer that

$$\begin{aligned} \|F_n(t, u_n(t), \mathcal{T}_n u_n(t))\|_X^2 &\leq 2\mathcal{L}(\Omega)\|r_n\|_\infty^2 \overline{g}^2 \overline{h}^2 + 2\|q_n(t,\cdot)\|_X^2 \\ &\leq C(1 + \|q_n(t,\cdot)\|_X^2). \end{aligned} \tag{3.24}$$

Thus, according to hypothesis (Hd_7), we deduce

$$\sup_{n\in\mathbb{N}} \int_0^T \|F_n(t, u_n(t), \mathcal{T}_n u_n(t))\|_X^2 dt < +\infty. \tag{3.25}$$

On the other hand, since u_n solves ($\mathcal{P}_n$) we have for a.e. $t \in (0,T)$,

$$\left\| \frac{du_n}{dt}(t) \right\|^2 + \left\langle D\Phi_n(u_n(t)), \frac{du_n}{dt}(t) \right\rangle = \left\langle F_n(t, u_n(t), \mathcal{T}_n u_n(t)), \frac{du_n}{dt}(t) \right\rangle,$$

and, by integrating this equality over $(0,T)$,

$$\begin{aligned} &\int_0^T \left\| \frac{du_n}{dt}(t) \right\|^2 dt + \int_0^T \left\langle D\Phi_n(u_n(t)), \frac{du_n}{dt}(t) \right\rangle dt \\ &= \int_0^T \left\langle F_n(t, u_n(t), \mathcal{T}_n u_n(t)), \frac{du_n}{dt}(t) \right\rangle dt. \end{aligned} \tag{3.26}$$

Since $u_n^0 = \eta_n(0) \in \operatorname{dom}(\Phi_n)$, we infer that $\frac{du_n}{dt}$ belongs to $L^2(0,T,X)$ and $t \mapsto \Phi_n(u_n(t))$ is absolutely continuous (see (Brezis, 1973, Theorem 3.6)). Therefore for a.e. $t \in (0,T)$, $\frac{d}{dt}\Phi(u_n(t)) = \langle D\Phi(u_n(t)), \frac{du_n}{dt}(t)\rangle$ (see (Attouch *et al.*, 2014,

Proposition 17.2.5)). From (3.26) we deduce

$$\begin{aligned}
&\int_0^T \left\|\frac{du_n}{dt}(t)\right\|^2 dt \\
&= -\Phi_n(u_n(T)) + \Phi_n(u_n^0) + \int_0^T \left\langle F_n(t, u_n(t), \mathcal{T}_n u_n(t)), \frac{du_n}{dt}(t)\right\rangle dt \\
&\leq - \inf_{v \in L^2(\Omega), n\in\mathbb{N}} \Phi_n(v) + \sup_{n\in\mathbb{N}} \Phi_n(u_n^0) \\
&\quad + \left(\int_0^T \|F_n(t, u_n(t), \mathcal{T}_n u_n(t))\|_X^2\right)^{\frac{1}{2}} \left(\int_0^T \left\|\frac{du_n}{dt}(t)\right\|_X^2\right)^{\frac{1}{2}}. \qquad (3.27)
\end{aligned}$$

Hence, from (3.25), (Hd$_1$), Lemma 2.2, and (Hd$_2$), we infer that there exists $C > 0$ such that

$$\int_0^T \left\|\frac{du_n}{dt}(t)\right\|_X^2 dt \leq C\left(1 + \left(\int_0^T \left\|\frac{du_n}{dt}(t)\right\|_X^2 dt\right)^{\frac{1}{2}}\right),$$

from which we deduce (3.23).

Step 2. We prove that there exist $u \in C([0,T], X)$, and a subsequence of $(u_n)_{n\in\mathbb{N}}$ not relabeled, satisfying $u_n \to u$ in $C([0,T], X)$ equipped with its uniform norm. Apply the Ascoli-Arzela compactness theorem and repeats point by point the proof of Theorem 2.6, ***Step 2***.

Step 3. We assert that $\frac{du_n}{dt} \rightharpoonup \frac{du}{dt}$ weakly in $L^2(0,T,X)$ for a non relabeled subsequence, and that $0 \leq u(t) \leq \sup_n \overline{y}_n(t)$ for all $t \in [0,T]$. The first claim is a straightforward consequence of (3.23) and ***Step 2***. The second one follows easily from inequality $0 \leq u_n(t) \leq \overline{y}_n(t)$ for all $t \in [0,T]$, and the convergence $u_n \to u$ in $C([0,T], X)$.

Step 4. We prove that u, extended by η in $(-\infty, 0]$, is the unique solution of $(\mathcal{P})$. The proof of

$$\frac{du}{dt}(t) + \partial\Phi(u(t)) \ni F(t, u(t), \mathcal{T}u(t)) \text{ for a.e. } t \in (0,T)$$

mimics that of Theorem 2.6. The only trickier result is the following version of Lemma 2.7 whose proof is postponed after the end of the proof.

Lemma 3.3. *The functional* $G_n = F_n(\cdot, u_n, \mathcal{T}_n u_n)$ *weakly converges in* $L^2(0,T,L^2(\Omega))$ *to the functional* G *defined by* $G(t) = F(t, u, \mathcal{T}u)$ *where* $F(t, u(t), \mathcal{T}u(t)) = r(t) \odot h(\mathcal{T}u(t)) \cdot g(u(t)) + q(t)$.

Step 5. We show that if $\Phi_n(u_n^0) \to \Phi(u^0)$, $r_n \to r$ strongly in $L^2(0,T,L^2(\Omega,\mathbb{R}^l))$ and $q_n \to q$ strongly in $L^2(0,T,X)$, then $\frac{du_n}{dt} \to \frac{du}{dt}$ strongly in $L^2(0,T,X)$. Repeat point by point the proof of Theorem 2.6, ***Step 5***.

Step 6. We establish the last assertion of Theorem 3.3. We start by establishing the following estimates on the total variations Var $(G_n, [0,T])$ of G_n in $[0,T]$, from which, with (3.23), we deduce that G_n belongs to $W^{1,1}(0,T,L^2(\Omega))$:

$$\mathrm{Var}(G_n,[0,T]) = \int_0^T \left\| \frac{G_n}{dt}(t) \right\|_{L^2(\Omega)} dt \le C\left(1 + \int_0^T \left\| \frac{du_n}{dt}(t)\right\|_{L^2(\Omega)} dt\right) \tag{3.28}$$

where $C > 0$ does not depend on n.

According to the structure of F_n, from (3.22) and (Hd$_6$), and noticing that from Lemma 3.1, $0 \le \mathcal{T}_n u_n \le \overline{y}_n(T) \le \sup_n \overline{y}_n(T)$, we have for $0 \le t \le s \le T$

$$\begin{aligned}
&\|G_n(t) - G_n(s)\|_X \le \overline{g}\overline{h}\|r_n(t) - r_n(s)\|_{L^2(\Omega,\mathbb{R}^l)} + \|q_n(t) - q_n(s)\|_{L^2(\Omega)} \\
&\quad + \sup_{n\in\mathbb{N}} \|r_n\|_\infty \|g_n(u_n(t)) \cdot h_n(\mathcal{T}_n u_n(t)) - g_n(u_n(s)) \cdot h_n(\mathcal{T}_n u_n(s))\|_{L^2(\Omega)} \\
&\le \overline{g}\overline{h} \int_s^t \left\| \frac{dr_n}{dt}(\tau,\cdot)\right\|_{L^2(\Omega,\mathbb{R}^l)} d\tau + \int_s^t \left\|\frac{dq_n}{dt}(\tau,\cdot)\right\|_{L^2(\Omega)} d\tau \\
&\quad + \sup_{n\in\mathbb{N}} \|r_n\|_\infty \|g_n(u_n(t))\cdot h_n(\mathcal{T}_n u_n(t)) - g_n(u_n(s))\cdot h_n(\mathcal{T}_n u_n(s))\|_{L^2(\Omega)}.
\end{aligned} \tag{3.29}$$

On the other hand, from (3.15), (3.17) and (3.22)), we infer that

$$\begin{aligned}
&\|g_n(u_n(t)) \cdot h_n(\mathcal{T}_n u_n(t)) - g_n(u_n(s)) \cdot h_n(\mathcal{T}_n u_n(s))\|_{L^2(\Omega)} \\
&\le \overline{g} L'_{[0,\sup_n \overline{y}_n(T)]} \|\mathcal{T}_n u_n(t) - \mathcal{T} u_n(s)\|_{L^2(\Omega)} + \overline{h} L_{[0,\sup_n \overline{y}_n(T)]} \|u_n(t) - u_n(s)\|_{L^2(\Omega)} \\
&\le g L'_{[0,\sup_n \overline{y}_n(T)]} \int_s^t \left(\left\|\frac{du_n}{d\sigma}(\cdot)\right\|_{L^2(\Omega)} \star \mu_n\right)(\sigma) d\sigma \\
&\quad + \overline{h} L_{[0,\sup_n \overline{y}_n(T)]} \int_s^t \left\| \frac{du_n}{d\sigma}(\sigma)\right\|_{L^2(\Omega)} d\sigma.
\end{aligned} \tag{3.30}$$

Combining (3.29) with (3.30), we deduce that

$$\begin{aligned}
&\int_0^T \left\| \frac{G_n}{d\sigma}(\sigma)\right\|_{L^2(\Omega)} d\sigma \\
&\le C\left(1 + \int_0^T \left(\left\| \frac{du_n}{d\sigma}(\cdot)\right\|_{L^2(\Omega)} \star \mu_n \right)(\sigma) d\sigma + \int_0^T \left\| \frac{du_n}{d\sigma}(\sigma)\right\|_{L^2(\Omega)} d\sigma \right).
\end{aligned} \tag{3.31}$$

Since $u_n \in C_{\eta_n}((-\infty, T], L^2(\Omega))$, according to Lemma 3.2, we have

$$\begin{aligned}
&\int_0^T \left(\left\| \frac{du_n}{d\sigma}(\cdot)\right\|_{L^2(\Omega)} \star \mu_n \right)(\sigma) d\sigma \\
&\qquad \le \int_{-\infty}^0 \left\| \frac{d\eta_n}{d\sigma}(\sigma)\right\|_{L^2(\Omega)} d\sigma + \int_0^T \left\| \frac{du_n}{d\sigma}(\sigma)\right\|_{L^2(\Omega)} d\sigma
\end{aligned}$$

so that, from (3.16),

$$\int_0^T \left(\left\| \frac{du_n}{d\sigma}(\cdot)\right\|_{L^2(\Omega)} \star \mu_n \right)(\sigma) d\sigma \le C\left(1 + \int_0^T \left\| \frac{du_n}{d\sigma}(\sigma)\right\|_{L^2(\Omega)} d\sigma \right). \tag{3.32}$$

Estimate (3.28) is obtained by Collecting (3.31) and (3.32).

According to Theorem 3.1, u_n satisfies (SD$_4$), then possesses a right derivative at each $t \in]0,T[$. Moreover,

$$\frac{d^+u_n}{dt}(t) + D\Phi_n(u_n(t)) = G_n(t) \text{ for all } t \in]0,T[. \tag{3.33}$$

Fix $t \in]0,T[$. We first establish successively

$$G_n(t) \rightharpoonup G(t) \text{ weakly in } L^2(\Omega); \tag{3.34}$$

$$\sup_{n\in\mathbb{N}} \left\| \frac{d^+u_n}{dt}(t) \right\|_X < +\infty; \tag{3.35}$$

$$\sup_{n\in\mathbb{N}} \|D\Phi_n(u_n(t))\|_X < +\infty. \tag{3.36}$$

Proof of (3.34). This convergence follows straightforwardly from $q_n(t,\cdot) \rightharpoonup q(t,\cdot)$, $r_n(t,\cdot) \rightharpoonup r(t,\cdot)$, and $h_n(\mathcal{T}_n u_n(t)) \odot g_n(u_n(t)) \to h(\mathcal{T}u(t)) \odot g(u(t))$ strongly in $L^2(\Omega,\mathbb{R}^l)$. This last point follows from (3.37) in the proof of Lemma 3.3 below, and the Lebesgue dominated convergence theorem.

Proof of (3.35). It follows by applying Proposition 2.5, and following the proof of (2.50). We deduce (3.36) from these two estimates and (3.33).

We end the proof by mimic the end of the proof of ***Step 6*** of Theorem 2.6: more precisely, it remains to identify the weak limit of $D\Phi_n(u_n(t))$ from the graph convergence $\partial\Phi_n \overset{G_{s,w}}{\to} \partial\Phi$, then to pass to the weak limit in (3.33) to obtain $\theta(t) + D\Phi(u(t)) = G(t)$. □

Proof of Lemma 3.3. The fact that $\mathcal{T}$ is well defined follows straightforwardly from the pointwise limit $\mathcal{T}_n u(t) \to \mathcal{T}u(t)$ in $L^2(\Omega)$ ensured by (Hd$_4$): indeed write $\mathcal{T}_n u(t) = \int_\mathbb{R} \mathbb{1}_{(-\infty,t]}\widetilde{u}_t d\mathbf{m}_t^n$ and $\mathcal{T}u(t) = \int_\mathbb{R} \mathbb{1}_{(-\infty,t]}\widetilde{u}_t d\mathbf{m}_t$ where $\widetilde{u}_t$ defined by (3.1) belongs to $C_c(\mathbb{R}, L^2(\Omega))$. Recall that from (3.14), $G_n(t) = H_n(t) + q_n(t)$ where

$$\begin{aligned} H_n(t)(x) &= r_n(t,x) \odot h_n(\mathcal{T}_n u_n(t,x)) \cdot g_n(u_n(t,x)) \\ &= r_n(t,x) \cdot h_n(\mathcal{T}_n u_n(t,x)) \odot g_n(u_n(t,x)). \end{aligned}$$

Since $q_n \rightharpoonup q$ in $L^2(0,T,X)$, we are reduced to prove that $H_n \rightharpoonup H$ in $L^2(0,T,X)$ where

$$H(t)(x) = r(t,x) \odot h(\mathcal{T}u(t,x)) \cdot g(u(t,x)) = r(t,x) \cdot h(\mathcal{T}u(t,x)) \odot g(u(t,x)).$$

Hence, since $r_n \rightharpoonup r$ in $L^2(0,T,X^l)$, it suffices to establish that

$$h_n(\mathcal{T}_n u_n) \odot g_n(u_n) \to h(\mathcal{T}u) \odot g(u) \text{ strongly in } L^2(0,T,X^l)$$

where X^l denotes the space $L^2(\Omega, \mathbb{R}^l)$. We have[5]

$$\begin{aligned}
&\|h_n(\mathcal{T}_n u_n(t)) \odot g_n(u_n(t)) - h(\mathcal{T}u(t)) \odot g(u(t))\|_{X^l} \\
\leq &\|h_n(\mathcal{T}_n u_n(t)) \odot g_n(u_n(t)) - h_n(\mathcal{T}_n u_n(t)) \odot g(u(t))\|_{X^l} \\
&+ \|h_n(\mathcal{T}_n u_n(t)) \odot g(u(t)) - h(\mathcal{T}u(t)) \odot g(u(t))\|_{X^l} \\
\leq &\overline{h}\|g_n(u_n(t)) - g(u(t))\|_{X^l} + \overline{g}\|h_n(\mathcal{T}_n u_n(t)) - h(\mathcal{T}u(t))\|_{X^l} \\
\leq &\overline{h}L_{[0,\sup_n \overline{y}_n(T)]}\|u_n(t) - u(t)\|_X + \overline{h}\|g_n(u(t)) - g(u(t))\|_{X^l} \\
&+ \overline{g}L'_{[0,\sup_n \overline{y}_n(T)]}\|\mathcal{T}_n u_n(t) - \mathcal{T}u(t)\|_X \\
&+ \overline{g}\|h_n(\mathcal{T}u(t)) - h(\mathcal{T}u(t))\|_{X^l}.
\end{aligned} \tag{3.37}$$

Hence, to prove the claim we need to establish that

$$\int_0^T \|g_n(u(t)) - g(u(t))\|^2_{X^l} dt \to 0, \int_0^T \|h_n(\mathcal{T}u(t)) - h(\mathcal{T}u(t))\|^2_{X^l} dt \to 0 \tag{3.38}$$

$$\int_0^T \|u_n(t) - u(t)\|^2_X dt \to 0 \tag{3.39}$$

$$\int_0^T \|\mathcal{T}_n u_n(t) - \mathcal{T}u(t)\|^2_X dt \to 0. \tag{3.40}$$

The two convergences in (3.38) follow easily from (Hd_5) and the Lebesgue dominated convergence theorem. Convergence (3.39) is straightforward. We establish (3.40). From (3.3), we have

$$\begin{aligned}
&\|\mathcal{T}_n u_n - \mathcal{T}u\|_{L^2(0,T,X)} \\
&\leq \|\mathcal{T}_n(u_n - u)\|_{L^2(0,T,X)} + \|\mathcal{T}_n u - \mathcal{T}u\|_{L^2(0,T,X)} \\
&\leq 2T^{\frac{1}{2}}\|u_n - u\|_{C_c((-\infty,T],X)} + \|\mathcal{T}_n u - \mathcal{T}u\|_{L^2(0,T,X)} \\
&\leq 2T^{\frac{1}{2}}\|u_n - u\|_{C([0,T],X)} + \|\eta_n - \eta\|_{C_c((-\infty,0],X)} + \|\mathcal{T}_n u - \mathcal{T}u\|_{L^2(0,T,X)}.
\end{aligned}$$

We are reduced to proving $\|\mathcal{T}_n u - \mathcal{T}u\|_{L^2(0,T,X)} \to 0$. The claim follows from the pointwise convergence $\mathcal{T}_n u(t) \to \mathcal{T}u(t)$ in $L^2(\Omega)$ for a.e. $t \in (0,T)$, (3.2), and the Lebesgue dominated convergence theorem. This completes the proof of Lemma 3.3. □

Remark 3.4. When F is a regular DCP-structured functional, the weak limit $F(t, u(t), \mathcal{T}u) - D\Phi(u(t))$ in the last part of Theorem 3.3, is equal to $\frac{du}{dt}(t)$ for a.e. $t \in (0,T)$. If the limit reaction functional G defined by $G(t) = F(t, u(t))$ belongs to $W^{1,1}(0,T,L^2(\Omega))$, then, from Lemma 2.1, $F(t, u(t), \mathcal{T}u) - D\Phi(u(t)) = \frac{d^+u}{dt}(t)$ for all $t \in]0,T[$; furthermore $u(t) \in \mathrm{dom}(\partial\Phi)$ for all $t \in]0,T]$.

By resuming the proof of ***Step 6***, estimate (3.28), one can show that if the limit operator $\mathcal{T}$ satisfies (M'_4), then $G \in W^{1,1}(0,T,L^2(\Omega))$, and $\frac{d^+u_n}{dt}(t) \to \frac{d^+u}{dt}(t)$ for each $t \in]0,T[$.

[5]To simplify the notation we write $g(v(t))$ for the function $x \mapsto g(v(t,x))$, idem for $h(v(t))$, $g_n(v(t))$ and $h_n(v(t))$.

It is not difficult to show that $\mathcal{T}$ satisfies (M'_4) under the following additional conditions:

- $\mu_n \rightharpoonup \mu$ weakly in $\mathcal{P}_+(\mathbb{R})$, and η_n does not depend on n;
- $\dfrac{du_n}{dt} \to \dfrac{du}{dt}$ strongly in $L^2(0,T,L^2(\Omega))$;
- $u \in C^1([0,T],L^2(\Omega))$.

Remark 3.5. It is easy to check that (Hd_4) holds for the measures described in the three Examples 3.1, 3.2 and 3.3 in the following situations

1) $\mathbf{m}_t^n = \sum_{k\in\mathbb{N}} \mathbf{d}_k \delta_{t-\tau_{k,n}}$ with $\tau_{k,n} \to \tau_k$ in $\mathbb{R}_+$; then $\mathbf{m}_t = \sum_{k\in\mathbb{N}} \mathbf{d}_k \delta_{t-\tau_k}$;
2) $\mathbf{m}_t^n = \dfrac{1}{\#\tau}\delta_{t-\tau_n(\cdot)}$ where $\#(\tau_n(\mathbb{R}))$ is assumed to be a constant, denoted by $\#\tau$, and $\tau_n \to \tau$ pointwise in Ω; then $\mathbf{m}_t = \frac{1}{\#\tau}\delta_{t-\tau(\cdot)}$;
3) $\mathbf{m}_t^n = \mathcal{K}_n(\cdot,t-\sigma)d\sigma$ where $\mathcal{K}_n(\cdot,\sigma) \to \mathcal{K}(\cdot,\sigma)$ in $L^2(\Omega)$, with the domination property: there exists $\mathcal{J} \in L^1(0,+\infty)$ such that $\|\mathcal{K}_n(\cdot,\sigma)\|_{L^2(\Omega)} \le \mathcal{J}(\sigma)$ a.e. Then $\mathbf{m}_t = \mathcal{K}(\cdot,t-\sigma)d\sigma$.

Remark 3.6. The sequence of probability measures $(\mu_n)_{n\in\mathbb{N}}$ can be constant although the family $(\mathbf{m}_t^n)_{t\ge0,n\in\mathbb{N}}$ is not. For some examples, see Chapter 8, Section 8.2.1.

3.3.2 *An alternative proof of Theorem 3.3 in the case of a single time delay*

We put ourself under the hypotheses of Theorem 3.3, when $\mu_t^n = \delta_{t-\tau}$, and $\tau > 0$ does not depend on the space variable. We develop an alternative shorter proof of Theorem 3.3 based on the following remark: splitting $[0,T]$ into sub-intervals of size τ, solving the problem

$$(\mathcal{P}_n)\begin{cases} \dfrac{du_n}{dt}(t) + D\Phi_n(u_n(t)) = F_n(t,u_n(t),u_n(t-\tau)) \text{ for a.e. } t \in (0,T) \\[2ex] u_n(t) = \eta_n(t) \text{ for all } t \in [-\tau,0], \end{cases}$$

is equivalent to solving each reaction-diffusion problem

$$(\mathcal{P}_{n,i})\begin{cases} \dfrac{du_n^i}{dt}(t) + D\Phi_n(u_n^i(t)) = F_n^i(t,u_n^i(t)) \text{ for a.e. } t \in ((i-1)\tau, i\tau) \\[2ex] u_n^i((i-1)\tau) = u_n^{i-1}((i-1)\tau), \end{cases}$$

for $i = 1,\dots,\left[\frac{T}{\tau}\right]+1$, where F_n^i is defined by $F_n^i(t,u_n^i(t)) := F_n(t,u_n^i(t),u_n^{i-1}(t))$, and $u_n^0 := \eta_n$ in $[-\tau,0]$.

Set $\overline{\rho}_n^i = \overline{y}_n(i\tau)$, with the initialization $\overline{\rho}_n^0 = \overline{\rho}_n$. It is worth noting that for all $n \in \mathbb{N}$, $(\overline{\rho}_n^i)_i$ is non decreasing with respect to i. In particular condition (3.19) yields that $0 \leq \phi_n \leq a_{0,n}\overline{\rho}_n^i$ on $\partial\Omega$ for $i = 0, \ldots, [\frac{T}{\tau}] + 1$. Furthermore, since $\overline{y}_n$ is non decreasing, we infer that $[0, \overline{\rho}_n^i] \subset [0, \overline{y}_n(T)]$ for all $i = 0, \ldots, [\frac{T}{\tau}] + 1$.

By using an inductive procedure for $i = 1, \ldots, [\frac{T}{\tau}] + 1$, we are going to establish that $(\mathcal{P}_{n,i})$ possesses a unique solution u_n^i satisfying

$$
\begin{aligned}
&0 \leq u_n^i \leq \overline{\rho}_n^i, \\
&\sup_{n\in\mathbb{N}} \Phi_n(u_n^i(i\tau)) < +\infty, \\
&u_n^i \to u^i \text{ in } C([(i-1)\tau, i\tau], L^2(\Omega)), \\
&\frac{du_n^i}{dt} \rightharpoonup \frac{du^i}{dt} \text{ in } L^2((i-1)\tau, i\tau, L^2(\Omega)),
\end{aligned}
$$

where u^i is the unique solution of

$$
(\mathcal{P}_i) \begin{cases} \dfrac{du^i}{dt}(t) + \partial\Phi(u^i(t)) \ni F^i(t, u^i(t)) \text{ for a.e. } t \in ((i-1)\tau, i\tau) \\ u^i((i-1)\tau) = u^{i-1}((i-1)\tau), \end{cases}
$$

which satisfies $0 \leq u^i \leq \sup_{n\in\mathbb{N}} \overline{\rho}_n^{i-1}$. The reaction functional F^i will be specified at each step. We will conclude by noting that the function u defined by $u = u^i$ on $[(i-1)\tau, i\tau]$ for $i = 1, \ldots, [\frac{T}{\tau}] + 1$, fulfills the conditions of Theorem 3.3. We implement this procedure below.

Proof. ***Step i = 1 (initialization).*** Since $u_n^0 = \eta_n$ in $[-\tau, 0]$, $(\mathcal{P}_{n,1})$ may be written

$$
(\mathcal{P}_{n,1}) \begin{cases} \dfrac{du_n^1}{dt}(t) + D\Phi_n(u_n^1(t)) = F_n^1(t, u_n^1(t)) \text{ for a.e. } t \in (0, \tau) \\ u_n^1(0) = \eta_n(0), \end{cases}
$$

with $0 \leq \eta_n(0) \leq \overline{\rho}_n$ from (SC2), and where $F_n^1(t, u_n^1(t)) := F_n(t, u_n^1(t), \eta_n(t-\tau))$. Set $f_n^1(t, x, \zeta) := r_n^1(t, x) \cdot g_n(\zeta) + q_n(t, x)$ where $r_n^1(t, x) := r_n(t, x) \odot h_n(\eta_n(t-\tau, x))$. According to (3.15), ($\text{Hd}_3$), ($\text{Hd}_5$) and ($\text{Hd}_6$), we easily deduce that $r_n^1 \rightharpoonup r^1$ weakly in $L^2(0, \tau, L^2(\Omega, \mathbb{R}^l))$ where $r^1(t, x) = r(t, x) \odot h(\eta(t-\tau, x))$. Finally we check that all the conditions of Theorem 2.6 are fulfilled with the time interval $(0, \tau)$ substitute for $(0, T)$. Therefore u_n^1 converges in $C([0, \tau], L^2(\Omega))$ to the unique solution u^1 of the problem

$$
(\mathcal{P}_1) \begin{cases} \dfrac{du^1}{dt}(t) + \partial\Phi(u^1(t)) \ni F^1(t, u^1(t)) \text{ for a.e. } t \in (0, \tau), \\ u^1(0) = \eta(0), \ 0 \leq \eta(0) \leq \sup\limits_{n\in\mathbb{N}} \overline{y}_n(\tau), \end{cases}
$$

where $F^1(t, v)(x) = f^1(t, x, v(x))$ and $f^1(t, x, \zeta) = r^1(t, x) \cdot g(\zeta) + q(t, x)$. Moreover $0 \leq u^1 \leq \sup\limits_{n\in\mathbb{N}} \overline{y}(\tau)$, i.e., $0 \leq u^1 \leq \sup\limits_{n\in\mathbb{N}} \overline{\rho}_n^1$, and $\dfrac{du_n^1}{dt} \rightharpoonup \dfrac{du^1}{dt}$ in $L^2(0, \tau, L^2(\Omega))$.

It remains to establish that $\sup_{n\in\mathbb{N}} \Phi_n(u_n^1(\tau)) < +\infty$. As we usually do to shorten the notation, we denote by X the space $L^2(\Omega)$. According to the proof of Theorem 2.6, we have

$$\sup_{n\in\mathbb{N}} \left\| \frac{du_n^1}{dt} \right\|_{L^2(0,\tau,X)} < +\infty, \tag{3.41}$$

$$\sup_{n\in\mathbb{N}} \int_0^\tau \|F_n^1(t,u_n^1(t))\|_X^2 dt < +\infty. \tag{3.42}$$

Let $t \in (0,T)$. Taking $\frac{du_n^1}{dt}(t)$ as a test function, from $(\mathcal{P}_{n,1})$ we obtain

$$\left\| \frac{du_n^1}{dt}(t) \right\|_X^2 + \left\langle \nabla\Phi_n(u_n^1(t)), \frac{du_n^1}{dt}(t) \right\rangle = \left\langle F_n^1(t,u_n^1(t)), \frac{du_n^1}{dt}(t) \right\rangle.$$

Thus (see Proposition (Attouch *et al.*, 2014, Proposition 17.2.5))

$$\frac{d}{dt}\Phi_n(u_n^1(t)) = \left\langle F_n^1(t,u_n^1(t)), \frac{du_n^1}{dt}(t) \right\rangle - \left\| \frac{du_n^1}{dt}(t) \right\|_X^2.$$

Integrating this equality over $(0,\tau)$ yields

$$\begin{aligned}
\Phi_n(u_n^1(\tau)) &= \Phi_n(u_n^1(0)) + \int_0^\tau \left\langle F_n^1(t,u_n^1(t)), \frac{du_n^1}{dt}(t) \right\rangle dt - \int_0^\tau \left\| \frac{du_n^1}{dt}(t) \right\|_X^2 dt \\
&= \Phi_n(\eta_n(0)) + \int_0^\tau \left\langle F_n^1(t,u_n^1(t)), \frac{du_n^1}{dt}(t) \right\rangle dt - \int_0^\tau \left\| \frac{du_n^1}{dt}(t) \right\|_X^2 dt \\
&\leq \Phi_n(\eta_n(0)) + \left(\int_0^\tau \|F_n^1(t,u_n^1(t))\|_X^2 dt \right)^{\frac{1}{2}} \left(\int_0^\tau \left\| \frac{du_n^1}{dt}(t) \right\|_X^2 dt \right)^{\frac{1}{2}}
\end{aligned}$$

and $\sup_{n\in\mathbb{N}} \Phi_n(u_n^1(\tau)) < +\infty$ follows from hypothesis (Hd_2) and (3.41), (3.42).

Step i starting from step i −1, i ≥ 2. We assume that $(\mathcal{P}_{n,i-1})$ possesses a unique solution u_n^{i-1} satisfying

$$0 \leq u_n^{i-1} \leq \overline{\rho}_n^{i-1},$$

$$\sup_{n\in\mathbb{N}} \Phi_n(u_n^{i-1}((i-1)\tau)) < +\infty, \tag{3.43}$$

$$u_n^{i-1} \to u^{i-1} \text{in } C([(i-2)\tau,(i-1)\tau], L^2(\Omega)), \tag{3.44}$$

$$\frac{du_n^{i-1}}{dt} \rightharpoonup \frac{du^{i-1}}{dt} \text{ in } L^2((i-2)\tau,(i-1)\tau, L^2(\Omega)),$$

where u^{i-1} is the unique solution of $(\mathcal{P}_{i-1})$.

Consider the problem

$$(\mathcal{P}_{n,i}) \begin{cases} \dfrac{du_n^i}{dt}(t) + D\Phi_n(u_n^i(t)) = F_n^i(t,u_n^i(t)) \text{ for a.e. } t \in ((i-1)\tau, i\tau) \\[2ex] u_n^i((i-1)\tau) = u_n^{i-1}((i-1)\tau) \end{cases}$$

where $F_n^i(t,u_n^i(t)) := F_n(t,u_n^i(t),u_n^{i-1}(t-\tau))$. Set $f_n^i(t,x,\zeta) := r_n^i(t,x)\cdot g_n(\zeta) + q_n(t,x)$ where $r_n^i(t,x) := r_n(t,x)\odot h_n(u^{i-1}(t-\tau,x))$. According to (3.15), (3.44), (Hd$_5$) and (Hd$_6$), we easily deduce that $r_n^i \rightharpoonup r^i$ weakly in $L^2((i-1)\tau,i\tau,L^2(\Omega,\mathbb{R}^l))$ where $r^i(t,x) = r(t,x)\odot h(u^{i-1}(t-\tau,x))$. Finally we check that all the conditions of Theorem 2.6 are fulfilled where the time interval $(0,T)$ is substitute for $((i-1)\tau,i\tau)$. Therefore u_n^i converges in $C([(i-1)\tau,i\tau],L^2(\Omega))$ to the unique solution u^i of the problem

$$(\mathcal{P}_i)\begin{cases}\dfrac{du^i}{dt}(t) + \partial\Phi(u^i(t)) \ni F^1(t,u^i(t)) \text{ for a.e. } t\in((i-1)\tau,i\tau),\\[2mm] u^i((i-1)\tau) = u^{i-1}((i-1)\tau),\ \inf\limits_n \underline{y}(i\tau) \le u^{i-1}((i-1)\tau) \le \sup\limits_{n\in\mathbb{N}} \overline{y}(i\tau),\end{cases}$$

where $F^i(t,v)(x) = f^i(t,x,v(x))$ and $f^i(t,x,\zeta) = r^i(t,x)\cdot g(\zeta) + q(t,x)$. Moreover $0 \le u^i \le \sup_{n\in\mathbb{N}} \overline{y}_n(i\tau)$, that is $0 \le u^i \le \sup_{n\in\mathbb{N}} \overline{\rho}_n^i$. The proof of $\sup_n \Phi_n(u_n^i(\tau)) < +\infty$ is similar to that of step $i=1$, by using estimates

$$\sup_{n\in\mathbb{N}} \left\| \frac{du_n^i}{dt} \right\|_{L^2((i-1)\tau,i\tau,X)} < +\infty,$$

$$\sup_{n\in\mathbb{N}} \int_{(i-1)\tau}^{i\tau} \|F_n^i(t,u_n^i(t))\|_X^2 dt < +\infty,$$

together with (3.43).

Last step. By finite induction we construct a function u defined by $u = u^i$ on $[(i-1)\tau,i\tau]$ for $i = 1\ldots[\frac{T}{\tau}]+1$, i.e.

$$u = \sum_{i=1}^{[\frac{T}{\tau}]+1} u^i \mathbb{1}_{[(i-1)T,iT]}.$$

By construction u fulfills all the conditions of the limit solution in Theorem 3.3. This completes the proof. □

3.3.3 *Non stability of the reaction functional: convergence with mixing effect between growth rates and time delays*

We substitute for (Hd$_4$) a global weak convergence (Hd$'_4$) for the product of the growth rate r_n with the time delays term $h_n(\mathcal{T}_n\cdot)$. The structure of the reaction functional is then no longer preserved at the limit. This phenomena occurs in stochastic homogenization and will be illustrated in Chapter 8. We make precise (SC2) and hypothesis (Hd$_3$) as follows: we assume that Support(η_n) and Support(η) are included in $[-M,0]$ for some $M>0$ which does not depend on n.

Theorem 3.4 (Mixing effect). *Assume that the sequence $(W_n)_{n\in\mathbb{N}}$ of functionals* (3.12) *or* (3.13) *satisfies* (D$_n$), *and that the sequence of DCP-structured reaction* *functionals* $(F_n)_{n\in\mathbb{N}}$ *satisfies* (SC1), (SC2), (SC3) *and* (SC4). *Let* u_n *be the unique solution of the delays problem*

$$(\mathcal{P}_n)\begin{cases} \dfrac{du_n}{dt}(t) + D\Phi_n(u_n(t)) = F_n(t, u_n(t), \mathcal{T}_n u_n(t)) \text{ for a.e. } t \in (0,T), \\ u_n(t) = \eta_n(t) \text{ for all } t \in (-\infty, 0]. \end{cases}$$

Assume hypotheses (Hd$_1$), (Hd$_2$), (Hd$_3$), (Hd$_7$) *and*

(Hd$'_4$) *there exists a function* $r_{\mathcal{T}} : C([-M,T], L^2(\Omega)) \to L^2(0,T,L^2(\Omega,\mathbb{R}^l))$ *such that* $r_n \odot h_n(\mathcal{T}_n\psi) \rightharpoonup r_{\mathcal{T}}(\psi)$[6] *weakly in* $L^2(0,T,L^2(\Omega,\mathbb{R}^l))$ *for all* $\psi \in C([-M,T], L^2(\Omega))$;

(Hd$'_5$) g_n *pointwise converges to* g *and* $\sup_{n\in\mathbb{N}} |h_n(0)| < +\infty$;

(Hd$'_6$) $\overline{r} = \sup_{n\in\mathbb{N}} \|r_n\|_{L^\infty([0,+\infty)\times\mathbb{R}^N,\mathbb{R}^l)} < +\infty$.

Then u_n *uniformly converges in* $C([0,T], L^2(\Omega))$ *to a function* u *whose extension by* η *in* $(-\infty, 0]$, *is the unique solution of the problem*

$$(\mathcal{P})\begin{cases} \dfrac{du}{dt}(t) + \partial\Phi(u(t)) \ni r_{\mathcal{T}}(u)(t)\cdot g(u(t)) + q(t) \text{ for a.e. } t \in (0,T) \\ u(t) = \eta(t),\ 0 \le \eta(t) \le \sup\limits_n \overline{\rho}_n \text{ for all } t \in (-\infty, 0],\ \eta(0) \in \mathrm{dom}(\Phi). \end{cases}$$

Moreover, all the conclusions of Theorem 3.3 hold.

Proof. All the proof of Theorem 3.3 remains valid, excepted Step 4, Lemma 3.3. We only have to substitute Lemma 3.4 below for Lemma 3.3. □

Lemma 3.4. *The functional* $G_n = F_n(\cdot, u_n, \mathcal{T}_n u_n)$ *weakly converges in* $L^2(0,T,X)$ *to the functional* G *defined for all* $t \in [0,T]$ *by* $G(t) = r_{\mathcal{T}}(u)(t)\cdot g(u(t)) + q(t)$.

Proof of Lemma 3.4. Recall that $G_n(t) = H_n(t) + q_n(t)$ where $H_n(t)(x) = r_n(t,x) \odot h_n(\mathcal{T}_n u_n(t,x)) \cdot g_n(u_n(t,x))$. Since $q_n \rightharpoonup q$ in $L^2(0,T,X)$, it remains to prove that $H_n \rightharpoonup H$ in $L^2(0,T,X)$ where $H(t) = r_{\mathcal{T}}(u)(t)\cdot g(u(t))$. From (Hd$'_5$), (3.15), (3.18), it is easy to show that $g_n(u_n) \to g(u)$ strongly in $L^2(0,T,L^2(\Omega;\mathbb{R}^l))$. To conclude, we prove that $r_n \odot h_n(\mathcal{T}_n u_n) \rightharpoonup r_{\mathcal{T}}(u)$ weakly in $L^2(0,T,L^2(\Omega;\mathbb{R}^l))$. We have

$$r_n \odot h_n(\mathcal{T}_n u_n) = r_n \odot h_n(\mathcal{T}_n u) + r_n(t) \odot (h_n(\mathcal{T}_n u_n) - h_n(\mathcal{T}_n u)) \tag{3.45}$$

From (3.15), (3.3), (Hd$'_6$), and the fact that $0 \le \mathcal{T}_n u_n \le \overline{\rho}$, $0 \le \mathcal{T}_n u \le \overline{\rho}$, we infer that

$$\|r_n(t) \odot (h_n(\mathcal{T}_n u_n) - h_n(\mathcal{T}_n u))\|_{L^2(0,T,X^l)} \le \overline{r}T^{\frac{1}{2}}\mathcal{L}_N(\Omega)^{\frac{1}{2}}L'_{[0,\overline{\rho}]}\|u_n - u\|_{C((-\infty,T],X)},$$

hence $r_n(t) \odot (h_n(\mathcal{T}_n u_n) - h_n(\mathcal{T}_n u))$ strongly converges to 0 in $L^2(0,T,L^2(\Omega;\mathbb{R}^l))$. The conclusion then follows from (3.45) and hypothesis (Hd$'_4$) by taking $\psi = u$ as a test function. □

[6]We write $r_n \odot h_n(\mathcal{T}_n\psi)$ to denote the function defined in $L^2(0,T,L^2(\Omega))$ by $(t,x) \mapsto r_n(t,x) \odot h_n(\mathcal{T}_n\psi(t,x))$.

Remark 3.7. In practice, the difficulty is to establish the convergence in hypothesis (Hd_4'). This condition is obtained in the context of stochastic homogenization in Chapter 7 for certain functions h_ε, then illustrating the interplay at the limit, between the growth rate r_ε, and the time delays modeled by $h_\varepsilon(\mathcal{T}_\varepsilon\psi)$.

Chapter 4

Variational convergence of two components nonlinear reaction-diffusion systems

Let $T > 0$ and Ω be a bounded domain in $\mathbb{R}^N$. This chapter is concerned with the variational convergence of sequences of reaction-diffusion systems in $L^2(0,T,L^2(\Omega)) \times L^2(0,T,L^2(\Omega))$ of the type

$$(\mathcal{S}_n)\begin{cases} \dfrac{du_n}{dt}(t) + \partial\Phi_{1,n}(u_n(t)) = F_{1,n}(t,u_n(t),v_n(t)) \text{ for a.e. } t \in (0,T) \\ \\ \dfrac{dv_n}{dt}(t) + \partial\Phi_{2,n}(v_n(t)) = F_{2,n}(t,u_n(t),v_n(t)) \text{ for a.e. } t \in (0,T) \\ \\ \underline{\rho}_{1,n} \le u_n(0) \le \overline{\rho}_{1,n},\ \underline{\rho}_{2,n} \le v_n(0) \le \overline{\rho}_{2,n} \\ \\ u_n(0) \in \overline{\mathrm{dom}(\partial\Phi_{1,n})},\ v_n(0) \in \overline{\mathrm{dom}(\partial\Phi_{2,n})}, \end{cases}$$

where, for $i = 1,2$, $\underline{\rho}_{i,n}$ and $\overline{\rho}_{i,n}$ are suitable constants depending on the reaction functional $F_{i,n}$. Problems $(\mathcal{S}_n)$ model various situations involving competition or symbiosis models, prey predator models in ecology, as well as heat mass transfer in chemical reactors and combustion theory, or gas-liquid interactions problems, etc. They are illustrated through Examples 4.1, 4.2, 4.3, 4.4. This study includes systems $(\mathcal{S}_n)$ coupling reaction-diffusion equations (r.d.e.) and non-diffusive reaction equations (n.d.r.e.), like the FitzHugh-Nagumo system in neurophysiology described in Example 4.5.

As in the previous chapters, every diffusion term is the subdifferential of a standard integral functional of the calculus of variations $\Phi_{i,n} : L^2(\Omega) \to \mathbb{R} \cup \{+\infty\}$, whose domain $\mathrm{dom}(\partial\Phi_{i,n})$ contains the boundary conditions. For shorten the notation and the proofs, it is assumed to be single valued. Recall that in their domain, the diffusion terms are of divergence form $-\mathrm{div}D_\xi W_{1,n}(x,\nabla u)$ and $-\mathrm{div}D_\xi W_{2,n}(x,\nabla v)$. We restrict our analysis to a special form of reactions functionals $F_{i,n}$: for fixed $v \in L^2(\Omega)$, $(t,u) \mapsto F_{n,1}(t,u,v)$, and for fixed $u \in L^2(\Omega)$, $(t,v) \mapsto F_{n,2}(t,u,v)$ are CP-structured reaction functionals as defined in Chapter 2.

Existence and uniqueness of bounded strong solutions in $C([0,T],L^2(\Omega)) \times C([0,T],L^2(\Omega))$ of systems $(\mathcal{S}_n)$ is established for any $T > 0$ in Section 4.1. The bounds of solutions are expressed in terms of the bounds of the initial functions and the reaction functionals. The proof is based on Corollary 2.1 combined with a suitable fixed point procedure (Theorem 4.1).

In Section 4.2, under the Mosco-convergence of functionals $\Phi_{i,n}$, and a suitable convergence of $F_{i,n}$ to F_i for $i = 1, 2$, we establish the first main result of the chapter, Theorem 4.3, which states the convergence of $(\mathcal{S}_n)$ toward a reaction-diffusion system $(\mathcal{S})$ of the same type. It can be seen as a stability result for the class of systems considered.

The convergence of systems $(\mathcal{S}_n)$ coupling a reaction-diffusion equation with a non-diffusive reaction equation is addressed in Section 4.3 and is discussed in Theorem 4.4.

4.1 Two components reaction-diffusion system associated with convex functionals of the calculus of variations and TCCP-structured reaction functionals

As the previous chapter, $\mathcal{L}_N$ denotes the Lebesgue measure in $\mathbb{R}^N$, and Ω is a domain of $\mathbb{R}^N$ of class C^1. We denote by Γ_i, $i = 1, 2$, two subsets of its boundary $\partial\Omega$ with positive $\mathcal{H}^{N-1}$-Hausdorff measure. To shorten the notation, we sometimes write X to denote the Hilbert space $L^2(\Omega)$ equipped with its standard scalar product and its associated norm, denoted by $\langle\ ,\ \rangle$ and $\|\ .\ \|_X$ respectively. All along the chapter we use the same notation $|\cdot|$ to denote the norms of the Euclidean spaces $\mathbb{R}^d$, $d \geq 1$, and by $\xi \cdot \xi'$ the standard scalar product of two elements ξ, ξ' in $\mathbb{R}^d$. We also denote by $\xi \odot \xi'$ the Hadamard (or Schur) product of two elements ξ and ξ' in $\mathbb{R}^d$. For any topological space $\mathbb{T}$, we denote by $\mathcal{B}(\mathbb{T})$ its Borel σ-algebra.

We are concerned with sequences of systems of reaction-diffusion Cauchy problems of the form

$$(\mathcal{S}) \begin{cases} \dfrac{du}{dt}(t) + \partial\Phi_1(u(t)) = F_1(t, u(t), v(t)) \text{ for a.e. } t \in (0,T) \\[2ex] \dfrac{dv}{dt}(t) + \partial\Phi_2(v(t)) = F_2(t, u(t), v(t)) \text{ for a.e. } t \in (0,T) \\[2ex] u(0) \in \overline{\operatorname{dom}(\partial\Phi_1)},\ v(0) \in \overline{\operatorname{dom}\partial\Phi_2}, \end{cases}$$

where, for $i = 1, 2$, $\partial\Phi_i$ denote the subdifferential of standard convex functionals Φ_i of the calculus of variations. More precisely $\Phi_i : L^2(\Omega) \to \mathbb{R} \cup \{+\infty\}$ is defined by

$$\Phi_i(u) = \begin{cases} \displaystyle\int_\Omega W_i(x, \nabla u(x))\, dx + \frac{1}{2}\int_{\partial\Omega} a_i u^2\, d\mathcal{H}_{N-1} - \int_{\partial\Omega} \phi_i u\, d\mathcal{H}_{N-1} & \text{if } u \in H^1(\Omega), \\[2ex] +\infty & \text{otherwise} \end{cases} \tag{4.1}$$

where $\phi_i \in L^2_{\mathcal{H}_{N-1}}(\partial\Omega)$, and $a_i \in L^\infty_{\mathcal{H}_{N-1}}(\partial\Omega)$ with

$$\begin{cases} a_i \geq 0 \quad \mathcal{H}_{N-1}\text{-a.e. in } \partial\Omega \\[1ex] \exists \sigma > 0 \quad a_i \geq \sigma \quad \mathcal{H}_{N-1}\text{-a.e. in } \Gamma_i \subset \partial\Omega, \end{cases}$$

or

$$\Phi_i(u) = \begin{cases} \displaystyle\int_\Omega W_i(x, \nabla u(x))\, dx & \text{if } u \in H^1_{\Gamma_i}(\Omega), \\ \\ +\infty & \text{otherwise.} \end{cases} \tag{4.2}$$

The density $W_i : \mathbb{R}^N \times \mathbb{R}^N \to \mathbb{R}$ is a Borel measurable function which satisfies (D) with the same constants α and β for $i = 1, 2$. As a consequence, for $i = 1, 2$, W_i fulfills (2.7) with the same constants, and the coercivity condition of Lemma 2.2: there exists two positive constants $C_1 = C(\alpha, \sigma, C_p, C_{trace}, \mathcal{L}_N(\Omega), \mathcal{H}_{N-1}(\Gamma))$ and $C_2 = C_2(\alpha, \sigma, C_p, C_{trace}, \mathcal{L}_N(\Omega), \mathcal{H}_{N-1}(\Gamma))$ such that

$$\Phi_i(v) \geq C_1 \|v\|^2_{H^1(\Omega)} - C_2 \|\phi_i\|^2_{L^2_{\mathcal{H}_{N-1}}(\partial\Omega)}. \tag{4.3}$$

According to Lemma 2.3, the subdifferential of the functional Φ_i (actually its Gâteaux derivative), whose domain captures the boundary condition, is given by:

$$\begin{cases} \operatorname{dom}(\partial\Phi_i) \\ = \{v \in H^1(\Omega) : \operatorname{div} D_\xi W_i(\cdot, \nabla v) \in L^2(\Omega),\ a_i v + D_\xi W_i(\cdot, \nabla v) \cdot \mathbf{n} = \phi_i \text{ on } \partial\Omega\} \\ \\ \partial\Phi_i(v) = -\operatorname{div} D_\xi W(\cdot, \nabla v) \text{ for } v \in \operatorname{dom}(\partial\Phi_i) \end{cases}$$

where $\mathbf{n}$ denotes the outer unit normal to $\partial\Omega$ and $a_i v + D_\xi W_i(\cdot, \nabla v) \cdot \mathbf{n}$ must be taken in the trace sense. In what follows, since $\partial\Phi_i$ are single valued, we denote them by $D\Phi_i$.

The pair (F_1, F_2) of reaction functionals belongs to a suitable class for which a comparison principle holds with respect to the initial and boundary data for lower and upper solutions. This class is defined in the next section.

4.1.1 *The class of TCCP-structured reaction functionals*

Reaction-diffusion systems which model a wide class of applications in the domain of ecosystems, and which gives rise to bounded or positive solutions, amenable to analytical calculation in homogenization (periodic or stochastic) involve a special class of pairs of reaction functionals that we define below.

Definition 4.1. A pair (F_1, F_2) of functionals $F_i : [0, +\infty) \times L^2(\Omega) \times L^2(\Omega) \to \mathbb{R}^\Omega$, $i = 1, 2$, is called a *TCCP-structured reaction functional*, if there exists a pair of Borel measurable functions (f_1, f_2), $f_i : [0, +\infty) \times \mathbb{R}^N \times \mathbb{R} \times \mathbb{R} \to \mathbb{R}$, $i = 1, 2$, such that for all $t \in [0, +\infty)$ and all $(u, v) \in L^2(\Omega) \times L^2(\Omega)$,

$$\begin{cases} F_1(t, u, v)(x) = f_1(t, x, u(x), v(x)), \\ F_2(t, u, v)(x) = f_2(t, x, u(x), v(x)), \end{cases}$$

and fulfilling the following structure conditions:[1]

- $\begin{cases} f_1(t,x,\zeta,\zeta') = r_1(t,x) \odot h_1(\zeta') \cdot g_1(\zeta) + q_1(t,x) \\ f_2(t,x,\zeta,\zeta') = r_2(t,x) \odot h_2(\zeta) \cdot g_2(\zeta') + q_2(t,x); \end{cases}$
- $h_i, g_i : \mathbb{R} \to \mathbb{R}^l$, $i = 1,2$, are locally Lipschitz continuous functions;
- $r_i \in L^\infty([0,T] \times \mathbb{R}^N, \mathbb{R}^l)$ for all $T > 0$;
- $q_i \in L^2(0,T,L^2_{loc}(\mathbb{R}^N))$ for all $T > 0$.

Furthermore (f_1, f_2) must satisfy the Two Components Comparison Principle condition $(TCCP)$: for $i = 1,2$, there exists a pair $(\underline{f}_i, \overline{f}_i)$ of functions $\underline{f}_i, \overline{f}_i$: $[0,+\infty) \times \mathbb{R} \to \mathbb{R}$ with $\underline{f}_i \le 0 \le \overline{f}_i$, and a pair $(\underline{\rho}_i, \overline{\rho}_i)$ in $\mathbb{R}^2$ with $\underline{\rho}_i \le \overline{\rho}_i$, such that each of the two ordinary differential equations

$$\underline{\mathrm{O}_{\mathrm{DE}_i}} \begin{cases} \underline{y}_i'(t) = \underline{f}_i(t, \underline{y}_i(t)) \text{ for } a.e.\ t \in (0,+\infty) \\ \underline{y}_i(o) = \underline{\rho}_i \end{cases}$$

$$\overline{\mathrm{O}_{\mathrm{DE}_i}} \begin{cases} \overline{y}_i'(t) = \overline{f}_i(t, \overline{y}_i(t)) \text{ for } a.e.\ t \in (0,+\infty) \\ \overline{y}_i(o) = \overline{\rho}_i \end{cases}$$

possesses at least one solution, such that for all $T > 0$, for a.e. $(t,x) \in (0,T) \times \mathbb{R}^N$, we have:

$$\begin{cases} f_1(t,x,\underline{y}_1(t),\zeta') \ge \underline{f}_1(t,\underline{y}_1(t)) \\ f_1(t,x,\overline{y}_1(t),\zeta') \le \overline{f}_1(t,\overline{y}_1(t)), \end{cases}$$

for all $\zeta' \in [\underline{y}_2(T), \overline{y}_2(T)]$, and

$$\begin{cases} f_2(t,x,\zeta,\underline{y}_2(t)) \ge \underline{f}_2(t,\underline{y}_2(t)) \\ f_2(t,x,\zeta,\overline{y}_2(t)) \le \overline{f}_2(t,\overline{y}_2(t)) \end{cases}$$

for all $\zeta \in [\underline{y}_1(T), \overline{y}_1(T)]$.

The pair (F_1, F_2) is called a *TCCP-structured reaction functional associated with* $(r_i, g_i, h_i, q_i)_{i=1,2}$, and (f_1, f_2) a *TCCP-structured reaction function associated with* $(r_i, g_i, h_i, q_i)_{i=1,2}$.

If furthermore, for all $T > 0$, and $i = 1,2$, $r_i \in W^{1,1}(0,T,L^2(\mathbb{R}^N,\mathbb{R}^l))$ and $q_i \in W^{1,1}(0,T,L^2_{\mathrm{loc}}(\mathbb{R}^N))$, the pair (F_1, F_2) is referred to as a *regular* TCCP-structured reaction functional and (f_1, f_2) as a *regular* TCCP-structured reaction function.

Remark 4.1. 1) Since $\underline{y}_i$ is nonincreasing, and $\overline{y}_i$ is nondecreasing, for any $T > 0$, and for $i = 1,2$ we have

$$\underline{y}_i(T) \le \underline{y}_i(0) = \underline{\rho}_i < \overline{\rho}_i = \overline{y}_i(0) \le \overline{y}_i(T).$$

2) It is worth noting that for each fixed ζ' in $[\underline{y}_2(T), \overline{y}_2(T)]$, the function $\zeta \mapsto f_1(t,x,\zeta,\zeta')$ is a CP-structured reaction function associated with $(r_1 \odot h_1(\zeta'), g_1, q_1)$ in the sense of Definition 2.1. Similarly for each fixed ζ in $[\underline{y}_1(T), \overline{y}_1(T)]$, the function $\zeta' \mapsto f_2(t,x,\zeta,\zeta')$ is a CP-structured reaction function associated with $(r_2 \odot h_2(\zeta), g_2, q_2)$.

[1]Using the coordinates of r_i, g_i and h_i we have $f_1(t,x,\zeta,\zeta') = \sum_{k=1}^l r_{1,k}(t,x) h_{1,k}(\zeta') g_{1,k}(\zeta) + q_1(t,x)$ and $f_2(t,x,\zeta,\zeta') = \sum_{k=1}^l r_{2,k}(t,x) h_{2,k}(\zeta) g_{2,k}(\zeta') + q_2(t,x)$.

4.1.2 *Examples*

In examples below, for any measurable function $a : [0, +\infty) \times \mathbb{R}^N \to \mathbb{R}$ we use the notation

$$\overline{a} := \operatorname*{ess\,sup}_{(t,x)\in\mathbb{R}_+\times\mathbb{R}^N} a(t,x), \ \underline{a} := \operatorname*{ess\,inf}_{(t,x)\in\mathbb{R}_+\times\mathbb{R}^N} a(t,x).$$

The proofs of propositions below are postponed to Section 4.4.

Example 4.1. *Example derived from competition models in ecology.*
Let $f_i : \mathbb{R}_+ \times \mathbb{R}^N \times \mathbb{R} \times \mathbb{R} \to \mathbb{R}$, $i = 1, 2$ be defined by

$$f_1(t,x,\zeta,\zeta') = \alpha_1(t,x)\zeta\left(1 - \frac{\zeta}{K_1(t,x)} - a_{1,2}\frac{\zeta'}{K_1(t,x)}\right)$$
$$f_2(t,x,\zeta,\zeta') = \alpha_2(t,x)\zeta'\left(1 - \frac{\zeta'}{K_2(t,x)} - a_{2,1}\frac{\zeta}{K_2(t,x)}\right),$$

where $\alpha_i \in L^\infty([0,T]\times\mathbb{R}^N)$ for all $T > 0$, $\alpha_i > 0$, K_i measurable, $K_i(t,x) \geq \underline{K}_i > 0$, $a_{1,2} > 0$ and $a_{2,1} > 0$.

Proposition 4.1. *The pair (f_1, f_2) is a TCCP-structured reaction function with for $i = 1, 2$,*

$$\begin{cases} \underline{\rho}_i = 0, \ \underline{y}_i = 0; \\ \overline{\rho}_i \text{ is any positive real number, } \overline{y}_i(t) = \overline{\rho}_i \exp(\overline{\alpha}_i t). \end{cases}$$

The pair (f_1, f_2) is associated with the *diffusive competition model* between two species

$$(\mathcal{S}) \begin{cases} \dfrac{du}{dt}(t) + D\Phi_1(u(t)) = \alpha_1(t,\cdot)u(t)\left(1 - \dfrac{u(t)}{K_1(t,\cdot)} - a_{1,2}\dfrac{v(t)}{K_1(t,\cdot)}\right) \\ \text{for a.e. } t \in (0,T) \\ \\ \dfrac{dv}{dt}(t) + D\Phi_2(v(t)) = \alpha_2(t,\cdot)v(t)\left(1 - \dfrac{v(t)}{K_2(t,\cdot)} - a_{2,1}\dfrac{u(t)}{K_2(t,\cdot)}\right) \\ \text{for a.e. } t \in (0,T) \\ \\ u(0) = u_0 \in \overline{\operatorname{dom}(D\Phi_1)}, \ v(0) = v_0 \in \overline{\operatorname{dom} D\Phi_2}, \end{cases}$$

where u and v denote the densities of two competing species having a logistic growth in the absence of the other. The α_i are the birth rates and the K_i the carrying capacities. The dimensionless coefficients $a_{1,2}$ and $a_{2,1}$ measure the competing effect of v to u and u to v respectively. In Theorem 4.1 we prove that for all $T > 0$, $(\mathcal{S})$ admits a unique solution $(u,v) \in C([0,T], L^2(\Omega))^2$. Under the initial conditions $0 \leq u_0 \leq \overline{\rho}_1$, $0 \leq v_0 \leq \overline{\rho}_2$, this solution fulfills for all $t \in [0,T]$ the bounds $0 \leq u(t) \leq \overline{\rho}_1 \exp(\overline{\alpha}_1 t)$, $0 \leq v(t) \leq \overline{\rho}_2 \exp(\overline{\alpha}_2 t)$. Furthermore, if we assume that the functions α_i and $\frac{\alpha_i}{K_i}$ belong to $W^{1,1}(0,T,L^2_{loc}(\mathbb{R}^N)) \cap L^\infty([0,T]\times\mathbb{R}^N)$, then u and v fulfill the boundary conditions for all $t \in]0,T]$ and possess a right derivative at each $t \in]0,T[$.

Example 4.2. *Example derived from symbiosis models in ecology.*

$$f_1(t,x,\zeta,\zeta') = \alpha_1(t,x)\zeta\left(1 - \frac{\zeta}{K_1(t,x)} + b_{1,2}\frac{\zeta'}{K_1(t,x)}\right)$$
$$f_2(t,x,\zeta,\zeta') = \alpha_2(t,x)\zeta'\left(1 - \frac{\zeta'}{K_2(t,x)} + b_{2,1}\frac{\zeta}{K_2(t,x)}\right),$$

where $\alpha_i \in L^\infty([0,T]\times\mathbb{R}^N)$ for all $T>0$, $\alpha_i > 0$, $K_i \in L^\infty([0,T]\times\mathbb{R}^N)$, $\overline{K}_i \geq K_i(t,x) \geq \underline{K}_i > 0$. We assume that

$$0 \leq b_{1,2} < \frac{\underline{K}_1}{\overline{K}_1} \text{ and } 0 \leq b_{2,1} < \frac{\underline{K}_2}{\overline{K}_2}. \tag{4.4}$$

Proposition 4.2. *The pair (f_1, f_2) is a TCCP-structured reaction function with for $i = 1,2$,*

$$\begin{cases} \underline{\rho}_i = 0, \ \underline{y}_i = 0; \\ \overline{\rho}_1 = \overline{\rho}_2 \geq \max\left(\dfrac{\underline{K}_1\overline{K}_1}{\underline{K}_1 - b_{1,2}\overline{K}_1}, \dfrac{\underline{K}_2\overline{K}_2}{\underline{K}_2 - b_{2,1}\overline{K}_2}\right); \\ \overline{y}_i = \overline{\rho}_i. \end{cases}$$

The pair (f_1, f_2) is associated with the *diffusive symbiosis model* between two species

$$(\mathcal{S})\begin{cases} \dfrac{du}{dt}(t) + D\Phi_1(u(t)) = \alpha_1(t)u(t)\left(1 - \dfrac{u(t)}{K_1(t)} + b_{1,2}\dfrac{v(t)}{K_1(t)}\right) \\ \text{for a.e. } t \in (0,T) \\ \dfrac{dv}{dt}(t) + D\Phi_2(v(t)) = \alpha_2(t)v(t)\left(1 - \dfrac{v(t)}{K_2(t)} + b_{2,1}\dfrac{u(t)}{K_2(t)}\right) \\ \text{for a.e. } t \in (0,T) \\ u(0) = u_0 \in \overline{\operatorname{dom}(D\Phi_1)}, \ v_0 = v_0 \in \overline{\operatorname{dom} D\Phi_2}, \end{cases}$$

where u and v denote the densities of two species having a logistic growth in the absence of the other. Like in Example 4.1, the α_i denote the birth rates and the K_i the carrying capacities. The dimensionless coefficients $b_{1,2}$ and $b_{2,1}$ measure the symbiosis effect of v to u and u to v respectively. By contrast with the competition model of two species described in Example 4.1, the signs preceding the b's are positive and reflect the fact that the interaction between the two species is to the advantage of all. Conditions (4.4) reflect the fact that symbiosis between both species must not be too large so that both populations grow while being bounded. It should be noted that the stability analysis of the system, for the model without diffusion and with constant carrying capacities, provides the less restrictive condition $b_{1,2}b_{2,1} < 1$ (see (Murray, 2002, Section 3.6)). In Theorem 4.1 we prove that for all $T > 0$, $(\mathcal{S})$ possesses a unique solution $(u,v) \in C([0,T], L^2(\Omega))^2$ which fulfills for all $t \in [0,T]$ the bounds $0 \le u(t) \le \overline{\rho}_1$, and $0 \le v(t) \le \overline{\rho}_2$ under the initial conditions $0 \le u_0 \le \overline{\rho}_1$, $0 \le v_0 \le \overline{\rho}_2$. Furthermore, if we assume that the functions α_i and $\frac{\alpha_i}{K_i}$ belong to $W^{1,1}(0,T,L^2_{loc}(\mathbb{R}^N)) \cap L^\infty([0,T]\times\mathbb{R}^N)$, then u and v fulfill the boundary conditions for all $t \in]0,T]$ and possess a right derivative at each $t \in]0,T[$.

Example 4.3. *Example derived from predator-prey models.*
Let $f_i : \mathbb{R}_+ \times \mathbb{R}^N \times \mathbb{R} \times \mathbb{R} \to \mathbb{R}$, $i = 1,2$ be defined by

$$\begin{aligned} f_1(t,x,\zeta,\zeta') &= \alpha_1(t,x)\zeta\left(1 - \frac{\zeta}{K_{car}(t,x)}\right) - a(x,t)\zeta'(1-\exp(-b\zeta)) \\ f_2(t,x,\zeta,\zeta') &= \alpha_2(t,x)\zeta'\left(1 - c\frac{\zeta'}{\zeta}\right), \end{aligned}$$

where

$\alpha_i \in L^\infty([0,T]\times\mathbb{R}^N)$ for all $T>0$, $\alpha_1(t,x) \ge \underline{\alpha}_1 > 0$, $+\infty > \overline{\alpha}_2 \ge \alpha_2(t,x) \ge \underline{\alpha}_2 > 0$,

$K_{car} \in L^\infty([0,T]\times\mathbb{R}^N)$ for all $T>0$, $K_{car}(t,x) \ge \underline{K} > 0$,

$a \in L^\infty([0,T]\times\mathbb{R}^N)$ for all $T>0$, $\overline{a} \ge a(t,x) > 0$,

b, c are positive constants.

Setting $\mu_{ext} := c\dfrac{\underline{\alpha}_1\underline{\alpha}_2}{\overline{a}\overline{\alpha}_2}$, we assume that $\mu_{ext} \ge 4$.

For $\delta > 0$, set

$$f_{2,\delta}(t,x,\zeta,\zeta') = \begin{cases} f_2(t,x,\zeta,\zeta') \text{ if } \zeta \ge \delta \\ f_2(t,x,\delta,\zeta') \text{ if } \zeta < \delta, \end{cases}$$

then we have,

Proposition 4.3. *The pair $(f_1, f_{2,\delta})$ is a TCCP-structured reaction function with*

$$\begin{cases} \delta = \underline{\rho}_1 \in \left[\underline{K}\dfrac{1-\sqrt{1-\frac{4}{\mu_{ext}}}}{2}, \underline{K}\dfrac{1+\sqrt{1-\frac{4}{\mu_{ext}}}}{2}\right], \ \underline{y}_1 = \underline{\rho}_1; \\ \overline{\rho}_1 = \theta\underline{\rho}_1, \ \theta \in \left[\dfrac{\underline{K}}{\underline{\rho}_1}, \mu_{ext}\left(1-\dfrac{\underline{\rho}_1}{\underline{K}}\right)\right], \ \overline{y}_1(t) = \overline{\rho}_1\exp(\overline{\alpha}_1 t); \\ \underline{\rho}_2 = 0, \ \underline{y}_2 = 0; \\ \overline{\rho}_2 \in \left[\dfrac{\overline{\alpha}_2}{c\underline{\alpha}_2}\overline{\rho}_1, \dfrac{\underline{\alpha}_1}{\overline{a}}\underline{\rho}_1\left(1-\dfrac{\underline{\rho}_1}{\underline{K}}\right)\right], \ \overline{y}_2 = \overline{\rho}_2. \end{cases} \tag{4.5}$$

As a consequence of Theorem 4.1, we obtain that under the initial conditions $\underline{\rho}_1 \le u_0 \le \overline{\rho}_1$ and $0 \le v_0 \le \overline{\rho}_2$, where $\underline{\rho}_1$, $\overline{\rho}_1$ and $\overline{\rho}_2$ fulfill condition (4.5), the diffusive predator-prey system

$$(\mathcal{S})\begin{cases} \dfrac{du}{dt}(t) + D\Phi_1(u(t)) \\ = \alpha_1(t,\cdot)u(t)\left(1 - \frac{u(t)}{K_{car}(t,\cdot)}\right) - a(t,\cdot)v(t)(1-\exp(-bu(t))) \\ \text{for a.e. } t \in (0,T) \\ \\ \dfrac{dv}{dt}(t) + D\Phi_2(v(t)) \\ = \alpha_2(t,\cdot)v(t)\left(1 - c\frac{v(t)}{u(t)}\right) \\ \text{for a.e. } t \in (0,T) \\ \\ u(0) = u_0 \in \overline{\mathrm{dom}(D\Phi_1)},\ v(0) = v_0 \in \overline{\mathrm{dom}\, D\Phi_2}, \end{cases}$$

admits for all $T > 0$ a unique solution $(u, v) \in C([0,T], L^2(\Omega))^2$ which satisfies for all $t \in [0,T]$, $\underline{\rho}_1 \leq u(t) \leq \overline{\rho}_1 \exp(\overline{\alpha}_1)$ and $0 \leq v(t) \leq \overline{\rho}_2$. Furthermore, if we assume that the functions α_i, $i = 1, 2$, $\frac{\alpha_1}{K_{car}}$ and a belong to $W^{1,1}(0,T,L^2_{loc}(\mathbb{R}^N)) \cap L^\infty([0,T] \times \mathbb{R}^N)$, then u and v fulfill the boundary conditions for all $t \in]0,T]$ and possess a right derivative at each $t \in]0,T[$. The system models the evolution of two species with density u and v of a prey and a predator, with birth growth rate α_1 and α_2 respectively. The prey population satisfies a logistic growth with a dependent time-space maximum carrying capacity K_{car} (the carrying capacity of the prey when the density of the predator is equal to zero), perturbed by a "predator term" $-a(t,\cdot)v(t)(1-\exp(-bu(t)))$ with a growth coefficient a. This term involves a saturation effect, i.e. $-a(t,\cdot)v(t)(1-\exp(-bu(t)))$ saturates to $-av(t)$ for $u(t)$ large, which reflects the limited capability of the predator when the prey is abundant. There exists many other choice of predator terms with saturation effects, and we refer the reader to (Murray, 2002, Section 3.3) for various examples in the context of o.d.e's. The predator population satisfies a logistic growth with a carrying capacity proportional to the prey density. The condition $\mu_{ext} \geq 4$ on the dimensionless coefficient μ_{ext}, prevents the extinction of the prey species since it ensures the existence of $\underline{\rho}_1 > 0$, so that $u(t) \geq \underline{\rho}_1 \geq \underline{K}\frac{1-\sqrt{1-\frac{4}{\mu_{ext}}}}{2}$. The coefficient μ_{ext} is referred to as *the extinction threshold*.

Example 4.4. *Example derived from thermo-chemical models.*
Let $f_i : \mathbb{R}_+ \times \mathbb{R}^N \times \mathbb{R} \times \mathbb{R} \to \mathbb{R}$, $i = 1, 2$ be defined by

$$f_1(t,x,\zeta,\zeta') = -\alpha_1(t,x)\zeta^p f_0(\zeta')$$
$$f_2(t,x,\zeta,\zeta') = \alpha_2(t,x)\zeta^p f_0(\zeta')$$

where

$$f_0(\zeta') = \begin{cases} \exp\left(\gamma - \dfrac{\gamma}{\zeta'}\right) & \text{if } \zeta' > 0 \\ \\ 0 & \text{otherwise,} \end{cases}$$

and $\alpha_i \in L^\infty([0,T] \times \mathbb{R}^N)$ for all $T > 0$, $\alpha_i > 0$, $p \geq 1$, and γ is a positive constant.

Proposition 4.4. *The pair (f_1, f_2) is a TCCP-structured reaction function with*

$$\begin{cases} \underline{\rho}_i = 0, \ \underline{y}_i = 0; \\ \overline{\rho}_i \text{ is any positive real number, } \overline{y}_1 = \overline{\rho}_1, \ \overline{y}_2(t) = \overline{\alpha}_2 \overline{\rho}_1^p t \exp(\gamma) + \overline{\rho}_2. \end{cases}$$

The pair (f_1, f_2) is associated with the diffusive system

$$(\mathcal{S}) \begin{cases} \dfrac{du}{dt}(t) + D\Phi_1(u(t)) = -\alpha_1(t,\cdot)u(t)^p f_0(v(t)) \text{ for a.e. } t \in (0,T) \\ \dfrac{dv}{dt}(t) + D\Phi_2(v(t)) \ni \alpha_2(t,\cdot)u(t)^p f_0(v(t)) \text{ for a.e. } t \in (0,T) \\ u(0) = u_0 \in \overline{\operatorname{dom}(D\Phi_1)}, \ v(0) = v_0 \in \overline{\operatorname{dom} D\Phi_2}, \end{cases}$$

where u and v denote a chemical concentration and the temperature respectively, in a non isothermal chemical reaction process; α_1 and $\frac{\alpha_2}{\alpha_1}$ are called Thiele number and Prater number respectively (see Pao (1992) and references therein). In Theorem 4.1 we prove that $(\mathcal{S})$ admits a unique solution $(u, v) \in C([0,T], L^2(\Omega))^2$ under the initial condition $0 \le u_0 \le \overline{\rho}_1$, $0 \le v_0 \le \overline{\rho}_2$, and that for all $t \in [0,T]$, the solution (u, v) satisfies the bounds $0 \le u(t) \le \overline{\rho}_1$, and $0 \le v(t) \le \overline{\alpha}_2 \overline{\rho}_1^p t \exp(\gamma) + \overline{\rho}_2$. Furthermore, if we assume that the functions α_i belong to $W^{1,1}(0,T, L^2_{loc}(\mathbb{R}^N)) \cap L^\infty([0,T] \times \mathbb{R}^N)$, then u and v fulfill the boundary conditions for all $t \in]0,T]$ and admit a right derivative at each $t \in]0,T[$.

Example 4.5. *Example derived from FitzHugh-Nagumo models.*
Let $f_i : \mathbb{R}_+ \times \mathbb{R}^N \times \mathbb{R} \times \mathbb{R} \to \mathbb{R}$, $i = 1, 2$ be defined by

$$f_1(t,x,\zeta,\zeta') = \alpha_1(t,x)\zeta(\zeta - a(t,x))(1-\zeta) - b(t,x)\zeta'$$
$$f_2(t,x,\zeta,\zeta') = \alpha_2(t,x)\zeta - c(t,x)\zeta'$$

where

$\alpha_i \in L^\infty([0,T] \times \mathbb{R}^N)$ for all $T > 0$, $\underline{\alpha}_i > 0$,
$b \in L^\infty([0,T] \times \mathbb{R}^N)$ for all $T > 0$, $\underline{b} > 0$,
$c \in L^\infty([0,T] \times \mathbb{R}^N)$ for all $T > 0$, $\underline{c} > 0$,
$a \in L^\infty([0,T] \times \mathbb{R}^N)$ for all $T > 0$, $\underline{a} > 0$, and $a < 1$.

Proposition 4.5. *Set* $\gamma = b\frac{\overline{\alpha}_2}{\underline{c}\underline{\alpha}_1}$. *Then the pair (f_1, f_2) is a TCCP-structured reaction function with*

$$\begin{cases} \overline{\rho}_1 \ge \max\left(\underline{\gamma} + 1, \dfrac{\sqrt{\frac{\overline{\gamma}}{\underline{\gamma}}} - \underline{a}}{\underline{\gamma}}\right); \\ \underline{\rho}_1 = -\underline{\gamma}\overline{\rho}_1; \\ \overline{\rho}_2 = \frac{\overline{\alpha}_2}{\underline{c}}\overline{\rho}_1; \\ \underline{\rho}_2 = \frac{\overline{\alpha}_2}{\underline{c}}\underline{\rho}_1; \\ \underline{f}_i = \overline{f}_i = 0, \ i = 1, 2. \end{cases} \tag{4.6}$$

The pair (f_1, f_2) is associated with the system in $L^2(0, T, L^2(\Omega)) \times H^1(\Omega)$

$$(\mathcal{S}) \begin{cases} \dfrac{du}{dt}(t) + D\Phi(u(t)) = \alpha_1(t,\cdot)u(t)(u(t) - a(t,\cdot))(1 - u(t)) - b(t,\cdot)v(t) \\ \text{for a.e. } t \in (0, T) \\ \\ \dfrac{dv}{dt}(t) = \alpha_2(t,\cdot)u(t) - c(t,\cdot)v(t) \text{ for a.e. } t \in (0, T) \\ \\ u(0) = u_0 \in \overline{\mathrm{dom}(D\Phi)}, \ v(0) = v_0 \in H^1(\Omega), \end{cases}$$

coupling a reaction-diffusion equation with a non-diffusive reaction equation. This coupling generalizes the FitzHugh-Nagumo model which describes the evolution of the electrical potential u across the axonal membrane. The variable v is a recovery variable obtained in the simplification of the Hodgkin-Huxley Theory of Nerve Membranes (see Murray (2002)). For a complete analysis of boundary value problems relating to FitzHugh-Nagumo equations in one space dimension, we refer the reader to Collins (1983), Dikansky (2005), Shonbek (1978). When the initial functions satisfy the bounds $\underline{\rho}_1 \leq u_0 \leq \overline{\rho}_1$ and $\underline{\rho}_2 \leq u_0 \leq \overline{\rho}_2$ where $\underline{\rho}_i$ and $\overline{\rho}_i$ are given by (4.6), existence and uniqueness of solutions fulfilling the same bounds are obtained according to Theorem 4.2. If we assume that the functions α_i, a, b, and c belong to $W^{1,1}(0, T, L^2_{loc}(\mathbb{R}^N)) \cap L^\infty([0, T] \times \mathbb{R}^N)$ then u fulfills the boundary condition for all $t \in]0, T]$ and u and v possess a right derivative at each $t \in]0, T[$. For bounds similar to those given by (4.6), in the case when the coefficients of the reaction functional are constants, we refer the reader to (Pao, 1992, Chapter 12, Section 12.7).

4.1.3 *Existence and uniqueness of a bounded solution*

Combining Corollary 2.1 with a suitable fixed point procedure, we establish the existence of a bounded unique solution to the Cauchy problem associated with TCCP-structured reaction functionals.

Theorem 4.1. *Let Φ_i, $i = 1, 2$, be standard functionals of the calculus of variations (4.1) or (4.2) and (F_1, F_2) a TCCP-structured reaction functional with $\underline{\rho}_i$, $\overline{\rho}_i$, $\underline{y}_i$ and $\overline{y}_i$ given by condition (TCCP). Assume that $a_i\underline{\rho}_i \leq \phi_i \leq a_i\overline{\rho}_i$ when Φ_i is of the form (4.1), or $\underline{\rho}_i \leq 0 \leq \overline{\rho}_i$ when Φ_i is of the form (4.2). Then the two component reaction-diffusion system*

$$(\mathcal{S}) \begin{cases} \dfrac{du}{dt}(t) + D\Phi_1(u(t)) = F_1(t, u(t), v(t)) \textit{ for a.e. } t \in (0, T) \\ \\ \dfrac{dv}{dt}(t) + D\Phi_2(v(t)) = F_2(t, u(t), v(t)) \textit{ for a.e. } t \in (0, T) \\ \\ \underline{\rho}_1 \leq u_0 = u(0) \leq \overline{\rho}_1, \quad \underline{\rho}_2 \leq v_0 = v(0) \leq \overline{\rho}_2 \\ \\ u_0 \in \overline{\mathrm{dom}(D\Phi_1)}, \quad v_0 \in \overline{\mathrm{dom}\, D\Phi_2}, \end{cases}$$

admits a unique solution $(u, v) \in C([0, T], L^2(\Omega)) \times C([0, T], L^2(\Omega))$ satisfying:

(SS$_1$) *$u(t) \in \mathrm{dom}(D\Phi_1)$ and $v(t) \in \mathrm{dom}(D\Phi_2)$ for a.e. $t \in]0,T[$,*
(SS$_2$) *u and v are almost everywhere derivable in $(0,T)$,*
(SS$_3$) *$u(t) \in \left[\underline{y}_1(t), \overline{y}_1(t)\right]$ and $v(t) \in \left[\underline{y}_2(t), \overline{y}_2(t)\right]$ for all $t \in [0,T]$.*
If moreover (F_1, F_2) is a regular TCCP-structured reaction functional, then u and v satisfy
(SS$_4$) *$u(t) \in \mathrm{dom}(D\Phi_1)$ and $v(t) \in \mathrm{dom}(D\Phi_2)$ for all $t \in]0,T]$, u and v possess a right derivative $\frac{d^+u}{dt}(t)$ and $\frac{d^+v}{dt}(t)$ at every $t \in]0,T[$, and*

$$\begin{cases} \dfrac{d^+u}{dt}(t) + D\Phi_1(u(t)) = F_1(t, u(t), v(t)), \\ \\ \dfrac{d^+v}{dt}(t) + D\Phi_1(v(t)) = F_2(t, u(t), v(t)). \end{cases}$$

Proof. ***Step 1 (local existence).*** We prove that there exists a unique solution of $(\mathcal{S})$ for T small enough. For $T > 0$ set

$$X_T := \{(u,v) \in C([0,T], X) \times C([0,T], X) : u \text{ and } v \text{ fulfill condition } (\mathrm{SS}_3)\}$$

which is clearly a closed subset of the space $C([0,T],X) \times C([0,T],X)$ equipped with the norm product defined by $\|(u,v)\|_{C\times C} := \|u\|_{C([0,T],X)} + \|v\|_{C([0,T],X)}$. Therefore X_T is a complete metric space when equipped with the metric associated with the norm $\|\cdot\|_{C\times C}$.

For each $(u,v) \in X_T$, we consider the two reaction-diffusion problems with unknown $\Lambda_1 v$ and $\Lambda_2 u$ respectively defined by

$$(\mathcal{P}_1) \begin{cases} \dfrac{d\Lambda_1 v}{dt}(t) + D\Phi_1(\Lambda_1 v(t)) = F_1(t, \Lambda_1 v(t), v(t)) \text{ for a.e. } t \in (0,T) \\ \\ \underline{\rho}_1 \le \Lambda_1 v(0) = u_0 \le \overline{\rho}_1, \end{cases}$$

$$(\mathcal{P}_2) \begin{cases} \dfrac{d\Lambda_2 u}{dt}(t) + D\Phi_2(\Lambda_2 u(t)) = F_2(t, u(t), \Lambda_2 u(t)) \text{ for a.e. } t \in (0,T) \\ \\ \underline{\rho}_2 \le \Lambda_2 u(0) = v_0 \le \overline{\rho}_2. \end{cases}$$

We first claim that $(\mathcal{P}_1)$ and $(\mathcal{P}_2)$ possess a unique solution $\Lambda_1 v$ and $\Lambda_2 u$ satisfying (SS$_1$), (SS$_2$) and (SS$_3$) where $\Lambda_1 v$ and $\Lambda_2 u$ are substituted for u and v respectively. Indeed, for fixed $(u,v) \in X_T$, set for every $(t,x,\zeta,\zeta') \in \mathbb{R}_+ \times \mathbb{R}^N \times \mathbb{R} \times \mathbb{R}$

$$r_v(t,x) = r_1(t,x) \odot h_1(v(t,x)), \ f_v(t,x,\zeta) = r_v(t,x) \cdot g_1(\zeta) + q_1(t,x),$$
$$r_u(t,x) = r_2(t,x) \odot h_2(u(t,x)), \ f_u(t,x,\zeta') = r_u(t,x) \cdot g_2(\zeta') + q_2(t,x),$$

and define F_v and F_u for $(U,V) \in L^2(\Omega) \times L^2(\Omega)$, by $F_v(t,U)(x) := f_v(t,x,U(x))$, $F_u(t,V)(x) := f_u(t,x,V(x))$. Therefore, $(\mathcal{P}_1)$ and $(\mathcal{P}_2)$ can be written as

$$(\mathcal{P}_1) \begin{cases} \dfrac{d\Lambda_1 v}{dt}(t) + D\Phi_1(\Lambda_1 v(t)) = F_v(t, \Lambda_1 v(t)) \text{ for a.e. } t \in (0,T) \\ \\ \underline{\rho}_1 \le \Lambda_1 v(0) = u_0 \le \overline{\rho}_1, \end{cases}$$

$$(\mathcal{P}_2)\begin{cases}\dfrac{d\Lambda_2 u}{dt}(t)+D\Phi_2(\Lambda_2 u(t))=F_u(t,\Lambda_2 u(t)) \text{ for a.e. } t\in(0,T)\\[2ex] \underline{\rho}_2\le\Lambda_2 u(0)=v_0\le\overline{\rho}_2.\end{cases}$$

The claim follows from Corollary 2.1, or Corollary 2.2, provided that we establish that F_v and F_u are CP-structured reaction functionals. For this, observe that each function f_v and f_u satisfies the structure condition of CP-structured reaction functions, and that condition (CP) is fulfilled because (f_1, f_2) is a TCCP-structured reaction function, and v and u satisfy (SS_3).

Let us consider the operator $\Lambda : X_T \to C([0,T],X)\times C([0,T],X)$ defined by $\Lambda(u,v)=(\Lambda_1 v,\Lambda_2 u)$. We are going to establish existence of a fixed point of Λ for $T>0$ small enough. Such a fixed point clearly furnishes a solution of $(\mathcal{S})$ fulfilling (SS_1)–(SS_4).

We claim that $\Lambda(X_T)\subset X_T$. Let $(u,v)\in X_T$, then $\Lambda(u,v)=(\Lambda_1 v,\Lambda_2 u)$. According to the considerations above, as $\Lambda_1 v$ and $\Lambda_2 u$ solve $(\mathcal{P}_1)$ and $(\mathcal{P}_2)$ respectively, we have $(\Lambda_1 v,\Lambda_2 u)\in C([0,T],X)\times C([0,T],X)$, and $\underline{y}_1(t)\le\Lambda_1 v(t)\le\overline{y}_1(t)$, $\underline{y}_2(t)\le\Lambda_2 u(t)\le\overline{y}_2(t)$. Therefore $(\Lambda_1 v,\Lambda_2 u)$ belongs to X_T.

We claim that Λ is a contraction for $T>0$ small enough. Let (u_1,v_1) and (u_2,v_2) in X_T. We first estimate

$$\|\Lambda(u_1,v_1)-\Lambda(u_2,v_2)\|_{C\times C}=\|(\Lambda_1 v_1-\Lambda_1 v_2)\|_X+\|(\Lambda_2 u_1-\Lambda_2 u_2)\|_X.$$

From $(\mathcal{P}_1)$, subtract the equation related to $\Lambda_1 v_1$ from the equation related to $\Lambda_1 v_2$ and take the scalar product in X with $\Lambda_1 v_1-\Lambda_1 v_2$. Using the fact that $D\Phi_1$ is a monotone operator, we obtain that for a.e. $t\in(0,T)$

$$\begin{aligned}&\frac{1}{2}\frac{d}{dt}\|(\Lambda_1 v_1-\Lambda_1 v_2)(t)\|_X^2\\&\le\langle F_1(t,\Lambda_1 v_1(t),v_1(t))-F_1(t,\Lambda_1 v_2(t),v_2(t)),\Lambda_1 v_1(t)-\Lambda_1 v_2(t)\rangle.\end{aligned}$$

Thus, for a.e. $t\in(0,T)$,

$$\begin{aligned}&\frac{d}{dt}\|(\Lambda_1 v_1-\Lambda_1 v_2)(t)\|_X^2\\&\le 2\|F_1(t,\Lambda_1 v_1(t),v_1(t))-F_1(t,\Lambda_1 v_2(t),v_2(t))\|_X\|\Lambda_1 v_1(t)-\Lambda_1 v_2(t)\|_X\\&\le\|F_1(t,\Lambda_1 v_1(t),v_1(t))-F_1(t,\Lambda_1 v_2(t),v_2(t))\|_X^2\\&\quad+\|\Lambda_1 v_1(t)-\Lambda_1 v_2(t)\|_X^2.\end{aligned}\tag{4.7}$$

According to the structure of the functional F_1, we have

$$\begin{aligned}&\|F_1(t,\Lambda_1 v_1(t),v_1(t))-F_1(t,\Lambda_1 v_2(t),v_2(t))\|_X^2\le\\&C(T,g_1,h_1)\|v_1(t)-v_2(t)\|_X^2+C'(T,g_1,h_1)\|\Lambda_1 v_1(t)-\Lambda_1 v_2(t)\|_X^2\end{aligned}\tag{4.8}$$

with

$$C(T,g_1,h_1)=2\sup_{\zeta\in[\underline{y}_1(T),\overline{y}_1(T)]}|g_1(\zeta)|^2\|r_1\|^2_{L^\infty(\mathbb{R}^N,\mathbb{R}^l)}L_{h_1,T},$$

$$C'(T,g_1,h_1)=2\sup_{\zeta'\in[\underline{y}_2(T),\overline{y}_2(T)]}|h_1(\zeta')|^2\|r_1\|^2_{L^\infty(\mathbb{R}^N,\mathbb{R}^l)}L_{g_1,T},$$

where $L_{g_1,T}$, $L_{h_1,T}$ denote the Lipschitz constants of the restrictions of g_1 and h_1 on $\left[\underline{y}_1(T), \overline{y}_1(T)\right]$ and $\left[\underline{y}_2(T), \overline{y}_2(T)\right]$ respectively. Combining (4.7) and (4.8) we infer that for a.e. $t \in (0,T)$

$$\frac{d}{dt}\|(\Lambda_1 v_1(t) - \Lambda_1 v_2(t))\|_X^2 \leq C(T, g_1, h_1)\|v_1(t) - v_2(t)\|_X^2 + (1 + C'(T, g_1, h_1))\|\Lambda_1 v_1(t) - \Lambda_1 v_2(t)\|_X^2. \tag{4.9}$$

By integrating this inequality over $(0,s)$ for $s \in [0,T]$ and noticing that $\Lambda_1 v_1(0) = \Lambda_1 v_2(0) = u_0$, we obtain

$$\|\Lambda_1 v_1(s) - \Lambda_1 v_2(s)\|_X^2 \leq C(T, g_1, h_1)\int_0^s \|v_1(t) - v_2(t)\|_X^2 dt + (1 + C'(T, g_1, h_1))\int_0^s \|\Lambda_1 v_1(t) - \Lambda_1 v_2(t)\|_X^2 dt \tag{4.10}$$

from which, according to Grönwall's lemma, we deduce that for all $s \in [0,T]$,

$$\|(\Lambda_1 v_1(s) - \Lambda_1 v_2(s))\|_X^2 \leq T\, C(T, g_1, h_1)\|v_1 - v_2\|_{C([0,T],X)}^2 \exp((1 + C'(T, g_1, h_1))T).$$

Proceeding similarly, we obtain, with suitable adapted notation, for all $s \in [0,T]$,

$$\|(\Lambda_2 u_1(s) - \Lambda_2 u_2(s))\|_X^2 \leq T\, C(T, g_2, h_2)\|u_1 - u_2\|_{C([0,T],X)}^2 \exp((1 + C'(T, g_2, h_2))T).$$

Consequently

$$\|\Lambda(u_1, v_1) - \Lambda(u_2, v_2)\|_{C\times C} \leq C(T)\|(u_1, v_1) - (u_2, v_2)\|_{C\times C}$$

where $C(T) = \max(C_1(T), C_2(T))$ with

$$C_1(T) = C(T, g_1, h_1)^{\frac{1}{2}} \exp((1 + C'(T, g_1, h_1)))\frac{T}{2}$$

and

$$C_2(T) = C(T, g_2, h_2)^{\frac{1}{2}} \exp((1 + C'(T, g_2, h_2)))\frac{T}{2}.$$

For $i = 1, 2$, the nonnegative constants $C(T, g_i, h_i)$ and $C'(T, g_i, h_i)$ are clearly nondecreasing so that $\lim_{T\to 0} C(T) = 0$. Consequently Λ is a contraction for T small enough and admits a fixed point (u, v), i.e. $(\Lambda_1 v, \Lambda_2 u) = (u, v)$ so that $\Lambda_1 v = u$ and $\Lambda_2 u = v$. This proves that (u, v) solves $(\mathcal{S})$.

To show (SS$_4$), it remains to prove that $t \mapsto r_v(t,\cdot)$ and $t \mapsto r_u(t,\cdot)$ from $[0,T]$ into $L^2(\Omega)$ are absolutely continuous. For $t \mapsto r_v(t,\cdot)$ the claim follows from the absolute continuity of r_1 and v, and the following estimate

$$\begin{aligned}\|r_v(t,\cdot) - r_v(s,\cdot)\|_{L^2(\Omega,\mathbb{R}^l)} &\leq \|r_1(t,\cdot) \odot h_1(v(t)) - r_1(s,\cdot) \odot h_1(v(t))\|_{L^2(\Omega,\mathbb{R}^l)} \\ &\quad + \|r_1(s,\cdot) \odot h_1(v(t)) - r_1(s,\cdot) \odot h_1(v(s))\|_{L^2(\Omega,\mathbb{R}^l)} \\ &\leq \|h_1\|_{L^\infty([\underline{y}_2(T),\overline{y}_2(T)],\mathbb{R}^l)}\|r_1(t,\cdot) - r_1(s,\cdot)\|_{L^2(\Omega,\mathbb{R}^l)} \\ &\quad + \|r_1\|_{L^\infty([0,T]\times\mathbb{R}^N,\mathbb{R}^l)} L_{h_1}\|v(t) - v(s)\|_{L^2(\Omega)}\end{aligned}$$

where L_{h_1} denotes the Lipschitz constant of h_1 in $[\underline{y}_2(T), \overline{y}_2(T)]$. For $t \mapsto r_u(t,\cdot)$ the proof is similar.

Step 2 (uniqueness). Let (u_1, v_1) and (u_2, v_2) be two solutions of $(\mathcal{S})$, then taking $\Lambda_1 v_1 = u_1$ and $\Lambda_1 v_2 = u_2$ in (4.10), we infer that for all $s \in [0,T]$

$$\|u_1(s) - u_2(s)\|_X^2 \leq C(T, g_1, h_1) \int_0^s \|v_1(t) - v_2(t)\|_X^2 dt + (1 + C'(T, g_1, h_1)) \int_0^s \|u_1(t) - u_2(t)\|_X^2 dt,$$

similarly

$$\|v_1(s) - v_2(s)\|_X^2 \leq C(T, g_2, h_2) \int_0^s \|u_1(t) - u_2(t)\|_X^2 dt + (1 + C'(T, g_2, h_2)) \int_0^s \|v_1(t) - v_2(t)\|_X^2 dt.$$

By summing these two inequalities, we obtain for a.e. $s \in [0,T]$,

$$\|u_1(s) - u_2(s)\|_X^2 + \|v_1(s) - v_2(s)\|_X^2 \leq C \int_0^s (\|u_1(t) - u_2(t)\|_X^2 + \|v_1(t) - v_2(t)\|_X^2) dt$$

for some nonnegative constant C. Hence, according to Grönwall's Lemma, for all $s \in [0,T]$,

$$\|u_1(s) - u_2(s)\|_X^2 + \|v_1(s) - v_2(s)\|_X^2 = 0,$$

which proves uniqueness.

Step 3 (existence of a global solution). Denote by $T^* > 0$ a small enough number obtained in ***Step 1*** so that $(\mathcal{S})$ admits a unique solution in $C([0,T^*], X) \times C([0,T^*], X)$.[2] By (Attouch *et al.*, 2014, Theorem 17.2.5) or (Brezis, 1973, Theorem 3.6)), we have $\sqrt{t}\frac{du}{dt} \in L^2(0, T^*, X)$. Hence, for $0 < \delta < T^*$, $\frac{du}{dt}$ belongs to $L^2(\delta, T^*, X)$. Set

$$E := \{T > \delta : \exists (u,v) \in C([0,T], X) \times C([0,T], X) \text{ solution of } (\mathcal{S})\}.$$

Since $T^* \in E$, we have $E \neq \emptyset$. Set $T_{Max} := \sup E$ in $\overline{\mathbb{R}}_+$ and denote by (u,v) the maximal solution of $(\mathcal{S})$ in $C([0, T_{Max}), X) \times C([0, T_{Max}), X)$. We argue by contradiction assuming that $T_{Max} < +\infty$.

a) We first prove the existence of the two limits $\lim_{t \to T_{Max}} u(t)$ and $\lim_{t \to T_{Max}} v(t)$ in X.

Let $T \in E$, then for a.e. $t \in (0,T)$ we have

$$\left\langle \frac{du}{dt}(t), \frac{du}{dt}(t) \right\rangle + \left\langle D\Phi_1 u(t), \frac{du}{dt}(t) \right\rangle = \left\langle F_1(t, u(t), v(t)), \frac{du}{dt}(t) \right\rangle, \quad (4.11)$$

[2] Recall that under the initial condition $u_0 \in \overline{\text{dom}(D\Phi_1)}$ we are not assured that the derivative $\frac{du}{dt}$ of the solution belongs to $L^2(0, T^*, X)$.

and

$$\left\langle \frac{dv}{dt}(t), \frac{dv}{dt}(t) \right\rangle + \left\langle D\Phi_2 v(t), \frac{dv}{dt}(t) \right\rangle = \left\langle F_2(t, u(t), v(t)), \frac{dv}{dt}(t) \right\rangle. \quad (4.12)$$

From (4.11), we infer that

$$\int_\delta^T \left\| \frac{du}{dt}(t) \right\|_X^2 dt + \Phi_1(u(t)) - \Phi_1(u(\delta))$$
$$\leq \left(\int_0^T \|F_1(t, u(t), v(t))\|_X^2 \, dt \right)^{\frac{1}{2}} \left(\int_\delta^T \left\| \frac{du}{dt}(t) \right\|_X^2 dt \right)^{\frac{1}{2}}. \quad (4.13)$$

For all $T \in E$, we have $[\underline{y}_1(T), \overline{y}_1(T)] \subset [\underline{y}_1(T_{max}), \overline{y}_1(T_{max})]$, and $[\underline{y}_2(T), \overline{y}_2(T)] \subset [\underline{y}_2(T_{max}), \overline{y}_2(T_{max})]$. Thus, according to the structure of F_1, there exists a constant $C = C\left(\|r_1\|_{L^\infty(\mathbb{R}^N, \mathbb{R}^l)}, \|g_1\|_{L^\infty([\underline{y}_1(T_{max}), \overline{y}_1(T_{max})], \mathbb{R}^l)}, \|h_1\|_{L^\infty([\underline{y}_2(T_{max}), \overline{y}_2(T_{max})], \mathbb{R}^l)} \right)$ such that

$$\|F_1(t, u(t), v(t))\|_X^2 \leq 2C^2 \mathcal{L}_N(\Omega) + 2\|q_1(t, \cdot)\|_X^2.$$

Therefore, since from Lemma 2.2

$$\inf_{v \in H^1(\Omega)} \Phi_1(v) \geq -C_2 \|\phi_1\|^2_{L^2_{\mathcal{H}_{N-1}}(\partial\Omega)}$$

and $q_1 \in L^2(0, T_{max}, L^2(\Omega))$, inequality (4.13) yields

$$\int_\delta^T \left\| \frac{du}{dt}(t) \right\|_X^2 dt \leq C \left(1 + \left(\int_\delta^T \left\| \frac{du}{dt}(t) \right\|_X^2 dt \right)^{\frac{1}{2}} \right)$$

where the new constant C does not depend on T. We infer that

$$\int_\delta^{T_{Max}} \left\| \frac{du}{dt}(t) \right\|_X^2 dt = \sup_{T \in E} \int_\delta^T \left\| \frac{du}{dt}(t) \right\|_X^2 dt < +\infty,$$

from which we deduce that $u : [\delta, T_{Max}) \to X$ is uniformly continuous. Indeed, for $s < t$ in $[\delta, T_{Max})$ we have

$$\|u(t) - u(s)\|_X \leq \int_s^t \left\| \frac{du}{d\tau}(\tau) \right\|_X^2 d\tau \leq (t-s)^{\frac{1}{2}} \left(\int_\delta^{T_{Max}} \left\| \frac{du}{d\tau}(t) \right\|_X^2 dt \right)^{\frac{1}{2}}$$

and u is more precisely $\frac{1}{2}$-Hölder continuous. According to the continuous extension principle in the Banach space X, u possesses a unique continuous extension $\overline{u}$ in $[\delta, T_{Max}]$ with $\lim_{t \to T_{Max}} u(t) = \overline{u}(T_{Max})$. Similarly, from (4.12), we deduce that v possesses a unique continuous extension $\overline{v}$ in $[\delta, T_{Max}]$ with $\lim_{t \to T_{Max}} v(t) = \overline{v}(T_{Max})$, which proves the claim.

b) Contradiction: For $T > 0$, consider the two component reaction-diffusion system

$$(\mathcal{S}')\begin{cases}\dfrac{dU}{dt}(t) + D\Phi_1(U(t)) = F_1(t, U(t), V(t)) \text{ for a.e. } t \in (0,T)\\[2ex] \dfrac{dV}{dt}(t) + D\Phi_2(V(t)) = F_2(t, U(t), V(t)) \text{ for a.e. } t \in (0,T)\\[2ex] U(0) = \overline{u}(T_{max}),\ V(0) = \overline{v}(T_{max})\\[2ex] \underline{\rho}_1' \le U(0) \le \overline{\rho}_1',\quad \underline{\rho}_2' \le V(0) \le \overline{\rho}_2'\end{cases}$$

where $\underline{\rho}_1' = \underline{y}_1(T_{max})$, $\overline{\rho}_1' = \overline{y}_1(T_{max})$, and $\underline{\rho}_2' = \underline{y}_2(T_{max})$, $\overline{\rho}_2' = \overline{y}_2(T_{max})$. Note that $U(0) \in \overline{\text{dom}(D\Phi_1)}$ and $V(0) \in \overline{\text{dom}(D\Phi_2)}$. Then from step 1, there exists $T^{**} > 0$ small enough such that $(\mathcal{S}')$ possesses a solution $(U, V) \in C([0, T^{**}], X) \times C([0, T^{**}], X)$. Set

$$\widetilde{u}(t) = \begin{cases} u(t) \text{ if } t \in [0, T_{Max}]\\ U(t - T_{Max}) \text{ if } t \in [T_{Max}, T_{Max} + T^{**}],\end{cases}$$

and

$$\widetilde{v}(t) = \begin{cases} v(t) \text{ if } t \in [0, T_{Max}]\\ V(t - T_{Max}) \text{ if } t \in [T_{Max}, T_{Max} + T^{**}].\end{cases}$$

Then $(\widetilde{u}, \widetilde{v}) \in C([0, T_{Max} + T^{**}], X) \times C([0, T_{Max} + T^{**}], X)$ is a solution of $(\mathcal{S})$, which leads to a contradiction with the maximality of T_{Max}. □

Theorem 4.1 remains valid for systems $(\mathcal{S})$ coupling a reaction-diffusion equation (r.d.e.) with a non-diffusive reaction equation (n.d.r.e.) (see Example 4.5), or two non-diffusive reaction equations (n.d.r.e.). As was noted in the preamble to Corollary 2.1, the non-diffusive reaction equation must be defined in $L^2(0, T, H^1(\Omega))$. We have

Theorem 4.2. *Let Φ be a standard functional of the calculus of variations (4.1) and (F_1, F_2) a TCCP-structured reaction functional with $\underline{\rho}_i$, $\overline{\rho}_i$, $\underline{y}_i$ and $\overline{y}_i$ given by condition (TCCP). Assume that $a\underline{\rho}_1 \le \phi \le a\overline{\rho}_1$, then the two component system defined in $L^2(0, T, L^2(\Omega)) \times L^2(0, T, H^1(\Omega))$ by*

$$(\mathcal{S})\begin{cases}\dfrac{du}{dt}(t) + D\Phi(u(t)) = F_1(t, u(t), v(t)) \textit{ for a.e. } t \in (0,T)\\[2ex] \dfrac{dv}{dt}(t) = F_2(t, u(t), v(t)) \textit{ for a.e. } t \in (0,T)\\[2ex] \underline{\rho}_1 \le u_0 = u(0) \le \overline{\rho}_1,\quad \underline{\rho}_2 \le v_0 = v(0) \le \overline{\rho}_2\\[2ex] u_0 \in \overline{\text{dom}(D\Phi)},\quad v_0 \in H^1(\Omega),\end{cases}$$

admits a unique solution $(u,v) \in C([0,T],L^2(\Omega)) \times C([0,T],H^1(\Omega))$ *satisfying* (SS_1), (SS_2), *and* (SS_3). *If moreover* (F_1,F_2) *is a regular TCCP-structured reaction functional, then* u *and* v *satisfy* (SS_4).[3]

Sketch of the proof. Set $X = L^2(\Omega)$ and $Y = H^1(\Omega)$. The proof consists in resuming that of Theorem 4.1 with some changes made precise below. Define

$$X_T := \{(u,v) \in C([0,T],X) \times C([0,T],Y) : u \text{ and } v \text{ fulfill condition } (SS_3)\}$$

which is a closed subset of the space $C([0,T],X) \times C([0,T],Y)$ equipped with the norm product defined by $\|(u,v)\| := \|u\|_{C([0,T],X)} + \|v\|_{C([0,T],Y)}$. For each $(u,v) \in X_T$, consider the two reaction problems with unknown $\Lambda_1 v$ and $\Lambda_2 u$ respectively defined by

$$(\mathcal{P}_1)\begin{cases} \dfrac{d\Lambda_1 v}{dt}(t) + D\Phi(\Lambda_1 v(t)) = F_1(t,\Lambda_1 v(t), v(t)) \text{ for a.e. } t \in (0,T) \\ \underline{\rho}_1 \le \Lambda_1 v(0) = u_0 \le \overline{\rho}_1, \end{cases}$$

$$(\mathcal{P}_2)\begin{cases} \dfrac{d\Lambda_2 u}{dt}(t) = F_2(t,u(t),\Lambda_2 u(t)) \text{ for a.e. } t \in (0,T) \\ \underline{\rho}_2 \le \Lambda_2 u(0) = v_0 \le \overline{\rho}_2. \end{cases}$$

By applying Corollary 2.3, and proceeding as in ***Step 1*** of the proof of Theorem 4.1, it is easily seen that $(\mathcal{P}_1)$ and $(\mathcal{P}_2)$ admit a unique solution $\Lambda_1 v$ and $\Lambda_2 u$ satisfying (SS_1), (SS_2), (SS_3), and furthermore (SS_4) when (F_1,F_2) is a regular TCCP-structured reaction functional. In the same way we prove that $\Lambda : X_T \to C([0,T],X) \times C([0,T],Y)$ defined by $\Lambda(u,v) = (\Lambda_1 v, \Lambda_2 u)$ possesses a fixed point of Λ for $T > 0$ small enough. Such a fixed point clearly furnishes a solution of $(\mathcal{S})$ fulfilling (SS_1)–(SS_4). The end of the proof is similar to that of Theorem 4.1. □

4.2 Convergence theorem of two components reaction-diffusion systems

For each $i = 1,2$, we consider a sequence $(\Phi_{i,n})_{n\in\mathbb{N}}$ of functional of the calculus of variations $\Phi_{i,n} : L^2(\Omega) \to \mathbb{R} \cup \{+\infty\}$, defined by

$$\Phi_{i,n}(u) = \begin{cases} \displaystyle\int_\Omega W_n(x,\nabla u(x))\,dx + \frac{1}{2}\int_{\partial\Omega} a_{i,n}u^2\,d\mathcal{H}_{N-1} - \int_{\partial\Omega} \phi_{i,n}u\,d\mathcal{H}_{N-1} & \text{if } u \in H^1(\Omega), \\ +\infty & \text{otherwise} \end{cases} \tag{4.14}$$

[3]By convention in (SS_4) we formally write $D\Phi_2$ for 0 and $\mathrm{dom}(D\Phi_2)$ for $H^1(\Omega)$.

where $a_{i,n} \in L^\infty_{\mathcal{H}_{N-1}}(\partial\Omega)$ with

$$\begin{cases} a_{i,n} \geq 0 \quad \mathcal{H}_{N-1}\text{-a.e. in } \partial\Omega \\ \exists \sigma > 0 \quad a_{i,n} \geq \sigma \quad \mathcal{H}_{N-1}\text{-a.e. in } \Gamma_i \subset \partial\Omega, \end{cases}$$

or

$$\Phi_{i,n}(u) = \begin{cases} \displaystyle\int_\Omega W_n(x, \nabla u(x))\, dx & \text{if } u \in H^1_{\Gamma_i}(\Omega), \\ +\infty & \text{otherwise.} \end{cases} \tag{4.15}$$

For $i = 1, 2$, the density $W_{i,n} : \mathbb{R}^N \times \mathbb{R}^N \to \mathbb{R}$ is a Borel measurable function which fulfills condition (D_n) of Chapter 2.

In the following we fix $T > 0$ and consider a sequence $((F_{1,n}, F_{2,n}))_{n\in\mathbb{N}}$ of TCCP-structured functionals. Each of them is associated with $(r_{i,n}, g_{i,n}, h_{i,n}, q_{i,n})$, i.e. $F_{i,n}(t, u, v)(x) = f_{i,n}(t, x, u(x)v(x))$ for all $t \in [0, T]$, a.e. $x \in \Omega$, and all $(u, v) \in L^2(\Omega)^2$, where for all $(t, x, \zeta, \zeta') \in \mathbb{R}_+ \times \mathbb{R}^N \times \mathbb{R} \times \mathbb{R}$,

$$\begin{aligned} f_{1,n}(t, x, \zeta, \zeta') &= r_{1,n}(t, x) \odot h_{1,n}(\zeta') \cdot g_{1,n}(\zeta) + q_{1,n}(t, x) \\ f_{2,n}(t, x, \zeta, \zeta') &= r_{2,n}(t, x) \odot h_{2,n}(\zeta) \cdot g_{2,n}(\zeta') + q_{2,n}(t, x). \end{aligned} \tag{4.16}$$

For $i = 1, 2$, we assume that $h_{i,n}$ and $g_{i,n}$ are locally Lipschitz functions, uniformly with respect to n, i.e. for all interval $I \subset \mathbb{R}$, there exists $L_I \geq 0$ and $L'_I \geq 0$ such that for all $(\zeta, \zeta') \in I \times I$,

$$\begin{aligned} &\sup_{n\in\mathbb{N}} |g_{i,n}(\zeta) - g_{i,n}(\zeta')| \leq L_I |\zeta - \zeta'|, \\ &\sup_{n\in\mathbb{N}} |h_{i,n}(\zeta) - h_{i,n}(\zeta')| \leq L'_I |\zeta - \zeta'|. \end{aligned} \tag{4.17}$$

We finally assume that

$$-\infty < \inf_n \underline{y}_{i,n}(T) \text{ and } \sup_n \overline{y}_{i,n}(T) < +\infty, \tag{4.18}$$

and, for all $n \in \mathbb{N}$,

$$a_{i,n}\underline{\rho}_{i,n} \leq \phi_{i,n} \leq a_{i,n}\overline{\rho}_{i,n} \text{ on } \partial\Omega \tag{4.19}$$

if $\Phi_{i,n}$ is of the form (4.14), or

$$\underline{\rho}_{i,n} \leq 0 \leq \overline{\rho}_{i,n} \tag{4.20}$$

if $\Phi_{i,n}$ is of the form (4.15). Recall that for $i = 1, 2$, $\underline{y}_{i,n}$, $\overline{y}_{i,n}$, $\underline{\rho}_{i,n}$ and $\overline{\rho}_{i,n}$ are given by condition $(TCCP)$ fulfilled by $(F_{1,n}, F_{2,n})$ where $\underline{y}_{i,n}$ and $\overline{y}_{i,n}$ are solution of $\underline{ODE_i}$ and $\overline{ODE_i}$ with two functions $\underline{f}_{i,n}$, $\overline{f}_{i,n}$ which do not depend on the space variable, and initial condition $\underline{\rho}_{i,n}$ and $\overline{\rho}_{i,n}$ respectively.

Theorem 4.3 (General convergence theorem). *Assume that for $i = 1, 2$, the sequence of densities $(W_{i,n})_{n\in\mathbb{N}}$ of functionals (4.14) or (4.15) satisfies (D_n), and that the sequence of TCCP-structured reaction functionals $(F_{1,n}, F_{2,n})_{n\in\mathbb{N}}$ satisfies conditions (4.17), (4.18), and (4.19) or (4.20). Let (u_n, v_n) be the unique solution of the system*

$$(\mathcal{S}_n)\begin{cases} \dfrac{du_n}{dt}(t) + D\Phi_{1,n}(u_n(t)) = F_{1,n}(t, u_n(t), v_n(t)) \text{ for a.e. } t \in (0, T) \\ \dfrac{dv_n}{dt}(t) + D\Phi_{2,n}(v_n(t)) = F_{2,n}(t, u_n(t), v_n(t)) \text{ for a.e. } t \in (0, T) \\ \underline{\rho}_{1,n} \le u_n^0 = u_n(0) \le \overline{\rho}_{1,n},\ \underline{\rho}_{2,n} \le v_n^0 = v_n(0) \le \overline{\rho}_{2,n}, \\ u_n^0 \in \mathrm{dom}(\Phi_{1,n}),\ v_n^0 \in \mathrm{dom}(\Phi_{2,n}). \end{cases}$$

For each $i = 1, 2$, assume that

(Hs$_1$) *$\Phi_{i,n} \overset{M}{\to} \Phi_i$ and $\sup_{n\in\mathbb{N}} \|\phi_{i,n}\|_{L^2_{\mathcal{H}_{N-1}}(\partial\Omega)} < +\infty$;*

(Hs$_2$) *$\sup_{n\in\mathbb{N}} \Phi_{1,n}(u_n^0) < +\infty$ and $\sup_{n\in\mathbb{N}} \Phi_{2,n}(v_n^0) < +\infty$;*

(Hs$_3$) *$u_n^0 \to u^0$ and $v_n^0 \to v^0$ strongly in $L^2(\Omega)$;*

(Hs$_4$) *$g_{i,n}$ and $h_{i,n}$ pointwise converge to g_i and h_i respectively;*

(Hs$_5$) *$\sup_n \|r_{i,n}\|_{L^\infty([0,T]\times\mathbb{R}^N,\mathbb{R}^l)} < +\infty$, and there exists $r_i \in L^2(0, T, L^2(\Omega, \mathbb{R}^l))$ with $r_i \in L^\infty([0, T] \times \mathbb{R}^N, \mathbb{R}^l)$, such that $r_{i,n} \rightharpoonup r_i$ in $L^2(0, T, L^2(\Omega, \mathbb{R}^l))$;*

(Hs$_6$) *there exists $q_i \in L^2(0, T, L^2(\Omega))$ such that $q_{i,n} \rightharpoonup q_i$ in $L^2(0, T, L^2(\Omega))$.*

Then (u_n, v_n) uniformly converges in $C([0, T], L^2(\Omega)) \times C([0, T], L^2(\Omega))$ to the unique solution (u, v) of the system

$$(\mathcal{S})\begin{cases} \dfrac{du}{dt}(t) + \partial\Phi_1(u(t)) \ni F_1(t, u(t), v(t)) \text{ for a.e. } t \in (0, T) \\ \dfrac{dv}{dt}(t) + \partial\Phi_2(v(t)) \ni F_2(t, u(t), v(t)) \text{ for a.e. } t \in (0, T) \\ \inf_n \underline{\rho}_{1,n} \le u^0 = u(0) \le \sup_n \rho_{1,n},\ \inf_n \underline{\rho}_{2,n} \le v^0 = v(0) \le \sup_n \overline{\rho}_{2,n}, \\ u^0 \in \mathrm{dom}(\Phi_1),\ v^0 \in \mathrm{dom}(\Phi_2). \end{cases}$$

The reaction functionals $F_i : [0, +\infty) \times L^2(\Omega) \times L^2(\Omega) \to \mathbb{R}^\Omega$, $i = 1, 2$, are defined for all $t \in [0, T]$, all $(U, V) \in L^2(\Omega) \times L^2(\Omega)$ and for a.e. $x \in \Omega$, by

$$\begin{aligned} &F_i(t, U, V)(x) = f_i(t, x, U(x), V(x)), \\ &f_1(t, x, \zeta, \zeta') = r_1(t, x) \odot h_1(\zeta') \cdot g_1(\zeta) + q_1(t, x), \\ &f_2(t, x, \zeta, \zeta') = r_2(t, x) \odot h_2(\zeta) \cdot g_2(\zeta') + q_2(t, x). \end{aligned}$$

Moreover $\inf_n \underline{y}_{1,n}(t) \le u(t) \le \sup \overline{y}_{1,n}(t)$, $\inf_n \underline{y}_{2,n}(t) \le v(t) \le \sup_n \overline{y}_{2,n}(t)$ *for all* $t \in [0,T]$, *and*

$$\left(\frac{du_n}{dt}, \frac{du_n}{dt}\right) \rightharpoonup \left(\frac{du}{dt}, \frac{dv}{dt}\right) \text{ weakly in } L^2(0,T,L^2(\Omega)) \times L^2(0,T,L^2(\Omega)).$$

If furthermore $(\Phi_{1,n}(u_n^0), \Phi_{2,n}(v_n^0)) \to (\Phi_1(u^0), \Phi_2(v^0))$, $r_{i,n} \to r_i$ *strongly in* $L^2(0,T,L^2(\Omega,\mathbb{R}^l))$, *and* $q_{i,n} \to q_i$ *strongly in* $L^2(0,T,L^2(\Omega))$, *then*

$$\left(\frac{du_n}{dt}, \frac{dv_n}{dt}\right) \to \left(\frac{du}{dt}, \frac{dv}{dt}\right) \text{ strongly in } L^2(0,T,L^2(\Omega)) \times L^2(0,T,L^2(\Omega)),$$

and $(\Phi_{1,n}(u_n(t)), \Phi_{2,n}(v_n(t)))$ *converges to* $(\Phi_1(u(t)), \Phi_2(v(t)))$ *for all* $t \in [0,T]$.

Assume that $(F_{1,n}, F_{2,n})_{n\in\mathbb{N}}$ *is a sequence of regular TCCP-structured reaction functionals, that* $\frac{dq_{i,n}}{dt} \rightharpoonup \frac{dq_i}{dt}$ *weakly in* $L^1(0,T,L^2(\Omega))$ *and* $\frac{dr_{i,n}}{dt} \rightharpoonup \frac{dr_i}{dt}$ *weakly in* $L^1(0,T,L^2(\Omega,\mathbb{R}^l))$. *Then the functional* G_i *defined by* $G_i(t) = F_i(t,u(t),v(t))$ *belongs to* $W^{1,1}(0,T,L^2(\Omega))$, *and* $(u(t),v(t)) \in \mathrm{dom}(\partial\Phi_1) \times \mathrm{dom}(\partial\Phi_2)$ *for all* $t \in]0,T]$. *Assume furthermore that for each* $t \in]0,T[$, $q_{i,n}(t,\cdot) \rightharpoonup q_i(t,\cdot)$ *weakly in* $L^2(\Omega)$, $r_{i,n}(t,\cdot) \rightharpoonup r_i(t,\cdot)$ *weakly in* $L^2(\Omega,\mathbb{R}^l)$, *and that* $\partial\Phi_i$ *is single valued. Then for each* $t \in]0,T[$

$$\left(\frac{d^+u_n}{dt}(t), \frac{d^+v_n}{dt}(t)\right) \rightharpoonup \left(\frac{d^+u}{dt}(t), \frac{d^+u}{dt}(t)\right) \text{ weakly in } L^2(\Omega).$$

Proof. Since for each $i = 1,2$, $\mathrm{dom}(\Phi_{i,n}) \subset \overline{\mathrm{dom}(D\Phi_{i,n})}$, we have $(u_n^0, v_n^0) \in \left(\overline{\mathrm{dom}(D\Phi_{1,n})}, \overline{\mathrm{dom}(D\Phi_{2,n})}\right)$, so that, according to Theorem 4.1, $(\mathcal{P}_n)$ admits a unique solution (u_n, v_n) which satisfies (SS_1)–(SS_4). We follow the strategy of the proof of Theorem 2.6.

Step 1. Let $i = 1,2$. Set $\underline{Y}_i := \inf_n \underline{y}_{i,n}$, and $\overline{Y}_i := \sup_n \overline{y}_{i,n}$. We establish

$$\underline{Y}_1 \le u_n \le \overline{Y}_1 \text{ and } \underline{Y}_2 \le v_n \le \overline{Y}_2; \tag{4.21}$$

$$\sup_{(\zeta,n)\in[\underline{Y}_i,\overline{Y}_i]\times\mathbb{N}} |g_{i,n}(\zeta)| < +\infty, \qquad \sup_{(\zeta,n)\in[\underline{Y}_{i+1},\overline{Y}_{i+1}]\times\mathbb{N}} |h_{i,n}(\zeta)| < +\infty; \tag{4.22}$$

$$\sup_{n\in\mathbb{N}} \left\|\frac{du_n}{dt}\right\|_{L^2(0,T,X)} < +\infty \text{ and } \sup_{n\in\mathbb{N}} \left\|\frac{dv_n}{dt}\right\|_{L^2(0,T,X)} < +\infty. \tag{4.23}$$

In what follows we set

$$\overline{g}_i := \sup_{(\zeta,n)\in[\underline{Y}_i,\overline{Y}_i]\times\mathbb{N}} |g_{i,n}(\zeta)|,$$

$$\overline{h}_i := \sup_{(\zeta,n)\in[\underline{Y}_{i+1},\overline{Y}_{i+1}]\times\mathbb{N}} |h_{i,n}(\zeta)|.$$

From (SS_3) the solution (u_n, v_n) of $(\mathcal{S}_n)$ satisfies $\underline{y}_{1,n}(T) \le u_n \le \overline{y}_{1,n}(T)$ and $\underline{y}_{2,n}(T) \le v_n \le \overline{y}_{2,n}(T)$, so that inequalities (4.21) follow directly from (4.18). We deduce (4.22) from (4.17), hypothesis (Hs_4) and estimate $|g_{i,n}(\zeta)| \le |g_{i,n}(0)| + L_{[\underline{Y}_i,\overline{Y}_i]}|\zeta|$; idem for $h_{i,n}$.[4]

[4]By convention $\underline{Y}_3 = \underline{Y}_1$, and $\overline{Y}_3 = \overline{Y}_1$.

Let us establish (4.23). In what follows the letter C denotes various constants which can vary from line to line. By using the structure of the TCCP-structured reaction functional F_n, and from (4.22) and hypothesis (Hs_5), we easily infer that

$$\begin{aligned}\|F_{i,n}(t,u_n(t),v_n(t))\|_X^2 &\leq 2\mathcal{L}_N(\Omega)\|r_{i,n}\|_\infty^2 \overline{g}_i^2 \overline{h}_i^{-2} + 2\|q_{i,n}(t,\cdot)\|_X^2 \\ &\leq C(1+\|q_{i,n}(t,\cdot)\|_X^2).\end{aligned} \tag{4.24}$$

Thus, according to (Hs_6), we deduce

$$\sup_n \int_0^T \|F_{i,n}(t,u_n(t),v_n(t))\|_X^2 \, dt < +\infty. \tag{4.25}$$

On the other hand, from $(\mathcal{S}_n)$ we infer that for a.e. $t \in (0,T)$,

$$\left\|\frac{du_n}{dt}(t)\right\|_X^2 + \left\langle D\Phi_{1,n}(u_n(t)), \frac{du_n}{dt}(t)\right\rangle = \left\langle F_{1,n}(t,u_n(t),v_n(t)), \frac{du_n}{dt}(t)\right\rangle$$

$$\left\|\frac{dv_n}{dt}(t)\right\|_X^2 + \left\langle D\Phi_{2,n}(u_n(t)), \frac{dv_n}{dt}(t)\right\rangle = \left\langle F_{2,n}(t,u_n(t),v_n(t)), \frac{dv_n}{dt}(t)\right\rangle.$$

By integrating over $(0,T)$, we obtain

$$\begin{aligned}&\int_0^T \left\|\frac{du_n}{dt}(t)\right\|_X^2 dt + \int_0^T \left\langle D\Phi_{1,n}(u_n(t)), \frac{du_n}{dt}(t)\right\rangle dt \\ &\quad = \int_0^T \left\langle F_{1,n}(t,u_n(t),v_n(t)), \frac{du_n}{dt}(t)\right\rangle dt, \\ &\int_0^T \left\|\frac{dv_n}{dt}(t)\right\|_X^2 dt + \int_0^T \left\langle D\Phi_{2,n}(v_n(t)), \frac{dv_n}{dt}(t)\right\rangle dt \\ &\quad = \int_0^T \left\langle F_{2,n}(t,u_n(t),v_n(t)), \frac{dv_n}{dt}(t)\right\rangle dt.\end{aligned} \tag{4.26}$$

Since $(u_n^0, v_n^0) \in (\text{dom}(\Phi_{1,n}), \text{dom}(\Phi_{2,n}))$, we deduce that $(\frac{du_n}{dt}, \frac{dv_n}{dt})$ belongs to $L^2(0,T,X) \times L^2(0,T,X)$ and $t \mapsto \Phi_{1,n}(u_n(t))$, $t \mapsto \Phi_{2,n}(v(t))$ are absolutely continuous (see (Brezis, 1973, Theorem 3.6)). Therefore for a.e. $t \in (0,T)$, $\frac{d}{dt}\Phi_{1,n}(u_n(t)) = \langle D\Phi_{1,n}(u_n(t)), \frac{du_n}{dt}(t)\rangle$, and $\frac{d}{dt}\Phi_{2,n}(v_n(t)) = \langle D\Phi_{2,n}(v_n(t)), \frac{dv_n}{dt}(t)\rangle$ (see (Attouch *et al.*, 2014, Proposition 17.2.5)). From the first equality in (4.26) we have

$$\begin{aligned}&\int_0^T \left\|\frac{du_n}{dt}(t)\right\|_X^2 dt \\ &= -\Phi_{1,n}(u_n(T)) + \Phi_{1,n}(u_n^0) + \int_0^T \left\langle F_{1,n}(t,u_n(t),v_n(t)), \frac{du_n}{dt}(t)\right\rangle dt \\ &\leq -\inf_{w\in L^2(\Omega)} \Phi_{1,n}(w) + \sup_n \Phi_{1,n}(u_n^0) \\ &\quad + \left(\int_0^T \|F_{1,n}(t,u_n(t),v_n(t))\|_X^2\right)^{\frac{1}{2}} \left(\int_0^T \left\|\frac{du_n}{dt}(t)\right\|_X^2\right)^{\frac{1}{2}}\end{aligned} \tag{4.27}$$

where from (4.3)

$$\inf_{w\in L^2(\Omega)} \Phi_{1,n}(w) \geq -C_2\|\phi_{1,n}\|^2_{L^2_{\mathcal{H}_{N-1}}(\partial\Omega)}.$$

From (Hs_1), (Hs_2), and (4.25), (4.27) yields that there exists a constant $C \geq 0$ such that

$$\int_0^T \left\|\frac{du_n}{dt}(t)\right\|_X^2 dt \leq C\left(1+\left(\int_0^T \left\|\frac{du_n}{dt}(t)\right\|_X^2 dt\right)^{\frac{1}{2}}\right).$$

Reasoning similarly with the second equality in (4.26), we obtain

$$\int_0^T \left\|\frac{dv_n}{dt}(t)\right\|_X^2 dt \leq C\left(1+\left(\int_0^T \left\|\frac{dv_n}{dt}(t)\right\|_X^2 dt\right)^{\frac{1}{2}}\right),$$

from which we deduce (4.23).

Step 2. We prove that there exist $(u,v) \in C([0,T],X) \times C([0,T],X)$, and a subsequence of $((u_n,v_n))_{n\in\mathbb{N}}$ not relabeled, satisfying $(u_n,v_n) \to (u,v)$ in $C([0,T],X) \times C([0,T],X)$ equipped with the norm $\|\cdot\|_{C\times C}$.

Basically we apply the Ascoli-Arzela compactness theorem for $(u_n)_{n\in\mathbb{N}}$ and $(v_n)_{n\in\mathbb{N}}$. We show the compactness only for $(u_n)_{n\in\mathbb{N}}$, the same reasoning holds for $(v_n)_{n\in\mathbb{N}}$, and we follow the proof of ***Step 2*** of Theorem 2.6.

From (4.21), $(u_n)_{n\in\mathbb{N}}$ is bounded in $C([0,T],X)$. Moreover for $(s,t) \in [0,T]^2$ with $s < t$, we have

$$\|u_n(t)-u_n(s)\|_X \leq \int_s^t \left\|\frac{du_n}{dt}(\tau)\right\|_X d\tau \leq (t-s)^{\frac{1}{2}} \sup_n \left\|\frac{du_n}{dt}\right\|_{L^2(0,T,X)}$$

which, from (4.23), yields the equicontinuity of the sequence $(u_n)_{n\in\mathbb{N}}$. It remains to establish for each $t \in [0,T]$, the relative compactness in X of the set $E_t := \{u_n(t) : n \in \mathbb{N}\}$. For $t = 0$ there is nothing to prove because of hypothesis (Hs_3) on the initial condition. Let $t \in]0,T]$, to establish the relative compactness of E_t, reproduce the proof of Theorem 2.6, ***Step 2*** by substituting t for T in (4.27). We obtain $\|\nabla u_n(t)\|_X \leq C$ where C does not depend on n and $t \in [0,T]$. The claim follows from the compact embedding $H^1(\Omega) \hookrightarrow L^2(\Omega)$.

Step 3. We assert that $(\frac{du_n}{dt}, \frac{dv_n}{dt}) \rightharpoonup (\frac{du}{dt}, \frac{dv}{dt})$ weakly in $L^2(0,T,X)\times L^2(0,T,X)$ for a non relabeled subsequence, and that $\inf_n \underline{y}_{1,n}(t) \leq u(t) \leq \sup_n \overline{y}_{1,n}(t)$, $\inf_n \underline{y}_{2,n}(t) \leq u(t) \leq \sup_n \overline{y}_{2,n}(t)$ for all $t \in [0,T]$. The first claim is a straightforward consequence of (4.23) and ***Step 2***. The second follows from inequality $\inf_n \underline{y}_{1,n} \leq u_n \leq \sup_n \overline{y}_{1,n}$, $\inf_n \underline{y}_{2,n} \leq v_n \leq \sup_n \overline{y}_{2,n}$, (4.18), and $(u_n,v_n) \to (u,v)$ in $C([0,T],X)$.

Step 4. We prove that (u,v) is the unique solution of $(\mathcal{S})$. The proof mimics that of Theorem 2.6. We give a sketch of the proof. According to the Fenchel extremality condition (see (Attouch *et al.*, 2014, Proposition 9.5.1)), (u_n, v_n) solves $(\mathcal{S}_n)$, is equivalent to

$$\int_0^T \left[\Phi_{1,n}(u_n(t)) + \Phi_{1,n}^*\left(G_{1,n}(t) - \frac{du_n}{dt}(t)\right)\right] dt + \frac{1}{2}\left(\|u_n(T)\|^2 - \|u_n^0\|^2\right) - \int_0^T \langle G_{1,n}(t), u_n(t)\rangle\, dt = 0,$$

$$\int_0^T \left[\Phi_{2,n}(u_n(t)) + \Phi_{2,n}^*\left(G_{2,n}(t) - \frac{dv_n}{dt}(t)\right)\right] dt + \frac{1}{2}\left(\|v_n(T)\|^2 - \|v_n^0\|^2\right) - \int_0^T \langle G_{2,n}(t), u_n(t)\rangle\, dt = 0,$$

where $G_{i,n}(t) = F_{i,n}(t, u_n(t), v_n(t))$. Observe that the functionals defined in $L^2(0, T, L^2(\Omega))$ by

$$w \mapsto \int_0^T \Phi_{i,n}(w(t))dt, \quad w \mapsto \int_0^T \Phi_{i,n}^*(w(t))dt$$

Mosco-converge to

$$w \mapsto \int_0^T \Phi_i(w(t))dt, \quad w \mapsto \int_0^T \Phi_i^*(w(t))dt$$

respectively (refer to Lemma 2.6). Thus going to the limit in two previous equalities, from ***Step 2***, ***Step 3***, and Lemma 4.1 postponed after the proof, we obtain

$$\int_0^T \left[\Phi_1(u(t)) + \Phi_1^*\left(G_1(t) - \frac{du}{dt}(t)\right)\right] dt + \frac{1}{2}\left(\|u(T)\|^2 - \|u^0\|^2\right) - \int_0^T \langle G_1(t), u(t)\rangle\, dt = 0,$$

$$\int_0^T \left[\Phi_2(u(t)) + \Phi_2^*\left(G_2(t) - \frac{dv}{dt}(t)\right)\right] dt + \frac{1}{2}\left(\|v(T)\|^2 - \|v^0\|^2\right) - \int_0^T \langle G_2(t), u(t)\rangle\, dt = 0,$$

where $G_i(t) = F_i(t, u(t), v(t))$. Observe that we applied the Legendre-Fenchel inequality in order to obtain equality above. This proves that (u, v) solves $(\mathcal{S})$.

Step 5. We assume furthermore that $(\Phi_{1,n}(u_n^0), \Phi_{1,n}(v_n^0)) \to (\Phi(u^0), \Phi(v^0))$, $r_{i,n} \to r_i$ strongly in $L^2(0, T, L^2(\Omega, \mathbb{R}^l))$, and $q_{i,n} \to q_i$ strongly in $L^2(0, T, L^2(\Omega))$. We claim that $(\frac{du_n}{dt}, \frac{dv_n}{dt}) \to (\frac{du}{dt}, \frac{dv}{dt})$ strongly in $L^2(0, T, L^2(\Omega)) \times L^2(0, T, L^2(\Omega))$ and $(\Phi_{1,n}(u_n(t)), \Phi_{2,n}(v_n(t)))$ converges to $(\Phi_1(u(t)), \Phi_2(v(t)))$ for all $t \in [0, T]$.

For the first claim it suffices to prove that $\|\frac{du_n}{dt}\|_{L^2(0,T,X)} \to \|\frac{du}{dt}\|_{L^2(0,T,X)}$ and $\|\frac{dv_n}{dt}\|_{L^2(0,T,X)} \to \|\frac{dv}{dt}\|_{L^2(0,T,X)}$ by replicating the proof of Theorem 2.6, ***Step 5***.

For the last claim, observe that in (4.27) T can be replaced by an arbitrary $t \in [0, T]$, which, using Lemma 4.1 and the strong convergence above, implies that $\lim_{n\to+\infty} \Phi_{1,n}(u_n(t)) = \Phi_1(u(t))$. The same reasoning leads to $\lim_{n\to+\infty} \Phi_{2,n}(v_n(t)) = \Phi_2(v(t))$.

Step 6. We assume that $(F_{1,n}, F_{2,n})$ is regular TCCP-structured reaction functional, and sketch the proof of the last assertion. We begin by establishing the following estimate on the total variation $\mathrm{Var}(G_i, [0,T])$ of G_i in $[0,T]$ from which we infer that $G_i \in W^{1,1}(0,T,L^2(\Omega))$.

$$\begin{aligned}\mathrm{Var}\,(G_i,[0,T]) &:= \int_0^T \left\| \frac{G_i}{dt}(t) \right\|_X dt \\ &\leq C\left(1 + \int_0^T \left\| \frac{dv}{dt}(t) \right\|_X dt + \int_0^T \left\| \frac{du}{dt}(t) \right\|_X dt\right) \end{aligned} \tag{4.28}$$

where C is a nonnegative constant which does not depend on n. To shorten the notation, we omit the index $i = 1$. According to the structure of F, to (4.22), and hypothesis (Hs_6), we have

$$\begin{aligned}\|G(t) - G(s)\|_X &\leq \overline{g}\overline{h}\|r(t) - r(s)\|_{L^2(\Omega,\mathbb{R}^l)} + \|q(t) - q(s)\|_X \\ &\quad + \|r\|_\infty \|g(u(t)) \cdot h(v(t)) - g(u(s)) \cdot h(v(s))\|_X \\ &\leq \overline{g}\overline{h} \int_s^t \left\| \frac{dr}{dt}(\tau,\cdot) \right\|_{L^2(\Omega,\mathbb{R}^l)} d\tau + \int_s^t \left\| \frac{dq}{dt}(\tau,\cdot) \right\|_X d\tau \\ &\quad + \|r\|_\infty \|g(u(t)) \cdot h(v(t)) - g(u(s)) \cdot h(v(s))\|_X \,. \end{aligned} \tag{4.29}$$

On the other hand, from (4.17) and (4.22), we infer that

$$\begin{aligned}&\|g(u(t)) \cdot h(v(t)) - g(u(s)) \cdot h_n(v(s))\|_X \\ &\leq \overline{g}L'_{[\underline{Y}_2,\overline{Y}_2]} \|v(t) - v(s)\|_X + \overline{h}L_{[\underline{Y}_1,\overline{Y}_1]} \|u(t) - u(s)\|_X \\ &\leq gL'_{[\underline{Y}_2,\overline{Y}_2]} \int_s^t \left\| \frac{dv}{d\sigma}(\sigma) \right\|_X d\sigma + \overline{h}L_{[\underline{Y}_1,\overline{Y}_1]} \int_s^t \left\| \frac{du}{d\sigma}(\sigma) \right\|_X d\sigma. \end{aligned} \tag{4.30}$$

Estimate (4.28) is then obtained by combining (4.29), (4.30).

Since $(F_{1,n}, F_{2,n})$ is regular TCCP-structured reaction functional by using the same calculation, we can easily show that $G_{i,n} \in W^{1,1}(0,T,L^2(\Omega))$. Hence, according to Theorem 4.1 u and v possess a right derivative $\frac{d^+u}{dt}(t)$ and $\frac{d^+v}{dt}(t)$ at every $t \in]0,T[$, satisfying

$$\begin{cases} \dfrac{d^+u_n}{dt}(t) + D\Phi_{1,n}(u_n(t)) = F_{1,n}(t, u_n(t), v_n(t)), \\[2ex] \dfrac{d^+v_n}{dt}(t) + D\Phi_{2,n}(v(t)) = F_{2,n}(t, u_n(t), v_n(t)). \end{cases} \tag{4.31}$$

Fix $t \in]0,T[$. Reasoning as in the proof of Theorem 2.6, ***Step 6***, we leave it to the reader to establish the following estimates:

$$F_{i,n}(t, u_n(t), v_n(t)) \rightharpoonup F_i(t, u(t), v(t)) \text{ weakly in } L^2(\Omega); \tag{4.32}$$

$$\sup_{n\in\mathbb{N}} \left\| \frac{d^+u_n}{dt}(t) \right\|_X < +\infty, \ \sup_{n\in\mathbb{N}} \left\| \frac{d^+v_n}{dt}(t) \right\|_X < +\infty; \tag{4.33}$$

$$\sup_{n\in\mathbb{N}} \|D\Phi_{1,n}(u_n(t))\|_X < +\infty, \ \sup_{n\in\mathbb{N}} \|D\Phi_{2,n}(v_n(t))\|_X < +\infty. \tag{4.34}$$

(Recall that (4.33) is obtained from Proposition 2.5.) According to the graph-convergence, the weak limits (for a subsequence) obtained from (4.34) can be identified, equal to $D\Phi_1(u(t))$ and $D\Phi_2(v(t))$. Hence, passing to the limit in (4.31), the weak limits $\theta_1(t)$ and $\theta_2(t)$ obtained from (4.33) solve the system

$$\begin{cases} \theta_1(t) + D\Phi_{1,n}(u_n(t)) = F_{1,n}(t, u_n(t), v_n(t)), \\ \\ \theta_2(t) + D\Phi_{2,n}(v(t)) = F_{2,n}(t, u_n(t), v_n(t)). \end{cases}$$

Since $G_i \in W^{1,1}(0, T, L^2(\Omega))$ we know that

$$\frac{d^+u}{dt}(t) + D\Phi_1(u(t)) = F_1(t, u(t), v(t)),$$

and

$$\frac{d^+v}{dt}(t) + D\Phi_2(v(t)) = F_2(t, u(t), v(t)).$$

Hence $(\theta(t), \theta_2(t)) = \left(\frac{d^+u}{dt}(t), \frac{d^+v}{dt}(t)\right)$. □

Lemma 4.1. *For each $i = 1, 2$, the functional $G_{i,n} = F_{i,n}(\cdot, u_n, v_n)$ weakly converges in $L^2(0, T, X)$ to the functional $G_i = F_i(\cdot, u, v)$.*

Proof of Lemma 4.1. We only prove the weak convergence of $G_{1,n}$ and omit the subscript 1. The proof of the weak convergence of $G_{2,n}$ is similar. Recall that from (4.16), $G_n(t) = H_n(t) + q_n(t)$ where

$$\begin{aligned} H_n(t)(x) &= r_n(t, x) \odot h_n(v_n(t, x)) \cdot g_n(u_n(t, x)) \\ &= r_n(t, x) \cdot h_n(v_n(t, x)) \odot g_n(u_n(t, x)) \end{aligned}$$

for all $(t, x) \in \mathbb{R}_+ \times \mathbb{R}^N$. Since $q_n \rightharpoonup q$ in $L^2(0, T, X)$, we are reduced to prove that $H_n \rightharpoonup H$ in $L^2(0, T, X)$ where

$$H(t)(x) = r(t, x) \odot h(v(t, x)) \cdot g(u(t, x)) = r(t, x) \cdot h(v(t, x)) \odot g(u(t, x))$$

for all $(t, x) \in \mathbb{R}_+ \times \mathbb{R}^N$. Hence, since $r_n \rightharpoonup r$ in $L^2(0, T, X)$, it suffices to establish that

$$h_n(v_n) \odot g_n(u_n) \to h(v) \odot g(u) \tag{4.35}$$

strongly in $L^2(0, T, X^l)$, where X^l denotes the space $L^2(\Omega, \mathbb{R}^l)$. We have for every $t \geq 0$[5]

$$\begin{aligned} &\|h_n(v_n(t)) \odot g_n(u_n(t)) - h(v(t)) \odot g(u(t))\|_{X^l} \\ &\leq \|h_n(v_n(t)) \odot g_n(u_n(t)) - h_n(v_n(t)) \odot g(u(t))\|_{X^l} \\ &\quad + \|h_n(v_n(t)) \odot g(u(t)) - h(v(t)) \odot g(u(t))\|_{X^l} \\ &\leq \overline{h}\|g_n(u_n(t)) - g(u(t))\|_{X^l} + \overline{g}\|h_n(v_n(t)) - h(v(t))\|_{X^l} \\ &\leq \overline{h}L_{[\underline{Y}_1, \overline{Y}_1]}\|u_n(t) - u(t)\|_X + \overline{h}\|g_n(u(t)) - g(u(t))\|_{X^l} \\ &\quad + \overline{g}L'_{[\underline{Y}_2, \overline{Y}_2]}\|v_n(t) - v(t)\|_X + \overline{g}\|h_n(v(t)) - h(v(t))\|_{X^l}. \end{aligned}$$

[5]To simplify the notation we write $g(v(t))$ for the function $x \mapsto g(v(t, x))$, idem for $h(v(t))$, $g_n(v(t))$ and $h_n(v(t))$.

Hence, to prove (4.35), it remains to establish that

$$\int_0^T \|g_n(u(t)) - g(u(t))\|^2_{X^l}\, dt \to 0, \int_0^T \|h_n(v(t)) - h(v(t))\|^2_{X^l}\, dt \to 0 \quad (4.36)$$

$$\int_0^T \|u_n(t) - u(t)\|^2_X\, dt \to 0 \quad (4.37)$$

$$\int_0^T \|v_n(t) - v(t)\|^2_X\, dt \to 0. \quad (4.38)$$

The two convergences in (4.36) are a straightforward consequence of hypothesis (Hs_4) and the Lebesgue dominated convergence theorem. Convergences (4.37) and (4.38) follow from ***Step 2***, this completes the proof of Lemma 4.1. □

4.3 Convergence theorem for problems coupling r.d.e. and n.d.r.e.

We keep the notation of the previous section and assume that $\Phi_{2,n} \equiv 0$. For obtaining the compactness of $(v_n)_{n\in\mathbb{N}}$ in $C([0,T], L^2(\Omega))$ (***Step 2*** in the proof of Theorem 4.3), we can no longer invoke the equi-coercivity of Φ_2. To overcome the difficulty, we assume additional regularity conditions on the reaction functional and the initial condition for the non-diffusive equation. To simplify the notation we denote by Φ_n the functional $\Phi_{1,n}$ and by W_n the density $W_{1,n}$. The theorem below provides a convergence result for FitzHugh-Nagumo like models (see Example 4.5).

In the following, for each $i = 1, 2$, we equip the spaces $C^1([\underline{\rho}_i, \overline{\rho}_i], \mathbb{R}^l)$, $i = 1, 2$, with the sup norm defined by

$$|||\varphi|||_{[\underline{\rho}_i, \overline{\rho}_i]} := \sup_{\zeta\in[\underline{\rho}_i, \overline{\rho}_i]} |\varphi(\zeta)| + \sup_{\zeta\in[\underline{\rho}_i, \overline{\rho}_i]} \left|\frac{d\varphi}{d\zeta}(\zeta)\right|.$$

The spaces

$$C^1\left([\underline{\rho}_1, \overline{\rho}_1], \mathbb{R}^l\right) \times C^1\left([\underline{\rho}_2, \overline{\rho}_2], \mathbb{R}^l\right)$$

and

$$C^1\left([\underline{\rho}_2, \overline{\rho}_2], \mathbb{R}^l\right) \times C^1\left([\underline{\rho}_1, \overline{\rho}_1], \mathbb{R}^l\right)$$

are endowed with their product norm.

Theorem 4.4. *Assume that the sequence of densities $(W_n)_{n\in\mathbb{N}}$ of functionals Φ_n of the form* (4.14) *or* (4.15) *satisfies* (D_n). *Assume that the sequence of TCCP-structured reaction functionals $(F_{1,n}, F_{2,n})_{n\in\mathbb{N}}$ satisfies conditions (4.17), (4.18), and that $F_{1,n}$ satisfies* (4.19) *or* (4.20). *Assume furthermore that $g_{2,n}$ and $h_{2,n}$*

belong to $C^1_{\mathrm{loc}}(\mathbb{R}, \mathbb{R}^l)$, *and that* $r_{2,n}$ *and* $q_{2,n}$ *do not depend on the space variable. Let* $(u_n, v_n) \in L^2(0, T, L^2(\Omega)) \times L^2(\Omega, H^1(\Omega))$ *be the unique solution of the system*

$$(\mathcal{S}_n) \begin{cases} \dfrac{du_n}{dt}(t) + D\Phi_n(u_n(t)) = F_{1,n}(t, u_n(t), v_n(t)) \text{ for a.e. } t \in (0,T) \\ \\ \dfrac{dv_n}{dt}(t) = F_{2,n}(t, u_n(t), v_n(t)) \text{ for a.e. } t \in (0,T) \\ \\ \underline{\rho}_{1,n} \leq u_n^0 = u_n(0) \leq \overline{\rho}_{1,n},\ \underline{\rho}_{2,n} \leq v_n^0 = v_n(0) \leq \overline{\rho}_{2,n}, \\ \\ u_n^0 \in \mathrm{dom}(\Phi_n),\ v_n^0 \in H^1(\Omega). \end{cases}$$

Under conditions (Hs$_1$), (Hs$_2$), (Hs$_5$), (Hs$_6$) *for* $i = 1$ *and*

(Hs$'_1$) $u_n^0 \to u^0$ *strongly in* $L^2(\Omega)$ *and* $v_n^0 \rightharpoonup v^0$ *weakly in* $H^1(\Omega)$;

(Hs$'_2$) $(g_{1,n}, h_{1,n})$ *pointwise converges to* (g_1, h_1),
$(g_{2,n}, h_{2,n})$ *converges to* (g_2, h_2) *in* $C([\underline{\rho}_2, \overline{\rho}_2], \mathbb{R}^l) \times C([\underline{\rho}_1, \overline{\rho}_1], \mathbb{R}^l)$,
$\sup_{n\in\mathbb{N}} \|(g_{2,n}, h_{2,n})\|_{C^1([\underline{\rho}_2,\overline{\rho}_2],\mathbb{R}^l)\times C^1([\underline{\rho}_1,\overline{\rho}_1],\mathbb{R}^l)} < +\infty$,

the solution (u_n, v_n) *uniformly converges in* $C([0,T], L^2(\Omega)) \times C([0,T], L^2(\Omega))$ *to the unique solution* $(u, v) \in L^2(0, T, L^2(\Omega)) \times L^2(\Omega, H^1(\Omega))$ *of the system*

$$(\mathcal{S}) \begin{cases} \dfrac{du}{dt}(t) + \partial\Phi(u(t)) \ni F_1(t, u(t), v(t)) \text{ for a.e. } t \in (0,T) \\ \\ \dfrac{dv}{dt}(t) = F_2(t, u(t), v(t)) \text{ for a.e. } t \in (0,T) \\ \\ \inf_n \underline{\rho}_{1,n} \leq u^0 = u(0) \leq \sup_n \overline{\rho}_{1,n},\ \inf_n \underline{\rho}_{2,n} \leq v^0 = v(0) \leq \sup_n \overline{\rho}_{2,n}, \\ \\ u^0 \in \mathrm{dom}(\Phi),\ v_0 \in H^1(\Omega). \end{cases}$$

The reaction functionals $F_i : [0, +\infty) \times L^2(\Omega) \times L^2(\Omega) \to \mathbb{R}^\Omega$, $i = 1, 2$, *are defined for all* $t \in [0, T]$, *all* $(U, V) \in L^2(\Omega) \times L^2(\Omega)$ *and for a.e.* $x \in \Omega$, *by*

$$\begin{aligned} F_i(t, U, V)(x) &= f_i(t, x, U(x), V(x)), \\ f_1(t, x, \zeta, \zeta') &= r_1(t, x) \odot h_1(\zeta') \cdot g_1(\zeta) + q_1(t, x), \\ f_2(t, \zeta, \zeta') &= r_2(t) \odot h_2(\zeta) \cdot g_2(\zeta') + q_2(t). \end{aligned}$$

Moreover all the conclusions of Theorem 4.3 hold.

Proof. The arguments of the proof of Theorem 4.3 remain valid, except those of ***Step 2*** and ***Step 4*** which have to be adapted.

Step 2. Because of the default of coercivity, the proof of the relative compactness in $L^2(\Omega)$ of the set $F_t = \{v_n(t) : n \in \mathbb{N}\}$ for $t \in]0, T]$ cannot be obtained by following the same arguments. In contrast, the proof of the relative compactness of $E_t = \{u_n(t) : n \in \mathbb{N}\}$ in $L^2(\Omega)$ for $t \in]0,T]$ remains the same. By taking advantage of the structure of F_2, we are going to estimate $\sup_{n\in\mathbb{N}} \|\nabla v_n(t)\|_{L^2(\Omega)}$ by using Grönwall's lemma, and will conclude to the compactness of F_t for all $t \in]0,T]$, according to the compact embedding $H^1(\Omega) \hookrightarrow L^2(\Omega)$. Take the distributional derivative of F_2 with respect to the space variable. We obtain for every $t \in]0,T]$

$$\begin{aligned}\nabla F_2(t, u_n(t), v_n(t)) = &\left(r_{2,n}(t) \odot \frac{dh_{2,n}}{d\zeta}(u_n(t)) \cdot g_{2,n}(v_n(t))\right) \nabla u_n(t)\\ &+ \left(r_{2,n}(t) \odot h_{2,n}(u_n(t)) \cdot \frac{dg_{2,n}}{d\zeta'}(v_n(t))\right) \nabla v_n(t).\end{aligned}$$

In the following, we set for a.e. $x \in \Omega$ and for all $t \in]0,T]$:

$$\begin{aligned}A_n(t,x) &:= r_{2,n}(t) \odot \frac{dh_{2,n}}{d\zeta}(u_n(t,x)) \cdot g_{2,n}(v_n(t,x)),\\ B_n(t,x) &:= r_{2,n}(t) \odot h_{2,n}(u_n(t,x)) \cdot \frac{dg_{2,n}}{d\zeta'}(v_n(t,x)).\end{aligned}$$

From (4.21), (Hs'_2) and (Hs_5), we deduce that

$$\overline{A} = \sup_{(t,x,n)\in[0,T]\times\Omega\times\mathbb{N}} |A_n(t,x)| < +\infty$$

and

$$\overline{B} = \sup_{(t,x,n)\in[0,T]\times\Omega\times\mathbb{N}} |B_n(t,x)| < +\infty.$$

Set $V_n(t) = \nabla v_n(t)$. Take the distributional derivative with respect to the space variable of each term of the second equation of $(\mathcal{S}_n)$. From the previous calculation, we infer that V_n solves the Cauchy problem

$$\begin{cases}\dfrac{dV_n}{dt}(t) = A_n(t)\nabla u_n(t) + B_n(t)V_n(t) \text{ for a.e. } t \in (0,T),\\[2ex] V_n(0) = \nabla v_n^0,\end{cases}$$

and belongs to $C([0,T], X^N)$ where $X^N := L^2(\Omega, \mathbb{R}^N)$ (see for instance Theorem 2.4 with $F(t,V) = A_n(t)\nabla u_n(t) + B_n(t)V$, and X^N substitute for X). Hence, for all $t \in]0,T]$,

$$V_n(t) = V_n(0) + \int_0^t (A_n(s)\nabla u_n(s) + B_n(t)V_n(s))\, ds,$$

from which we deduce

$$\|V_n(t)\|_{X^N} \le \|\nabla v_n^0\|_{X^N} + \overline{A}\int_0^t \|\nabla u_n(s)\|_{X^N}\, ds + \overline{B}\int_0^t \|V_n(s)\|_{X^N}\, ds.$$

From Gronwall's lemma, we infer that for all $t \in]0,T]$ (note that $s \mapsto \|V_n(s)\|_{X^N}$ is continuous in $[0,T]$)

$$\|V_n(t)\|_{X^N} \le \left(\|\nabla v_n^0\|_{X^N} + \overline{A}\int_0^T \|\nabla u_n(s)\|_{X^N}\, ds\right)\exp(\overline{B}T).$$

From (Hs_1'), the fact that $\|\nabla u_n(t)\|_X \le C$ (see ***Step 2*** in the proof of Theorem 4.3), and from the estimate above, we infer that

$$\sup_{n\in\mathbb{N},t\in[0,T]} \|\nabla v_n(t)\|_{L^2(\Omega,\mathbb{R}^N)} < +\infty. \tag{4.39}$$

Therefore $v_n(t)$ is bounded in $H^1(\Omega)$ for all $t \in [0,T]$, which completes the proof of the relative compactness of the set F_t.

According to the Ascoli-Arzela compactness theorem, there exists a subsequence and $v \in C([0,T], L^2(\Omega))$ such that $v_n \to v$ in $C([0,T], L^2(\Omega))$. Furthermore, from (4.39), $(\nabla v_n)_{n\in\mathbb{N}}$ is bounded in $L^2(0,T,L^2(\Omega,\mathbb{R}^N))$, thus, for a subsequence, weakly converges to some Vin $L^2(0,T,L^2(\Omega,\mathbb{R}^N))$. Since $v_n \to v$ in $L^2(0,T,L^2(\Omega,\mathbb{R}^N))$, we infer that $V = \nabla v$ so that $v \in L^2(0,T,H^1(\Omega))$.

Step 4. We prove that (u,v) is solution of $(\mathcal{S})$. To establish that u solves the first equation, the proof of Theorem 4.3 remains valid. We only have to establish that v solves the second equation of $(\mathcal{S})$ which is easier: write

$$\int_0^T \left\langle \frac{dv_n}{dt}(t), \varphi(t) \right\rangle \, dt = \int_0^T \langle F_{2,n}(t,u_n(t),v_n(t)), \varphi(t)\rangle \, dt$$

for all $\varphi \in L^2(0,T,L^2(\Omega))$, then pass to the limit. □

In the specific case of a coupling between two non-diffusive reaction equations, we have the following convergence of $(\mathcal{S}_n)$ to $(\mathcal{S})$. The proof which is an easy adaptation of the proof above can be find in Anza Hafsa *et al.* (2021d), under slightly different assumptions.

Theorem 4.5 (Convergence theorem for problems coupling two n.d.r.e.). *Assume that for each $i = 1,2$, the sequence of TCCP-structured reaction functionals $(F_{i,n})_{n\in\mathbb{N}}$ satisfies conditions (4.17), (4.18), that g_i and h_i belong to $C^1_{loc}(\mathbb{R},\mathbb{R}^l)$, and that $r_{i,n}$ and $q_{i,n}$ do not depend on the space variable. Let (u_n, v_n) be the unique solution of the system of n.d.r.e.*

$$(\mathcal{S}_n)\begin{cases} \dfrac{du_n}{dt}(t) = F_{1,n}(t,u_n(t),v_n(t)) \text{ for a.e. } t \in (0,T) \\[2ex] \dfrac{dv_n}{dt}(t) = F_{2,n}(t,u_n(t),v_n(t)) \text{ for a.e. } t \in (0,T) \\[2ex] \underline{\rho}_{1,n} \le u_n^0 = u_n(0) \le \overline{\rho}_{1,n},\ \underline{\rho}_{2,n} \le v_n^0 = v_n(0) \le \overline{\rho}_{2,n}, \\[2ex] u_n^0 \in H^1(\Omega),\ v_n^0 \in H^1(\Omega). \end{cases}$$

Assume (Hs_5), (Hs_6), *and*

(Hs_1'') *$u_n^0 \rightharpoonup u^0$ and $v_n^0 \rightharpoonup v^0$ weakly in $H^1(\Omega)$;*

(Hs$_2''$) *for* $i = 1,2$, $(g_{i,n}, h_{i,n})$ *converges to* (g_i, h_i) *in* $C^1\left([\underline{\rho}_1, \overline{\rho}_1], \mathbb{R}^l\right) \times C^1\left([\underline{\rho}_2, \overline{\rho}_2], \mathbb{R}^l\right)$ *(*$i=1$*)*, $C^1\left([\underline{\rho}_2, \overline{\rho}_2], \mathbb{R}^l\right) \times C^1\left([\underline{\rho}_1, \overline{\rho}_1], \mathbb{R}^l\right)$ *(*$i=2$*), and*

$$\sup_{n\in\mathbb{N}} \|(g_{1,n}, h_{1,n})\|_{C^1([\underline{\rho}_1,\overline{\rho}_1],\mathbb{R}^l)\times C^1([\underline{\rho}_2,\overline{\rho}_2],\mathbb{R}^l)} < +\infty,$$

$$\sup_{n\in\mathbb{N}} \|(g_{2,n}, h_{2,n})\|_{C^1([\underline{\rho}_2,\overline{\rho}_2],\mathbb{R}^l)\times C^1([\underline{\rho}_1,\overline{\rho}_1],\mathbb{R}^l)} < +\infty.$$

Then (u_n, v_n) *uniformly converges in* $C([0,T], L^2(\Omega)) \times C([0,T], L^2(\Omega))$ *to the unique solution* (u, v) *of the system*

$$(\mathcal{S})\begin{cases} \dfrac{du}{dt}(t) = F_1(t, u(t), v(t)) \text{ for a.e. } t \in (0,T) \\[2ex] \dfrac{dv}{dt}(t) = F_2(t, u(t), v(t)) \text{ for a.e. } t \in (0,T) \\[2ex] \inf_n \underline{\rho}_{1,n} \leq u^0 = u(0) \leq \sup_n \overline{\rho}_{1,n},\ \inf_n \underline{\rho}_{2,n} \leq v^0 = v(0) \leq \sup_n \overline{\rho}_{2,n}, \\[2ex] u^0 \in H^1(\Omega),\ v_0 \in H^1(\Omega). \end{cases}$$

The reaction functionals $F_i : [0,+\infty) \times L^2(\Omega) \times L^2(\Omega) \to \mathbb{R}^\Omega$, $i = 1,2$, *are defined for all* $t \in [0,T]$, *all* $(U,V) \in L^2(\Omega) \times L^2(\Omega)$, *for a.e.* $x \in \Omega$ *and for all* $(\zeta, \zeta') \in \mathbb{R}$ *by*

$$\begin{aligned} F_i(t, U, V)(x) &= f_i(t, x, U(x), V(x)), \\ f_1(t, \zeta, \zeta') &= r_1(t) \odot h_1(\zeta') \cdot g_1(\zeta) + q_1(x), \\ f_2(t, \zeta, \zeta') &= r_2(t) \odot h_2(\zeta) \cdot g_2(\zeta') + q_2(t). \end{aligned}$$

Moreover all the conclusions of Theorem 4.3 hold.

4.4 Proofs of Propositions 4.1–4.5

4.4.1 *Proof of Proposition 4.1*

Recall that for all $(t, x, \zeta, \zeta') \in \mathbb{R}_+ \times \mathbb{R}^N \times \mathbb{R} \times \mathbb{R}$,

$$\begin{aligned} f_1(t, x, \zeta, \zeta') &= \alpha_1(t,x)\zeta\left(1 - \frac{\zeta}{K_1(t,x)} - a_{1,2}\frac{\zeta'}{K_1(t,x)}\right) \\ f_2(t, x, \zeta, \zeta') &= \alpha_2(t,x)\zeta'\left(1 - \frac{\zeta'}{K_2(t,x)} - a_{2,1}\frac{\zeta}{K_2(t,x)}\right). \end{aligned}$$

Clearly each function f_i satisfies the structure condition of TCCP-structured reaction functions with $l = 3$ and for all $(t, x, \zeta, \zeta') \in \mathbb{R}_+ \times \mathbb{R}^N \times \mathbb{R} \times \mathbb{R}$,

$$\begin{aligned} &r_i(t,x) = \left(\alpha_i(t,x), -\frac{\alpha_i(t,x)}{K_i(t,x)}, -a_{i,\overline{i+1}}\frac{\alpha_i(t,x)}{K_i(t,x)}\right) \quad (\overline{i+1} \in \mathbb{Z}/2\mathbb{Z}), \\ &h_1(\zeta') = (1, 1, \zeta'),\ h_2(\zeta) = (1, 1, \zeta), \\ &g_1(\zeta) = (\zeta, \zeta^2, \zeta),\ g_2(\zeta') = (\zeta', \zeta'^2, \zeta'). \end{aligned}$$

Let us show that condition (TCCP) is fulfilled. From now on, all the estimates below hold for all $t \in \mathbb{R}_+$ and all $x \in \mathbb{R}^N$. For each $i = 1, 2$, take $\underline{f}_i = 0$, and $\underline{\rho}_i = 0$, then $\underline{y}_i = 0$ and $f_1\left(t, x, \underline{y}_1(t), \zeta'\right) = 0 \geq \underline{f}_1\left(t, \underline{y}_1(t)\right)$ for all $\zeta' \in \mathbb{R}$. Similarly $f_2\left(t, x, \zeta, \underline{y}_2(t)\right) = 0 \geq \underline{f}_1\left(t, \underline{y}_2(t)\right)$ for all $\zeta \in \mathbb{R}$.

On the other hand take $\overline{f}_1(t, \zeta) = \overline{\alpha_1}\zeta$, where $\overline{\rho}_1 > 0$ is any real number. Similarly $\overline{f}_2(t, \zeta') = \overline{\alpha_2}\zeta'$, where $\overline{\rho}_2 > 0$ is any real number. Then for each $i = 1, 2$, $\overline{y}_i(t) = \overline{\rho}_i \exp(\overline{\alpha}_i t)$ and $f_1(t, x, \overline{y}_1(t), \zeta') \leq \overline{\alpha}_1\overline{y}_1(t) = \overline{f}_1(t, \overline{y}_1(t))$ for all $\zeta' \geq 0$. Similarly $f_2(t, x, \zeta, \overline{y}_2(t)) \leq \overline{\alpha}_2\overline{y}_2(t) = \overline{f}_2(t, \overline{y}_2(t))$ for all $\zeta \geq 0$. This proves the claim since $[\underline{y}_i(T), \overline{y}_i(T)] = [0, \overline{\rho}_i \exp(\overline{\alpha}_i T)]$ for $i = 1, 2$.

4.4.2 *Proof of Proposition 4.2*

Recall that for all $(t, x, \zeta, \zeta') \in \mathbb{R}_+ \times \mathbb{R}^N \times \mathbb{R} \times \mathbb{R}$,

$$f_1(t, x, \zeta, \zeta') = \alpha_1(t, x)\zeta\left(1 - \frac{\zeta}{K_1(t, x)} + b_{1,2}\frac{\zeta'}{K_1(t, x)}\right)$$
$$f_2(t, x, \zeta, \zeta') = \alpha_2(t, x)\zeta'\left(1 - \frac{\zeta'}{K_2(t, x)} + b_{2,1}\frac{\zeta}{K_2(t, x)}\right).$$

As in the previous example, each function f_i satisfies the structure condition of TCCP-structured reaction functions with $l = 3$ and for all (t, x, ζ, ζ'),

$$r_i(t, x) = \left(\alpha_i(t, x), -\frac{\alpha_i(t, x)}{K_i(t, x)}, b_{i,\overline{i+1}}\frac{\alpha_i(t, x)}{K_i(t, x)}\right) \cdot \quad (\overline{i+1} \in \mathbb{Z}/2\mathbb{Z}),$$
$$h_1(\zeta') = (1, 1, \zeta'),\ h_2(\zeta) = (1, 1, \zeta),$$
$$g_1(\zeta) = (\zeta, \zeta^2, \zeta),\ g_2(\zeta') = (\zeta', \zeta'^2, \zeta').$$

Let us show that condition (TCCP) is fulfilled. For each $i = 1, 2$ take $\underline{f}_i = 0$, and $\underline{\rho}_i = 0$, then $\underline{y}_i = 0$ and $f_1\left(t, x, \underline{y}_1(t), \zeta'\right) = 0 \geq \underline{f}_1\left(t, \underline{y}_1(t)\right)$ for all $\zeta' \in \mathbb{R}$. Similarly $f_2\left(t, x, \zeta, \underline{y}_2(t)\right) = 0 \geq \underline{f}_1\left(t, \underline{y}_2(t)\right)$ for all $\zeta \in \mathbb{R}$.

Because of the signs positive preceding the b's, we cannot proceed as in the previous example for completing the condition (TCCP). We take $\overline{f}_i = 0$ and look for $\overline{y}_i$ in the form of constants $\overline{\rho}_i > 0$. For all $\zeta' \in [0, \overline{\rho}_2]$ and for all $(t, x) \in \mathbb{R}_+ \times \mathbb{R}^N$ we have

$$\begin{aligned} f_1(t, x, \overline{\rho}_1, \zeta') &= \alpha_1(t, x)\overline{\rho}_1\left(1 - \frac{\overline{\rho}_1}{K_1(t, x)} + b_{1,2}\frac{\zeta'}{K_1(t, x)}\right) \\ &\leq \alpha_1(t, x)\overline{\rho}_1\left(1 - \frac{\overline{\rho}_1}{\overline{K}_1} + b_{1,2}\frac{\overline{\rho}_2}{\underline{K}_1}\right). \end{aligned} \tag{4.40}$$

Take $\overline{\rho}_1$ and $\overline{\rho}_2$ positive, satisfying

$$1 - \frac{\overline{\rho}_1}{\overline{K}_1} + b_{1,2}\frac{\overline{\rho}_2}{\underline{K}_1} \leq 0. \tag{4.41}$$

With this choice, from (4.40) and (4.41), we infer that for all $\zeta' \in [0, \overline{\rho}_2]$

$$f_1(t, x, \overline{y}_1(t), \zeta') \leq 0 = \overline{f}_1(t, \overline{y}_1(t)).$$

Similarly, with $\overline{\rho}_1$ and $\overline{\rho}_2$ positive, satisfying

$$1 - \frac{\overline{\rho}_2}{\overline{K}_2} + b_{2,1}\frac{\overline{\rho}_1}{\underline{K}_2} \leq 0, \tag{4.42}$$

for all $\zeta \in [0, \overline{\rho}_1]$, we have

$$f_2(t, x, \zeta, \overline{\rho}_2) \leq 0 = \overline{f}_2(t, \overline{y}_2(t)).$$

Therefore, $\overline{\rho}_i > 0$, $i = 1, 2$, satisfying the system of two inequalities (4.41) and (4.42), i.e.

$$\overline{\rho}_1 \geq \frac{\underline{K}_1\overline{K}_1}{\underline{K}_1 - b_{1,2}\overline{K}_1},$$

$$\overline{\rho}_2 \geq \frac{\underline{K}_2\overline{K}_2}{\underline{K}_2 - b_{2,1}\overline{K}_2},$$

are suitable for (TCCP) to be fulfilled. It is easy to see that $\overline{\rho}_1 = \overline{\rho}_2 \geq \max\left(\frac{\underline{K}_1\overline{K}_1}{\underline{K}_1 - b_{1,2}\overline{K}_1}, \frac{\underline{K}_2\overline{K}_2}{\underline{K}_2 - b_{2,1}\overline{K}_2}\right)$ is a solution provided that $\frac{\underline{K}_1\overline{K}_1}{\underline{K}_1 - b_{1,2}\overline{K}_1} > 0$ and $\frac{\underline{K}_2\overline{K}_2}{\underline{K}_2 - b_{2,1}\overline{K}_2} > 0$. These two conditions are ensured by (4.4).

4.4.3 *Proof of Proposition 4.3*

Recall that for all $(t, x, \zeta, \zeta') \in \mathbb{R}_+ \times \mathbb{R}^N \times \mathbb{R} \times \mathbb{R}$,

$$f_1(t, x, \zeta, \zeta') = \alpha_1(t, x)\zeta\left(1 - \frac{\zeta}{K_{car}(t, x)}\right) - a(x, t)\zeta'(1 - \exp(-b\zeta))$$

$$f_2(t, x, \zeta, \zeta') = \alpha_2(t, x)\zeta'\left(1 - c\frac{\zeta'}{\zeta}\right).$$

Fix $\delta := \underline{\rho}_1 > 0$ satisfying $0 < \underline{\rho}_1 < \underline{K}$, chosen later. Set for all $(t, x, \zeta, \zeta') \in \mathbb{R}_+ \times \mathbb{R}^N \times \mathbb{R} \times \mathbb{R}$

$$f_{2,\delta}(t, x, \zeta, \zeta') = \begin{cases} f_2(t, x, \zeta, \zeta') \text{ if } \zeta \geq \delta \\ f_2(t, x, \delta, \zeta') \text{ if } \zeta < \delta. \end{cases}$$

At the end of the proof, we show that $f_{2,\delta} = f_2$ for ζ and ζ' in suitable intervals. We claim that the pair $(f_1, f_{2,\delta})$ satisfies the structure condition of TCCP-structured reaction functions with $l = 3$: indeed take

$$r_1(t, x) = \left(\alpha_1(t, x), \frac{\alpha_1(t, x)}{K_{car}(t, x)}, -a(t, x)\right),$$
$$h_1(\zeta') = (1, 1, \zeta'),$$
$$g_1(\zeta) = (\zeta, -\zeta^2, 1 - \exp(-b\zeta));$$

and

$$r_2(t, x) = (\alpha_2(t, x), -c\alpha_2(t, x), 0),$$

$$h_{2,\delta}(\zeta) = \begin{cases} \left(1, \frac{1}{\zeta}, 0\right) \text{ if } \zeta \geq \delta \\ \left(1, \frac{1}{\delta}, 0\right) \text{ if } \zeta < \delta, \end{cases}$$

$$g_2(\zeta') = (\zeta', \zeta'^2, 0)$$

for all $(t,x,\zeta,\zeta') \in \mathbb{R}_+ \times \mathbb{R}^N \times \mathbb{R} \times \mathbb{R}$. It remains to show that $(f_1, f_{2,\delta})$ fulfills condition (TCCP). First take $\underline{\rho}_2 = 0$, $\underline{f}_{2,\delta} = 0$, then $\underline{y}_2 = \underline{\rho}_2$ and $f_{2,\delta}\left(t,x,\zeta,\underline{y}_2(t)\right) = 0 \geq \underline{f}_{2,\delta}\left(t,\underline{y}_2(t)\right)$ for all $\zeta \in \mathbb{R}$. To complete condition (TCCP), we look for

$$\begin{cases} \overline{\rho}_2 > 0,\ \overline{f}_{2,\delta} = 0,\ \overline{y}_2 = \overline{\rho}_2; \\ \underline{\rho}_1,\ \underline{f}_1 = 0,\ \underline{y}_1 = \underline{\rho}_1; \\ \overline{\rho}_1 > \underline{\rho}_1,\ \overline{f}_1(t,\xi) = \overline{\alpha}_1\xi,\ \overline{y}_1(t) = \overline{\rho}_1\exp(\overline{\alpha}_1 t). \end{cases}$$

We first look for $\overline{\rho}_2$ satisfying

$$\begin{aligned} f_{2,\delta}(t,x,\zeta,\overline{\rho}_2) &\leq \overline{\rho}_2\left(\overline{\alpha}_2 - c\underline{\alpha}_2\frac{\overline{\rho}_2}{\overline{\rho}_1}\right) \\ &\leq 0 = \overline{f}_2(t,\overline{\rho}_2), \end{aligned}$$

for all $\zeta \geq \underline{\rho}_1$, which furnishes the first condition:

$$\frac{\overline{\alpha}_2}{c\underline{\alpha}_2}\overline{\rho}_1 \leq \overline{\rho}_2. \tag{4.43}$$

Secondly, we look for $\underline{\rho}_1$, $0 < \underline{\rho}_1 < \underline{K}$, satisfying

$$\begin{aligned} f_1\left(t,x,\underline{\rho}_1,\zeta'\right) &= \alpha_1(t,x)\underline{\rho}_1\left(1 - \frac{\underline{\rho}_1}{K_{car}(t,x)}\right) - a(x,t)\zeta'\left(1 - \exp\left(-b\underline{\rho}_1\right)\right) \\ &\geq \underline{\alpha}_1\underline{\rho}_1\left(1 - \frac{\underline{\rho}_1}{\underline{K}}\right) - \overline{a}\zeta' \geq 0 \end{aligned}$$

for all $\zeta' \in [0,\overline{\rho}_2]$, which requires the following second condition

$$\overline{\rho}_2 \leq \frac{\underline{\alpha}_1}{\overline{a}}\underline{\rho}_1\left(1 - \frac{\underline{\rho}_1}{\underline{K}}\right). \tag{4.44}$$

Combining (4.43) and (4.44), we infer that the choice of $\overline{\rho}_2$, $\overline{\rho}_1$ and $\underline{\rho}_1$ is conditioned by the following inequality

$$\frac{\overline{\alpha}_2}{c\underline{\alpha}_2}\overline{\rho}_1 \leq \frac{\underline{\alpha}_1}{\overline{a}}\underline{\rho}_1\left(1 - \frac{\underline{\rho}_1}{\underline{K}}\right),$$

or equivalently

$$\overline{\rho}_1 \leq \mu_{ext}\ \underline{\rho}_1\left(1 - \frac{\underline{\rho}_1}{\underline{K}}\right), \tag{4.45}$$

where $\mu_{ext} := c\frac{\underline{\alpha}_1\underline{\alpha}_2}{\overline{a}\overline{\alpha}_2}$. Hence, we can choose $\overline{\rho}_2$ with

$$\overline{\rho}_2 \in \left[\frac{\overline{\alpha}_2}{c\underline{\alpha}_2}\overline{\rho}_1, \frac{\underline{\alpha}_1}{a}\underline{\rho}_1\left(1 - \frac{\underline{\rho}_1}{\underline{K}}\right)\right].$$

which is the last condition in (4.5).

Set $\overline{\rho}_1 = \theta \underline{\rho}_1$ where $\theta \geq \frac{\underline{K}}{\underline{\rho}_1}$ $\left(\text{recall that by the choice of } \underline{\rho}_1, \text{ we have } \frac{\underline{K}}{\underline{\rho}_1} > 1\right)$. Then (4.45) and the previous condition on θ are equivalent to

$$\frac{\underline{K}}{\underline{\rho}_1} \leq \theta \leq \mu_{ext}\left(1 - \frac{\underline{\rho}_1}{\underline{K}}\right),$$

so that the possible choice of θ and $0 < \underline{\rho}_1 \leq \underline{K}$ is governed by

$$\frac{1}{\mu_{ext}} \leq \frac{\underline{\rho}_1}{\underline{K}}\left(1 - \frac{\underline{\rho}_1}{\underline{K}}\right) \tag{4.46}$$

Since $\mu_{ext} \geq 4$, condition (4.46) is fulfilled by any

$$\underline{\rho}_1 \in \left[\underline{K}\frac{1 - \sqrt{1 - \frac{4}{\mu_{ext}}}}{2}, \underline{K}\frac{1 + \sqrt{1 - \frac{4}{\mu_{ext}}}}{2}\right]$$

which is the first condition in (4.5). The choice of θ is then given by

$$\theta \in \left[\frac{\underline{K}}{\underline{\rho}_1}, \mu_{ext}\left(1 - \frac{\underline{\rho}_1}{\underline{K}}\right)\right]. \tag{4.47}$$

which is the second condition in (4.5).

Since $f_1(t, x, \zeta, \zeta') \leq \overline{\alpha}_1 \zeta$ for all $\zeta' \geq 0$, with our choice of $\overline{f}_1$ and $\overline{\rho}_1 = \theta \underline{\rho}_1$, we have

$$f_1(t, x, \overline{y}_1(t), \zeta') \leq \overline{\alpha}_1 \overline{y}_1(t) = \overline{f}_1(t, \overline{y}_1(t))$$

for all $\zeta' \in [\underline{y}_2(T) = 0, \overline{\rho}_2 = \overline{y}_2(T)]$ and all $(t, x) \in \mathbb{R}_+ \times \mathbb{R}^N$. Note that $f_{2,\delta} = f_2$ for $(\zeta, \zeta') \in [\underline{\rho}_1, \overline{\rho}_1 \exp(\overline{\alpha}_1 t)] \times [0, \overline{\rho}_2]$. This ends the proof.

4.4.4 *Proof of Proposition 4.4*

Recall that for all $(t, x, \zeta, \zeta') \in \mathbb{R}_+ \times \mathbb{R}^N \times \mathbb{R} \times \mathbb{R}$,

$$f_1(t, x, \zeta, \zeta') = -\alpha_1(t, x)\zeta^p f_0(\zeta')$$
$$f_2(t, x, \zeta, \zeta') = \alpha_2(t, x)\zeta^p f_0(\zeta')$$

where

$$f_0(\zeta') = \begin{cases} \exp\left(\gamma - \frac{\gamma}{\zeta'}\right) & \text{if } \zeta' > 0 \\ 0 & \text{otherwise.} \end{cases}$$

Clearly the pair (f_1, f_2) satisfies the structure condition of TCCP-structured reaction functions with $l = 1$. Let us show that (f_1, f_2) fulfills condition (TCCP). From the fact that $f_1(t, x, 0, \zeta') = f_2(t, x, \zeta, 0) = 0$ we see that $\underline{f}_1 = \underline{f}_2 = 0$ and $\underline{\rho}_1 = \underline{\rho}_2 = 0$, $\underline{y}_1 = \underline{y}_2 = 0$ are suitable. Take $\overline{\rho}_1 > 0$ arbitrary and $\overline{f}_1 = 0$, $\overline{y}_1 = \overline{\rho}_1$. We have

$$f_1(t, x, \overline{y}_1(t), \zeta') \leq 0 = \overline{f}_1(t, \overline{y}_1(t))$$

for all $\zeta' \geq 0$. Finally, from inequality

$$\begin{aligned} f_2(t,x,\zeta,\zeta') &= \alpha_2(t,x)\zeta^p \exp\left(\gamma - \frac{\gamma}{\zeta'}\right) \\ &\leq \overline{\alpha}_2\overline{\rho}_1^p \exp(\gamma) \end{aligned}$$

fulfilled for all $\zeta' > 0$ and all $\zeta \in [0, \overline{y}_1(T) = \overline{\rho}_1]$, we deduce that the constant function $\overline{f}_2(t,\zeta') = \overline{\alpha}_2\overline{\rho}_1^p \exp(\gamma)$, and the affine function $\overline{y}_2(t) = \overline{\alpha}_2\overline{\rho}_1^p t \exp(\gamma) + \overline{\rho}_2$, with $\overline{\rho}_2 > 0$ arbitrary chosen, are suitable to complete condition (TCCP).

4.4.5 *Proof of Proposition 4.5*

Recall that for all $(t,x,\zeta,\zeta') \in \mathbb{R}_+ \times \mathbb{R}^N \times \mathbb{R} \times \mathbb{R}$,

$$\begin{aligned} f_1(t,x,\zeta,\zeta') &= \alpha_1(t,x)\zeta(\zeta - a(t,x))(1-\zeta) - b(t,x)\zeta' \\ f_2(t,x,\zeta,\zeta') &= \alpha_2(t,x)\zeta - c(t,x)\zeta'. \end{aligned}$$

The pair (f_1, f_2) satisfies the structure condition of TCCP-structured reaction functions with $l = 3$: indeed, take

$$\begin{aligned} r_1(t,x) &= (\alpha_1(t,x), -\alpha_1(t,x)a(t,x), -b(t,x)), \\ h_1(\zeta') &= (1, 1, \zeta'), \\ g_1(\zeta) &= (\zeta^2(1-\zeta), \zeta(1-\zeta), 1); \end{aligned}$$

and

$$\begin{aligned} r_2(t,x) &= (\alpha_2(t,x), -c(t,x), 0), \\ h_2(\zeta) &= (\zeta, 1, 0), \\ g_2(\zeta') &= (1, \zeta', 0). \end{aligned}$$

Let us show that (f_1, f_2) fulfills condition (TCCP) with $\underline{\rho}_i$, $\overline{\rho}_i$, and $\underline{f}_i$, $\overline{f}_i$ given by (4.6). For each $i = 1, 2$, we take $\underline{y}_i = \underline{\rho}_i$, $\overline{y}_i = \overline{\rho}_i$ and $\underline{\rho}_i$, $\overline{\rho}_i$ must satisfy for all $(\zeta, \zeta') \in \left[\underline{\rho}_1, \overline{\rho}_1\right] \times \left[\underline{\rho}_2, \overline{\rho}_2\right]$,

$$f_1\left(t, x, \underline{\rho}_1, \zeta'\right) \geq 0, \tag{4.48}$$

$$f_1\left(t, x, \overline{\rho}_1, \zeta'\right) \leq 0, \tag{4.49}$$

$$f_2\left(t, x, \zeta, \underline{\rho}_2\right) \geq 0, \tag{4.50}$$

$$f_2\left(t, x, \zeta, \overline{\rho}_2\right) \leq 0. \tag{4.51}$$

We look for $\underline{\rho}_i \leq 0$ for $i = 1, 2$, $\overline{\rho}_1 \geq 1$, and $\overline{\rho}_2 > 0$. Let $(t,x) \in \mathbb{R}^+ \times \mathbb{R}^N$. For all $\zeta \in \left[\underline{\rho}_1, \overline{\rho}_1\right]$ we have

$$\begin{aligned} f_2\left(t, x, \zeta, \underline{\rho}_2\right) &= \alpha_2\zeta - c\underline{\rho}_2 \\ &\geq \overline{\alpha}_2\underline{\rho}_1 - \underline{c}\underline{\rho}_2, \end{aligned}$$

then, for obtaining (4.50), it suffices to set

$$\underline{\rho}_2 = \frac{\overline{\alpha}_2}{\underline{c}} \underline{\rho}_1.$$

Similarly, for all $\zeta \in [\underline{\rho}_1, \overline{\rho}_1]$ we have

$$\begin{aligned} f_2(t, x, \zeta, \overline{\rho}_2) &= \alpha_2 \zeta - c \overline{\rho}_2 \\ &\leq \overline{\alpha}_2 \overline{\rho}_1 - \underline{c} \overline{\rho}_2, \end{aligned}$$

so that, for obtaining (4.51), it suffices to set

$$\overline{\rho}_2 = \frac{\overline{\alpha}_2}{\underline{c}} \overline{\rho}_1.$$

On the other hand, for all $\zeta' \in \left[\underline{\rho}_2, \overline{\rho}_2\right] = \left[\frac{\overline{\alpha}_2}{\underline{c}} \underline{\rho}_1, \frac{\overline{\alpha}_2}{\underline{c}} \overline{\rho}_1\right]$, we have (recall that $\overline{\rho}_1 \geq 1 > a$)

$$\begin{aligned} f_1(t, x, \overline{\rho}_1, \zeta') &= \alpha_1 \overline{\rho}_1 (\overline{\rho}_1 - a)(1 - \overline{\rho}_1) - b\zeta' \\ &\leq \underline{\alpha}_1 \overline{\rho}_1 (\overline{\rho}_1 - a)(1 - \overline{\rho}_1) - \underline{b} \frac{\overline{\alpha}_2}{\underline{c}} \underline{\rho}_1 \end{aligned}$$

so that, for obtaining (4.49), it suffices to take $\underline{\rho}_1$ and $\overline{\rho}_1$ satisfying

$$\overline{\rho}_1 (\overline{\rho}_1 - a)(1 - \overline{\rho}_1) \leq \underline{\gamma}\, \underline{\rho}_1. \tag{4.52}$$

By a similar calculation, for obtaining (4.48), it suffices to take $\underline{\rho}_1$ and $\overline{\rho}_1$ satisfying

$$\underline{\rho}_1 \left(\underline{\rho}_1 - a\right) \left(1 - \underline{\rho}_1\right) \geq \overline{\gamma}\, \overline{\rho}_1. \tag{4.53}$$

Set $\underline{\rho}_1 = -\underline{\gamma} \overline{\rho}_1$. Then (4.52) and (4.53) yield

$$\begin{cases} (\overline{\rho}_1 - a)(\overline{\rho}_1 - 1) \geq \underline{\gamma}^2; \\ \\ \left(\underline{\rho}_1 - a\right)\left(\underline{\rho}_1 - 1\right) \geq \frac{\overline{\gamma}}{\underline{\gamma}}. \end{cases}$$

The first inequality is fulfilled for $\overline{\rho}_1 \geq \underline{\gamma} + 1$, the second for $\underline{\rho}_1 \leq a - \sqrt{\frac{\overline{\gamma}}{\underline{\gamma}}}$, i.e. for $\overline{\rho}_1 \geq \frac{\sqrt{\frac{\overline{\gamma}}{\underline{\gamma}}} - a}{\underline{\gamma}}$. Hence (4.48) and (4.49) are satisfied for $\overline{\rho}_1$ satisfying the first condition in (4.6).

Chapter 5

Variational convergence of integrodifferential reaction-diffusion equations

In the spirit of Chapter 2, under suitable variational convergences on the classes of functionals Φ and Ψ, we investigate the stability in terms of convergence of integrodifferential reaction-diffusion problems defined in $L^2(0,T,X)$ by

$$(\mathcal{P})\begin{cases} \dfrac{du}{dt}(t) + \partial\Phi(u(t)) + \displaystyle\int_0^t K(t-s)\partial\Psi(u(s))ds \ni F(t,u(t))\text{a.e. } t \in (0,T) \\ \\ u(0) = u^0 \in \text{dom}(\partial\Phi). \end{cases}$$

The functionals $\Phi, \Psi : X \to \mathbb{R} \cup \{+\infty\}$ are proper, convex, lower semicontinuous, and their domain is a subspace V compactly embedded in a Hilbert space X. Unlike the previous chapters, we treat the problem in an abstract framework where Φ, Ψ are not necessarily integral functionals of the calculus of variations. The integral in the first member is taken in the sense of Bochner. The class of functionals Φ is equipped with the Mosco-convergence and the class of the restrictions to V of functionals Ψ, with the sequential Γ-convergence associated with the weak topology of V. When dealing with concrete integral functionals Φ and Ψ of the Calculus of Variations, problems of type $(\mathcal{P})$ arise from the conservation of mass when the flux is splitted into two terms: the Fickian flux whose divergence is $\partial\Phi(u(t,\cdot))$ and the non Fickian flux which takes time memory effects into account, and whose divergence is $\int_0^t K(t-s)\partial\Psi(u(s))ds$; we refer the reader to Chapter 10 for various examples in the framework of stochastic homogenization.

We treat the problem with smooth kernels K and in a strong formulation: we say that u is a solution of $(\mathcal{P})$ if $u \in L^2([0,T],X)$ is absolutely continuous in time and satisfies $(\mathcal{P})$. Well posedness in the sense of existence of solutions has been extensively studied using maximal monotone operator techniques under a condition involving the scalar product of $\partial\Phi$ with the Yosida approximation of $\partial\Psi$, not easy to handle, even for elementary examples; see (Crandall *et al.*, 1978, Example 2). We refer the reader to the pioneers Barbu (1976), Barbu and Malik (1979), Crandall *et al.* (1978), Rennolet (1979) and references therein when F is a source without reaction term. In Theorems 5.2, 5.3 we establish existence of a strong solution of $(\mathcal{P})$ with a right derivative at each $t \in [0,T[$ under a coerciveness condition on

$\partial\Phi \circ (\partial\Psi)^{-1}$, which is more stable for the product of the Mosco-convergence with the Γ-convergence.

For existence of solutions of nonlinear Volterra integrodifferental equations under a weak formulation but with nonsmooth kernels refer to Grasselli and Lorenzi (1991). For existence results related integrodifferential equations of noncovolution type or for integrodifferential equations whose source includes a delay term refer to Chang (2007), Chang *et al.* (2010), Rennolet (1979). For recent developments in non-Fickian diffusion and its applications to viscoelastic materials, refer to Gudino Rojas (2014) and references therein.

The main result of this chapter is stated in Theorem 5.4 where we establish the convergence of sequences in the class of problems $(\mathcal{P})$, when the classes of functionals Φ and Ψ are equipped with the two variational convergences mentioned above. This theorem can be seen as a compactness result for the class of problems $(\mathcal{P})$. In the concrete case when $X = L^2(\Omega)$, we extend the convergence to reaction-diffusion problems (see Theorem 5.5).

5.1 The general analysis framework

As in the previous chapters X denotes a Hilbert space endowed with a scalar product denoted by $\langle \cdot, \cdot \rangle$ and its associated norm $\|\cdot\|_X$. In Section 5.4.2, $X = L^2(\Omega)$ where Ω is a C^1-regular domain in $\mathbb{R}^N$. We denote by V a reflexive Banach space compactly embedded in X, by $\|\cdot\|_V$ its norm, and by V' its dual. Therefore, for $T > 0$ fixed all along the chapter, the following compact embeddings hold:

$$V \hookrightarrow X \hookrightarrow V'$$

$$L^2(0,T,V) \hookrightarrow L^2(0,T,X) \hookrightarrow L^2(0,T,V')$$

where X and $L^2(0,T,X)$ are identified with their duals. For a proof and general results concerning compact embeddings for vector valued spaces, we refer the reader to Amann (2000) and references therein. We assume that $\langle u, v\rangle_{V',V} = \langle u, v\rangle$ whenever $u \in X$ and $v \in V$.

5.1.1 *Structure of the first member of* $(\mathcal{P})$

We are given two lower-semicontinuous (lsc in short) convex proper functionals $\Phi, \Psi : X \to]-\infty, +\infty]$ with domain V, satisfying $\inf_X \Phi > -\infty$, $\inf_X \Psi > -\infty$, and such that $\mathrm{dom}(\partial\Phi) \subset \mathrm{dom}(\partial\Psi)$. We denote by $\Psi_{\lfloor V}$ the restriction of Ψ to V.

Lemma 5.1. *The subdifferentials of* $\Psi_{\lfloor V}$ *and* Ψ *are connected through the relation*

$$\partial\Psi = \partial(\Psi_{\lfloor V}) \cap X.$$

Proof. Inclusion $\partial\Psi \subset \partial(\Psi_{\lfloor V}) \cap X$ is obvious. The converse inclusion $\partial(\Psi_{\lfloor V}) \cap X \subset \partial\Psi$ is checked as follows: take $u^* \in \partial(\Psi_{\lfloor V}(u)) \cap X$, then, from Definition B.5, $u^* \in X$, $u \in V$, and

$$\Psi_{\lfloor V}(u) \leq \langle u^*, u - v\rangle_{V',V} + \Psi_{\lfloor V}(v) \quad \forall v \in V,$$

that is, since $u^* \in X$, and $(u,v) \in V \times V$,

$$\Psi(u) \leq \langle u^*, u-v\rangle + \Psi(v) \quad \forall v \in V.$$

Since $\mathrm{dom}(\Psi) = V$, the right member is $+\infty$ for $v \in X \setminus V$, so that inequality

$$\Psi(u) \leq \langle u^*, u-v\rangle + \Psi(v)$$

still holds for all $v \in X$. Hence $u^* \in \partial\Psi(u)$. □

When there is no ambiguity, to simplify the notation, for $u \in \mathrm{dom}(\partial\Phi)$ we write $\partial\Phi(u)$ and $\partial\Psi(u)$ to denote any element of the sets $\partial\Phi(u)$ and $\partial\Psi(u)$ respectively. We assume that $0 \in \mathrm{dom}(\partial\Psi)$ with $\partial\Psi(0) = \{0\}$.

The functional $\Psi_{\lfloor V}$ is assumed to be *strongly convex* in the following sense: for all $(u,v) \in \mathrm{dom}(\partial(\Psi_{\lfloor V}))^2$,

$$\langle \partial(\Psi_{\lfloor V})(u) - \partial(\Psi_{\lfloor V})(v), u-v\rangle_{V',V} \geq \alpha_\Psi \|u-v\|_V^2. \tag{5.1}$$

Note that when $(u,v) \in \mathrm{dom}(\partial\Psi)^2$, then (5.1) yields

$$\langle \partial\Psi(u) - \partial\Psi(v), u-v\rangle \geq \alpha_\Psi \|u-v\|_V^2,$$

hence, since $\partial\Psi(0) = \{0\}$, for all $u \in \mathrm{dom}(\partial\Psi)$, we infer that

$$\langle \partial\Psi(u), u\rangle \geq \alpha_\Psi \|u\|_V^2.$$

We assume that the subdifferentials $\partial\Phi$ and $\partial\Psi$ are connected via the following coercivity condition on $\partial\Phi \circ \partial\Psi^{-1}$: there exist two constants $\alpha_{\Phi,\Psi} > 0$ and $\beta_{\Phi,\Psi} \geq 0$ such that for all $u^* \in R_{\partial\Phi}(\partial\Psi)$,

$$\left\langle \partial\Phi\left((\partial\Psi)^{-1}(u^*)\right), u^*\right\rangle \geq \alpha_{\Phi,\Psi}\|u^*\|_X^2 - \beta_{\Phi,\Psi}. \tag{5.2}$$

For the definition of $(\partial\Psi)^{-1}$ and the relative range $R_{\partial\Phi}(\partial\Psi)$ of Ψ with respect to $\partial\Phi$, refer to Appendix B.4. This condition must be understood in the sense of a set relation, i.e. for every $\xi \in \mathrm{dom}(\partial\Phi)$ every $u^* \in R_{\partial\Phi}(\partial\Psi)$ such that $u^* \in \partial\Psi(\xi)$, we have

$$\langle \xi^*, u^*\rangle \geq \alpha_{\Phi,\Psi}\|u^*\|_X^2 - \beta_{\Phi,\Psi} \text{ for all } \xi^* \in \partial\Phi(\xi).$$

Condition (5.2) replaces condition

$$\langle \partial\Phi(u), \partial\Psi_\lambda(u)\rangle \geq -\beta(\|\partial\Phi(u)\|_X^2 + \|u\|_X^2 + 1) \text{ for all } u \in \mathrm{dom}(\partial\Phi) \tag{5.3}$$

in Barbu and Malik (1979), Crandall *et al.* (1978), Rennolet (1979), or (Barbu, 1976, (d) page 253), which links $\partial\Phi$ and $\partial\Psi_\lambda$, the subdifferential of the Moreau-Yosida envelope at $\lambda > 0$ of $\partial\Psi$. We say that (5.1) and (5.2) hold uniformly if the constants α_Ψ, $\alpha_{\Phi,\Psi}$ and $\beta_{\Phi,\Psi}$ do not depend on the functionals Φ and Ψ.

Remark 5.1. It is easily seen that Condition 5.2 is equivalent to

$$\langle \partial\Phi(u), \partial\Psi(u)\rangle \geq \alpha_{\Phi,\Psi}\|\partial\Psi(u)\|_X^2 - \beta_{\Phi,\Psi}$$

for all $u \in \mathrm{dom}(\partial\Phi)$, which, using elements, must be understood as

for all u^* in $\mathrm{dom}(\partial\Phi)$ for all ξ^* in $\partial\Phi(u)$ for all u in $\partial\Psi(u)$,

$$\langle \xi^*, u^*\rangle \geq \alpha_{\Phi,\Psi}\|u^*\|_X^2 - \beta_{\Phi,\Psi}.$$

Example 5.1. Consider $\Psi : X \to]-\infty, +\infty]$ lsc convex proper, and let $\chi : X \to]-\infty, +\infty[$ be a lsc convex functional, continuous at a point of V. Assume that there exists $\beta_{\Psi,\chi} \geq 0$ such that

$$\inf_{u \in \mathrm{dom}(\partial\Psi) \cap \mathrm{dom}(\partial\chi)} \langle \partial\Psi(u), \partial\chi(u) \rangle \geq -\beta_{\Psi,\chi} \tag{5.4}$$

in the sense that $\inf_{u \in \mathrm{dom}(\partial\Psi)\cap\mathrm{dom}(\partial\chi)} \inf_{\xi \in \partial\Phi(u), \zeta \in \partial\chi(u)} \langle \xi, \zeta \rangle > -\infty$. Then the functional Ψ and its perturbation $\Phi = \Psi + \chi$ by χ, satisfy (5.2).

Indeed, $\mathrm{dom}(\partial\Phi) = \mathrm{dom}(\partial\Psi) \cap \mathrm{dom}(\partial\chi) \subset \mathrm{dom}(\partial\Psi)$, and from (Attouch *et al.*, 2014, Theorem 9.5.4), $\partial\Phi = \partial\Psi + \partial\chi$. Hence for all $u \in \mathrm{dom}(\partial\Phi)$

$$\begin{aligned}
\langle \partial\Phi(u), \partial\Psi(u) \rangle &= \|\partial\Psi(u)\|_X^2 + \langle \partial\Psi(u), \partial\chi(u) \rangle \\
&\geq \|\partial\Psi(u)\|_X^2 + \inf_{u \in \mathrm{dom}(\partial\Psi)\cap\mathrm{dom}(\partial\chi)} \langle \partial\Psi(u), \partial\chi(u) \rangle \\
&\geq \|\partial\Psi(u)\|_X^2 - \beta_{\Psi,\chi}.
\end{aligned}$$

As a particular case, take Ψ satisfying (5.1), and $\chi = b\|\cdot\|_X^2$ where $b \geq 0$. For all $u \in \mathrm{dom}(\partial\Psi)$ we have from (5.1)

$$\langle \partial\Psi(u), \partial\chi(u) \rangle = 2b \langle \partial\Psi(u), u \rangle \geq 2b\|u\|_V^2.$$

Consequently, Ψ and χ satisfy (5.4) since

$$\inf_{u \in \mathrm{dom}(\partial\Psi)\cap\mathrm{dom}(\partial\chi)} \langle \partial\Psi(u), \partial\chi(u) \rangle \geq 0.$$

Therefore the functionals Ψ and $\Phi = \Psi + b\|\cdot\|_X^2$ satisfy (5.2). Existence in the case when $b = 0$, i.e. $\Phi = \Psi$ has been established in MacCamy (1976). Other examples are provided in Section 10.2.

Define the class $\mathcal{F}$ of pairs of functionals (Φ, Ψ) by

$$(\Phi, \Psi) \in \mathcal{F} \iff \begin{cases} \Phi, \Psi : X \to]-\infty, +\infty] \text{ are lsc convex proper,} \\[1ex] \mathrm{dom}(\Phi) = \mathrm{dom}(\Psi) = V, \\[1ex] \mathrm{dom}(\partial\Phi) \subset \mathrm{dom}(\partial\Psi), \\[1ex] 0 \in \mathrm{dom}(\partial\Psi) \text{ and } \partial\Psi(0) = \{0\}, \end{cases}$$

and let $(\Phi_n, \Psi_n)_{n\in\mathbb{N}}$ be a sequence of $\mathcal{F}$. Then we write

$$\Phi_n \overset{\mathrm{M}}{\to} \Phi,$$

$$\Psi_{n\lfloor V} \overset{\Gamma_{w\text{-}V}}{\to} \Psi_{\lfloor V},$$

to denote the Mosco-convergence of the sequence $(\Phi_n)_{n\in\mathbb{N}}$ to some lsc convex proper functional $\Phi : X \to]-\infty, +\infty]$ and the $\Gamma_{w\text{-}V}$-convergence of the sequence $(\Psi_{n\lfloor V})_{n\in\mathbb{N}}$ to the restriction to V of some lsc convex proper functional $\Psi : X \to]-\infty, +\infty]$

with $\mathrm{dom}(\Psi) = V$. Endow $X \times X$ with the strong product topology, the space $V \times V'$ with the product of the weak topology of V with the strong topology of V', and denote by $G_{s,s}$ and $G_{w,s}$ the associated graph convergence (see Appendix B). Then the following implications hold (see Theorem B.4):

$$\Phi_n \overset{\mathrm{M}}{\to} \Phi \Longrightarrow \partial\Phi_n \overset{G_{s,s}}{\to} \partial\Phi,$$
$$\Psi_{n\lfloor V} \overset{\Gamma_{w\text{-}V}}{\to} \Psi_{\lfloor V} \Longrightarrow \partial(\Psi_{n\lfloor V}) \overset{G_{w,s}}{\to} \partial(\Psi_{\lfloor V}). \tag{5.5}$$

According to the above considerations, we equip $\mathcal{F}$ with the $M \times \Gamma_{w\text{-}V}$-convergence. Although $\mathcal{F}$ is not closed for this convergence, the proposition below shows that conditions (5.1) and (5.2), which are essential in establishing existence in Sections 5.2, 5.3, are in some sense stable in $\mathcal{F}$, then well suited to the convergence analysis of Section 5.4 (see Remark 5.3).

Proposition 5.1. *Assume that* (5.1) *and* (5.2) *are satisfied uniformly with respect to all elements of* $\mathcal{F}$. *For every sequence* $(\Phi_n, \Psi_n)_{n\in\mathbb{N}}$ *of* $\mathcal{F}$ *and every lsc convex proper functionals* $\Phi, \Psi : X \to]-\infty, +\infty]$, *if* $\mathrm{dom}(\Psi) = V$, $\Phi_n \overset{M}{\to} \Phi$ *and* $\Psi_{n\lfloor V} \overset{\Gamma_{w\text{-}V}}{\to} \Psi_{\lfloor V}$, *then* (Φ, Ψ) *satisfies* (5.1), (5.2) *and* $\mathrm{dom}(\partial\Phi) \subset \mathrm{dom}(\partial\Psi)$.

Proof. We denote by α_Ψ and $\alpha_{\Phi,\Psi}$ the two uniform constants in (5.1) and (5.2).

Stability of (5.1). According to (5.5), $\partial(\Psi_{n\lfloor V}) \overset{G_{w,s}}{\to} \partial(\Psi_{\lfloor V})$. Hence, from Proposition B.7, for $(u, v) \in \mathrm{dom}(\partial(\Psi_{\lfloor V}))^2$, there exists $(u_n, v_n) \in \mathrm{dom}(\partial(\Psi_{n\lfloor V}))^2$ such that

$$\begin{cases} u_n \rightharpoonup u \text{ weakly in } V, \\ \partial(\Psi_{n\lfloor V})(u_n) \to \partial(\Psi_{\lfloor V})(u) \text{ strongly in } V', \\ v_n \rightharpoonup v \text{ weakly in } V, \\ \partial(\Psi_{n\lfloor V})(v_n) \to \partial(\Psi_{\lfloor V})(v) \text{ strongly in } V'. \end{cases}$$

The claim then follows from the convergences above by passing to the limit on

$$\langle \partial(\Psi_{n\lfloor V})(u_n) - \partial(\Psi_{n\lfloor V})(v_n), u_n - v_n \rangle_{V',V} \geq \alpha_\Psi \|u_n - v_n\|_V^2.$$

Stability of (5.2) *associated with* (5.1). Take $u \in \mathrm{dom}(\partial\Phi)$. From $\partial\Phi_n \overset{G_{s,s}}{\to} \partial\Phi$ and Proposition B.7, there exists $u_n \in \mathrm{dom}(\partial\Phi_n)$ such that

$$\begin{cases} u_n \to u \text{ strongly in } X, \\ \partial\Phi_n(u_n) \to \partial\Phi(u) \text{ strongly in } X. \end{cases} \tag{5.6}$$

For all $n \in \mathbb{N}$ we have

$$\langle \partial\Phi_n(u_n), \partial\Psi_n(u_n) \rangle \geq \alpha_{\Phi,\Psi} \|\partial\Psi_n(u_n)\|_X^2 - \beta_{\Phi,\Psi}. \tag{5.7}$$

From (5.6) and (5.7) we deduce that

$$\sup_{n\in\mathbb{N}} \|\partial\Psi_n(u_n)\|_X < +\infty. \tag{5.8}$$

From (5.1) we have for all $n \in \mathbb{N}$

$$\langle \partial\Psi_n(u_n), u_n \rangle \geq \alpha_\Psi \|u_n\|_V^2$$

which combined with (5.8) gives

$$\sup_{n\in\mathbb{N}} \|u_n\|_V < +\infty.$$

Hence, there exist a (non relabeled) subsequence $(u_n)_{n\in\mathbb{N}}$ and $\xi \in X$ such that

$$\begin{cases} u_n \rightharpoonup u \text{ weakly in } V \quad \text{(and strongly in } X), \\ \partial\Psi_n(u_n) \rightharpoonup \xi \text{ weakly in } X \text{ thus } \partial\Psi_n(u_n) \to \xi \text{ strongly in } V'. \end{cases} \tag{5.9}$$

Since $\partial(\Psi_{n\lfloor V}) \overset{G_{w,s}}{\to} \partial(\Psi_{\lfloor V})$, we conclude from above and Lemma 5.1, that $\xi \in \partial(\Psi_{\lfloor V})(u) \cap X = \partial\Psi(u)$. From (5.6), (5.9) and passing to the limit in (5.7), we obtain

$$\begin{aligned} \langle \partial\Phi(u), \partial\Psi(u) \rangle &\geq \alpha_{\Phi,\Psi} \liminf_{n\to+\infty} \|\partial\Psi_n(u_n)\|_X^2 + \beta_{\Phi,\Psi} \\ &\geq \alpha_{\Phi,\Psi} \|\partial\Psi(u)\|_X^2 + \beta_{\Phi,\Psi} \end{aligned}$$

where ξ is denoted by $\partial\Psi(u)$. This completes the proof. □

Remark 5.2. It is not clear that (5.3) is stable in the following sense: let $((\Phi_n, \Psi_n)_{n\in\mathbb{N}}, (\Phi, \Psi))$ be a sequence of lsc convex proper functionals from X into $]-\infty, +\infty]$ with $\mathrm{dom}(\Psi_n) = \mathrm{dom}(\Psi) = V$ such that (Φ_n, Ψ_n) satisfies (5.3) and converges to (Φ, Ψ) for the product $\mathrm{M} \times \Gamma_{w\text{-}V}$-convergence, then (Φ, Ψ) satisfies (5.3). However under an additional equi-coerciveness condition, one can establish this stability. For a proof we refer the reader to Anza Hafsa *et al.* (2021e).

Regarding the kernel of the Bochner integral, we assume that $K : [0,T] \to \mathbb{R}_+$ belongs to $C^1([0,T])$. For every $v \in L^2(0,T,X)$ we adopt the notation

$$K \star v(t) := \int_0^t K(t-s)v(s)ds.$$

In case $\partial\Psi$ is linear, to obtain the uniqueness of the solution we assume that K satisfies the additional conditions

$$K \in C^2(]0,T]), \; K(0) > 0, \text{ and } (-1)^k K^k(t) \geq 0 \text{ for } t > 0 \text{ and } k = 0,1,2, \tag{5.10}$$

which imply that for every $v \in L^2(0,T,X)$ such that $v(t) \in \mathrm{dom}(\partial\Psi)$ for a.e. $t \geq 0$,

$$\int_0^t \langle v(s), K \star \partial\Psi(v(s)) \rangle ds \geq \frac{K(t)}{2} \left\langle \partial\Psi\left(\int_0^t v(s)ds\right), \int_0^t v(s)ds \right\rangle, \text{ for all } t \geq 0. \tag{5.11}$$

For a proof see Barbu and Malik (1979), Londen (1977).

5.1.2 *Structure of the reaction functional*

The reaction functional $F : [0,T] \times X \to X$ is a Borel measurable map satisfying:

(CI_1) there exists $L \in L^2(0,T)$ such that

$$\|F(t,u) - F(t,v)\|_X \leq L(t)\|u - v\|_X$$

for all $(u,v) \in X^2$ and all $t \in [0,T]$;

(CI_2) the map $t \mapsto \|F(t,0)\|_X$ belongs to $L^2(0,T)$;

(CI_3) L belongs to $L^2(0,T) \cap W^{1,1}(0,T)$ and there exists $\Theta : [0,T] \to \mathbb{R}_+$, $\Theta \in L^1(0,T)$ such that for all $s < t$ and all $u \in X$,

$$\|F(t,u) - F(s,u)\|_X \leq \int_s^t \Theta(\sigma)d\sigma.$$

When Ω is a bounded domain of $\mathbb{R}^N$ and $X = L^2(\Omega)$, we specify F as follows: let $l \in \mathbb{N}^*$, then for all $u \in L^2(\Omega)$ and a.e. $x \in \Omega$, $F(t,u)(x) = r(t,x) \cdot g(u(x)) + q(t,x)$ where

- $r \in L^\infty\left((0,T) \times \mathbb{R}^N, \mathbb{R}^l\right) \cap W^{1,1}\left(0,T, L^2_{loc}\left(\mathbb{R}^N, \mathbb{R}^l\right)\right)$,
- $q \in L^2\left(0,T, L^2_{loc}\left(\mathbb{R}^N\right)\right) \cap W^{1,1}\left(0,T, L^2_{loc}\left(\mathbb{R}^N\right)\right)$,
- $g : \mathbb{R} \to \mathbb{R}^l$ is bounded and L_g-Lipschitz continuous.

It is easy to check that F fulfills the conditions (CI_1), (CI_2) and (CI_3) with $L = \|r\|_{L^\infty((0,T)\times\mathbb{R}^N,\mathbb{R}^l)} Lg$ and $\Theta(\tau) = M_g\|\frac{dr}{dt}(\tau,\cdot)\|_{L^2(\Omega,\mathbb{R}^l)} + \|\frac{dq}{dt}(\tau,\cdot)\|_{L^2(\Omega)}$ where $M_g = \sup_{r\in\mathbb{R}} |g(r)|$. See Section 5.4.2 for sequences of functionals of this type, and Chapter 10 for more details when F is a random reaction functional.

5.2 Existence of a local solution

From now on, to simplify the presentation, we assume that Φ and Ψ are Gâteaux-differentiable, i.e. $\partial\Phi$ and $\partial\Psi$ are single valued. We follow the standard strategy of Barbu (1976), Barbu and Malik (1979), Crandall *et al.* (1978), Rennolet (1979) consisting in regularizing the non Fickian term $\int_0^t K(t-s)\partial\Psi(u(s))ds$ by means of the Yosida approximation of $\partial\Psi$ (cf. Definition B.3).

5.2.1 *The regularized problem* $(\mathcal{P}_\lambda)$

Let $\lambda > 0$. Consider the Cauchy problem

$$(\mathcal{P})\begin{cases} \dfrac{du}{dt}(t) + \partial\Phi(u(t)) + K \star \partial\Psi(u)(t) = F(t,u(t)) \text{ for a.e. } t \in (0,T) \\ \\ u(0) = u^0 \in \text{dom}(\partial\Phi), \end{cases}$$

and denote by Ψ_λ the Moreau-Yosida approximation of index λ of Ψ (see Definition B.3). We begin by establishing the global existence and uniqueness of a strong solution for the approximate problem expressed in $L^2(0,T,X)$

$$(\mathcal{P}_\lambda)\begin{cases}\dfrac{du_\lambda}{dt}(t)+\partial\Phi(u_\lambda(t))+K\star\partial\Psi_\lambda(u_\lambda)(t)=F(t,u_\lambda(t)) \text{ for a.e. } t\in(0,T)\\ \\ u_\lambda(0)=u^0,\ u^0\in\operatorname{dom}(\partial\Phi)\end{cases}$$

where λ is intended to tend towards 0. Set for all $t\in\mathbb{R}_+$

$$G_\lambda(t,u_\lambda):=F(t,u_\lambda(t))-K\star\partial\Psi_\lambda(u_\lambda)(t).$$

We write $(\mathcal{P}_\lambda)$ as

$$\begin{cases}\dfrac{du_\lambda}{dt}(t)+\partial\Phi(u_\lambda(t))=G_\lambda(t,u_\lambda) \text{ for a.e. } t\in(0,T)\\ \\ u_\lambda(0)=u^0,\ u^0\in\operatorname{dom}(\partial\Phi).\end{cases}\tag{5.12}$$

Lemma 5.2. *Assume that* (CI_1), (CI_2), *and* (CI_3) *hold. Then, there exists a unique solution* $u_\lambda\in C([0,T],X)$ *of* $(\mathcal{P}_\lambda)$. *Furthermore* $\frac{du_\lambda}{dt}\in L^2(0,T,X)$, $\partial\Phi(u_\lambda)\in L^2(0,T,X)$, *and*

(S_1) $u_\lambda(t)\in\operatorname{dom}(\partial\Phi)$ *for all* $t\in[0,T]$, *and admits a right derivative* $\dfrac{d^+u_\lambda}{dt}(t)$ *which satisfies for every* $t\in[0,T[$

$$\frac{d^+u_\lambda}{dt}(t)+\partial\Phi(u_\lambda(t))=G_\lambda(t,u_\lambda).$$

Proof. According to (Attouch *et al.*, 2014, Proposition 17.2.1), $\partial\Psi_\lambda$ is Lipschitz continuous with Lipschitz constant $\frac{1}{\lambda}$. Therefore, it is easy to show that for all $(u,v)\in C([0,T],X)\times C([0,T],X)$,

$$\|K\star\partial\Psi_\lambda(u)-K\star\partial\Psi_\lambda(v)\|_{C([0,T],X)}\leq C_{\lambda,T}\|u-v\|_{C([0,T],X)}\tag{5.13}$$

where $C_{\lambda,T}:=\frac{1}{\lambda}\left(\int_0^T K(s)ds\right)$. For each $u\in C([0,T],X)$, denote by Λu the unique solution in $C([0,T],X)$ with $\frac{d\Lambda u}{dt}\in L^2(0,T,X)$ of the Cauchy problem

$$(\mathcal{P}_u)\begin{cases}\dfrac{d\Lambda u}{dt}(t)+\partial\Phi(\Lambda u(t))=G_\lambda(t,u) \text{ for a.e. } t\in(0,T)\\ \\ \Lambda u(0)=u_0\in\operatorname{dom}(\partial\Phi).\end{cases}$$

For existence and uniqueness of Λu, we only have to check that $G_\lambda\in L^2(0,T,X)$ (refer to (Attouch *et al.*, 2014, Theorem 17.2.5), or (Brezis, 1973, Theorem 3.7)). The claim follows straightforwardly from (CI_1), (CI_2) and (5.13).

The method is to show that the iterated map Λ^n is a strict contraction for n large enough. Indeed, from existence of a unique fixed point u_λ for Λ^n we will deduce that Λu_λ is a fixed point too. Thus, from uniqueness $\Lambda u_\lambda=u_\lambda$, so that u_λ

is a fixed point for Λ which clearly solves $(\mathcal{P}_\lambda)$ and satisfies $\frac{du_\lambda}{dt} \in L^2(0,T,X)$ and $\partial\Phi(u_\lambda) \in L^2(0,T,X)$.

Let $(u,v) \in C([0,T],X) \times C([0,T],X)$ satisfying for a.e. $t \in (0,T)$

$$\frac{d\Lambda u}{dt}(t) + \partial\Phi(\Lambda u(t)) = G_\lambda(t,u),$$
$$\frac{d\Lambda v}{dt}(t) + \partial\Phi(\Lambda v(t)) = G_\lambda(t,v).$$

From the monotonicity of $\partial\Phi$, we infer that for a.e. $\sigma \in (0,T)$

$$\left\langle \frac{d\Lambda v}{dt}(\sigma) - \frac{d\Lambda u}{dt}(\sigma), \Lambda v(\sigma) - \Lambda u(\sigma) \right\rangle \leq \langle G_\lambda(\sigma,u) - G_\lambda(\sigma,v), \Lambda v(\sigma) - \Lambda u(\sigma)\rangle,$$

hence

$$\frac{1}{2}\frac{d}{dt}\|\Lambda v(\sigma) - \Lambda u(\sigma)\|_X^2 \leq \langle G_\lambda(\sigma,u) - G_\lambda(\sigma,v), \Lambda v(\sigma) - \Lambda u(\sigma)\rangle.$$

By integration we have for all $t \in [0,T]$

$$\begin{aligned}\frac{1}{2}\|\Lambda v(t) - \Lambda u(t)\|_X^2 &\leq \int_0^t \langle G_\lambda(\sigma,u) - G_\lambda(\sigma,v), \Lambda v(\sigma) - \Lambda u(\sigma)\rangle\, d\sigma \\ &\leq \int_0^t \|G_\lambda(\sigma,u) - G_\lambda(\sigma,v)\|_X \|\Lambda v(\sigma) - \Lambda u(\sigma)\|_X d\sigma.\end{aligned}$$

Thus, according to Lemma A.2 with $p = 2$, it follows that for all $t \in [0,T]$

$$\|\Lambda v(t) - \Lambda u(t)\|_X \leq \int_0^t \|G_\lambda(\sigma,u) - G_\lambda(\sigma,v)\|_X d\sigma.$$

From (5.13) and (CI_1) we infer that for all $t \in [0,T]$

$$\|\Lambda v - \Lambda u\|_{C([0,t],X)} \leq \int_0^t L_{\lambda,T}(\sigma)\|u - v\|_{C([0,\sigma],X)} d\sigma \tag{5.14}$$

where $L_{\lambda,T}(\sigma) := C_{\lambda,T} + L(\sigma)$. By iterating (5.14), and according to the formula

$$\int_0^t L_{\lambda,T}(\sigma_1) \int_0^{\sigma_1} L_{\lambda,T}(\sigma_2) \ldots \int_0^{\sigma_{n-1}} L_{\lambda,T}(\sigma_n) d\sigma_n \ldots d\sigma_1 = \frac{\left(\int_0^t L_{\lambda,T}(\sigma)d\sigma\right)^n}{n!}$$

obtained by a standard calculus for multiple integrals, we obtain

$$\|\Lambda^n v - \Lambda^n u\|_{C([0,T],X)} \leq \frac{\left(\int_0^T L_{\lambda,T}(\sigma)d\sigma\right)^n}{n!}\|u - v\|_{C([0,T],X)}.$$

The claim follows for n sufficiently large.

For proving that u_λ satisfies (S_1), we have to establish that $G_\lambda \in W^{1,1}(0,T,X)$ (see (Attouch *et al.*, 2014, Theorem 17.2.6), or (Brezis, 1973, Theorem 3.7)). We first claim that $K \star \partial\Psi_\lambda(u_\lambda)$ belongs to $W^{1,2}(0,T,X)$. This follows from

$$\left\|\frac{d}{dt}K \star \partial\Psi_\lambda(u_\lambda)\right\|_{L^2(0T,X)} \leq \left(K(0) + T^{\frac{1}{2}}\|K'\|_{L^2(0,T)}\right)\|\partial\Psi_\lambda(u_\lambda)\|_{L^2(0,T,X)}$$

which is obtained from the formula

$$\frac{d}{dt}K \star \partial\Psi_\lambda(u_\lambda)(t) = K(0)\partial\Psi_\lambda(u_\lambda)(t) + K' \star \partial\Psi_\lambda(u_\lambda)(t) \tag{5.15}$$

for a.e. $t \in (0,T)$. It remains to establish that $F(\cdot, u_\lambda) \in W^{1,1}(0,T,X)$. This follows from (CI_3), and the following calculation: for every $(s,t) \in [0,T]^2$ with $s \le t$ we have

$$\begin{aligned}
&\|F(t,u_\lambda(t)) - F(s,u_\lambda(s))\|_X \\
&\le \|F(t,u_\lambda(t)) - F(s,u_\lambda(t))\|_X + |L(s)|\|u_\lambda(t) - u_\lambda(s)\|_X \\
&\le \int_s^t \Theta(\sigma)d\sigma + \left(L(0) + \int_0^T \left|\frac{dL}{d\sigma}(\sigma)\right| d\sigma\right)\int_s^t \left\|\frac{du_\lambda}{d\sigma}(\sigma)\right\|_X d\sigma,
\end{aligned} \tag{5.16}$$

which proves that $F(\cdot, u_\lambda)$ is absolutely continuous. The proof is complete. □

5.2.2 *Convergence of $(\mathcal{P}_\lambda)$ to $(\mathcal{P})$: existence of a local solution of $(\mathcal{P})$*

The following lemma provides local estimates for the solution of $(\mathcal{P}_\lambda)$, in order to establish the convergence of $(\mathcal{P}_\lambda)$ to $(\mathcal{P})$.

Lemma 5.3. *Assume that* (5.1), (5.2) *and* (CI_1), (CI_2), (CI_3) *hold. Then for every* $\widetilde{T}$ *in* $]0,T]$ *satisfying* $\widetilde{T}^{\frac{1}{2}}\|K\|_{L^2(0,T)} + \alpha_\Psi\|L\|_{L^2(0,\widetilde{T})} < \alpha_{\Phi,\Psi}$, *the following estimates hold:*

$$\sup_{\lambda>0} \|\partial\Psi(u_\lambda)\|_{L^2(0,\widetilde{T},X)} < +\infty, \tag{5.17}$$

$$\sup_{\lambda>0} \|u_\lambda\|_{L^2(0,\widetilde{T},V)} < +\infty, \tag{5.18}$$

$$\sup_{\lambda>0} \left\|\frac{du_\lambda}{dt}\right\|_{L^2(0,\widetilde{T},X)} < +\infty, \tag{5.19}$$

$$\sup_{\lambda>0} \|\partial\Phi(u_\lambda)\|_{L^2(0,\widetilde{T},X)} < +\infty, \tag{5.20}$$

$$\sup_{\lambda>0} \|u_\lambda\|_{C(0,\widetilde{T},X)} < +\infty, \tag{5.21}$$

$$\sup_{\lambda>0} \left\|\frac{d^+u_\lambda}{dt}(t)\right\|_X < +\infty \textit{ for each } t \in]0,\widetilde{T}] \tag{5.22}$$

$$\sup_{\lambda>0} \|\partial\Psi(u_\lambda(t))\|_X < +\infty \textit{ for each } t \in [0,\widetilde{T}], \tag{5.23}$$

$$\sup_{\lambda>0} \|\partial\Phi(u_\lambda(t))\|_X < +\infty \textit{ for each } t \in [0,\widetilde{T}]. \tag{5.24}$$

Proof. Let $\lambda > 0$. ***Step 1.*** Proof of (5.17) and (5.18).

Observe that from Lemma 5.2, for all $t \in [0,T]$, $u_\lambda(t) \in \mathrm{dom}(\partial\Phi) \subset \mathrm{dom}(\partial\Psi)$. For a.e. $t \in [0,\widetilde{T}]$, form the scalar product in X of $\partial\Psi(u_\lambda(t))$ with the approximate equation (5.12) and integrate over $[0, \widetilde{T}]$. This yields

$$\int_0^{\widetilde{T}} \frac{d}{dt}\Psi(u_\lambda(t))dt + \int_0^{\widetilde{T}} \langle \partial\Phi(u_\lambda(t)), \partial\Psi(u_\lambda(t))\rangle \, dt + \int_0^{\widetilde{T}} \langle K \star \partial\Psi_\lambda(u_\lambda)(t), \partial\Psi(u_\lambda(t))\rangle \, dt = \int_0^{\widetilde{T}} \langle F(t, u_\lambda(t)), \partial\Psi(u_\lambda(t))\rangle \, dt. \quad (5.25)$$

We have used the fact that $\partial\Phi(u_\lambda) \in L^2(0,T,X)$ yields from (5.2), that $\partial\Psi(u_\lambda) \in L^2(0,T,X)$, hence $\langle \frac{du_\lambda}{dt}(t), \partial\Psi(u_\lambda)(t)\rangle = \frac{d}{dt}\Psi(u_\lambda)(t)$ (cf. (Attouch *et al.*, 2014, Proposition 17.2.5)). An easy calculation gives

$$\|K \star \partial\Psi_\lambda(u_\lambda)\|_{L^2(0,\widetilde{T},X)} \leq \widetilde{T}^{\frac{1}{2}} \|K\|_{L^2(0,T)} \|\partial\Psi_\lambda(u_\lambda)\|_{L^2(0,\widetilde{T},X)}. \quad (5.26)$$

Since for all $\lambda > 0$,

$$\|\partial\Psi_\lambda(u_\lambda(t))\|_X \leq \|\partial\Psi(u_\lambda(t))\|_X \quad (5.27)$$

(see (Attouch *et al.*, 2014, Proposition 17.2.2)), we infer that

$$\left| \int_0^{\widetilde{T}} \langle K \star \partial\Psi_\lambda(u_\lambda)(t), \partial\Psi(u_\lambda(t))\rangle \, dt \right| \leq \widetilde{T}^{\frac{1}{2}} \|K\|_{L^2(0,T)} \|\partial\Psi(u_\lambda)\|^2_{L^2(0,\widetilde{T},X)}. \quad (5.28)$$

On the other hand from (CI_1) and (5.1)

$$\begin{aligned}\|F(\cdot, u_\lambda)\|_{L^2(0,\widetilde{T},X)} &\leq \|F(\cdot,0)\|_{L^2(0,T,X)} + \|L\|_{L^2(0,\widetilde{T})} \|u_\lambda\|_{L^2(0,\widetilde{T},X)} \\ &\leq \|F(\cdot,0)\|_{L^2(0,T,X)} + \alpha_\Psi \|L\|_{L^2(0,\widetilde{T})} \|\partial\Psi(u_\lambda)\|_{L^2(0,\widetilde{T},X)}. \quad (5.29)\end{aligned}$$

Hence from (5.29)

$$\begin{aligned}\int_0^{\widetilde{T}} \langle F(t, u_\lambda(t)), \partial\Psi(u_\lambda(t))\rangle \, dt &\leq \|F(\cdot,0)\|_{L^2(0,T,X)} \|\partial\Psi(u_\lambda)\|_{L^2(0,\widetilde{T},X)} \\ &\quad + \alpha_\Psi \|L\|_{L^2(0,\widetilde{T})} \|\|\partial\Psi(u_\lambda)\|^2_{L^2(0,\widetilde{T},X)}. \quad (5.30)\end{aligned}$$

Combining (5.25), (5.2), (5.28), (5.30) yields that

$$\begin{aligned}&\left(\alpha_{\Phi,\Psi} - \left(\widetilde{T}^{\frac{1}{2}} \|K\|_{L^2(0,T)} + \alpha_\Psi \|L\|_{L^2(0,\widetilde{T})}\right)\right) \|\partial\Psi(u_\lambda)\|^2_{L^2(0,\widetilde{T},X)} \\ &\leq T\beta_{\Phi,\Psi} + \Psi(u_0) - \inf_X \Psi + \|F(\cdot,0)\|_{L^2(0,\widetilde{T},X)} \|\partial\Psi(u_\lambda)\|_{L^2(0,\widetilde{T},X)}\end{aligned}$$

from which we deduce (5.17) provided that $\widetilde{T}^{\frac{1}{2}} \|K\|_{L^2(0,T)} + \alpha_\Psi \|L\|_{L^2(0,\widetilde{T})} < \alpha_{\Phi,\Psi}$. Estimate (5.18) follows by combining (5.17) with (5.1).

Step 2. Proof of (5.19), (5.20) and (5.21).

For a.e. $t \in \left(0, \widetilde{T}\right)$, form the scalar product in X of $\frac{du_\lambda}{dt}(t)$ with the approximate equation and integrate over $\left(0, \widetilde{T}\right)$. This yields

$$\begin{aligned} &\left\| \frac{du_\lambda}{dt}(t) \right\|^2_{L^2(0,\widetilde{T},X)} \\ &\le \Phi(u_0) - \inf_X \Phi + \left(\|K \star \partial\Psi_\lambda(u_\lambda)\|_{L^2(0,\widetilde{T},X)} + \|F(\cdot, u_\lambda)\|_{L^2(0,\widetilde{T},X)} \right) \left\| \frac{du_\lambda}{dt} \right\|_{L^2(0,\widetilde{T},X)} \end{aligned}$$

and (5.19) follows from (5.26), (5.17), (5.29) and (5.18). Estimate (5.20) follows straightforwardly from the approximate equation, (5.26), (5.19), (5.17), and (5.18). Estimate (5.21) is obtained from (5.19), according to

$$\|u_\lambda(t)\|_X \le \|u_0\|_X + \widetilde{T}^{\frac{1}{2}} \left\| \frac{du_\lambda}{dt} \right\|_{L^2(0,\widetilde{T},X)}.$$

Step 3. Proof of (5.22), (5.23) and (5.24). First observe that

$$\sup_{\lambda > 0} \|K \star \partial\Psi_\lambda(u_\lambda)\|_{W^{1,2}(0,\widetilde{T},X)} < +\infty. \tag{5.31}$$

which follows from (5.17) and the two inequalities:

$$\begin{aligned} &\|K \star \partial\Psi_\lambda(u_\lambda)\|_{L^2(0\widetilde{T},X)} \le \widetilde{T}^{\frac{1}{2}} \|K\|_{L^2(0,T)} \|\partial\Psi(u_\lambda)\|_{L^2(0,\widetilde{T},X)}, \\ &\left\| \frac{d}{dt} K \star \partial\Psi_\lambda(u_\lambda) \right\|_{L^2(0\widetilde{T},X)} \le \left(K(0) + \widetilde{T}^{\frac{1}{2}} \|K'\|_{L^2(0,T)} \right) \|\partial\Psi(u_\lambda)\|_{L^2(0,\widetilde{T},X)} \end{aligned}$$

(the second inequality follows from (5.15)). Next, from (5.16) and (5.19), we have

$$\sup_\lambda \left\| \frac{dF(\cdot, u_\lambda)}{dt} \right\|_{L^1(0,\widetilde{T},X)} \le +\infty. \tag{5.32}$$

From (5.31) and (5.32) we deduce that

$$\sup_\lambda \left\| \frac{dG_\lambda}{dt} \right\|_{L^1(0,\widetilde{T},X)} < +\infty.$$

Hence (5.22) is a straightforward consequence of 5.19 and Lemma 2.5. which states that for each $t \in]0, \widetilde{T}]$

$$\left\| \frac{d^+u_\lambda}{dt}(t) \right\|_X \le \frac{1}{t} \int_0^t \left\| \frac{du_\lambda}{dt}(s) \right\|_X ds + \int_0^t \left\| \frac{dG_\lambda}{dt}(s) \right\|_X ds.$$

To establish (5.23), form the scalar product of the approximate equation (S_1)

$$\frac{d^+u_\lambda}{dt}(t) + \partial\Phi(u_\lambda(t)) = G_\lambda(t, u_\lambda)(t)$$

with $\partial\Psi(u_\lambda(t))$ for all $t \in]0, \widetilde{T}]$. This yields from (5.2)

$$\left\langle \frac{d^+u_\lambda}{dt}(t), \partial\Psi(u_\lambda(t)) \right\rangle + \alpha_{\Phi,\Psi} \|\partial\Psi(u_\lambda(t))\|^2_X \le \langle G_\lambda(t, u_\lambda)(t), \partial\Psi(u_\lambda(t)) \rangle - \beta_{\Phi,\Psi}$$

from which we deduce

$$\alpha_{\Phi,\Psi}\|\partial\Psi(u_\lambda(t))\|_X^2 \le \left(\left\|\frac{d^+u_\lambda}{dt}(t)\right\|_X + \|G_\lambda(t,u_\lambda)(t)\|_X\right)\|\partial\Psi(u_\lambda(t))\|_X - \beta_{\Phi,\Psi}.$$

The claim follows from (5.17), (5.21), (5.22), and $\sup_\lambda \|G_\lambda(t,u_\lambda(t))\|_X < +\infty$ obtained according to

$$\|G_\lambda(t,u_\lambda(t))\|_X \le \|K\|_{L^2(0,T)}\|\partial\Psi(u_\lambda)\|_{L^2(0,\widetilde{T},X)} + \|F(t,0)\|_X + L(t)\|u_\lambda(t)\|_X.$$

For $t = 0$, $\partial\Psi(u_\lambda(t)) = \partial\Psi(u_0)$ which does not depend on λ. To obtain (5.24), take the scalar product of the approximate equation with $\partial\Phi(u_\lambda(t))$ and follows the same calculation. □

Theorem 5.1 (Local solution). *Assume that* (5.1), (5.2), (CI$_1$), (CI$_2$), (CI$_3$) *hold, and let* $\widetilde{T} > 0$ *satisfying* $\widetilde{T}^{\frac{1}{2}}\|K\|_{L^2(0,T)} + \alpha_\Psi\|L\|_{L^2(0,\widetilde{T})} < \alpha_{\Phi,\Psi}$. *Then* $(\mathcal{P})$ *admits a solution* $u_{\widetilde{T}}$ *in* $C([0,\widetilde{T}],X)$ *which satisfies* $u_{\widetilde{T}}(t) \in \mathrm{dom}(\partial\Phi)$ *for all* $t \in [0,\widetilde{T}]$.

Proof. To shorten the notation we write u for $u_{\widetilde{T}}$. The proof falls into four steps.

Step 1. (Compactness in $C([0,\widetilde{T}],X)$). We establish existence of $u \in C(0,\widetilde{T},X)$ and a subsequence of $(u_\lambda)_{\lambda>0}$ (not relabeled), such that

$$u_\lambda \to u \text{ in } C([0,\widetilde{T}],X), \tag{5.33}$$

$$J_\lambda u_\lambda(t) \to u(t) \text{ in } X, \text{ for all } t \in [0,\widetilde{T}], \tag{5.34}$$

where $J_\lambda := (I + \lambda\partial\Psi)^{-1} : X \to X$ is the resolvent of index λ of $\partial\Psi$ (for the properties of J_λ see (Attouch *et al.*, 2014, Proposition 17.2.1)).

In standard way, to prove (5.33), the method consists in applying Ascoli's theorem. From (5.19) and (5.21) we have

$$\sup_{\lambda>0}\|u_\lambda\|_{C(0,\widetilde{T},X)} < +\infty \text{ (equiboundedness)},$$

$$\|u_\lambda(t) - u_\lambda(s)\|_X \le (t-s)^{\frac{1}{2}}\sup_{\lambda>0}\left\|\frac{du_\lambda}{dt}\right\|_{L^2(0,\widetilde{T},X)} \qquad \text{(equicontinuity)}$$

for all $(s,t) \in [0,\widetilde{T}]^2$ with $s \le t$. It remains to show that for each $t \in]0,\widetilde{T}]$, the set $E_t := \{u_\lambda(t) : \lambda > 0\}$ is relatively compact in X (for $t = 0$, E_t is reduced to $\{u_0\}$). Let $t \in]0,\widetilde{T}]$. From the compact embedding $V \hookrightarrow X$, it suffices to establish that $\sup_{\lambda>0}\|u_\lambda(t)\|_V < +\infty$. The claim follows from (5.1) and (5.23) which yields

$$\begin{aligned}\|u_\lambda(t)\|_V^2 &\le \frac{1}{\alpha_\Psi}\langle\partial\Psi(u_\lambda(t)), u_\lambda(t)\rangle \le \frac{1}{\alpha_\Psi}\|\partial\Psi(u_\lambda(t))\|_X\|u_\lambda(t)\|_X \\ &\le \frac{1}{\alpha_\Psi}\|\partial\Psi(u_\lambda(t))\|_X\|u_\lambda(t)\|_V.\end{aligned}$$

Estimate (5.34) is established as follows: from the definition of J_λ, we have $J_\lambda u_\lambda(t) - u_\lambda(t) = \lambda\partial\Psi_\lambda(u_\lambda(t))$ so that, from (5.27),

$$\|J_\lambda u_\lambda(t) - u_\lambda(t)\|_X \le \lambda\|\partial\Psi_\lambda(u_\lambda(t))\|_X \le \lambda\|\partial\Psi(u_\lambda(t))\|_X.$$

Hence, from (5.23), $J_\lambda u_\lambda(t) - u_\lambda(t) \to 0$ in X for all $t \in]0,\widetilde{T}]$.

Step 2. We prove that $u(t) \in \mathrm{dom}(\partial\Phi)$ for all $t \in \left[0, \widetilde{T}\right]$, and that (5.33), (5.34) hold in V equipped with its norm $\|\cdot\|_V$. More precisely

$$u_\lambda \to u \text{ in } C\left(\left[0, \widetilde{T}\right], V\right), \tag{5.35}$$

$$J_\lambda u_\lambda(t) \to u(t) \text{ in } V \text{ for all } t \in [0, T]. \tag{5.36}$$

Fix $t \in \left[0, \widetilde{T}\right]$. From (5.24), there exist $B(t) \in X$ and a subsequence such that

$$\partial\Phi(u_\lambda(t)) \rightharpoonup B(t) \text{weakly in } X.$$

From (5.35), $u_\lambda(t) \to u(t)$ strongly in X, and since the maximal monotone operator $\partial\Phi$ is demi-closed (see (Attouch *et al.*, 2014, Proposition 17.2.4)), we deduce that $u(t) \in \mathrm{dom}(\partial\Phi)$ and $B(t) = \partial\Phi(u(t))$.

Observe that $u_\lambda(t)$ and $u(t)$ belong to $\mathrm{dom}(\partial\Phi) \subset \mathrm{dom}(\partial\Psi)$ for all $t \in \left[0, \widetilde{T}\right]$. Hence from (5.1), we deduce

$$\begin{aligned}\|u_\lambda(t) - u(t)\|_V^2 &\leq \langle \partial\Psi(u_\lambda(t)) - \partial\Psi(u(t)), u_\lambda(t) - u(t)\rangle \\ &\leq \left(\sup_{\lambda>0} \|\partial\Psi(u_\lambda(t))\|_X + \|\partial\Psi(u(t))\|_X\right) \|u_\lambda(t) - u(t)\|_X.\end{aligned}$$

Hence (5.35) follows from (5.23) and (5.33). The proof of (5.36) follows the same scheme. More precisely for all $t \in [0, \widetilde{T}]$,

$$\begin{aligned}&\|J_\lambda u_\lambda(t) - u_\lambda(t)\|_V^2 \\ &\leq \langle \partial\Psi(J_\lambda u_\lambda(t)) - \partial\Psi(u_\lambda(t)), J_\lambda u_\lambda(t) - u_\lambda(t)\rangle \\ &\leq \sup_{\lambda>0}(\|\partial\Psi_\lambda(u_\lambda(t))\|_X + \|\partial\Psi(u_\lambda(t))\|_X)\|J_\lambda u_\lambda(t) - u_\lambda(t)\|_X\end{aligned}$$

(recall that $\partial\Psi_\lambda(u_\lambda(t)) = \partial\Psi(J_\lambda u_\lambda(t))$, see (Attouch *et al.*, 2014, Proposition 17.2.1)). From (5.27) and (5.23)

$$\sup_{\lambda>0} \|\partial\Psi_\lambda(u_\lambda(t))\|_X \leq \sup_{\lambda>0} \|\partial\Psi(u_\lambda(t))\|_X < +\infty,$$

hence

$$\|J_\lambda u_\lambda(t) - u_\lambda(t)\|_V^2 \leq 2 \sup_{\lambda>0} \|\partial\Psi(u_\lambda(t))\|_X \|J_\lambda u_\lambda(t) - u_\lambda(t)\|_X$$

so that (5.36) follows from (5.23) and (5.34).

Step 3. We establish that $G_\lambda(\cdot, u_\lambda) \rightharpoonup G(\cdot, u)$ in $L^2\left(0, \widetilde{T}, X\right)$ where $G(\cdot, u)$ is defined by $G(t, u(t)) := F(t, u(t)) - K \star \partial\Psi(u)(t)$.

From (CI_1), (CI_2) and (5.33), $F(\cdot, u_\lambda)$ strongly converges to $F(\cdot, u)$ in $L^2\left(0, \widetilde{T}, X\right)$. We claim that $\partial\Psi_\lambda(u_\lambda) \rightharpoonup \partial\Psi(u)$ in $L^2\left(0, \widetilde{T}, X\right)$, from which we easily deduce that $K \star \partial\Psi_\lambda(u) \rightharpoonup K \star \partial\Psi(u)$ in $L^2\left(0, \widetilde{T}, X\right)$. From (5.17) we have

$$\sup_{\lambda>0} \|\partial\Psi_\lambda(u_\lambda(t))\|_{L^2(0,\widetilde{T},X)} \leq \sup_{\lambda>0} \|\partial\Psi(u_\lambda(t))\|_{L^2(0,\widetilde{T},X)} < +\infty.$$

Thus, using the compact embedding $L^2\left(0, \widetilde{T}, X\right) \hookrightarrow L^2\left(0, \widetilde{T}, V'\right)$, we infer that there exist a subsequence (not relabeled) and $C \in L^2\left(0, \widetilde{T}, X\right)$ such that successively,

$$\partial\Psi_\lambda(u_\lambda) \rightharpoonup C \text{ weakly in } L^2\left(0, \widetilde{T}, X\right),$$

$$\partial\Psi_\lambda(u_\lambda) \to C \text{ strongly in } L^2\left(0, \widetilde{T}, V'\right),$$

$$\partial\Psi_\lambda(u_\lambda(t)) \to C(t) \text{ in } V' \text{ for a.e. } t \in \left(0, \widetilde{T}\right).$$

Since $\partial\Psi_\lambda(u_\lambda) = \partial\Psi(J_\lambda u_\lambda)$, we deduce from above that $\partial\Psi(J_\lambda u_\lambda(t)) \to C(t)$ in V' for a.e. $t \in (0, T)$. As from (5.36), $J_\lambda u_\lambda(t) \to u(t)$ in V, from the maximality of $\partial\Psi$ we infer that $C(t) = \partial\Psi(u(t))$ for a.e. $t \in (0, T)$. This proves the claim.

Step 4. (u solves $(\mathcal{P})$). To shorten the notation, we write $G_\lambda(t)$ for $G_\lambda(t, u_\lambda(t))$. Denote by Φ^* the Legendre-Fenchel conjugate of Φ. According to the Fenchel extremality condition (see (Attouch *et al.*, 2014, Proposition 9.5.1)), equation (5.12) is equivalent to

$$\Phi(u_\lambda(t)) + \Phi^*\left(G_\lambda(t) - \frac{du_\lambda}{dt}(t)\right) + \left\langle \frac{du_\lambda}{dt}(t) - G_\lambda(t), u_\lambda(t)\right\rangle = 0$$

for a.e. $t \in \left(0, \widetilde{T}\right)$, which, from the Legendre-Fenchel inequality, is in turn equivalent to

$$\int_0^{\widetilde{T}} \left[\Phi(u_\lambda(t)) + \Phi^*\left(G_\lambda(t) - \frac{du_\lambda}{dt}(t)\right) + \left\langle \frac{du_\lambda}{dt}(t) - G_\lambda(t), u_\lambda(t)\right\rangle\right] dt = 0.$$

Therefore, (5.12) is equivalent to

$$\int_0^{\widetilde{T}} \left[\Phi(u_\lambda(t)) + \Phi^*\left(G_\lambda(t) - \frac{du_\lambda}{dt}(t)\right) + \frac{d}{dt}\frac{1}{2}\|u_\lambda(t)\|^2 - \langle G_\lambda(t), u_\lambda(t)\rangle\right] dt = 0,$$

hence to

$$\int_0^{\widetilde{T}} \left[\Phi(u_\lambda(t)) + \Phi^*\left(G_\lambda(t) - \frac{du_\lambda}{dt}(t)\right)\right] dt + \frac{1}{2}\left(\left\|u_\lambda(\widetilde{T})\right\|^2 - \|u_0\|^2\right)$$
$$- \int_0^{\widetilde{T}} \langle G_\lambda(t), u_\lambda(t)\rangle \; dt = 0.$$

Equivalently

$$I_\Phi(u_\lambda) + I_{\Phi^*}\left(G_\lambda(t) - \frac{du_\lambda}{dt}\right) + \frac{1}{2}\left(\left\|u_\lambda(\widetilde{T})\right\|^2 - \|u_0\|^2\right)$$
$$- \int_0^{\widetilde{T}} \langle G_\lambda(t), u_\lambda(t)\rangle \; dt = 0 \tag{5.37}$$

where the integral functionals I_Φ and I_{Φ^*} are respectively defined in $L^2\left(0,\widetilde{T},X\right)$ by $I_\Phi(v)=\int_0^{\widetilde{T}}\Phi(v(t))dt$ and $I_{\Phi^*}(v)=\int_0^{\widetilde{T}}\Phi^*(v(t))dt$.

Combining $u_\lambda\left(\widetilde{T}\right)=u_0+\int_0^{\widetilde{T}}\frac{du_\lambda}{dt}(t)dt$ with $\frac{du_\lambda}{dt}\rightharpoonup\frac{du}{dt}$ in $L^2(0,T,X)$ which is obtained from (5.19), we infer that

$$\liminf_{\lambda\to+\infty}\left\|u_\lambda\left(\widetilde{T}\right)\right\|^2\geq\left\|u\left(\widetilde{T}\right)\right\|^2. \tag{5.38}$$

By passing to the lower limit in (5.37), from (5.38), (5.33), ***Step 3***, and noticing that I_Φ and I_{Φ^*} are lower semicontinuous for the weak topology of $L^2\left(0,\widetilde{T},X\right)$, we obtain

$$\int_0^T\left[\Phi(u(t))+\Phi^*\left(G(t)-\frac{du}{dt}(t)\right)\right]dt+\frac{1}{2}(\|u(T)\|^2-\|u^0\|^2)$$
$$-\int_0^T\langle G(t),u(t)\rangle\ dt\leq 0$$

or equivalently,

$$\int_0^T\left[\Phi(u(t))+\Phi^*\left(G(t)-\frac{du}{dt}(t)\right)+\left\langle\frac{du}{dt}(t)-G(t),u(t)\right\rangle\right]dt\leq 0. \tag{5.39}$$

From the Legendre-Fenchel inequality, we have $\Phi(u(t))+\Phi^*(G(t)-\frac{du}{dt}(t))+\langle\frac{du}{dt}(t)-G(t),u(t)\rangle\geq 0$, so that (5.39) yields that for a.e. $t\in(0,T)$, $\Phi(u(t))+\Phi^*(G(t)-\frac{du}{dt}(t))+\langle\frac{du}{dt}(t)-G(t),u(t)\rangle=0$ which is equivalent to

$$\frac{du}{dt}(t)+\partial\Phi(u(t))=G(t)\text{ for a.e. }t\in(0,T).$$

This completes the proof. □

5.3 Existence of solutions in $C([0,T],X)$

5.3.1 *Existence of a global solution in $C([0,T],X)$: translation-induction method*

Any local solution obtained in Theorem 5.1 can be continued on $[0,T]$ as follows: cover $[0,T]$ by the translated segments of $[0,\widetilde{T}]$, and stick together the $\widetilde{T}$-translated local solutions in $C([0,\widetilde{T}],X)$ of each suitably modified problem $(\mathcal{P})$. This process is a generalization of a standard method, see for instance (Barbu, 1976, page 243).

Theorem 5.2. *Given $\widetilde{T}>0$ satisfying $\widetilde{T}^{\frac{1}{2}}\|K\|_{L^2(0,T)}+\alpha_\Psi\|L\|_{L^2(0,\widetilde{T})}<\alpha_{\Phi,\Psi}$, any local solution $u_{\widetilde{T}}$ of $(\mathcal{P})$ in $C([0,\widetilde{T}],X)$ obtained in Theorem 5.1 can be extended to a solution of $(\mathcal{P})$ in $C([0,T],X)$.*

Proof. For each $i = 1, \ldots, \ell$ where $\ell := \max\{k \in \mathbb{N} : k\widetilde{T} \leq T\}$, set $T_i := i\widetilde{T}$, $T_{\ell+1} = T$, and denote by u_0 the solution $u_{\widetilde{T}}$ of $(\mathcal{P})$ on $(0, \widetilde{T})$ whose existence has been established in Theorem 5.1. For each $i = 1, \ldots, \ell$ consider the Cauchy problem defined by induction:

$$(\mathcal{P}_i)\begin{cases} \dfrac{du_i}{dt}(t) + \partial\Phi(u_i(t)) + K \star \partial\Psi(u_i)(t) = F(t + T_i, u_i(t)) - R_i(t) \\ \text{for a.e. } t \in (0, \widetilde{T}) \\ \\ u_i(0) = u_{i-1}(\widetilde{T}) \end{cases}$$

where for every $t \in [0, T]$

$$R_i(t) := \sum_{k=1}^{i} \int_{T_{k-1}}^{T_k} K(t + T_i - s)\partial\Psi(u_{k-1}(s - T_{k-1}))ds.$$

Let $i = 1 \ldots \ell$. Existence of u_i can be obtained as in the proof of Theorem 5.1: substitute $F_i(t, u_i(t)) = F(t + T_i, u_i(t)) - R_i(t)$ for $F(t, u(t))$, and observe that $R_i \in W^{1,1}(0, \widetilde{T}, X)$ so that F_i satisfies (CI_1), (CI_2), (CI_3). Note that $u_{i-1}(\widetilde{T}) \in \mathrm{dom}(\partial\Phi)$ (repeat the first part of ***Step 2*** in the proof of Theorem 5.1 and reason by induction).

Finally we show that the function u defined for all $t \in [0, T]$ by

$$u(t) = \sum_{i=1}^{\ell} u_i(t - T_i)\mathbb{1}_{[T_i, T_{i+1}]}(t)$$

solves $(\mathcal{P})$. Indeed, for $t \in [T_i, T_{i+1}]$ the following calculation holds:

$$\begin{aligned} &\frac{du}{dt}(t) + \partial\Phi(u(t)) + \int_0^t K(t - s)\partial\Psi(u(s))ds - F(t, u(t)) \\ &= \frac{du_i}{dt}(t - T_i) + \partial\Phi(u_i(t - T_i)) + \sum_{k=1}^{i} \int_{T_{k-1}}^{T_k} K(t - s)\partial\Psi(u_{k-1}(s - T_{k-1}))ds \\ &\quad + \int_{T_i}^{t} K(t - s)\partial\Psi(u_i(s - T_i))ds - F(t, u_i(t - T_i)). \end{aligned}$$

Since $\sigma := t - T_i \in [0, T_{i+1} - T_i] = [0, \widetilde{T}]$, the second member is equal to

$$\begin{aligned} &\frac{du_i}{dt}(\sigma) + \partial\Phi(u_i(\sigma)) + \sum_{k=1}^{i} \int_{T_{k-1}}^{T_k} K(\sigma + T_i - s)\partial\Psi(u_{k-1}(s - T_{k-1}))ds \\ &\quad + \int_{T_i}^{\sigma + T_i} K(\sigma + T_i - s)\partial\Psi(u_i(s - T_i))ds - F(\sigma + T_i, u_i(\sigma)) \\ &= \frac{du_i}{dt}(\sigma) + \partial\Phi(u_i(t)) + R_i(\sigma) + \int_0^{\sigma} K(\sigma - s)\partial\Psi(u_i(s))ds - F(\sigma + T_i, u_i(\sigma)) \\ &= \frac{du_i}{dt}(\sigma) + \partial\Phi(u_i(t)) + K \star \partial\Psi(u_i)(\sigma) + R_i(\sigma) - F(\sigma + T_i, u_i(\sigma)) \end{aligned}$$

which, from $(\mathcal{P}_i)$, is equal to 0. Moreover $u(T_i^-) = u_{i-1}(\widetilde{T})$ and $u(T_i^+) = u_i(0) = u_{i-1}(\widetilde{T})$ so that $u \in C([0, T], X)$. □

5.3.2 *Existence and uniqueness when Ψ is a quadratic functional*

Proposition 5.2. *Under the conditions of Theorem 5.1, assume further that Ψ is a quadratic form in V. Then $(\mathcal{P})$ admits a unique solution.*

Proof. Let u_1 and u_2 be two solutions of $(\mathcal{P})$. This yields for a.e. $s \in (0,T)$,

$$\frac{d(u_1-u_2)}{dt}(s) + (\partial\Phi(u_1(s)) - \partial\Phi(u_2(s))) + K \star \partial\Psi(u_1-u_2)(s)$$
$$= F(s,u_1(s)) - F(s,u_2(s)). \tag{5.40}$$

Fix $t \in [0,T]$. Form the scalar product of (5.40) with $u_1(s)-u_2(s)$, and integrate over $(0,t)$. Taking into account the monotonicity of $\partial\Phi$ and (5.11), this gives

$$\frac{1}{2}\|u_1(t)-u_2(t)\|^2 \leq \int_0^t L(s)\|u_1(s)-u_2(s)\|^2 ds.$$

We conclude by applying the standard Grönwall's lemma □

From Proposition 5.2 and Theorem 5.2, we have

Corollary 5.1. *When Ψ is a quadratic functional, then $(\mathcal{P})$ admits a unique solution in $C([0,T],X)$.*

5.3.3 *Existence of a right derivative of the solutions at each $t \in [0,T[$*

Theorem 5.3 below is crucial to establish the convergence in Section 5.4. For its proof, condition (5.2) does not play any role.

Theorem 5.3. *Any solution u of $(\mathcal{P})$ admits a right derivative at every $t \in [0,T[$ which satisfies the equation:*

$$\frac{d^+u}{dt}(t) + \partial\Phi(u(t)) + K \star \partial\Psi(u)(t) = F(t,u(t)), \ t \in [0,T[. \tag{5.41}$$

Proof. ***Step 1.*** Fix t_0 in $[0,T[$ and write h to denote a sequence $(h_n)_{n\in\mathbb{N}}$ of positive numbers decreasing to 0. This step is devoted to the following estimate:

$$\limsup_{h\to 0}\left\|\frac{1}{h}(u(t_0+h)-u(t_0))\right\|_X \leq \|-\partial\Phi(u(t_0)) - K\star\partial\Psi(u)(t_0) + F(t_0,u(t_0))\|_X. \tag{5.42}$$

Observe that the constant function $v := u(t_0)$ satisfies

$$\frac{dv}{dt}(t) + \partial\Phi(v(t)) = \partial\Phi(u(t_0)) \tag{5.43}$$

for all $t \in (0,T)$. Subtract (5.43) from

$$\frac{du}{dt}(t) + \partial\Phi(u(t)) = -K\star\partial\Psi(u)(t) + F(t,u(t)),$$

form the scalar product with $u(t) - v(t)$ and integrate over $(t_0, t_0 + h)$. This yields

$$\frac{1}{2}\|u(t_0+h) - u(t_0)\|_X^2$$
$$\leq \int_{t_0}^{t_0+h} \langle -\partial\Phi(u(t_0)) - K \star \partial\Psi(u)(s) + F(s, u(s)), u(s) - u(t_0)\rangle \, ds.$$

According to the Grönwall type lemma, Lemma A.2 with $p = 2$, it follows that

$$\left\| \frac{1}{h}(u(t_0+h) - u(t_0)) \right\|_X$$
$$\leq \frac{1}{h}\int_{t_0}^{t_0+h} \| -\partial\Phi(u(t_0)) - K \star \partial\Psi(u)(s) + F(s, u(s))\|_X ds.$$

The conclusion follows by passing to the upper limit when $h \to 0^+$.

Step 2. We prove that

$$\frac{1}{h}(u(t_0+h) - u(t_0)) \rightharpoonup -\partial\Phi(u(t_0)) - K \star \partial\Psi(u)(t_0) + F(t_0, u(t_0)) \tag{5.44}$$

weakly in X. From ***Step 1*** a subsequence not relabeled of $\frac{1}{h}(u(t_0+h) - u(t_0))$ weakly converges to some w in X. For identifying w, take $(\xi, \xi^*) \in \partial\Phi$, i.e. $\xi^* = \partial\Phi(\xi)$. Then ξ satisfies equation

$$\frac{dv}{dt}(t) + \partial\Phi(v(t)) = \xi^* \tag{5.45}$$

for all $t \in (0, T)$. Subtract (5.45) from

$$\frac{du}{dt}(t) + \partial\Phi(u(t)) = -K \star \partial\Psi(u)(t) + F(t, u(t)),$$

form the scalar product with $u(t_0) - \xi$ and integrate over $(t_0, t_0 + h)$. This yields

$$\frac{1}{2}\|u(t_0+h) - \xi\|_X^2 - \frac{1}{2}\|u(t_0) - \xi\|_X^2$$
$$\leq \int_{t_0}^{t_0+h} \langle -K \star \partial\Psi(u)(s) + F(s, u(s)) - \xi^*, u(s) - \xi\rangle \, ds.$$

From the elementary inequality $2\langle a - b, b\rangle \leq \|a\|_X^2 - \|b\|_X^2$ we infer that

$$\left\langle \frac{1}{h}(u(t_0+h) - u(t_0)), u(t_0) - \xi \right\rangle$$
$$\leq \frac{1}{h}\int_{t_0}^{t_0+h} \langle -K \star \partial\Psi(u)(s) + F(s, u(s)) - \xi^*, u(s) - \xi\rangle \, ds.$$

Passing to the limit $h \to 0$ we find

$$\langle w, u(t_0) - \xi\rangle \leq \langle -K \star \partial\Psi(u)(t_0) + F(t_0, u(t_0)) - \xi^*, u(t_0) - \xi\rangle ,$$

thus

$$\langle -K \star \partial\Psi(u)(t_0) + F(t_0, u(t_0)) - w - \xi^*, u(t_0) - \xi\rangle \geq 0$$

for all $(\xi, \xi^*) \in \partial\Phi$, i.e.

$$(u(t_0), -K \star \partial\Psi(u)(t_0) + F(t_0, u(t_0)) - w)$$

is monotonically related to $\partial\Phi$ (see Definition B.4). Since $\partial\Phi$ is maximal monotone, from Proposition B.6, we deduce that $-K\star\partial\Psi(u)(t_0)+F(t_0,u(t_0))-w = \partial\Phi(u(t_0))$.

Step 3. (End of the proof). Combining (5.44), the lower semicontinuity of the norm, and (5.42), we deduce that

$$\lim_{h\to 0}\left\|\frac{1}{h}(u(t_0+h)-u(t_0))\right\|_X = \|-\partial\Phi(u(t_0)) - K\star\partial\Psi(u)(t_0) + F(t_0,u(t_0))\|_X.$$

Hence

$$\frac{1}{h}(u(t_0+h)-u(t_0)) \to -\partial\Phi(u(t_0)) - K\star\partial\Psi(u)(t_0) + F(t_0,u(t_0)) \text{ strongly in } X,$$

i.e.

$$\frac{d^+u}{dt}(t_0) + \partial\Phi(u(t_0)) + K\star\partial\Psi(u)(t_0) = F(t_0,u(t_0))$$

which completes the proof. □

5.4 Convergence under Mosco×Γ-convergence

5.4.1 *The abstract case*

In this section we use the framework of Section 5.1. Let $(\mathcal{P}_n)_{n\in\mathbb{N}}$ be a sequence of integrodifferential diffusion problems in $L^2(0,T,X)$ defined by

$$(\mathcal{P}_n)\begin{cases}\dfrac{du_n}{dt}(t) + \partial\Phi_n(u_n(t)) + \displaystyle\int_0^t K(t-s)\partial\Psi_n(u_n(s))\,ds = F_n(t)\\ \text{for a.e. } t\in(0,T)\\ \\ u_n(0) = u_n^0 \in \operatorname{dom}(\partial\Phi_n),\end{cases}$$

where $F_n \in L^2(0,T,X)\cap W^{1,1}(0,T,X)$ fulfills (CI_1), (CI_2), and (CI_3) with $F_n(t,u) = F_n(t)$ for all $t\in[0,T]$. In next Section 5.4.2, $X = L^2(\Omega)$ and the source F_n is structured as a reaction functional $F_n(t,u_n(t))$, as defined in Section 5.1.2. Since $F_n(t,u) = F_n(t)$, we have $L_n = 0$. Recall that $\Phi_n, \Psi_n : X \to]-\infty,+\infty]$ are lsc convex proper functionals with domain V. We assume that

- $\inf_{n,X} \Phi_n \geq 0$ and $\inf_{n,X} \Psi_n \geq 0$;
- the subdifferentials $\partial\Phi_n$ and $\partial\Psi_n$ are single valued (observe that this hypothesis is not closed under the Mosco and the $\Gamma_{w\text{-}V}$ convergence of $(\Phi_n)_{n\in\mathbb{N}}$ and $(\Psi_{n\lfloor V})_{n\in\mathbb{N}}$ respectively);
- $\operatorname{dom}(\partial\Phi_n) \subset \operatorname{dom}(\partial\Psi_n)$;
- conditions (5.1) and (5.2) hold uniformly in the sense that α_{ψ_n} and α_{Φ_n,Ψ_n} do not depend on n. We denote it by α_Ψ and $\alpha_{\Phi,\Psi}$ respectively.

Let $\widetilde{T} > 0$ satisfying $\widetilde{T}^{\frac{1}{2}} \|K\|_{L^2(0,T)} < \alpha_{\Phi,\Psi}$, by a particular solution of $(\mathcal{P}_n)$, we mean any solution in $C([0,T],X)$ obtained by translation-induction of a local solution $u_{n,\widetilde{T}} \in C([0,\widetilde{T}],X)$, whose existence is established in Theorem 5.2. Note that when Ψ_n is quadratic, according to Proposition 5.2, a particular solution is nothing but the unique solution of $(\mathcal{P}_n)$.

Theorem 5.4. *Under the general conditions above, assume furthermore that*

(STAB$_1$) $F_n \rightharpoonup F$ *in* $L^2(0,T,X)$, $\sup_{n\in\mathbb{N}} \|F_n(t)\|_X < +\infty$ *for all* $t \in [0,T]$, *and* $\sup_{n\in\mathbb{N}} \|\frac{dF_n}{dt}\|_{L^1(0,T,X)} < +\infty$;

(STAB$_2$) $\sup_{n\in\mathbb{N}} \Phi_n(u_n^0) < +\infty$;

(STAB$_3$) $u_n^0 \to u^0$ *strongly in* X;

(STAB$_4$) *there exists* $\Phi : X \to]-\infty,+\infty]$ *such that* $\Phi_n \overset{M}{\to} \Phi$;

(STAB$_5$) *there exists* $\Psi : V \to]-\infty,+\infty]$ *lsc convex proper, such that* $\Psi_{n\lfloor V} \overset{\Gamma_{w-V}}{\to} \Psi$.

Then any particular solution u_n *of* $(\mathcal{P}_n)$ *admits a subsequence which converges to* u *in* $C([0,T],X)$, *which solves the differential inclusion*

$$(\mathcal{P})\begin{cases} \dfrac{du}{dt}(t) + \partial\Phi(u(t)) + \displaystyle\int_0^t K(t-s)\big[\partial\Psi(u(s)) \cap X\big]\, ds \ni F(t) \ \text{a.e. } t \in (0,T) \\ u(0) = u^0 \in \mathrm{dom}(\partial\Phi). \end{cases}$$

Proof. We use the notation of the proof of Theorem 5.2, and do not relabel the various subsequences. Take $\widetilde{T} > 0$ satisfying $\widetilde{T}^{\frac{1}{2}} \|K\|_{L^2(0,T)} < \alpha_{\Phi,\Psi}$. Set $T_i := i\widetilde{T}$ for $i = 1,\dots,\ell$ where $\ell := \max\{k \in \mathbb{N} : k\widetilde{T} \le T\}$, and $T_{\ell+1} := \widetilde{T}$. Let $n \in \mathbb{N}$. According to Theorem 5.2, for $i = 0,\dots,\ell$, the restriction of u_n to $[T_i, T_{i+1}]$ is given by $u_n(t) = u_{i,n}(t - T_i)$ where $u_{i,n}$ is a solution in $C\left([0,\widetilde{T}],X\right)$ of

$$(\mathcal{P}_{i,n})\begin{cases} \dfrac{du_{i,n}}{dt}(t) + \partial\Phi_n(u_{i,n}(t)) + K \star \partial\Psi_n(u_{i,n})(t) = F_{i,n}(t) \text{ for a.e. } t \in \left(0,\widetilde{T}\right) \\ u_{i,n}(0) = u_{i-1,n}\left(\widetilde{T}\right) \in \mathrm{dom}(\partial\Phi_n), \end{cases}$$

with

$$F_{i,n}(t) := F_n(t+T_i) - \sum_{k=0}^{i-1} \int_{T_k}^{T_{k+1}} K(t+T_i-s)\partial\Psi_n(u_{k,n}(s-T_k))ds,$$

and by convention, $u_{-1,n}\left(\widetilde{T}\right) = u_n^0$ and $\sum_{k=0}^{-1} = 0$. We set

$$G_{i,n}(t) := F_{i,n}(t) - K \star \partial\Psi_n(u_{i,n})(t).$$

Then, $(\mathcal{P}_{i,n})$ may be written as

$$(\mathcal{P}_{i,n}) \begin{cases} \dfrac{du_{i,n}}{dt}(t) + \partial\Phi_n(u_{i,n}(t)) = G_{i,n}(t) \text{ for a.e. } t \in \left(0, \widetilde{T}\right) \\ u_{i,n}(0) = u_{i-1,n}\left(\widetilde{T}\right) \in \operatorname{dom}(\partial\Phi_n). \end{cases}$$

Our strategy is the following: for each $i = 0, \ldots \ell$, we show that $u_{i,n}$ converges to some u_i in $C\left(\left[0, \widetilde{T}\right], X\right)$, next we claim that u defined by $u(t) = u_i(t - T_i)$ for $t \in [T_i, T_{i+1}]$ solves $(\mathcal{P})$ and that $u_n \to u$ in $C([0,T], X)$. We proceed in this way to check the uniform estimates similar to those of Lemma 5.3 which require $\widetilde{T} > 0$ small enough, i.e. $\widetilde{T}^{\frac{1}{2}} \|K\|_{L^2(0,T)} < \alpha_{\Phi,\Psi}$.

Step 1. Reasoning by induction for $i = 0, \ldots, \ell$, we prove the following three assertions:

a) The following bounds hold for each $i = 0, \ldots, \ell$

$$\sup_{n\in\mathbb{N}} \|\partial\Psi_n(u_{i,n})\|_{L^2(0,\widetilde{T},X)} < +\infty, \tag{5.46}$$

$$\sup_{n\in\mathbb{N}} \left\| \frac{du_{i,n}}{dt} \right\|_{L^2(0,\widetilde{T},X)} < +\infty, \tag{5.47}$$

$$\sup_{n\in\mathbb{N}} \|u_{i,n}\|_{C(0,\widetilde{T},X)} < +\infty, \tag{5.48}$$

$$\sup_{n\in\mathbb{N}} \|\partial\Psi_n(u_{i,n}(t))\|_X < +\infty \text{ for all } t \in \left]0, \widetilde{T}\right], \tag{5.49}$$

$$\sup_{n\in\mathbb{N}} \|\partial\Phi_n(u_{i,n}(t))\|_X < +\infty \text{ for all } t \in \left]0, \widetilde{T}\right]. \tag{5.50}$$

b) For each $i = 0, \ldots, \ell$, there exists a subsequence of $(u_{i,n})_{n\in\mathbb{N}}$ which uniformly converges to some u_i in $C\left(\left[0, \widetilde{T}\right], X\right)$.

c) For each $i = 0, \ldots, \ell$ and for each $k = 0, \ldots, i$, there exists $\xi_k \in L^2\left(0, \widetilde{T}, X\right)$ with $\xi_k(t) \in \partial\Psi(u_k(t)) \cap X$ such that

$$\partial\Psi_n(u_{k,n}) \rightharpoonup \xi_k \text{ weakly in } L^2\left(0, \widetilde{T}, X\right).$$

Step $i = 0$. *Proof of a).* According to the uniform bounds (5.1), (5.2), (STAB$_1$), (STAB$_2$), $\inf_X \Phi_n \geq 0$, $\inf_X \Psi_n \geq 0$, and finally to the existence of a right derivative of $u_{i,n}$ at each $t \in \left]0, \widetilde{T}\right]$ (cf. Theorem 5.3), assertion a) is obtained by reproducing the proof of (5.17)–(5.24) with F_n substitute for F, Φ_n for Φ, and Ψ_n for Ψ_λ (unlike (5.23) and (5.24), we cannot claim that (5.49) and (5.50) hold for $t = 0$ because of the dependance on n of $u_{n,0}(0) = u_n^0$). We only establish (5.46) to highlight the importance of condition (5.2), and (5.49) to underline the need for hypothesis (STAB$_1$). For a.e. $t \in \left(0, \widetilde{T}\right)$, form the scalar product in X of $\partial\Psi_n(u_{0,n}(t))$ with the equation of the first formulation of $(\mathcal{P}_{i,n})$ and integrate over $\left(0,\widetilde{T}\right)$. This yields

$$\int_0^{\widetilde{T}} \frac{d}{dt}\Psi_n(u_{0,n}(t))dt + \int_0^{\widetilde{T}} \langle \partial\Phi_n(u_{0,n}(t)), \partial\Psi_n(u_{0,n}(t))\rangle\, dt$$
$$+ \int_0^{\widetilde{T}} \langle K \star \partial\Psi_n(u_{0,n})(t), \partial\Psi(u_{0,n}(t))\rangle\, dt$$
$$= \int_0^{\widetilde{T}} \langle F_n(t), \partial\Psi_n(u_{0,n}(t))\rangle\, dt. \tag{5.51}$$

An easy calculation gives

$$\|K \star \partial\Psi_n(u_{0,n})\|_{L^2(0,\widetilde{T},X)} \leq \widetilde{T}^{\frac{1}{2}} \|K\|_{L^2(0,T)} \|\partial\Psi_n(u_{0,n})\|_{L^2(0,\widetilde{T},X)}. \tag{5.52}$$

Combining (5.51), (5.52) and (5.2) we conclude that

$$\left[\alpha_{\Phi,\Psi} - \widetilde{T}^{\frac{1}{2}}\|K\|_{L^2(0,T)}\right] \|\partial\Psi_n(u_{0,n})\|^2_{L^2(0,\widetilde{T},X)}$$
$$\leq T\beta_{\Phi,\Psi} + \sup_{n\in\mathbb{N}} \Psi_n(u_{0,n}) + \sup_{n\in\mathbb{N}} \|F_n(t)\|_{L^2(0,\widetilde{T},X)} \|\partial\Psi_n(u_{0,n})\|_{L^2(0,\widetilde{T},X)}.$$

We deduce (5.46) from (STAB$_1$), (STAB$_2$), provided that $\widetilde{T}^{\frac{1}{2}}\|K\|_{L^2(0,T)} < \alpha_{\Phi,\Psi}$.

For establishing (5.49) first observe that

$$\sup_{n\in\mathbb{N}} \|K \star \partial\Psi_n(u_{0,n})\|_{W^{1,2}(0\widetilde{T},X)} < +\infty \tag{5.53}$$

(reproduce the proof of (5.31)). Next, from (5.53) and (STAB$_1$) we deduce that

$$\sup_{n\in\mathbb{N}} \left\|\frac{dG_{0,n}}{dt}\right\|_{L^1(0,\widetilde{T},X)} < +\infty. \tag{5.54}$$

Hence, combining

$$\left\|\frac{d^+u_{0,n}}{dt}(t)\right\|_X \leq \frac{1}{t}\int_0^t \left\|\frac{du_{0,n}}{dt}(s)\right\|_X ds + \int_0^t \left\|\frac{dG_{0,n}}{dt}(s)\right\|_X ds$$

for each $t \in]0,\widetilde{T}]$ (see Lemma 2.5), with (5.47) and (5.54), we obtain that for each $t \in]0,\widetilde{T}]$,

$$\sup_{n\in\mathbb{N}} \left\|\frac{d^+u_{0,n}}{dt}(t)\right\|_X < +\infty. \tag{5.55}$$

Form the scalar product of the equation

$$\frac{d^+u_{0,n}}{dt}(t) + \partial\Phi_n(u_{0,n}(t)) = G_{0,n}(t)$$

with $\partial\Psi_n(u_{0,n}(t))$ for each $t \in]0,\widetilde{T}]$. This yields

$$\left\langle \frac{d^+u_{0,n}}{dt}(t), \partial\Psi(u_{0,n}(t))\right\rangle + \alpha_{\Phi,\Psi}\|\partial\Psi(u_{0,n}(t))\|^2_X$$
$$\leq \langle G_{0,n}(t), \partial\Psi_n(u_{0,n}(t))\rangle - \beta_{\Phi,\Psi}$$

from which we deduce

$$\alpha_{\Phi,\Psi}\|\partial\Psi_n(u_{0,n}(t))\|_X^2 \leq \left(\left\|\frac{d^+u_{0,n}}{dt}(t)\right\|_X + \|G_{0,n}(t)\|_X\right)\|\partial\Psi_n(u_{0,n}(t))\|_X - \beta_{\Phi,\Psi}.$$

The claim follows from (5.55), and $\sup_{n\in\mathbb{N}}\|G_{0,n}(t)\|_X < +\infty$ obtained according to

$$\|G_{0,n}(t)\|_X \leq \|K\|_{L^2(0,T)}\|\partial\Psi_n(u_{0,n})\|_{L^2(0,\widetilde{T},X)} + \|F_n(t)\|_X$$

and (5.46) and (STAB_1).

Proof of b). From (5.47) and (5.48) we infer that the sequence $(u_{0,n})_{n\in\mathbb{N}}$ is bounded and uniformly equicontinuous in $C\left(\left[0,\widetilde{T}\right],X\right)$. Assertion b) then follows from the Ascoli compactness theorem provided that for each fixed $t \in [0,T]$, we establish that the set $E_0(t) := \{u_{0,n}(t) : n \in \mathbb{N}\}$ is relatively compact in X. For $t = 0$ we have $E_0(0) = \{u_n^0 : n \in \mathbb{N}\}$ so that the claim follows directly from (STAB_3). For $t \in]0,\widetilde{T}]$, (5.1) yields

$$\begin{aligned}\|u_{0,n}(t)\|_V^2 &\leq \frac{1}{\alpha_\Psi}\langle\partial\Psi_n(u_{0,n}(t)),u_{0,n}(t)\rangle \leq \frac{1}{\alpha_\Psi}\|\partial\Psi_n(u_{0,n}(t))\|_X\|u_{0,n}(t)\|_X\\ &\leq \frac{1}{\alpha_\Psi}\|\partial\Psi_n(u_{0,n}(t))\|_X\|u_{0,n}(t)\|_V\end{aligned}$$

and the claim follows from (5.49) and the compact embedding $V \hookrightarrow X$.

Proof of c). We have to establish the existence of $\xi_0 \in L^2\left(0,\widetilde{T},X\right)$ with $\xi_0 \in \partial\Psi(u_0(t))$ such that $\partial\Psi_n(u_{0,n}) \rightharpoonup \xi_0$ in $L^2\left(0,\widetilde{T},X\right)$. From (5.46) and the compact embedding $L^2\left(0,\widetilde{T},X\right) \hookrightarrow L^2\left(0,\widetilde{T},V'\right)$, there exist a subsequence of $(\partial\Psi_n(u_{0,n}))_{n\in\mathbb{N}}$ and $\xi_0 \in L^2\left(0,\widetilde{T},X\right)$ such that successively

$$\begin{aligned}&\partial\Psi_n(u_{0,n}) \rightharpoonup \xi_0 \text{ weakly in } L^2\left(0,\widetilde{T},X\right),\\ &\partial\Psi_n(u_{0,n}) \to \xi_0 \text{ strongly in } L^2\left(0,\widetilde{T},V'\right),\\ &\partial\Psi_n(u_{0,n}(t)) \to \xi_0(t) \text{ strongly in } V' \text{ for a.e. } t \in \left(0,\widetilde{T}\right),\\ &u_{0,n}(t) \rightharpoonup u_0(t) \text{ weakly in } V \text{ for each } t \in]0,\widetilde{T}]\end{aligned}$$

(the last convergence follows from (5.1), (5.49), and b) to identify the weak limit). According to (STAB_5) and the implication (cf. Theorem B.4)

$$\Psi_{n\lfloor V} \overset{\Gamma_{w-V}}{\to} \Psi \Longrightarrow \partial\Psi_{n\lfloor V} \overset{G_{w,s}}{\to} \partial\Psi,$$

the two last convergences above yield that for a.e. $t \in \left(0,\widetilde{T}\right)$, $u_0(t) \in \operatorname{dom}(\partial\Psi)$ and $\xi_0(t) \in \partial\Psi(u_0(t)) \cap X$.

Step $i > 1$ from steps $0,\cdots,i-1$. We assume that assertions a), b) and c) hold for all $k = 0,\dots i-1$ and we establish that they hold for index i. We do not relabel the subsequence obtained by successively extracting each subsequence from steps $k = 0,\dots i-1$.

Proof of a). We first claim that $F_{i,n}$ satisfies (STAB$_1$) for all $t \in [0, \widetilde{T}]$. From c) for $k = 0, \cdots i-1$, we easily deduce that

$$\sum_{k=0}^{i-1} \int_{T_k}^{T_{k+1}} K(\cdot + T_i - s)\partial\Psi_n(u_{k,n}(s - T_k))ds$$
$$\rightharpoonup \sum_{k=0}^{i-1} \int_{T_k}^{T_{k+1}} K(\cdot + T_i - s)\xi_k(s - T_k)ds. \tag{5.56}$$

in $L^2\left(0, \widetilde{T}, X\right)$. Hence from (STAB$_1$) and (5.56), $F_{i,n} \rightharpoonup F_i$ in $L^2\left(0, \widetilde{T}, X\right)$. On the other hand, from (STAB$_1$) and (5.49) for $k = 0, \cdots i-1$, we easily deduce that $\sup_{n\in\mathbb{N}} \|F_{i,n}(t)\|_X$ for all $t \in [0, \widetilde{T}]$. Finally from (5.46), (STAB$_1$) and

$$\frac{dF_{i,n}}{dt}(t) := \frac{dF_n}{dt}(t + T_i) - \sum_{k=0}^{i-1} \int_{T_k}^{T_{k+1}} K'(t + T_i - s)\partial\Psi_n(u_{k,n}(s - T_k))ds$$

for all $t \in [0, T]$, we infer that

$$\sup_{n\in\mathbb{N}} \left\| \frac{dF_{i,n}}{dt} \right\|_{L^1(0,\widetilde{T},X)} < +\infty,$$

which proves the claim. By repeating the arguments of the proof of a) at index $i = 0$ where $F_{i,n}$ is substitute for F_n, we obtain the estimates of i) provided that $\sup_{n\in\mathbb{N}} \Phi_n(u_{i,n}(0)) < +\infty$, that is to say $\sup_{n\in\mathbb{N}} \Phi_n\left(u_{i-1,n}\left(\widetilde{T}\right)\right) < +\infty$ (this condition replace (STAB$_2$)). For that, first note that from b) at index $i-1$

$$\sup_{n\in\mathbb{N}} \left\| u_{i-1,n}\left(\widetilde{T}\right) \right\|_X < +\infty. \tag{5.57}$$

Next, fix arbitrary $v \in \mathrm{dom}(\Phi)$. From refstab3 there exists a sequence $(v_n)_{n\in\mathbb{N}}$ such that $v_n \to v$ strongly in X and $\Phi_n(v_n) \to \Phi(v)$. The thesis then follows, from the convexity inequality

$$\Phi_n\left(u_{i-1,n}\left(\widetilde{T}\right)\right) \leq \Phi_n(v_n) + \left\langle \partial\Phi_n\left(u_{i-1,n}\left(\widetilde{T}\right)\right), u_{i-1,n}\left(\widetilde{T}\right) - v_n \right\rangle$$

and (5.50), (5.57).

Proof of b). The proof is exactly that of b) at index $i = 0$, by establishing that $E_i(t) := \{u_{i,n}(t) : n \in \mathbb{N}\}$ is relatively compact in X. Observe that for $t = 0$, $E_i(0) = \{u_{i-1,n}\left(\widetilde{T}\right) : n \in \mathbb{N}\}$ so that the claim follows directly from b) at index $i-1$.

Proof of c). The proof is exactly that of Step $i = 0$ by using estimates obtained in a).

Step 2. By using a method similar to that of the proof of Theorem 5.1, and from the convergences obtained in ***Step 1***, we are going to prove that u_i defined in b), ***Step 1***, solves the Cauchy problem

$$(\mathcal{P}_i) \begin{cases} \dfrac{du_i}{dt}(t) + \partial\Phi(u_i(t)) + K \star \partial\Psi(u_i)(t) \ni F_i(t) \text{ for a.e. } t \in \left(0, \widetilde{T}\right) \\ \\ u_i(0) = u_{i-1}\left(\widetilde{T}\right) \in \mathrm{dom}(\partial\Phi). \end{cases}$$

We will infer that the function u defined by $u(t) = u_i(t - T_i)$ for $t \in [T_i, T_{i+1}]$ converges toward u in $C([0,T], X)$ and, according to Theorem 5.2, solves $(\mathcal{P})$.

By using the Fenchel extremality condition, the equation of $(\mathcal{P}_{i,n})$ written with $G_{i,n}$ as second member, is equivalent to

$$\int_0^{\widetilde{T}} \left[\Phi_n(u_{i,n}(t)) + \Phi_n^*\left(G_{i,n}(t) - \frac{du_{i,n}}{dt}(t)\right) + \left\langle \frac{du_{i,n}}{dt}(t) - G_{i,n}(t), u_{i,n}(t)\right\rangle\right] dt$$
$$= 0,$$

where we have denoted by Φ_n^* the Legendre-Fenchel conjugate of Φ_n. Equivalently

$$I_{\Phi_n}(u_{i,n}) + I_{\Phi_n^*}\left(G_{i,n}(t) - \frac{du_{i,n}}{dt}\right) + \frac{1}{2}\left(\|u_{i,n}\left(\widetilde{T}\right)\|^2 - \|u_{i,n}(0)\|^2\right)$$
$$- \int_0^{\widetilde{T}} \langle G_{i,n}(t), u_{i,n}(t)\rangle \; dt = 0 \quad (5.58)$$

where the integral functionals I_{Φ_n} and $I_{\Phi_n^*}$ are respectively defined in $L^2\left(0, \widetilde{T}, X\right)$ by $I_{\Phi_n}(v) = \int_0^{\widetilde{T}} \Phi_n(v(t))dt$ and $I_{\Phi_n^*}(v) = \int_0^{\widetilde{T}} \Phi_n^*(v(t))dt$. From (STAB$_4$) and Lemma 2.6 we have

$$I_{\Phi_n} \overset{M}{\to} I_{\Phi}. \quad (5.59)$$

On the other hand, combining $u_{i,n}\left(\widetilde{T}\right) = u_{i,n}^0 + \int_0^{\widetilde{T}} \frac{du_{i,n}}{dt}(t)dt$ with $\frac{du_{i,n}}{dt} \rightharpoonup \frac{du_i}{dt}$ in $L^2\left(0, \widetilde{T}, X\right)$ which is obtained from (5.47), we infer that

$$\liminf_{n\to+\infty} \|u_{i,n}\left(\widetilde{T}\right)\|^2 \geq \|u_i\left(\widetilde{T}\right)\|^2. \quad (5.60)$$

Finally, from (STAB$_1$) and assertion c) of ***Step 1***

$$G_{i,n} \rightharpoonup G_i := F_i - K \star \xi_i \text{ weakly in } L^2\left(0, \widetilde{T}, X\right) \quad (5.61)$$

where

$$F_i(t) = F(t + T_i) - \sum_{k=0}^{i-1} \int_{T_k}^{T_{k+1}} K(\cdot + T_i - s)\xi_k(s - T_k)ds.$$

Hence, by passing to the lim inf in (5.58), from (5.59), (5.60), ***Step 1***, b) and (5.61), we infer that

$$\int_0^T \left[\Phi(u_i(t)) + \Phi^*\left(G_i(t) - \frac{du_i}{dt}(t)\right)\right] dt + \frac{1}{2}(\|u_i(T)\|^2 - \|u_i^0\|^2)$$
$$- \int_0^T \langle G_i(t), u_i(t)\rangle \; dt \leq 0$$

or equivalently,

$$\int_0^T \left[\Phi(u_i(t)) + \Phi^*\left(G_i(t) - \frac{du_i}{dt}(t)\right) + \left\langle \frac{du_i}{dt}(t) - G_i(t), u_i(t)\right\rangle\right] dt \leq 0, \quad (5.62)$$

from which we conclude that

$$\frac{du_i}{dt}(t) + \partial\Phi(u_i(t)) \ni G_i(t) \text{ for a.e. } t \in (0, T).$$

The initial condition $u_i(0) = u_{i-1}\left(\widetilde{T}\right)$ is obtained from

$$u_i(0) = \lim_{n\to+\infty} u_{i,n}(0) = \lim_{n\to+\infty} u_{i-1,n}\left(\widetilde{T}\right) = u_{i-1}\left(\widetilde{T}\right).$$

Finally we claim that $u_{i-1}\left(\widetilde{T}\right) \in \mathrm{dom}(\partial\Phi)$. The claim follows from

$$u_{i-1,n}\left(\widetilde{T}\right) \in \mathrm{dom}(\partial\Phi_n),$$
$$\lim_{n\to+\infty} u_{i-1,n}\left(\widetilde{T}\right) = u_{i-1}\left(\widetilde{T}\right) \text{ strongly in } X,$$

($STAB_4$) and Theorem B.4. This completes the proof. □

Remark 5.3. Let us strengthen ($STAB_5$) by:

($STAB'_5$) there exists $\Psi : X \to]-\infty, +\infty]$ lsc convex proper, such that $\mathrm{dom}(\Psi) = V$ and $\Psi_{n\lfloor V} \overset{\Gamma_{w-V}}{\to} \Psi_{\lfloor V}$.

Then, with the notation above, we can assert that $\partial\Psi_{\lfloor V}(u(s)) \cap X = \partial\Psi$. The limit problem then becomes

$$(\mathcal{P})\begin{cases} \dfrac{du}{dt}(t) + \partial\Phi(u(t)) + \displaystyle\int_0^t K(t-s)\partial\Psi(u(s))ds \ni F(t) \text{ for a.e. } t \in (0,T) \\ u(0) = u^0, \ u^0 \in \mathrm{dom}(\partial\Phi). \end{cases}$$

Moreover, from Proposition 5.1, $\mathrm{dom}(\partial\Phi) \subset \mathrm{dom}(\partial\Psi)$ and Φ and Ψ fulfill conditions (5.1), (5.2). Therefore, under condition ($STAB'_5$), Theorem 5.4 may be considered as a stability result although $\partial\Phi$ and $\partial\Psi$ are not single valued in general.

5.4.2 *The case $X = L^2(\Omega)$*

In this section Ω is a bounded domain of $\mathbb{R}^N$, $X = L^2(\Omega)$, and $V = H_0^1(\Omega)$. We keep the same conditions on $(\Phi_n)_{n\in\mathbb{N}}$, $(\Psi_n)_{n\in\mathbb{N}}$ and $(u_n^0)_{n\in\mathbb{N}}$ but we further specify the structure of the source F_n. Given a positive integer l, $r_n \in L^\infty((0,T)\times\mathbb{R}^N, \mathbb{R}^l) \cap W^{1,1}(0,T,L^2_{loc}(\mathbb{R}^N,\mathbb{R}^l))$, $q_n \in L^2(0,T,L^2_{loc}(\mathbb{R}^N)) \cap W^{1,1}(0,T,L^2_{loc}(\mathbb{R}^N))$, and $g_n : \mathbb{R} \to \mathbb{R}^l$ a uniformly bounded and Lipschitz continuous function, we consider the reaction functional F_n defined for all $v \in L^2(\Omega)$ and all $(t,x) \in \mathbb{R}_+ \times \Omega$ by

$$F_n(t,u)(x) = r_n(t,x) \cdot g_n(u(x)) + q_n(t,x)$$

for all $n \in \mathbb{N}$. We assume that

$$\begin{cases} \sup\limits_{n\in\mathbb{N}} \displaystyle\int_0^T \left\| \frac{dr_n}{dt} \right\|_{L^2(\Omega,\mathbb{R}^l)} dt < +\infty, \\ \sup\limits_{n\in\mathbb{N}} \displaystyle\int_0^T \left\| \frac{dq_n}{dt} \right\|_{L^2(\Omega)} dt < +\infty. \end{cases} \tag{5.63}$$

Denote by L_g the uniform Lipschitz constant of the functions g_n and by $M_g = \sup_{r\in\mathbb{R}} |g_n(r)|$ their uniform norm. Then, as noticed in Section 5.1, F_n satisfies (CI_1), (CI_2) and (CI_3) with $L_n = \|r_n\|_{L^\infty((0,T)\times\mathbb{R}^N,\mathbb{R}^l)} Lg$ and $\Theta_n(\tau) = M_g \|\frac{dr_n}{dt}(\tau,\cdot)\|_{L^2(\Omega,\mathbb{R}^l)} + \|\frac{dq_n}{dt}(\tau,\cdot)\|_{L^2(\Omega)}$. Theorem 5.5 below states a concrete version of Theorem 5.4.

Theorem 5.5. *In addition to* (STAB_2)–(STAB_5), *assume that*

$\sup_{n\in\mathbb{N}} \|r_n\|_{L^\infty([0,T]\times\mathbb{R}^N,\mathbb{R}^l)} < +\infty$ *and*

$r_n \rightharpoonup r$ *weakly for the* $\sigma(L^\infty(0,T,L^2(\Omega,\mathbb{R}^l)), L^1(0,T,L^2(\Omega,\mathbb{R}^l)))$ *topology;*

$g_n \to g$ *pointwise in* $\mathbb{R}^l$;

$\sup_{n\in\mathbb{N}} \|q_n(t,\cdot)\|_{L^2(\Omega)} < +\infty$, *and* $q_n \rightharpoonup q$ *weakly in* $L^2(0,T,L^2(\Omega))$ *for each* $t \in [0,T]$.

Then any particular solution u_n *of* $(\mathcal{P}_n)$ *admits a subsequence which converges to* u *in* $C([0,T],X)$, *solution of*

$$(\mathcal{P})\begin{cases} \dfrac{du}{dt}(t) + \partial\Phi u(t) + \displaystyle\int_0^t K(t-s)\big[\partial(\Psi)(u(s)) \cap L^2(\Omega)\big]\, ds \ni F(t,u(t)) \\ \text{for a.e. } t \in (0,T) \\ \\ u(0) = u^0,\ u^0 \in \mathrm{dom}(\partial\Phi), \end{cases}$$

where $F(t,v)(x) = r(t,x)\cdot g(v(x)) + q(t,x)$ *for all* $(t,x) \in [0,T]\times\Omega$ *and all* $v \in L^2(\Omega)$.

Proof. We use the notation of the proof of Theorem 5.4. Clearly, for all $t \in [0,T]$,

$$\sup_{n\in\mathbb{N}} \|F_n(t,u_n(t))\|_{L^2(\Omega)} < +\infty.$$

On the other hand from (CI_3) and the uniform bounds (5.63) we easily deduce that

$$\sup_{n\in\mathbb{N}} \left\| \frac{dF_n}{dt} \right\|_{L^1(0,T,L^2(\Omega))} < +\infty.$$

Therefore ***Step 1*** of the proof of Theorem 5.4 is still valid. The rest of the proof mimics that of ***Step 2***. We only have to establish that $F_n(\cdot,u_{i,n}(\cdot)) \rightharpoonup F(\cdot,u_i(\cdot))$ in $L^2(0,\widetilde{T},L^2(\Omega))$ for all $i = 0,\dots,\ell$. This convergence is a straightforward consequence of the weak convergences $r_n \rightharpoonup r$, $q_n \rightharpoonup q$ and the pointwise convergence $g_n \to g$ together with the uniform bound of g_n. □

Chapter 6

Variational convergence of a class of functionals indexed by Young measures

The topic of this chapter, of more abstract aspect, differs significantly from the previous; however, on a strictly mathematical level, it has its own interest. Although it is essentially preparatory to Chapter 11, in the spirit of the general framework of Part 1 of the book, we investigate in Section 6.2 the convergence with emergence of nonlocal effects in non-diffusive reaction differential equations of the type

$$\begin{cases} -\dfrac{\partial u_\varepsilon}{\partial t}(t,x) = \dfrac{\partial \psi_\varepsilon}{\partial s}(t,x,u_\varepsilon(t,x)), & \text{for a.e. } (t,x) \in (0,T) \times \Omega, \\ u_\varepsilon(0,x) = u_0, \end{cases} \tag{6.1}$$

where $\zeta \mapsto \psi_\varepsilon(t,x,\zeta)$ is a convex potential. The results are obtained as a direct consequence of the abstract study of Section 6.1. For other applications and a discussion on nonlocal effects related to Young measure, we refer the reader to Anza Hafsa *et al.* (2020c) and references therein. For nonlocal diffusion problems, we refer the reader to Andreu-Vaillo *et al.* (2010), Chipot and Lovat (2001), Chang and Chipot (2003, 2004) and references therein.

Let us start by fixing the notation and by defining the various classes of functions and functionals involved in this chapter. Let T be a positive real number fixed all along the chapter. Given (α,β) in $\mathbb{R}^2$, with $0 < \alpha \leq \beta < +\infty$, and a continuous function $\gamma : \mathbb{R}_+ \to \mathbb{R}_+$ satisfying $\lim_{r\to 0} \gamma(r) = 0$, we denote by $\mathbb{E}_{\alpha,\beta,\gamma}$ the set of functions $\Lambda : [0,T] \times \mathbb{R}^N \times \mathbb{R}^2 \to \mathbb{R}$, $\left(t,x,\zeta,\dot\zeta\right) \mapsto \Lambda\left(t,x,\zeta,\dot\zeta\right)$ satisfying the following conditions:

- for all $(t,x) \in [0,T] \times \mathbb{R}^N$, $\left(\zeta,\dot\zeta\right) \mapsto \Lambda\left(t,x,\zeta,\dot\zeta\right)$ is convex;
- for all $\left(t,x,\zeta,\dot\zeta\right) \in [0,T] \times \mathbb{R}^N \times \mathbb{R}^2$,

$$\alpha\left(\left|\left(\zeta,\dot\zeta\right)\right|^2 - 1\right) \leq \Lambda\left(t,x,\zeta,\dot\zeta\right) \leq \beta\left(1 + \left|\left(\zeta,\dot\zeta\right)\right|^2\right); \tag{6.2}$$

- for all $\left(\zeta,\dot{\zeta}\right)\in\mathbb{R}^2$, all $(t_1,x_1)\in[0,T]\times\mathbb{R}^N$, and all $(t_2,x_2)\in[0,T]\times\mathbb{R}^N$,

$$\left|\Lambda\left(t_2,x_2,\zeta,\dot{\zeta}\right)-\Lambda\left(t_1,x_1,\zeta,\dot{\zeta}\right)\right| \leq \gamma\left(|t_2-t_1|+|x_2-x_1|\right)\left(1+\left|\left(\zeta,\dot{\zeta}\right)\right|^2\right). \tag{6.3}$$

In addition, we are given a convex function $\Theta:\mathbb{R}\to\mathbb{R}_+$ of at most 2-polynomial growth, i.e. there exists $\beta'\geq 0$ such that for all $\zeta\in\mathbb{R}$,

$$0\leq\Theta(\zeta)\leq\beta'(1+|\zeta|^2). \tag{6.4}$$

Note that Θ is locally Lipschitz, more precisely, there exists a constant L_Θ depending on β' such that, for all ζ and all ζ' in $\mathbb{R}$,

$$|\Theta(\zeta)-\Theta(\zeta')|\leq L_\Theta|\zeta-\zeta'|(1+|\zeta|+|\zeta'|). \tag{6.5}$$

Using standard topological arguments, Lemma 6.2 below states that $\mathbb{E}_{\alpha,\beta,\gamma}$ is a compact metrizable space when equipped with the topology of uniform convergence on compact subsets of $[0,T]\times\mathbb{R}^N\times\mathbb{R}^2$, and that the evaluation map $\mathcal{E}:\left(\Lambda,t,x,\zeta,\dot{\zeta}\right)\mapsto\Lambda\left(t,x,\zeta,\dot{\zeta}\right)$ from $\mathbb{E}_{\alpha,\beta,\gamma}\times[0,T]\times\mathbb{R}^N\times\mathbb{R}^2$ into $\mathbb{R}$ is continuous.

Let Ω be a bounded domain in $\mathbb{R}^N$, $N\in\mathbb{N}^*$, and $\mathcal{X}$ a Polish subspace of $(\mathbb{E}_{\alpha,\beta,\gamma},\mathbf{d})$, namely a separable and complete subspace for the induced metric $\mathbf{d}$. We denote by $\mathcal{Y}(\Omega,\mathcal{X})$ the topological space of Young measures on $\Omega\times\mathcal{X}$ equipped with the narrow topology (see Appendix E.2). For each $\mu\in\mathcal{Y}(\Omega,\mathcal{X})$ we introduce the following spaces:

- $\mathcal{V}(0,T,L^2(\Omega))$ the subspace of $H^1(0,T,L^2(\Omega))$ made up of all the functions u satisfying $u(0,\cdot)=0$, endowed with the weak topology of $H^1(0,T,L^2(\Omega))$,
- $\mathcal{V}(0,T,L^2_\mu(\Omega\times\mathcal{X}))$ the subspace of $H^1(0,T,L^2_\mu(\Omega\times\mathcal{X}))$ made up of all the functions U satisfying $U(0,\cdot,\cdot)=0$.

We write $\dot{u}$, $\dot{U}$ rather than $\frac{du}{dt}$, $\frac{dU}{dt}$, to denote the distributional derivative with respect to t of $u\in H^1(0,T,L^2(\Omega))$ and $U\in\mathcal{V}(0,T,L^2_\mu(\Omega\times\mathcal{X}))$ respectively. We write each μ of $\mathcal{Y}(\Omega,\mathcal{X})$ in the form $\mu=dx\otimes\mu_x$ to indicate its disintegration. Recall that $(\mu_x)_{x\in\Omega}$ is a family of probability measures on $\mathcal{X}$ (see Theorem E.11). Then, for every $\mu\in\mathcal{Y}(\Omega,\mathcal{X})$, we define the functional $\Phi_\mu:\mathcal{V}(0,T,L^2(\Omega))\to\mathbb{R}$ by

$$\Phi_\mu(u) = \inf_{U\in X_\mu(u)}\left\{\int_{(0,T)\times\Omega\times\mathcal{X}}\mathcal{E}\left(\Lambda,t,x,U(t,x,\Lambda),\dot{U}(t,x,\Lambda)\right)dt\otimes d\mu(x,\Lambda)+K_\mu(U)\right\}$$

where

$$K_\mu(U)=\int_{\Omega\times\mathcal{X}}\Theta(U(T,x,\Lambda))d\mu(x,\Lambda),$$

and

$$X_\mu(u) = \left\{ U \in \mathcal{V}(0,T,L^2_\mu(\Omega \times \mathcal{X})) : \int_{\mathcal{X}} U(t,x,\Lambda)\, d\mu_x(\Lambda) = u(t,x) \text{ for a.e. } (t,x) \right\}.$$

Finally we set $\mathcal{F} = \{\Phi_\mu : \mu \in \mathcal{Y}(\Omega, \mathcal{X})\}$. The main objective of this chapter is to establish the sequential continuity of the map $\mu \mapsto \Phi_\mu$ from $\mathcal{Y}(\Omega, \mathcal{X})$ endowed with the narrow topology into $\mathcal{F}$ equipped with the Γ-convergence (Theorem 6.1). For a precise notation and relevant definitions, we refer to Section 6.1 and Appendix E.2.

Given for each $\varepsilon > 0$ a measurable map $f_\varepsilon : \Omega \to \mathcal{X}$, as a straightforward consequence, in Corollary 6.1 we infer that the integral functional defined for every u in $\mathcal{V}(0,T,L^2(\Omega))$ by

$$\Phi_{\mu_\varepsilon}(u) = \int_{(0,T)\times\Omega} f_\varepsilon(x)(x,t,u(t,x),\dot{u}(t,x))\, dt \otimes dx + \int_\Omega \Theta(u((T,x)))\, dx \tag{6.6}$$

Γ-converges as $\varepsilon \to 0$ to the functional Φ_μ whenever $\mu_\varepsilon := dx \otimes \delta_{f_\varepsilon(x)}$ narrow converges to some Young measure μ. When the disintegration of the limit measure μ is a family of Dirac measures, i.e. when there exists $f : \Omega \to \mathcal{X}$ such that $\mu = dx \otimes \delta_{f(x)}$, then Φ_μ is local with the integral representation of the form (6.6) where f is substituted for f_ε. The convergence in measure of f_ε to f is a sufficient condition which ensures that the limit functional Φ_μ is local (see Corollary 6.2). For a discussion about the local property of Φ_μ in relation to the atomic character of μ, refer to (Anza Hafsa *et al.*, 2020c, Remarks 2.10, 2.11). It is worthwhile to note that a particular case of integral functional (6.6) arises in the context of non-diffusive reaction differential equations (6.1). Indeed the solution of (6.1) in $H^1(0,T,L^2(\Omega))$ is a minimizer of Φ_{μ_ε} with

$$f_\varepsilon(x)\left(t,y,\zeta,\dot{\zeta}\right) = \psi_\varepsilon(t,x,\zeta) + \psi^*_\varepsilon\left(t,x,-\dot{\zeta}\right), \Theta(\zeta) = \frac{1}{2}|\zeta|^2.$$

In this specific case, $f_\varepsilon(x)$ is a function, element of $\mathcal{X}$, which does not depend on the y-variable. Corollary 6.3 states a characterization of the weak cluster points of the solutions of (6.1) in terms of the Young measure generated by the sequence $(f_\varepsilon(x))_{\varepsilon>0}$. Furthermore, when ψ does not depend on t and $\zeta \mapsto \psi_\varepsilon(x,\zeta)$ pointwise converges toward a C^1-convex function for a.e. $x \in \mathbb{R}^N$, Proposition 6.3 states that the sequence of problems (6.1) is stable.

6.1 The main continuity result

We take back the notation and definitions of Appendix B.1, E.2 and focus on the continuity of the map $\mu \mapsto \Phi_\mu$ from $\mathcal{Y}(\Omega, \mathcal{X})$ equipped with the narrow convergence into $\mathcal{F}$ equipped with the Γ-convergence, when the space $\mathcal{V}(0,T,L^2(\Omega))$ is equipped with the weak convergence of $H^1(0,T,L^2(\Omega))$. To shorten the notation, we sometimes denote by $\mathcal{O}$ the product space $(0,T)\times\Omega$, by dt and dx (or $\mathcal{L}_N$) the Lebesgue measures on $(0,T)$ and Ω respectively, and we sometimes write S to denote the couple $\left(\zeta,\dot{\zeta}\right)$ of $\mathbb{R}^2$. It must be notice that each element $u \in H^1(0,T,L^2(\Omega))$, resp.

$U \in H^1(0,T,L^2_\mu(\Omega\times\mathcal{X}))$ can be identified with an element in $L^2((0,T)\times\Omega)$, respectively in $L^2_{dt\otimes\mu}((0,T)\times\Omega\times\mathcal{X})$; idem for $\dot{u}$ and $\dot{U}$. In all the chapter, we do not distinguish u, respectively U, $\dot{u}$, respectively $\dot{U}$ from their identifications. We systematically apply Fubini's theorem, without the basic justifications.

Lemma 6.1. *There exists a positive constant $L\in\mathbb{R}$ such that for all Λ in $\mathbb{E}_{\alpha,\beta,\gamma}$, all (t_1,x_1,S_1) and all (t_2,x_2,S_2) in $[0,T]\times\mathbb{R}^N\times\mathbb{R}^2$,*

$$\begin{aligned}|\Lambda(t_1,x_1,S_1)-\Lambda(t_2,x_2,S_2)| &\le \gamma(|t_2-t_1|+|x_2-x_1|)(1+|S_1|^2)\\ &\quad + L|S_2-S_1|(1+|S_1|+|S_2|).\end{aligned}\tag{6.7}$$

Proof. From the uniform growth condition (6.2), and the convexity of $S\mapsto\Lambda(t,x,S)$, it is easy to prove that there exists a positive constant $L\in\mathbb{R}$ such that for all $\Lambda\in\mathbb{E}_{\alpha,\beta,\gamma}$, all $(t,x)\in[0,T]\times\mathbb{R}^N$ and all (S_1,S_2) in $\mathbb{R}^2\times\mathbb{R}^2$,

$$|\Lambda(t,x,S_1)-\Lambda(t,x,S_2)|\le L|S_2-S_1|(1+|S_1|+|S_2|).\tag{6.8}$$

Estimate (6.7) is obtained by combining (6.8) and (6.3). □

Lemma 6.2. *The space $\mathbb{E}_{\alpha,\beta,\gamma}$ equipped with the topology of uniform convergence on compact subsets of $[0,T]\times\mathbb{R}^N\times\mathbb{R}^2$ is compact metrizable. Moreover, the evaluation map $\mathcal{E}:(\Lambda,(t,x,S))\mapsto\Lambda(t,x,S)$ from $\mathbb{E}_{\alpha,\beta,\gamma}\times[0,T]\times\mathbb{R}^N\times\mathbb{R}^2$ into $\mathbb{R}$ is continuous.*

Proof. Let $n\in\mathbb{N}$. Let denote by $\|\cdot\|_{\infty,n}$ the uniform norm on $[0,T]\times\overline{B}_n$, where $\overline{B}_n$ is the closed ball of $\mathbb{R}^{N+2}$, centered at 0, with radius n. Define the metric $\mathbf{d}:\mathbb{E}_{\alpha,\beta,\gamma}\times\mathbb{E}_{\alpha,\beta,\gamma}\to[0,+\infty[$ by

$$\mathbf{d}(\Lambda_1,\Lambda_2)=\sum_{n=0}^{\infty}\frac{\inf(1,\|\Lambda_1-\Lambda_2\|_{\infty,n})}{2^{n+1}}\tag{6.9}$$

for all Λ_1,Λ_2 in $\mathbb{E}_{\alpha,\beta,\gamma}\times\mathbb{E}_{\alpha,\beta,\gamma}$. On the other hand, for each $n\in\mathbb{N}$, consider the following set of restrictions of the functions Λ:

$$X_n=\left\{\Lambda_{\lfloor[0,T]\times\overline{B}_n}:\Lambda\in\mathbb{E}_{\alpha,\beta,\gamma}\right\},$$

that we equip with the uniform norm $\|\cdot\|_{\infty,n}$. Then, $\mathbb{E}_{\alpha,\beta,\gamma}$ equipped with the distance $\mathbf{d}$ is homeomorphic to the product $\prod_{n=0}^{+\infty}X_n$. Moreover each normed spaces $(X_n,\|\cdot\|_{\infty,n})$ is compact according to the Ascoli compactness theorem: the relative pointwise compactness follows from (6.2), and the equicontinuity, from (6.7). The first claim then follows from the Tychonov compactness theorem.

Let (Λ_1,t_1,x_1,S_1) and (Λ_2,t_2,x_2,S_2) be two elements of $\mathbb{E}_{\alpha,\beta,\gamma}\times[0,T]\times\mathbb{R}^N\times\mathbb{R}^2$. For n large enough, (t_i,x_i,S_i), $i=1,2$, belong to $[0,T]\times\overline{B}_n$ so that, for Λ_1 closed to Λ_2 for the distance $\mathbf{d}$, and from (6.7), we have

$$\begin{aligned}&|\mathcal{E}(\Lambda_1,t_1,x_1,S_1)-\mathcal{E}(\Lambda_2,t_2,x_2,S_2)|\\ &\le|\Lambda_1(t_1,x_1,S_1)-\Lambda_1(t_2,x_2,S_2)|+|(\Lambda_1-\Lambda_2)(t_2,x_2,S_2)|\\ &\le C_n(\gamma(|t_2-t_1|+|x_2-x_1|)+L|S_2-S_1|)+2^{n+1}d(\Lambda_1,\Lambda_2)\end{aligned}$$

where $C_n=\sup\{\max(1+|S_1|^2,|S_1|+|S_2|):(x,S_i)\in\overline{B}_n,\ i=1,2\}$. This completes the proof. □

From Lemma 6.2, and the general theory of Young measures (see Appendix E.2), for any Polish subspace $\mathcal{X}$ of $(\mathbb{E}_{\alpha,\beta,\gamma}, \mathbf{d})$, we can consider the space of Young measures $\mathcal{Y}(\Omega, \mathcal{X})$, and for each $\mu \in \mathcal{Y}(\Omega, \mathcal{X})$, the functional Φ_μ defined in the introduction. Observe that $X_\mu(u)$ is non empty since $u \in X_\mu(u)$. Note also that each two integrals expressing Φ_μ are well defined. Indeed the measurability of the map $(t, x, \Lambda) \mapsto \Lambda\left(t, x, U(t, x, \Lambda), \dot{U}(t, x, \Lambda)\right)$ is a straightforward consequence of the continuity of the evaluation map $\mathcal{E} : (\Lambda, t, x, S) \mapsto \Lambda(t, x, S)$, and the measurability of $(t, x, \Lambda) \mapsto \left(U(t, x, \Lambda), \dot{U}(t, x, \Lambda)\right)$. Moreover, from the growth condition (6.2), and since U belongs to $\mathcal{V}(0, T, L^2_\mu(\Omega \times \mathcal{X}))$, we have

$$\int_{\mathcal{O}\times\mathcal{X}} \Lambda\left(t, x, U(t, x, \Lambda), \dot{U}(t, x, \Lambda)\right) d\mu(x, \Lambda)dt$$
$$\leq \beta \int_{\mathcal{O}\times\mathcal{X}} \left(1 + \left|(U, \dot{U})\right|^2\right) d\mu(x, \Lambda)dt < +\infty.$$

Similarly, from the continuity of Θ which fulfills the growth condition (6.4), the second integral is clearly well defined.

Remark 6.1. The functional, Φ_μ satisfies the growth conditions

$$\alpha \int_{\mathcal{O}} (|(u, \dot{u})|^2 - 1)\, dt \otimes dx \leq \Phi_\mu(u) \leq \beta \int_{\mathcal{O}} (1 + |(u, \dot{u})|^2)\, dt \otimes dx$$
$$+ \beta' \int_\Omega (1 + |u(T, x)|^2)\, dx. \tag{6.10}$$

Indeed, the upper bound follows from (6.2) after taking $U = u$ as admissible function in $X_\mu(u)$. The lower bound follows from

$$\begin{aligned}\Phi_\mu(u) &\geq \alpha \inf \left\{\int_{\mathcal{O}\times\mathcal{X}} \left|\left(U, \dot{U}\right)\right|^2 dt \otimes d\mu : U \in X_\mu(u)\right\} - \alpha T \mathcal{L}_N(\Omega) \\ &\geq \alpha \inf \left\{\int_{\mathcal{O}} \left|\int_{\mathcal{X}} \left(U, \dot{U}\right) d\mu_x\right|^2 dt \otimes dx : U \in X_\mu(u)\right\} - \alpha T \mathcal{L}_N(\Omega) \\ &= \alpha \int_{\mathcal{O}} (|(u, \dot{u})|^2 - 1) dt \otimes dx.\end{aligned}$$

Theorem 6.1. *The map $\mu \mapsto \Phi_\mu$ is sequentially continuous from $\mathcal{Y}(\Omega, \mathcal{X})$ equipped with the narrow topology into $\mathcal{F}$ equipped with the Γ-convergence.*

According to the definition of the Γ-convergence (see Appendix B.1), the proof is the consequence of the two propositions, Proposition 6.1 and Proposition 6.2 below, and is postpone to Section 6.1.3.

6.1.1 *The lower bound*

Proposition 6.1. *Assume that the sequence $(\mu_n)_{n\in\mathbb{N}}$ narrow converges to μ in $\mathcal{Y}(\Omega, \mathcal{X})$. Then, for all u_n weakly converging to u in $\mathcal{V}(0, T, L^2(\Omega))$, we have*

$$\Phi_\mu(u) \leq \liminf_{n\to+\infty} \Phi_{\mu_n}(u_n).$$

Proof. In what follows, the symbol $\otimes$ denotes the product measure operation. Without loss of generality, we may assume that $\liminf_{n\to+\infty}\Phi_{\mu_n}(u_n)<+\infty$. By using the direct method of the calculus of variations we can assert that there exists a minimizer $U_n\in X_{\mu_n}(u_n)$ of Φ_{μ_n}. Then

$$\sup_{n\in\mathbb{N}}\left\|\left(U_n,\dot{U}_n\right)\right\|_{L^2(\mu_n\otimes dt(\mathcal{O}\times\mathcal{X},\mathbb{R}^2))}<+\infty. \tag{6.11}$$

First consider the Borel measurable map $\mathcal{G}_n:\mathcal{O}\times\mathcal{X}\to\mathcal{O}\times\mathcal{X}\times\mathbb{R}^2$ defined by

$$(t,x,\Lambda)\mapsto\Big(t,x,\Lambda,U_n(t,x,\Lambda),\dot{U}_n(t,x,\Lambda)\Big),$$

and the push-forward measure $\nu_n=\mathcal{G}_n\#dt\otimes\mu_n$ of $dt\otimes\mu_n$ by $\mathcal{G}_n$. Clearly $\nu_n\in\mathcal{Y}(\mathcal{O},\mathcal{X}\times\mathbb{R}^2)$ with the projection measure $dt\otimes dx$ on $\mathcal{O}$. We claim that the sequence $(\nu_n)_{n\in\mathbb{N}}$ is tight in $\mathcal{Y}(\mathcal{O},\mathcal{X}\times\mathbb{R}^2)$ (see Definition E.12 in Appendix E). Indeed, for any compact subsets $\mathcal{K}$ and K of $\mathcal{X}$ and $\mathbb{R}^2$ respectively, we have

$$\begin{aligned}
&\nu_n(\mathcal{O}\times(\mathcal{K}\cap K)^c)\\
&=dt\otimes\mu_n\left(\left\{(t,x,\Lambda)\in\mathcal{O}\times\mathcal{X}:\Big(\Lambda,U_n(t,x,\Lambda),\dot{U}_n(t,x,\Lambda)\Big)\in(\mathcal{K}\cap K)^c\right\}\right)\\
&\le dt\otimes\mu_n(\mathcal{O}\times\mathcal{K}^c)\\
&\quad+dt\otimes\mu_n\left(\left\{(t,x,\Lambda)\in\mathcal{O}\times\mathcal{X}:\Big(U_n(t,x,\Lambda),\dot{U}_n(t,x,\Lambda)\Big)\in K^c\right\}\right)\\
&=T\mu_n(\mathcal{O}\times\mathcal{K}^c)\\
&\quad+dt\otimes\mu_n\left(\left\{(t,x,\Lambda)\in\mathcal{O}\times\mathcal{X}:\Big(U_n(t,x,\Lambda),\dot{U}_n(t,x,\Lambda)\Big)\in K^c\right\}\right).
\end{aligned}$$

The claim is then a consequence of the following considerations: the sequence $(\mu_n)_{n\in\mathbb{N}}$ is tight because it narrow converges, then $\mu_n(\mathcal{O}\times\mathcal{K}^c)<\frac{\varepsilon}{2T}$ for a suitable choice of $\mathcal{K}$; on the other hand, from (6.11) and Markov's inequality, the second term is less than $\frac{\varepsilon}{2}$ when K_2 is a closed ball of $\mathbb{R}^2$ with radius large enough.

Secondly consider the Borel measurable map $\mathcal{H}_n:\Omega\times\mathcal{X}\to\Omega\times\mathcal{X}\times\mathbb{R}$ defined by

$$(x,\Lambda)\mapsto\mathcal{H}_n(x,\Lambda)=(x,\Lambda,U_n(T,x,\Lambda)),$$

and the push-forward measure $\eta_n=\mathcal{H}_n\#\mu_n$ of μ_n by $\mathcal{H}_n$. Then $\eta_n\in\mathcal{Y}(\Omega,\mathcal{X}\times\mathbb{R})$ whose measure projection on Ω is the measure dx. By using similar arguments, it is easy to establish that the sequence $(\eta_n)_{n\in\mathbb{N}}$ is tight.

Therefore, according to Theorem E.9, there exists $\nu\in\mathcal{Y}(\mathcal{O},\mathcal{X}\times\mathbb{R}^2)$ and $\eta\in\mathcal{Y}(\Omega,\mathcal{X}\times\mathbb{R})$ such that, up to the extraction of a subsequence, not relabeled, $\nu_n\overset{\text{nar}}{\rightharpoonup}\nu$ and $\eta_n\overset{\text{nar}}{\rightharpoonup}\eta$. It is easy to deduce that

$$\begin{aligned}
\pi_{\mathcal{O}\times\mathcal{X}}\#\nu&=w\text{-}\lim_{n\to+\infty}\pi_{\mathcal{O}\times\mathcal{X}}\#\nu_n=dt\otimes\mu,\\
\pi_{\Omega\times\mathcal{X}}\#\eta&=w\text{-}\lim_{n\to+\infty}\pi_{\Omega\times\mathcal{X}}\#\eta_n=\mu,
\end{aligned}$$

where w-lim denotes the standard narrow limit of Borel measures, i.e. the limit associated with the weak $\sigma(C_b', C_b)$ topology (cf. (Attouch *et al.*, 2014, Definition 4.2.2)), and $\pi_{\mathcal{O}\times\mathcal{X}}$, $\pi_{\Omega\times\mathcal{X}}$ denote the canonical projections of $\mathcal{O}\times\mathcal{X}\times\mathbb{R}$ onto $\mathcal{O}\times\mathcal{X}$ and $\Omega\times\mathcal{X}\times\mathbb{R}$ onto $\Omega\times\mathcal{X}$ respectively. From the disintegration theorem, Theorem E.11, the disintegration of ν on the product $(\mathcal{O}\times\mathcal{X})\times\mathbb{R}^2$ provides a family $(\nu_{t,x,\Lambda})_{t,x,\Lambda}$ of probability measures on $\mathbb{R}^2$ with $\nu=(dt\otimes\mu)\otimes\nu_{t,x,\Lambda}$. Similarly, the disintegration of η on the product $(\Omega\times\mathcal{X})\times\mathbb{R}$ provides a family $(\eta_{x,\Lambda})_{x,\Lambda}$ of probability measures on $\mathbb{R}$ with $\eta=\mu\otimes\eta_{x,\Lambda}$. Let denote by $U(t,x,\Lambda)$, $V(t,x,\Lambda)$ and $W_T(x,\Lambda)$ the first moments of the probability measures $\nu_{t,x,\Lambda}$ and $\eta_{x,\Lambda}$, namely[1]

$$\begin{cases} U(t,x,\Lambda) := \displaystyle\int_{\mathbb{R}^2} \zeta \, d\nu_{t,x,\Lambda}\left(\zeta,\dot\zeta\right), \\[2ex] V(t,x,\Lambda) := \displaystyle\int_{\mathbb{R}^2} \dot\zeta \, d\nu_{t,x,\Lambda}\left(\zeta,\dot\zeta\right), \\[2ex] W_T(x,\Lambda) := \displaystyle\int_{\mathbb{R}} \zeta_T d\eta_{x,\Lambda}(\zeta_T). \end{cases}$$

The proof of the following result is postponed to Section 6.1.4.

Lemma 6.3. *$U\in X_\mu(u)$, $V=\dot U$ and $W_T=U(T,\cdot,\cdot)$.*

From now on, we write $\overset{\text{Jensen}}{\leq}$, $\overset{\text{s.c.i}}{\leq}$, $\overset{\text{U.I.}}{=}$, and converse inequalities, to mean that the inequality, the equality or the convergence, is justified thank's to Jensen's inequality, to the lower semicontinuity, Theorem E.10 (i) or to the uniform integrability, Theorem E.10 (ii). Justifications are left to the reader. We are in a position to establish the lower bound $\Phi_\mu(u)\leq\liminf_{n\to+\infty}\Phi_{\mu_n}(u_n)$, which is a straightforward consequence of Theorem E.10, i), the definition of the measures ν_n, ν, η_n, η, Theorem E.11, Jensen's inequality, and Lemma 6.3. Indeed we have

[1] We choose to denote by $\left(\zeta,\dot\zeta\right)$ and ζ_T the integration variables for the disintegrations measures of ν and η respectively. Note that η depends on T but we choose to shorten the notation and to write η for η_T.

$$
\begin{aligned}
&\liminf_{n\to+\infty} \Phi_{\mu_n}(u_n) \\
&\geq \liminf_{n\to+\infty} \int_{\mathcal{O}\times\mathcal{X}} \Lambda\left(t,x,U_n(t,x,\Lambda),\dot{U}_n(t,x,\Lambda)\right) dt\otimes d\mu_n(x,\Lambda) \\
&\quad + \liminf_{n\to+\infty} \int_{\Omega\times\mathcal{X}} \Theta(U_n(T,x,\Lambda))d\mu_n(x,\Lambda) \\
&= \liminf_{n\to+\infty} \int_{\mathcal{O}\times\mathcal{X}\times\mathbb{R}^2} \Lambda(t,x,S)\ d\nu_n(t,x,\Lambda,S) + \liminf_{n\to+\infty} \int_{\Omega\times\mathcal{X}\times\mathbb{R}} \Theta(\zeta_T)\ d\eta_n(x,\Lambda,\zeta_T) \\
&\overset{\text{s.c.i}}{\geq} \int_{\mathcal{O}\times\mathcal{X}\times\mathbb{R}^2} \Lambda(t,x,S)\ d\nu(t,x,\Lambda,S) + \int_{\Omega\times\mathcal{X}\times\mathbb{R}} \Theta(\zeta_T)\ d\eta(x,\Lambda,\zeta_T) \\
&\overset{\text{Jensen}}{\geq} \int_{\mathcal{O}\times\mathcal{X}} \Lambda\left(t,x,\int_{\mathbb{R}^2} S\ d\nu_{t,x,\Lambda}(S)\right) dt\otimes d\mu(x,\Lambda) \\
&\quad + \int_{\Omega\times\mathcal{X}} \Theta\left(\int_{\mathbb{R}} \zeta_T\ d\eta_{x,\Lambda}(\zeta_T)\right) d\mu(x,\Lambda) \\
&= \int_{\mathcal{O}\times\mathcal{X}} \Lambda(t,x,U(t,x,\Lambda),V(t,x,\Lambda))dt\otimes d\mu(x,\Lambda) + \int_{\Omega\times\mathcal{X}} \Theta(W_T(x,\Lambda))d\mu(x,\Lambda) \\
&= \int_{\mathcal{O}\times\mathcal{X}} \Lambda\left(t,x,U(t,x,\Lambda),\dot{U}(t,x,\Lambda)\right) dt\otimes d\mu(x,\Lambda) + \int_{\Omega\times\mathcal{X}} \Theta(U(T,x,\Lambda))d\mu(x,\Lambda) \\
&= \Phi_\mu(u)
\end{aligned}
$$

which completes the proof. □

6.1.2 *The upper bound*

Proposition 6.2. *Assume that the sequence $(\mu_n)_{n\in\mathbb{N}}$ narrow converges to μ in $\mathcal{Y}(\Omega,\mathcal{X})$. Then, for every $u\in\mathcal{V}(0,T,L^2(\Omega))$, there exists a subsequence $(\mu_{\sigma(n)})_{n\in\mathbb{N}}$ of $(\mu_n)_{n\in\mathbb{N}}$, possibly depending on u, and $(v_n)_{n\in\mathbb{N}}$ weakly converging to u in $\mathcal{V}(0,T,L^2(\Omega))$ such that*

$$\Phi_\mu(u) \geq \limsup_{n\to+\infty} \Phi_{\mu_{\sigma(n)}}(v_n).$$

Proof. Fix $u\in\mathcal{V}(0,T,L^2(\Omega))$ and let $U\in X_{\mu,u}$ be a minimizer of $\Phi_\mu(u)$. Observe that the space $C_c(0,T,C_c(\Omega\times\mathcal{X}))$ is dense in $\mathcal{V}(0,T,L^2_\mu(\Omega\times\mathcal{X}))$. For $\delta>0$ ($\delta=\frac{1}{m}$, $m\in\mathbb{N}^*$), consider $U_\delta\in C_c(0,T,C_c(\Omega\times\mathcal{X}))$ satisfying

$$\int_{\mathcal{O}\times\mathcal{X}} |U_\delta-U|^2\, dt\otimes d\mu < \delta \tag{6.12}$$

$$\int_{\mathcal{O}\times\mathcal{X}} \left|\dot{U}_\delta-\dot{U}\right|^2 dt\otimes d\mu < \delta, \tag{6.13}$$

and set

$$u_{\delta,n}(t,x) := \int_{\mathcal{X}} U_\delta(t,x,\Lambda)\ d\mu_x^n(\Lambda)$$

where $(\mu_x^n)_{x\in\Omega}$ is a disintegration of the Young measure μ_n. For each $\delta > 0$, there exists $N_\delta \in \mathbb{N}$ such that, for all $n \geq N_\delta$, the following estimate holds:

$$\begin{aligned}
\int_{(0,T)\times\Omega} |u_{n,\delta}|^2 \, dt \otimes dx &= \int_{(0,T)\times\Omega} \left| \int_{\mathcal{X}} U_\delta \, d\mu_x^n \right|^2 \, dt \otimes dx \\
&\leq \int_{\mathcal{O}\times\mathcal{X}} |U_\delta|^2 \, dt \otimes d\mu_n \\
&\leq 2\left(\frac{1}{\alpha}\Phi_\mu(u) + T\mathcal{L}_N(\Omega) + \delta \right) + 1 \\
&\leq \frac{2}{\alpha}\Phi_\mu(u) + C
\end{aligned}$$

where $C > 0$ depends only on T and $\mathcal{L}_N(\Omega)$. The same calculation leads to

$$\int_{(0,T)\times\Omega} |\dot{u}_{n,\delta}|^2 \, dt \otimes dx \leq \frac{2}{\alpha}\Phi_\mu(u) + C.$$

Set

$$\widetilde{u}_{n,\delta}(t,x) := \int_{\mathcal{X}} U_\delta(t,x,\Lambda) \, d\mu_x^{n+N_\delta}(\Lambda). \tag{6.14}$$

From above $\widetilde{u}_{n,\delta}$ belongs to the ball $\mathbb{B}(0, R_u)$ of $\mathcal{V}(0,T,L^2(\Omega))$ with $R_u := \frac{2}{\alpha}\Phi_\mu(u)+C$, that we equip with the metrizable weak topology τ_w. According to the definition of the narrow convergence in $\mathcal{Y}(\Omega,\mathcal{X})$, which is in duality with the space $\mathcal{G}_\mathcal{C}(\Omega,\mathcal{X})$ of Carathéodory integrands, and applying the Lebesgue dominated convergence theorem, we infer that

$$\lim_{\delta\to 0} \lim_{n\to+\infty} \widetilde{u}_{n,\delta} = \lim_{\delta\to 0} \int_{\mathcal{X}} U_\delta \, d\mu_x = u \tag{6.15}$$

where the limits are taken in the sense of the weak convergence in $\mathcal{V}(0,T,L^2(\Omega))$. Moreover, by using the same arguments together with (6.12), (6.13), and the local Lipshitz regularity (6.8), (6.5) of Λ and Θ, we obtain

$$\begin{aligned}
\lim_{\delta\to 0} \lim_{n\to+\infty} \int_{\mathcal{O}\times\mathcal{X}} \Lambda\left(t, x, U_\delta(t,x,\Lambda), \dot{U}_\delta(t,x,\Lambda)\right) \, dt \otimes d\mu_{n+N_\delta} \\
+ \int_{\Omega\times\mathcal{X}} \Theta(U_\delta(T,x,\Lambda)) \, d\mu_{n+N_\delta} = \Phi_\mu(u).
\end{aligned} \tag{6.16}$$

From (6.15), (6.16), and a standard diagonalisation argument in the metric space $(\mathbb{B}(0,R_u),\tau_w) \times (\mathbb{R}, |\cdot|)$, we can assert that there exists an increasing map $\delta \mapsto n(\delta)$ such that

$$\lim_{\delta\to 0} \widetilde{u}_{n(\delta),\delta} = u \text{ weakly in } \mathcal{V}(0,T,L^2(\Omega));$$

$$\begin{aligned}
&\lim_{\delta\to 0} \int_{\mathcal{O}\times\mathcal{X}} \Lambda\left(t, x, U_\delta(t,x,\Lambda), \dot{U}_\delta(t,x,\Lambda)\right) \, dt \otimes d\mu_{n(\delta)+N_\delta} \\
&+ \int_{\Omega\times\mathcal{X}} \Theta(U_\delta(T,x,\Lambda)) \, d\mu_{n(\delta)+N_\delta} = \Phi_\mu(u).
\end{aligned}$$

Therefore, since from (6.14), $U_\delta \in X_{\mu_{n(\delta)+N_\delta}}(\widetilde{u}_{n(\delta),\delta})$, we deduce from above that

$$\limsup_{\delta\to 0} \Phi_{\mu_{n(\delta)+N_\delta}}(\widetilde{u}_{n(\delta),\delta}) \le \Phi_\mu(u).$$

The map $\delta \mapsto n(\delta) + N_\delta$ is increasing since one may assume that $(N_\delta)_{\delta>0}$ is an increasing sequence. Hence, $(\mu_{n(\delta)+N_\delta})_{\delta>0}$ is a subsequence of $(\mu_\delta)_{\delta>0}$. The claim follows by setting $v_\delta = \widetilde{u}_{n(\delta),\delta}$. □

6.1.3 *Proof of Theorem 6.1*

According to the compactness theorem for the Γ-convergence, there exists a subsequence of $(\Phi_{\mu_n})_{n\in\mathbb{N}}$, that we do not relabel, and $\Psi : \mathcal{V}(0,T,L^2(\Omega)) \to \overline{\mathbb{R}}$ such that

$$\Gamma\text{-}\lim_{n\to+\infty} \Phi_{\mu_n} = \Psi$$

(see (Attouch, 1984, Theorem 2.2)). Suppose for the moment that we have proved that $\Psi = \Phi_\mu$. The same result would hold for any given subsequence of the initial sequence $(\Phi_{\mu_n})_{n\in\mathbb{N}}$, so that $(\Phi_{\mu_n})_{n\in\mathbb{N}}$ possesses a unique limit point Φ_μ, then Γ-converges to Φ_μ (see (Attouch, 1984, Proposition 2.73)).

We are going to show that $\Psi = \Phi_\mu$. Fix $u \in \mathcal{V}(0,T,L^2(\Omega))$. Applying Proposition 6.1 with $\mu_{\sigma(n)} := \widetilde{\mu}_n$ obtained in Proposition 6.2 (recall that $\widetilde{\mu}_n$ depends on u), and substituted for μ_n, for all $u_n \in \mathcal{V}(0,T,L^2(\Omega))$ with $u_n \rightharpoonup u$ in $\mathcal{V}(0,T,L^2(\Omega))$, we have

$$\liminf_{n\to+\infty} \Phi_{\widetilde{\mu}_n}(u_n) \ge \Phi_\mu(u). \tag{6.17}$$

On the other hand, from Proposition 6.2, there exists v_n such that $v_n \rightharpoonup u$ in $\mathcal{V}(0,T,L^2(\Omega))$, and

$$\limsup_{n\to+\infty} \Phi_{\widetilde{\mu}_n}(v_n) \le \Phi_\mu(u). \tag{6.18}$$

Combining (6.17) and (6.18), we infer that $\Gamma\text{-}\lim_{n\to+\infty} \Phi_{\widetilde{\mu}_n}(u) = \Phi_\mu(u)$. According to the fact that $(\Phi_{\widetilde{\mu}_n})_{n\in\mathbb{N}}$ is a subsequence of $(\Phi_{\mu_n})_{n\in\mathbb{N}}$, we deduce that $\Gamma\text{-}\lim_{n\to+\infty} \Phi_{\widetilde{\mu}_n} = \Psi$. Thus $\Gamma\text{-}\lim_{n\to+\infty} \Phi_{\widetilde{\mu}_n}(u) = \Phi_\mu(u) = \Psi(u)$. Since the reasoning holds for all $u \in \mathcal{V}(0,T,L^2(\Omega))$, we conclude to $\Psi = \Phi_\mu$.

6.1.4 *Proof of Lemma 6.3*

We refer the reader to Anza Hafsa *et al.* (2020c) for a complete proof which consists in applying Theorem E.10 to suitable test functions, together with a disintegration process (Theorem E.11). To familiarize the reader with the computational technique, we establish that $V = \dot{U}$ (equality in $L^2(0,T,L^2_\mu(\Omega\times\mathcal{X}))$), and $U \in X_\mu(u)$.

We begin by proving that for $dt\otimes dx$-a.e. $(t,x) \in (0,T)\times\Omega$,

$$\int_{\mathcal{X}} U(t,x,\Lambda)\, d\mu_x(\Lambda) = u(t,x).$$

Let $\varphi \in \mathcal{D}((0,T)\times\Omega)$, and set $\psi\left(t,x,\Lambda,\zeta,\dot{\zeta}\right) = \varphi(t,x)\zeta$ for all $\left(t,x,\Lambda,\zeta,\dot{\zeta}\right) \in [0,T]\times\Omega\times\mathcal{X}\times\mathbb{R}\times\mathbb{R}$. Then, since $u_n \rightharpoonup u$ in $\mathcal{V}(0,T,L^2(\Omega))$, according to the definition of the measure ν_n, together with a uniform integrability argument (Theorem E.10, (ii)) and disintegration process (Theorem E.11), we have

$$\begin{aligned}
\int_{(0,T)\times\Omega} \varphi u \, dt\otimes dx &= \lim_{n\to+\infty}\int_{(0,T)\times\Omega} \varphi u_n \, dt\otimes dx \\
&= \lim_{n\to+\infty}\int_{\mathcal{O}\times\mathcal{X}\times\mathbb{R}^2} \psi \, d\nu_n \\
&\overset{\text{U.I.}}{=} \int_{\mathcal{O}\times\mathcal{X}\times\mathbb{R}^2} \psi \, d\nu \\
&= \int_{\mathcal{O}\times\mathcal{X}} \varphi(t,x)\left(\int_{\mathbb{R}}^2 \zeta \, d\nu_{t,x,\Lambda}\left(\zeta,\dot{\zeta}\right)\right) dt\otimes d\mu(x,\Lambda) \\
&= \int_{(0,T)\times\Omega} \varphi(t,x)\left(\int_{\mathcal{X}} U(t,x,\Lambda)\, d\mu(x,\Lambda)\right) dt\otimes dx,
\end{aligned}$$

which proves the thesis. Let us prove that $V=\dot{U}$. Take $\varphi\in C_c^1(0,T)$, $\psi\in C_c(\Omega)$, and set $\Psi\left(t,x,\Lambda,\zeta,\dot{\zeta}\right) := \varphi(t)\dot{\zeta}\psi(x)$ and $\Xi\left(t,x,\zeta,\dot{\zeta}\right) = \varphi'(t)\zeta\psi(x)$. Using the arguments above, we have

$$\begin{aligned}
&\int_{\mathcal{O}\times\mathcal{X}} \varphi(t)\psi(x)V(t,x,\Lambda)\, dt\otimes d\mu(x,\Lambda) \\
&= \int_{\mathcal{O}\times\mathcal{X}} \varphi(t)\psi(x)\left(\int_{\mathbb{R}^2} \dot{\zeta}\, d\nu_{t,x,\Lambda}\left(\zeta,\dot{\zeta}\right)\right) dt\otimes d\mu(x,\Lambda) \\
&= \int_{\mathcal{O}\times\mathcal{X}\times\mathbb{R}^2} \Psi d\nu \\
&\overset{\text{U.I.}}{=} \lim_{n\to+\infty}\int_{\mathcal{O}\times\mathcal{X}\times\mathbb{R}^2} \Psi d\nu_n \\
&= \lim_{n\to+\infty}\int_{\mathcal{O}\times\mathcal{X}} \Psi\left(t,x,\Lambda,U_n(t,x,\Lambda),\dot{U}_n(t,x,\Lambda)\right) dt\otimes d\mu(x,\Lambda) \\
&= \lim_{n\to+\infty}\int_{\mathcal{O}\times\mathcal{X}} \varphi(t)\psi(x)\dot{U}_n(t,x,\Lambda)dt\otimes d\mu_n(x,\Lambda) \\
&= -\lim_{n\to+\infty}\int_{\mathcal{O}\times\mathcal{X}} \varphi'(t)\psi(x)U_n(t,x,\Lambda)dt\otimes d\mu_n(x,\Lambda) \\
&= -\lim_{n\to+\infty}\int_{\mathcal{O}\times\mathcal{X}\times\mathbb{R}^2} \Xi\left(t,x,\Lambda,\zeta,\dot{\zeta}\right) d\nu_n \\
&\overset{\text{U.I.}}{=} -\int_{\mathcal{O}\times\mathcal{X}\times\mathbb{R}^2} \Xi\left(t,x,\Lambda,\zeta,\dot{\zeta}\right) d\nu \\
&= -\int_{\mathcal{O}\times\mathcal{X}} \varphi'(t)\psi(x)\left(\int_{\mathbb{R}^2} \zeta\, d\nu_{t,x,\Lambda}\left(\zeta,\dot{\zeta}\right)\right) dt\otimes d\mu(x,\Lambda) \\
&= \int_{\mathcal{O}\times\mathcal{X}} \varphi'(t)\psi(x)U(t,x,\Lambda)\, dt\otimes d\mu(x,\Lambda),
\end{aligned}$$

which proves the thesis. To establish that $U(0,x,\Lambda)=0$ for μ-a.e. $(x,\Lambda)\in\Omega\times\mathcal{X}$, take $\varphi\in C^1([0,T])$ with $\varphi(T)=0$, $\varphi(0)\neq 0$, and $\psi\in C_c(\Omega\times\mathcal{X})$. Then, on a one hand

$$\begin{aligned}
&\int_{\mathcal{O}\times\mathcal{X}}\varphi(t)\dot{U}(t,x,\Lambda)\psi(x,\Lambda)dt\otimes d\mu(x,\Lambda)\\
&=\int_{\Omega\times\mathcal{X}}\psi(x,\Lambda)\left(\int_0^T\varphi(t)\dot{U}(t,x,\Lambda)\,dt\right)d\mu(x,\Lambda)\\
&=\int_{\Omega\times\mathcal{X}}\psi(x,\Lambda)\left(-\int_0^T\varphi'(t)U(t,x,\Lambda)\,dt\right)d\mu(x,\Lambda)\\
&\quad-\int_{\Omega\times\mathcal{X}}\varphi(0)U(0,x,\Lambda)\psi(x,\Lambda)d\mu(x,\Lambda).
\end{aligned}\tag{6.19}$$

On the other hand, it is easy to show that

$$\begin{aligned}
&\int_{\mathcal{O}\times\mathcal{X}}\varphi(t)\dot{U}(t,x,\Lambda)\psi(x,\Lambda)dt\otimes d\mu(x,\Lambda)\\
&\overset{\text{U.I.}}{=}\lim_{n\to+\infty}\int_{\mathcal{O}\times\mathcal{X}\times\mathbb{R}^2}\varphi(t)\psi(x,\Lambda)\dot{\zeta}\,d\nu_n\\
&=\int_{\Omega\times\mathcal{X}}\psi(x,\Lambda)\left(-\int_0^T\varphi'(t)U(t,x,\Lambda)\,dt\right)d\mu(x,\Lambda).
\end{aligned}\tag{6.20}$$

Collecting (6.19) and (6.20), we infer that $\varphi(0)U(0,x,\Lambda)=0$ for μ-a.e. $(x,\Lambda)\in\Omega\times\mathcal{X}$, i.e., $U(0,x,\Lambda)=0$ for μ-a.e. $(x,\Lambda)\in\Omega\times\mathcal{X}$.

6.2 The case of integral functionals

In this section, for each Borel measurable map $h:\Omega\to\mathcal{X}$, we focus on the measure $\mu\in\mathcal{Y}(\Omega,\mathcal{X})$ associated with h, i.e. $\mu=dx\otimes\delta_{h(x)}$. In this specific case, it is easily seen that the functional $\Phi_\mu:\mathcal{V}(0,T,L^2(\Omega))\to\mathbb{R}$ is the integral functional given for every $u\in\mathcal{V}(0,T,L^2(\Omega))$ by

$$\Phi_\mu(u)=\int_{(0,T)\times\Omega}h(x)(x,t,u(t,x),\dot{u}(t,x))\,dt\otimes dx+\int_\Omega\Theta(u(T,x))\,dx.$$

From now on, we write ε to denote a sequence $(\varepsilon_n)_{n\in\mathbb{N}}$ of positive numbers ε_n going to zero when $n\to+\infty$, and we briefly write $\varepsilon\to 0$ instead of $\lim_{n\to+\infty}\varepsilon_n=0$. As a straightforward corollary of Theorem 6.1, we have

Corollary 6.1. *Let $(f_\varepsilon)_{\varepsilon>0}$ be a sequence of Borel measurable maps $f_\varepsilon:\Omega\to\mathcal{X}$, and assume that the sequence of young measures $(\mu_\varepsilon=dx\otimes\delta_{f_\varepsilon(x)})_{\varepsilon>0}$ narrow converges to some $\mu\in\mathcal{Y}(\Omega,\mathcal{X})$. Then, the integral functional Φ_{μ_ε} defined for every $u\in\mathcal{V}(0,T,L^2(\Omega))$ by*

$$\Phi_{\mu_\varepsilon}(u)=\int_{(0,T)\times\Omega}f_\varepsilon(t,x,u(t,x),\dot{u}(t,x))\,dt\otimes dx+\int_\Omega\Theta(u(T,x))\,dx,$$

Γ-converges to the functional Φ_μ defined for every $u \in \mathcal{V}(0,T,L^2(\Omega))$ by

$$\Phi_\mu(u) = \inf_{U \in X_\mu(u)} \Psi(U)$$

where

$$\Psi(U) = \int_{(0,T)\times\Omega\times\mathcal{X}} \Lambda\left(t, x, U(t,x,\Lambda), \dot{U}(t,x,\Lambda)\right) dt \otimes d\mu(x,\Lambda) + \int_{\Omega\times\mathcal{X}} \Theta(U(T,x,\Lambda)) d\mu(x,\Lambda).$$

Note that, in general, the limit functional Φ_μ is non local even if μ is purely atomic (see (Anza Hafsa *et al.*, 2020c, Remarks 2.10, 2.11)). In Corollary 6.2 below, we give a sufficient condition which ensures that the Γ-limit Φ_μ is local in the sense that Φ_μ possesses an integral representation.

Corollary 6.2. *Assume that the sequence $(f_\varepsilon)_{\varepsilon>0}$ converges in measure to a Borel measurable function $f : \Omega \to \mathcal{X}$, i.e. $\lim_{\varepsilon\to 0} \mathcal{L}_N(\{x \in \Omega : \mathbf{d}(f_\varepsilon(x), f(x)) > r\}) = 0$ for all $r > 0$, where $\mathbf{d}$ is the distance defined by (6.9). Then the sequence $(\Phi_{\mu_\varepsilon})_{\varepsilon>0}$ Γ-converges to the integral functional Φ defined for every $u \in \mathcal{V}(0,T,L^2(\Omega))$ by*

$$\Phi(u) = \int_{(0,T)\times\Omega} f(x)(x, t, u(t,x), \dot{u}(t,x))\, dt \otimes dx + \int_\Omega \Theta(u(T,x))\, dx.$$

Proof. According to (Attouch *et al.*, 2014, Proposition 4.3.8), or (Valadier, 1990, Proposition 6) in the general theory of Young measures in $\mathcal{Y}(\Omega,\mathcal{X})$, where $\mathcal{X}$ is a Souslin space, the Young measure $\mu_\varepsilon = dx \otimes \delta_{f_\varepsilon(x)}$ narrow converges to the Young measure $dx \otimes \delta_{f(x)}$ where $f : \Omega \to \mathcal{X}$ is some Borel measurable function, if and only if the sequence $(f_\varepsilon)_{\varepsilon>0}$ converges in measure to f. The claim then follows by applying Corollary 6.1. □

From now on, we denote indifferently by $\dot{v}$ or $\frac{\partial v}{\partial t}$ the time derivative of any element of $\mathcal{V}(0,T,L^2(\Omega))$. For each $T > 0$ and each $\varepsilon > 0$, consider the non-diffusive reaction differential equation (6.1) defined in $H^1(0,T,L^2(\Omega))$ by

$$\begin{cases} -\dfrac{\partial u_\varepsilon}{\partial t}(t,x) = \dfrac{\partial \psi_\varepsilon}{\partial \zeta}(t,x,\cdot,u_\varepsilon(t)) \text{ for a.e. } (t,x) \in (0,T)\times\Omega \\ u_\varepsilon(0,\cdot) = u_0 \in L^2(\Omega). \end{cases} \tag{6.21}$$

We assume that for a.e. $(t,x) \in (0,T)\times\mathbb{R}^N$, $\zeta \mapsto \psi_\varepsilon(t,x,\zeta)$ is convex of class $C^1(\mathbb{R})$ and fulfills the standard growth conditions

$$\alpha(|\zeta|^2 - 1) \le \psi_\varepsilon(t,x,\zeta) \le \beta(1 + |\zeta|^2) \tag{6.22}$$

for all $\zeta \in \mathbb{R}$, where $0 < \alpha \le \beta < +\infty$. We also assume that ψ and its Fenchel conjugate with respect to the third variable satisfy the equicontinuity condition with respect to t: there exist $\gamma : \mathbb{R}_+ \to \mathbb{R}_+$ and $\gamma^* : \mathbb{R}_+ \to \mathbb{R}_+$ with $\lim_{r\to 0} \gamma(r) = \lim_{r\to 0} \gamma^*(r)0$, such that, for all $(x,\zeta) \in \mathbb{R}^{N+1}$ and all t_1, t_2 in $[0,T]$,

$$\begin{aligned} |\psi_\varepsilon(t_2,x,\zeta) - \psi_\varepsilon(t_1,x,\zeta)| &\leq \delta(|t_2-t_1|)(1+|\zeta|^2), \\ |\psi^*_\varepsilon(t_2,x,\zeta) - \psi^*_\varepsilon(t_1,x,\zeta)| &\leq \delta^*(|t_2-t_1|)(1+|\zeta|^2). \end{aligned} \tag{6.23}$$

Finally, to obtain existence of a solution, the function $\zeta \mapsto \frac{\partial \psi_\varepsilon}{\partial \zeta}(t,x,\zeta)$ is assumed to be Lipschitz continuous, uniformly with respect to (t,x), i.e. there exists $L_\varepsilon > 0$ such that for all $(\zeta_1,\zeta_2) \in \mathbb{R}^2$,

$$\left| \frac{\partial \psi_\varepsilon}{\partial \zeta}(\cdot,\cdot,\zeta_1) - \frac{\partial \psi_\varepsilon}{\partial \zeta}(\cdot,\cdot,\zeta_2) \right| \leq L_\varepsilon |\zeta_1 - \zeta_2|. \tag{6.24}$$

We know that (6.21) admits a unique solution u_ε (apply for instance Theorem 2.3). Without loss of generality we can assume that $u_0 = 0$. Indeed, setting $\widetilde{u}_\varepsilon = u_\varepsilon - u_0$, and $\widetilde{\psi}_\varepsilon(t,x,\zeta) = \psi_\varepsilon(t,x,\zeta + u_0(x))$, we obtain that u_ε solves (6.21) if and only if $\widetilde{u}_\varepsilon$ solves in $\mathcal{V}(0,T,L^2(\Omega))$.

$$\begin{cases} -\dot{\widetilde{u}}_\varepsilon(t) = \dfrac{\partial \widetilde{\psi}_\varepsilon}{\partial \zeta}(t,\cdot,\widetilde{u}_\varepsilon(t)) \text{ for a.e. } t \in (0,T) \\ \widetilde{u}_\varepsilon(0) = 0 \end{cases}$$

where $\widetilde{\psi}_\varepsilon$ satisfies all the conditions fulfilled by ψ_ε. From now on we assume $u_0 = 0$.

Remark 6.2. When the function ψ_ε takes the form $\psi_\varepsilon(t,x,\zeta) = r_\varepsilon(t,x)g_\varepsilon(\zeta) + q_\varepsilon(t,x)$, then the reaction functional of (6.21), i.e. the second member, is given by $F_\varepsilon(t,v)(x) = r_\varepsilon(t,x)\frac{dg_\varepsilon}{d\zeta}(v(x)) + q_\varepsilon(t,x)$. Therefore F_ε is a CP-structured reaction functional associated with $\left(r_\varepsilon, \frac{dg_\varepsilon}{d\zeta}, q_\varepsilon\right)$ as defined in Chapter 2, Section 2.2.2. Consequently, using the notation of Section 2.2.2, we infer that (6.21) admits a unique solution which satisfies $\underline{y}(T) \leq u(t) \leq \overline{y}(T)$ under the initial condition $\underline{\rho} \leq u_0 \leq \overline{\rho}$.

Consider $\Psi_\varepsilon : (0,T) \times L^2(\Omega) \to \mathbb{R}$ defined for every $(t,v) \in (0,T) \times L^2(\Omega)$ by

$$\Psi_\varepsilon(t,v) = \int_\Omega \psi_\varepsilon(t,x,v(x))\, dx.$$

Then (6.21) is equivalent to the differential inclusion in $\mathcal{V}(0,T,L^2(\Omega))$:

$$\begin{cases} -\dot{u}_\varepsilon(t) \in \partial\Psi_\varepsilon(t,u_\varepsilon(t)) \text{ for a.e. } t \in (0,T) \\ u_\varepsilon(0) = 0, \end{cases} \tag{6.25}$$

where $\partial\Psi_\varepsilon$ denotes the subdifferential of $v \mapsto \Psi_\varepsilon(t,v)$. According to the Fenchel extremality condition, (see (Attouch *et al.*, 2014, Proposition 9.5.1)), (6.25) is equivalent to

$$\Psi_\varepsilon(t,u_\varepsilon(t)) + \Psi^*_\varepsilon(t,-\dot{u}_\varepsilon(t)) - \langle -\dot{u}_\varepsilon(t), u_\varepsilon(t)\rangle = 0, \text{ a.e. } t \in (0,T)$$

then to

$$\int_0^T \left(\Psi_\varepsilon(t, u_\varepsilon(t)) + \Psi_\varepsilon^*(t, -\dot{u}_\varepsilon(t)) + \frac{1}{2}\frac{d}{dt}\|u_\varepsilon(t)\|^2_{L^2(\Omega)} \right) dt = 0, \tag{6.26}$$

where Ψ_ε^* denotes the Legendre-Fenchel conjugate of $v \mapsto \Psi_\varepsilon(t, v)$. (For the last equivalence, apply the Legendre-Fenchel inequality

$$\Psi_\varepsilon(t, v) + \Psi_\varepsilon^*(t, -\dot{v}) - \langle -\dot{v}(t), v \rangle \geq 0$$

valid for all $v \in L^2(\Omega)$ and all $t \in [0, T]$.) Consequently u_ε solves (6.21) if and only if u_ε is a minimizer in $\mathcal{V}(0, T, L^2(\Omega))$ of the functional

$$u \mapsto \int_0^T (\Psi_\varepsilon(t, u(t)) + \Psi_\varepsilon^*(t, -\dot{u}(t)))dt + \frac{1}{2}\int_\Omega |u(T, x)|^2 dx.$$

Therefore, by using the standard Legendre-Fenchel calculus, u_ε minimizes the functional defined by

$$\Phi_{\mu_\varepsilon}(u) = \int_{(0,T)\times\Omega} f_\varepsilon(x)(t, u(t, x), \dot{u}(t, x))\, dx\, dt + \int_\Omega \Theta(u(T, x))\, dx,$$

where

$$f_\varepsilon(x)\left(t, \zeta, \dot{\zeta}\right) = \psi_\varepsilon(t, x, \zeta) + \psi_\varepsilon^*\left(t, x, -\dot{\zeta}\right), \Theta(\zeta_T) = \frac{1}{2}|\zeta_T|^2.$$

According to the definition of the Legendre-Fenchel conjugate, it is easy to prove that ψ_ε^* satisfies condition (6.22) (with other constants α and β) so that, from (6.23), $f_\varepsilon(x)$ belongs to $\mathbb{E}_{\alpha,\beta,\gamma}$ with suitable α and β, and $\gamma = \delta + \delta^*$ (note that the function $f_\varepsilon(x)$ of $\mathbb{E}_{\alpha,\beta,\gamma}$ does not depend on the x-variable). Moreover $f_\varepsilon : \Omega \to \mathcal{X} = \mathbb{E}_{\alpha,\beta,\gamma}$ is clearly Borel measurable. The corollary below states a characterization of the weak cluster points of $(u_\varepsilon)_{\varepsilon>0}$ as minimizers of the functional Φ_μ where μ is the narrow limit of the Young measure $dx \otimes \delta_{f_\varepsilon(x)}$.

Corollary 6.3. *Assume that* (6.22), (6.23), (6.26) *hold and that the sequence* $(dx \otimes \delta_{f_\varepsilon(x)})_{\varepsilon>0}$ *of Young measures associated with the Borel measurable map* f_ε*, narrow converges to some* $\mu \in \mathcal{Y}(\Omega, \mathcal{X})$*. Then, any cluster point of the sequence of solutions* u_ε *of (6.21) for the weak convergence in* $\mathcal{V}(0, T, L^2(\Omega))$*, is a minimizer of* Φ_μ *in* $\mathcal{V}(0, T, L^2(\Omega))$.

Proof. Apply Proposition B.2 together with Corollary 6.1 after noticing that, from (6.10), the sequence $(u_\varepsilon)_{\varepsilon>0}$ is relatively compact for the weak convergence in $\mathcal{V}(0, T, L^2(\Omega))$. □

Under further conditions, u_ε weakly converges to the solution u of a reaction differential equation of the same type, reflecting a stability result. More precisely, we have

Proposition 6.3. *Assume that (6.22), (6.23), (6.26) hold, that ψ does not depend on t, and $\zeta \mapsto \psi_\varepsilon(x,\zeta)$ pointwise converges toward a C^1-convex function $\zeta \mapsto \psi(x,\zeta)$ for a.e. $x \in \mathbb{R}^N$ as $\varepsilon \to 0$. Then, the solution u_ε of (6.21) weakly converges in $\mathcal{V}(0,T,L^2(\Omega))$ to the solution u of*

$$\begin{cases} -\dfrac{\partial u}{\partial t}(t,x) = \dfrac{\partial \psi}{\partial \zeta}(x,u(t)) \text{ for a.e. } (t,x) \in (0,T)\times\Omega \\ \\ u(0) = 0. \end{cases} \tag{6.27}$$

Proof. The proof falls into two steps.

Step 1. We claim that Ψ_ε Mosco-converges to the functional Ψ defined in $L^2(\Omega)$ by $\Psi(v) = \int_\Omega \psi(x,v(x))dx$. Indeed for each $\lambda > 0$, let denote by Ψ_ε^λ, Ψ^λ, the Moreau-Yosida approximations of Ψ_ε, Ψ, and, for a.e. $x \in \mathbb{R}^N$, by $\psi_\varepsilon^\lambda(x,\cdot)$, $\psi^\lambda(x,\cdot)$, the Moreau-Yosida approximations of $\psi_\varepsilon(x,\cdot)$, $\psi(x,\cdot)$. Then, by using an interchange argument of infimum and integral (for a proof see Lemma 2.6), we easily infer that

$$\Psi_\varepsilon^\lambda(v) = \int_\Omega \psi_\varepsilon^\lambda(x,v(x))dx,\ \Psi^\lambda(v) = \int_\Omega \psi^\lambda(x,v(x))dx. \tag{6.28}$$

According to (Attouch, 1984, Theorem 3.24), for proving the claim, it remains to establish that for each $v \in L^2(\Omega)$, $\Psi_\varepsilon^\lambda(v) \to \Psi^\lambda(v)$, thus from (6.28) and the Lebesgue dominated convergence theorem, that for each $\zeta \in \mathbb{R}$ and all $\lambda > 0$, and for a.e. $x \in \mathbb{R}^N$, $\psi_\varepsilon^\lambda(x,\zeta) \to \psi^\lambda(x,\zeta)$. This last convergence follows directly from (Attouch, 1984, Theorem 3.24) and the following observation: for all $x \in \Omega$

$$\psi_\varepsilon(x,\cdot) \text{ pointwise converges to } \psi(x,\cdot) \Longrightarrow \psi_\varepsilon(x,\cdot) \text{ Mosco-converge to } \psi(x,\cdot),$$

which is the straightforward consequence of the following calculation: for all $x \in \Omega$, and all $\zeta_\varepsilon \to \zeta$ in $\mathbb{R}$,

$$\psi_\varepsilon(x,\zeta_\varepsilon) \geq \psi_\varepsilon(x,\zeta) - L|\zeta_\varepsilon - \zeta|(1 + |\zeta_\varepsilon| + |\zeta|),$$

so that $\liminf_{\varepsilon\to 0} \psi_\varepsilon(x,\zeta_\varepsilon) \geq \psi(x,\zeta)$. Moreover, for the constant sequence $(\zeta_\varepsilon)_{\varepsilon>0}$, $\zeta_\varepsilon = \zeta$, we have $\lim_{\varepsilon\to 0} \psi_\varepsilon(x,\zeta_\varepsilon) = \psi(x,\zeta)$.

Step 2. In what follows, we still denote by $(u_\varepsilon)_{\varepsilon>0}$ one of its subsequence which weakly converges to some u in $\mathcal{V}(0,T,L^2(\Omega))$, whose existence is ensured by (6.10) satisfied by Φ_{μ_ε}. From Step 1 and Lemma 2.6, the functional defined in $L^2(0,T,L^2(\Omega))$ by

$$v \mapsto \int_0^T \Psi_\varepsilon(v(t))dt$$

Mosco-converges toward the functional

$$v \mapsto \int_0^T \Psi(v(t))dt.$$

Therefore, going to the limit on (6.26), we obtain

$$\int_0^T \left(\Psi(u(t)) + \Psi^*(-\dot{u}(t)) + \frac{1}{2}\frac{d}{dt}\|u(t)\|^2_{L^2(\Omega)} \right) dt \leq 0,$$

thus, according to Legendre-Fenchel inequality,

$$\int_0^T \left(\Psi(u(t)) + \Psi_\varepsilon^*(-\dot{u}(t)) + \frac{1}{2}\frac{d}{dt}\|u(t)\|^2_{L^2(\Omega)} \right) dt = 0. \tag{6.29}$$

Note that we have used the fact that

$$\int_0^T \frac{1}{2}\frac{d}{dt}\|u_\varepsilon(t)\|^2_{L^2(\Omega)}dt = \frac{1}{2}\|u_\varepsilon(T)\|^2_{L^2(\Omega)}, \int_0^T \frac{1}{2}\frac{d}{dt}\|u(t)\|^2_{L^2(\Omega)}dt = \frac{1}{2}\|u(T)\|^2_{L^2(\Omega)},$$

and $u_\varepsilon(T) = \int_0^T \frac{du_\varepsilon}{dt}(t)dt$ with $\frac{du_\varepsilon}{dt} \rightharpoonup \frac{du}{dt}$ in $L^2(0, T, L^2(\Omega))$, to infer that

$$\begin{aligned}\int_0^T \frac{1}{2}\frac{d}{dt}\|u(t)\|^2_{L^2(\Omega)}dt = \|u(T)\|^2_{L^2(\Omega)} &\leq \liminf_{\varepsilon\to 0} \|u_\varepsilon(T)\|^2_{L^2(\Omega)} \\ &= \liminf_{\varepsilon\to 0} \int_0^T \frac{1}{2}\frac{d}{dt}\|u_\varepsilon(t)\|^2_{L^2(\Omega)}dt.\end{aligned}$$

From (6.29), we infer that u solves (6.27). □

PART 2

Sequences of reaction-diffusion problems: Stochastic homogenization

Chapter 7

Stochastic homogenization of nonlinear reaction-diffusion equations

In the modeling of biological invasion or for example in food-limited models, the interplay between environment heterogeneities in the individual evolution of propagation species, plays an essential role. Indeed, growth rates, or various thresholds appearing in the models, are mostly influenced by the environment, and vary in each small habitats (forests, marshes, hedges, etc.) which are, in most of the cases, statistically homogeneously distributed. These heterogeneities appear very small compared with the dimension of the domain. Therefore both diffusion and reaction terms in the problems modeling the propagation, present random coefficients and a small parameter ε which accounts for the dimension of heterogeneities. In order to characterize the effective coefficients (effective growth rate, various effective thresholds etc.), the purpose of this chapter is to identify the homogenized problem when ε goes to zero, by suitably applying Theorem 2.6 of Chapter 2. Theorem 7.1 provides a stochastic homogenization framework in nonlinear reaction-diffusion problems. In its domain, the homogenized operator is in divergence form $-\text{div}\partial_\xi W^{hom}(\omega, \cdot)$. The subdifferential $\partial_\xi W^{hom}(\omega, \cdot)$ is the almost sure graph limit of a sequence of subdifferentials $(\partial_\xi W_n(\omega, \cdot))_{n\in\mathbb{N}^*}$ where the random functions W_n is of the form $\frac{S_{(0,n)^N}}{n^N}$, defined from a subbaditive process $A \mapsto S_A$. The graph limit is in some cases an almost sure pointwise limit (see Proposition 7.2). Some precision on the rate of convergence of the law of W_n is expressed in Proposition 7.1 by using the Large Deviations principle.

As an example, we treat in two situations the stochastic homogenization of the reaction-diffusion problem describing the food-limited population model, whose reaction functional is that of the Fisher logistic growth model with Allee effect. In the first situation the spatial heterogeneities are distributed following a random patch model. From a probabilistic point of view, this means that the dynamic probability system which models heterogeneities is that of a random checkerboard-like environment. In the second situation, the probability dynamical system describes spatial heterogeneities distributed following a Poisson point process. We have completed this example by the homogenization of a problem stemming from a hydrogeological model described in Examples 2.5 and 2.6. For homogenization of convection-diffusion equations, and parabolic problems in perforated domains or in periodic or random environments, we refer the reader to Allaire *et al.* (2010, 2012a,b), Armstrong and Souganidis (2013), Cioranescu and Piatnitski (2006) and references therein. For homogenization of a Fokker-Planck equation with space-time periodic potential, we refer to Perthame and Souganidis (2011).

7.1 Probabilistic setting

The notation used here is local to this preamble and should not be confused with the notation of the other sections. For any topological space $\mathcal{X}$, we denote by $\mathcal{B}(\mathcal{X})$ its Borel σ-algebra, and we refer to the basic concepts of Appendix C (see also (Attouch *et al.*, 2014, Section 12.4.3)) concerning probability dynamic systems. Let $(\Sigma, \mathcal{A}, \mathbb{P})$ be a probability space, we consider a group $(T_z)_{z\in\mathbb{Z}^N}$ of $\mathbb{P}$-preserving transformations on $(\Sigma, \mathcal{A})$, i.e.

- (measurability) $T_z : \Sigma \to \Sigma$ is $\mathcal{A}$-measurable for all $z \in \mathbb{Z}^N$;
- (group property) $T_z \circ T_{z'} = T_{z+z'}$ and $T_{-z} = T_z^{-1}$ for all $z, z' \in \mathbb{Z}^N$;
- (mass invariance) $T_z\#\mathbb{P} = \mathbb{P}$ for all $z \in \mathbb{Z}^N$,

where we use the standard notation $T_z\#\mathbb{P}$ to denote the image measure (or push-forward) of $\mathbb{P}$ by T_z. The probability space $(\Sigma, \mathcal{A}, \mathbb{P})$ equipped with the group $(T_z)_{z\in\mathbb{Z}^N}$[1] is denoted by $(\Sigma, \mathcal{A}, \mathbb{P}, (T_z)_{z\in\mathbb{Z}^N})$, and referred to as a probability dynamical system.

We denote by $\mathcal{I}$ the σ-algebra of invariant sets of $\mathcal{A}$ by the group $(T_z)_{z\in\mathbb{Z}^N}$ and, for every $\mathbf{h}$ in the space $L^1_{\mathbb{P}}(\Sigma)$ of $\mathbb{P}$-integrable functions, by $\mathbb{E}^{\mathcal{I}}\mathbf{h}$ the conditional expectation of $\mathbf{h}$ with respect to $\mathcal{I}$, i.e. the unique $\mathcal{I}$-measurable function in $L^1_{\mathbb{P}}(\Sigma)$ satisfying for every $E \in \mathcal{I}$

$$\int_E \mathbb{E}^{\mathcal{I}}\mathbf{h}(\omega)d\mathbb{P}(\omega) = \int_E \mathbf{h}(\omega)d\mathbb{P}(\omega).$$

Note that a function $\mathbf{h}$ in $L^1_{\mathbb{P}}(\Sigma)$ is $\mathcal{I}$-measurable if and only if it is invariant under the group $(T_z)_{z\in\mathbb{Z}}$, i.e. $h \circ T_z = h$ for all $z \in \mathbb{Z}^N$ (see Proposition C.8). If $\mathcal{I}$ is made up of sets with probability 0 or 1, then we say that the probability dynamical system $(\Sigma, \mathcal{A}, \mathbb{P}, (T_z)_{z\in\mathbb{Z}^N})$ is ergodic. Under this condition, we have $\mathbb{E}^{\mathcal{I}}\mathbf{h} = \mathbb{E}\mathbf{h}$ where $\mathbb{E}\mathbf{h} = \int_\Sigma h(\omega)d\mathbb{P}(\omega)$ is the mathematical expectation of $\mathbf{h}$.

A sufficient condition to ensure ergodicity is the so called mixing condition which expresses an asymptotic independence: for all sets E and F of $\mathcal{A}$

$$\lim_{|z|\to+\infty} \mathbb{P}(T_z E \cap F) = \mathbb{P}(E)\mathbb{P}(F).$$

For more definitions and properties we refer to Appendix C or to (Attouch *et al.*, 2014, Section 12.4) and references therein. In the two sections below, $(\Sigma, \mathcal{A}, \mathbb{P}, (T_z)_{z\in\mathbb{Z}^N})$ is a general probability dynamical system.

[1]We could consider continuous groups or semi-groups $(T_x)_{x\in\mathbb{R}^N}$. But groups indexed by $\mathbb{Z}^N$ are sufficient for the applications considered in this book.

7.1.1 *The random diffusion part*

Given $\alpha > 0$, $\beta > 0$, we denote by $\text{Conv}_{\alpha,\beta}$ the class of functions $\mathbf{G} : \mathbb{R}^N \times \mathbb{R}^N \to \mathbb{R}$, $(x,\xi) \mapsto \mathbf{G}(x,\xi)$, satisfying conditions (D) defined in Section 2.2.1, endowed with the σ-algebra $\mathcal{T}_{\text{Conv}_{\alpha,\beta}}$, trace of the product σ-algebra of $\mathbb{R}^{\mathbb{R}^N\times\mathbb{R}^N}$. Recall that $\mathcal{T}_{\text{Conv}_{\alpha,\beta}}$ is the smallest σ-algebra on $\text{Conv}_{\alpha,\beta}$ such that all the evaluation maps

$$\left\{ e_{(x,\xi)} : \begin{array}{l} \text{Conv}_{\alpha,\beta} \longrightarrow \mathbb{R} \\ \qquad \mathbf{G} \longmapsto e_{(x,\xi)}(\mathbf{G}) = \mathbf{G}(x,\xi) \end{array} \right\}_{(x,\xi)\in\mathbb{R}^N\times\mathbb{R}^N}$$

are measurable.

We consider a random convex integrand $W : \Sigma \times \mathbb{R}^N \times \mathbb{R}^N \to \mathbb{R}$, i.e. a $(\mathcal{A} \otimes \mathcal{B}(\mathbb{R}^N) \otimes \mathcal{B}(\mathbb{R}^N), \mathcal{B}(\mathbb{R}))$-measurable function such that for every $\omega \in \Sigma$, the function $W(\omega,\cdot,\cdot)$, belongs to $\text{Conv}_{\alpha,\beta}$. Since $W(\cdot,x,\zeta)$ is $(\mathcal{A},\mathcal{B}(\mathbb{R}))$-measurable for all $(x,\xi) \in \mathbb{R}^N \times \mathbb{R}^N$, the map $\widetilde{W} : \Sigma \to \text{Conv}_{\alpha,\beta}$, $\omega \mapsto W(\omega,\cdot,\cdot)$, is $(\mathcal{A}, \mathcal{T}_{\text{Conv}_{\alpha,\beta}})$-measurable, and we denote by $\widetilde{\mathbb{P}}$ its law, that is $\widetilde{\mathbb{P}} = \widetilde{W}\#\mathbb{P}$.

We assume that W satisfies the so called *covariance* property with respect to the probability dynamical system $(\Sigma, \mathcal{A}, \mathbb{P}, (T_z)_{z\in\mathbb{Z}^N})$: for all $z \in \mathbb{Z}^N$

$$W(T_z\omega, x, \xi) = W(\omega, x+z, \xi)$$
$$\text{for a.e. } x \in \mathbb{R}^N, \text{ for all } \xi \in \mathbb{R}^N, \text{ and for } \mathbb{P}\text{-a.e. } \omega \in \Sigma.$$

For all $\mathbf{G}$ in $\text{Conv}_{\alpha,\beta}$ all $x \in \mathbb{R}^N$, and all $z \in \mathbb{Z}^N$, set $\widetilde{T}_z\mathbf{G}(x,\cdot) = \mathbf{G}(x+z,\cdot)$. This defines a group $\left(\widetilde{T}_z\right)_{z\in\mathbb{Z}^N}$ which acts on $\text{Conv}_{\alpha,\beta}$: for all $z \in \mathbb{Z}^N$, $\widetilde{T}_z : \text{Conv}_{\alpha,\beta} \to \text{Conv}_{\alpha,\beta,\gamma}$ is $\mathcal{T}_{\text{Conv}_{\alpha,\beta}}$ measurable. Then it is easy to show that the *covariance* property implies that the law $\widetilde{\mathbb{P}}$ of $\widetilde{W}$ is invariant under the group $\left(\widetilde{T}_z\right)_{z\in\mathbb{Z}^N}$, i.e. $\widetilde{T}_z\#\widetilde{\mathbb{P}} = \widetilde{\mathbb{P}}$ for all $z \in \mathbb{Z}^N$. The random function W is said to be *periodic in law*.

We write ε to denote a sequence $(\varepsilon_n)_{n\in\mathbb{N}}$ of positive numbers with $\lim_{n\to+\infty} \varepsilon_n = 0$. We briefly write $\varepsilon \to 0$. The following random functional $\widetilde{\Phi}_\varepsilon : \Sigma \times L^2(\Omega) \longrightarrow \mathbb{R}_+ \cup \{+\infty\}$ defined by

$$\widetilde{\Phi}_\varepsilon(\omega,u) = \begin{cases} \displaystyle\int_\Omega W\left(\omega, \frac{x}{\varepsilon}, \nabla u\right) dx & \text{if } u \in H^1(\Omega) \\ \\ +\infty & \text{otherwise,} \end{cases} \tag{7.1}$$

models a random energy concerning various steady-states situations, where the small parameter ε accounts for the size of small and randomly distributed heterogeneities in the context of a statistically homogeneous media. The measurability of $\omega \mapsto \widetilde{\Phi}_\varepsilon(\omega,u)$ for all $u \in H^1(\Omega)$ may be obtained by standard arguments (see for instance (Attouch *et al.*, 2014, Section 12.4.3 and Proposition 12.4.1)).

Under the hypotheses above on W with respect to the probability dynamical system $(\Sigma, \mathcal{A}, \mathbb{P}, (T_z)_{z\in\mathbb{Z}^N})$, it is a basic result, using the subadditive ergodic theorem, Theorem C.6, that for $\mathbb{P}$-a.e. ω in Σ the sequence $\left(\widetilde{\Phi}_\varepsilon(\omega,\cdot)\right)_{\varepsilon>0}$ Γ-converges to the integral functional $\widetilde{\Phi}^{hom}(\omega, \cdot)$, where $\widetilde{\Phi}^{hom} : \Sigma \times L^2(\Omega) \longrightarrow \mathbb{R}_+ \cup \{+\infty\}$ is given by

$$\widetilde{\Phi}^{hom}(\omega, u) = \begin{cases} \displaystyle\int_\Omega W^{hom}(\omega, \nabla u)dx & \text{if } u \in H^1(\Omega) \\ +\infty & \text{otherwise,} \end{cases} \tag{7.2}$$

when $L^2(\Omega)$ is equipped with its strong convergence.

The density W^{hom} is defined as follows: denote by Y the unit cell $]0,1[^N$ of $\mathbb{R}^N$, then there exists a set Σ' of full probability such that for all $\omega \in \Sigma'$ and all $\xi \in \mathbb{R}^N$

$$\begin{aligned} W^{hom}(\omega, \xi) &= \lim_{n \to +\infty} \inf \left\{ \frac{1}{n^N} \int_{nY} W(\omega, y, \xi + \nabla u(y))dy : u \in H_0^1(nY) \right\} \qquad (7.3) \\ &= \inf_{n \in \mathbb{N}^*} \mathbb{E}^{\mathcal{I}} \inf \left\{ \frac{1}{n^N} \int_{nY} W(\cdot, y, \xi + \nabla u(y))dy : u \in H_0^1(nY) \right\}. \end{aligned}$$

If $(\Sigma, \mathcal{A}, \mathbb{P}, (T_z)_{z \in \mathbb{Z}^N})$ is ergodic, then W^{hom} is deterministic and given for all $\omega \in \Sigma'$ and all $\xi \in \mathbb{R}^N$ by

$$\begin{aligned} W^{hom}(\xi) &= \lim_{n \to +\infty} \inf \left\{ \frac{1}{n^N} \int_{nY} W(\omega, y, \xi + \nabla u(y))dy : u \in H_0^1(nY) \right\} \\ &= \inf_{n \in \mathbb{N}^*} \mathbb{E} \inf \left\{ \frac{1}{n^N} \int_{nY} W(\cdot, y, \xi + \nabla u(y))dy : u \in H_0^1(nY) \right\}. \end{aligned}$$

For a proof we refer the reader to (Attouch *et al.*, 2014, Proposition 12.4.3, Theorem 12.4.7) or, in the ergodic case, to Dal Maso and Modica (1986).

For all $\omega \in \Sigma'$, all $\xi \in \mathbb{R}^N$, and every $n \in \mathbb{N}^*$ let us set

$$W_n(\omega, \xi) := \inf \left\{ \frac{1}{n^N} \int_{nY} W(\omega, y, \xi + \nabla u(y))\, dy : u \in H_0^1(nY) \right\} \tag{7.4}$$

which defines for each $\xi \in \mathbb{R}^N$ the real valued process $\omega \mapsto W_n(\omega, \xi)$. According to the above, for every fixed ξ, this process almost surely converges to $\omega \mapsto W^{hom}(\omega, \xi)$. The next result stated in Proposition 7.1 below, concerns the rate of convergence of the law $\mu_n := \mathbb{P} \# W_n(\cdot, \xi)$ of $W_n(\cdot, \xi)$, towards the law $\mu := \mathbb{P} \# W^{hom}(\xi)$ of $W^{hom}(\xi)$ in the independent case. Note that in this situation, $(\Sigma, \mathcal{A}, \mathbb{P}, (T_z)_{z \in \mathbb{Z}^N})$ is ergodic, therefore μ is the Dirac measure $\delta_{W^{hom}(\xi)}$. This question is within the probabilistic scope of the theory of Large Deviations (see Appendix D) which provides exponential rates for decay of random processes structured as $\frac{S_{nY}}{n^N}$. For more applications in the field of stochastic homogenization, we refer the reader to Michaille *et al.* (1998). For the structure of fluctuations, or optimal quantitative estimates in stochastic homogenization consult Gloria *et al.* (2019), Duerinckx *et al.* (2020), Gloria and Otto (2021).

Proposition 7.1. *Let denote by $\mathcal{B}_b(\mathbb{R}^N)$ the family of bounded Borel subsets of $\mathbb{R}^N$. Assume that $(W(\cdot, x, \xi))_{x \in A, \xi \in \mathbb{R}^N}$ and $(W(\cdot, x, \xi))_{x \in B, \xi \in \mathbb{R}^N}$ are two independent random vector value functions whenever A and B are two disjoint subsets in $\mathcal{B}_b(\mathbb{R}^N)$. Fix $\xi \in \mathbb{R}^N$, then*

i) for all $\lambda \in \mathbb{R}$ the following limit exists in $\mathbb{R}$

$$\Lambda_\xi(\lambda) := \lim_{n\to+\infty} \frac{1}{n^N} \mathrm{Log}(\mathbb{E}(\exp(n^N \lambda W_n(\cdot,\xi)))),$$

and $\Lambda_\xi : \mathbb{R} \to \mathbb{R}$ is convex, lower semicontinuous, and satisfies $|\Lambda_\xi(\lambda)| \le C_\xi |\lambda|$ where $C_\xi = \beta(1+|\xi|^2)$;

ii) the Legendre-Fenchel conjugate Λ_ξ^* of Λ_ξ is a convex, lower semicontinuous function from $\mathbb{R}$ into $[0,+\infty]$, with domain included in $[-C_\xi, C_\xi]$, and is a good rate function in the sense of Definition D.9;

iii) for every closed subset F in $\mathbb{R}$,

$$\limsup_{n\to+\infty} \frac{1}{n^N} \mathrm{Log}(\mu_n(F)) \le -\Lambda_\xi^*(F);$$

for every open set G in $\mathbb{R}$,

$$\liminf_{n\to+\infty} \frac{1}{n^N} \mathrm{Log}(\mu_n(G)) \ge -\Lambda_\xi^*(G \cap E_{\mathrm{ex}})$$

where E_{ex} is the set of exposed points of Λ_ξ^*, and $\Lambda_\xi^*(B) = \inf_{x\in B} \Lambda_\xi^*(x)$ for any Borel measurable set of $\mathbb{R}^N$;

iv) if Λ_ξ^* is strictly convex, then $W_n(\cdot,\xi)$ satisfies the full large deviations property: for all $B \in \mathcal{B}(\mathbb{R})$

$$-\Lambda_\xi^*\left(\overset{0}{B}\right) \le \liminf_{n\to+\infty} \frac{1}{n^N} \mathrm{Log}(\mu_n(B)) \le \limsup_{n\to+\infty} \frac{1}{n^N} \mathrm{Log}(\mu_n(B)) \le -\Lambda_\xi^*\left(\bar{B}\right).$$

Proof. Proof of i). To shorten the notation, we do not indicate the dependance with respect to fixed $\xi \in \mathbb{R}^N$. For all $A \in \mathcal{B}_b(\mathbb{R}^N)$, set

$$S_A(\omega) = \inf\left\{ \int_A W(\omega, y, \xi + \nabla u(y))\, dy : u \in H_0^1\left(\overset{\circ}{A}\right) \right\}.$$

Then $A \mapsto S_A(\cdot)$ is a subadditive process, covariant with respect to $(T_z)_{z\in\mathbb{Z}^N}$ (see Appendix C and (Attouch *et al.*, 2014, Proposition 12.4.3, Theorem 12.4.7)). For all $A \in \mathcal{B}_b(\mathbb{R}^N)$, set

$$\Lambda_A(\lambda) := \frac{1}{\mathcal{L}_N(A)} \mathrm{Log}(\mathbb{E}(\exp(\lambda S_A))).$$

First, consider the case $\lambda \ge 0$, and take A, B in $\mathcal{B}_b(\mathbb{R}^N)$ with $A \cap B = \emptyset$. From the independence hypothesis, using standard arguments, it is easily seen that S_A and S_B are independent. Thanks to the subadditivity of S and independence, we have

$$\begin{aligned}
\mathcal{L}_N(A\cup B)\Lambda_{A\cup B}(\lambda) &= \mathrm{Log}(\mathbb{E}(\exp(\lambda S_{A\cup B}))) \\
&\le \mathrm{Log}(\mathbb{E}(\exp(\lambda S_A + \lambda S_B))) \\
&= \mathrm{Log}(\mathbb{E}(\exp(\lambda S_A)\exp(\lambda S_B))) \\
&= \mathrm{Log}(\mathbb{E}(\exp(\lambda S_A))\mathbb{E}(\exp(\lambda S_B))) \\
&= \mathrm{Log}(\mathbb{E}(\exp(\lambda S_A))) + \mathrm{Log}(\mathbb{E}(\exp(\lambda S_B))) \\
&= \mathcal{L}_N(A)\Lambda_A(\lambda) + \mathcal{L}_N(B)\Lambda_B(\lambda).
\end{aligned}$$

Hence $A \mapsto \mathcal{L}_N(A)\Lambda_A(\lambda)$ is a subadditive (deterministic) function so that

$$\lim_{n\to+\infty} \Lambda_{[0,n]^N} = \lim_{n\to+\infty} \frac{1}{n^N}\text{Log}(\mathbb{E}(\exp(n^N\lambda W_n(\cdot,\xi))))\Lambda_\xi(\lambda)$$

exists in $\mathbb{R}$. Invoking this time the superadditivity of $A \mapsto \mathcal{L}_N(A)\Lambda_A(\lambda)$, the same conclusion holds with $\lambda \leq 0$. The proof of the growth condition fulfilling by Λ_ξ follows easily from the growth conditions satisfied by W. This completes the proof of assertion i).

Proof of ii). For all $\lambda^* \in \mathbb{R}$ we have

$$\begin{aligned}
\Lambda_\xi^*(\lambda^*) &= \sup_{\lambda\in\mathbb{R}}(\lambda^*\lambda - \Lambda_\xi(\lambda)) \\
&\geq \sup_{\lambda\in\mathbb{R}}(\lambda^*\lambda - C_\xi|\lambda|) \\
&\geq \sup_{\lambda\in\mathbb{R}^*} \lambda\left(\lambda^* - C_\xi\frac{|\lambda|}{\lambda}\right) \\
&\geq \max\left(\sup_{\lambda>0}\lambda(\lambda^* - C_\xi), \sup_{\lambda<0}\lambda(\lambda^* + C_\xi)\right) = +\infty
\end{aligned}$$

if $\lambda^* \notin [-C_\xi, C_\xi]$. Clearly $\Lambda_\xi^*(\lambda^*) \geq 0$ for $\lambda^* \in [-C_\xi, C_\xi]$. It remains to establish that Λ^* is a good rate function. Convexity and lsc of Λ^* are straightforwardly checked. The fact that the level sets $[\Lambda^* \leq r]$, $r \in \mathbb{R}$, are compact follows directly from $\lim_{|\lambda^*|\to+\infty} \frac{\Lambda^*(\lambda^*)}{|\lambda^*|} = +\infty$.

Proof of iii). This assertion follows directly from i), ii) and Gärtner-Ellis's theorem, Theorem D.7 quoted in Appendix D.

Proof of iv). When Λ^* is strictly convex, then $E = \mathbb{R}$ and the claim follows directly from iii). □

Remark 7.1. Roughly speaking, assertion iv) says that the probabilities $\mu_n(B)$ behave like $\exp\left(-n^N\Lambda_\xi^*(B)\right)$. We say that $W_n(\cdot,\xi)$ satisfies a Large Deviation Principle (LDP) with the rate function $I = \Lambda_\xi^*$ also called entropy (see Appendix D). The set E_{ex} of exposed points of Λ_ξ^* is also the range of the derivative of Λ (see (Dembo and Zeitouni, 1998, lemma 2.3.9)). If one could get the exact value of Λ, its differentiability would also yield a full large deviations property as defined in Definition D.9.

Remark 7.2. The main feature of Large Deviations principle is to measure large rare events. Let us illustrate this fact when the domain of Λ_ξ^* is not reduced to the set $\{W^{hom}(\xi)\}$. First assume that there exists $\lambda^* \in \text{dom}\left(\Lambda_\xi^*\right)$ such that $W^{hom}(\xi) < \lambda^*$ and choose $\delta > 0$ such that $W^{hom}(\xi) + \delta \leq \lambda^*$. Consider the large closed set $F_\delta = [r \geq W^{hom}(\xi) + \delta]$. Thanks to the choice of δ we have $F_\delta \cap \text{dom}\left(\Lambda_\xi^*\right) \neq \emptyset$. Since the limit probability measure μ is the Dirac probability measure with mass

$W^{hom}(\xi)$, we have $\mu(F_\delta) = 0$ so that F_δ is a large rare event which "deviates" from $W^{hom}(\xi)$. Moreover

$$\Lambda_\xi^*(F_\delta) := \inf_{r \in F_\delta} \Lambda_\xi^*(r) \neq +\infty.$$

According to Proposition 7.1, iii), we infer that for n large enough

$$\mu_n([r \geq W^{hom}(\xi) + \delta]) \leq \exp(-n^N \gamma)$$

for any strict lower bound γ of $\Lambda_\xi^*(F_\delta)$. Thus μ_n measures the rate of decay to 0 of rare large events as F_δ. This explains the terminology *Large Deviations.* Note that if Λ is differentiable everywhere, from iv) and Remark 7.1, the rate of decay can be refined by the estimate from below

$$\exp(-n^N \Gamma) \leq \mu_n([r \geq W^{hom}(\xi) + \delta])$$

where Γ is any strict upper bound of $\Lambda_\xi^*(F_\delta)$.

If there is no $\lambda^* \in \mathrm{dom}\left(\Lambda_\xi^*\right)$ such that $W^{hom}(\xi) < \lambda^*$, choose $\lambda^* \in \mathrm{dom}\left(\Lambda_\xi^*\right)$ such that $W^{hom}(\xi) > \lambda^*$, set $F_\delta = [r \leq W^{hom}(\xi) - \delta]$ for $\delta > 0$ small enough and follow the same reasoning.

Diffusions terms of nonlinear reaction-diffusion problems considered in this chapter as well as in the following two, are the subdifferentials of random functionals of one of the two forms below:

$$\Phi_\varepsilon(\omega, u) = \begin{cases} \widetilde{\Phi}_\varepsilon(\omega, u) + \dfrac{1}{2} \displaystyle\int_{\partial\Omega} a_0 u \, d\mathcal{H}_{N-1} - \int_{\partial\Omega} \phi u \, d\mathcal{H}_{N-1} \\ \qquad\qquad\qquad\qquad\qquad\qquad \text{if } u \in H^1(\Omega) \\ +\infty \qquad\qquad\qquad\qquad\quad \text{otherwise}, \end{cases} \tag{7.5}$$

where $a_0 \in L^\infty_{\mathcal{H}_{N-1}}(\partial\Omega)$, $a_0 \geq 0$, $a_0 \geq \sigma > 0$ on $\Gamma \subset \partial\Omega$ with $\mathcal{H}_{N-1}(\Gamma) > 0$; or

$$\Phi_\varepsilon(\omega, u) = \begin{cases} \widetilde{\Phi}_\varepsilon(\omega, u) \text{ if } u \in H^1_\Gamma(\Omega) \\ +\infty \quad \text{otherwise}, \end{cases} \tag{7.6}$$

where $\widetilde{\Phi}_\varepsilon$ is the random functional defined in (7.1).

According to Lemma 2.3, for $\mathbb{P}$-a.e. $\omega \in \Sigma$, the subdifferential of the functional (7.5), actually its Gâteaux derivative since $W(\omega, \cdot)$ satisfies (D), is the operator $\partial\Phi_\varepsilon(\omega) : L^2(\Omega) \to 2^{L^2(\Omega)}$ defined for every $\omega \in \Sigma$ by

$$\begin{aligned} &\mathrm{dom}(\partial\Phi_\varepsilon(\omega)) \\ &= \left\{ v \in H^1(\Omega) : \mathrm{div} D_\xi W\left(\omega, \frac{\cdot}{\varepsilon}, \nabla v\right) \in L^2(\Omega), a_0 v + D_\xi W\left(\omega, \frac{\cdot}{\varepsilon}, \nabla v\right) \cdot \mathbf{n} = \phi \text{ on } \partial\Omega \right\} \end{aligned}$$

and, for all $v \in \mathrm{dom}\, \partial\Phi_\varepsilon(\omega)$,

$$\partial\Phi_\varepsilon(\omega) v = -\mathrm{div} D_\xi W\left(\omega, \frac{\cdot}{\varepsilon}, \nabla v\right).$$

Similarly the subdifferential of the functional (7.6), has the same expression in its domain, which is: for every $\omega \in \Sigma$

$$\begin{aligned}&\operatorname{dom}(\partial\Phi_\varepsilon(\omega))\\&=\Big\{v \in H^1_\Gamma(\Omega) : \operatorname{div} D_\xi W\Big(\omega, \frac{\cdot}{\varepsilon}, \nabla v\Big) \in L^2(\Omega), D_\xi W\Big(\omega, \frac{\cdot}{\varepsilon}, \nabla v\Big) \cdot \mathbf{n} = 0 \text{ on } \partial\Omega \setminus \Gamma\Big\}.\end{aligned}$$

We also consider one of the two functionals

$$\Phi^{hom}(\omega, u) = \begin{cases} \widetilde{\Phi}^{hom}(\omega, u) + \dfrac{1}{2}\displaystyle\int_{\partial\Omega} a_0 u \, d\mathcal{H}_{N-1} - \int_{\partial\Omega} \phi u \, d\mathcal{H}_{N-1} & \text{if } u \in H^1(\Omega) \\ +\infty & \text{otherwise;} \end{cases} \tag{7.7}$$

or

$$\Phi^{hom}(\omega, u) = \begin{cases} \widetilde{\Phi}^{hom}(\omega, u) \text{ if } u \in H^1_\Gamma(\Omega) \\ +\infty \quad \text{otherwise.} \end{cases} \tag{7.8}$$

where $\widetilde{\Phi}^{hom}$ are the random functionals defined in (7.2). The subdifferential of (7.7) is the operator $\partial\Phi^{hom}(\omega) : L^2(\Omega) \to 2^{L^2(\Omega)}$ defined for every $\omega \in \Sigma$ by

$$\begin{aligned}&\operatorname{dom}(\partial\Phi^{hom}(\omega))\\&=\{v \in H^1(\Omega) : \operatorname{div}\partial_\xi W^{hom}(\omega, \nabla v) \in L^2(\Omega), a_0 v + \partial_\xi W^{hom}(\omega, \nabla v) \cdot \mathbf{n} \ni \phi \text{ on } \partial\Omega\}\end{aligned}$$

and, for all $v \in \operatorname{dom}(\partial\Phi^{hom}(\omega))$,

$$\partial\Phi^{hom}(\omega)v = -\operatorname{div}\partial_\xi W^{hom}(\omega, \nabla v).$$

When W is ergodic, then $\partial\Phi^{hom}$ is deterministic and

$$\partial\Phi^{hom} v = -\operatorname{div}\partial_\xi W^{hom}(\nabla v).$$

The subdifferential of the functional (7.8), has the same expression in its domain, which is:

$$\begin{aligned}&\operatorname{dom}(\partial\Phi^{hom}(\omega))\\&= \{v \in H^1_\Gamma(\Omega) : \operatorname{div} D_\xi W^{hom}(\omega, \nabla v) \in L^2(\Omega), D_\xi W^{hom}(\omega, \nabla v) \cdot \mathbf{n} \ni 0 \text{ on } \partial\Omega \setminus \Gamma\}.\end{aligned}$$

Note that we cannot guarantee that $W^{hom}(\omega, \cdot)$ is Gâteaux-differentiable. Hence $\partial_\xi W^{hom}(\omega, \cdot)$ is possibly multivalued. Nevertheless, to shorten the notation (see also Remark 2.7), we write $\partial_\xi W^{hom}(\omega, \cdot)$ to denote any element of the subdifferential $\partial_\xi W^{hom}(\omega, \cdot)$. In the proposition below where the density $\xi \mapsto W(\omega, \xi)$ is not necessarily assumed to be Gâteaux-differentiable, we establish that we can express $\partial_\xi W^{hom}(\omega, .)$ as a Graph limit. More precisely,

Proposition 7.2. *Let W_n be the random function defined in (7.4), where $W(\omega, x, \cdot)$ is not necessarily Gâteaux-differentiable. Then the following assertions hold for $\mathbb{P}$-a.e. $\omega \in \Sigma$:*

(i) $W_n(\omega,.) \overset{M}{\to} W^{hom}(\omega,.)$;

(ii) $\partial_\xi W_n(\omega,.) \overset{G}{\to} \partial_\xi W^{hom}(\omega,.)$ *where* $\partial_\xi W_n(\omega,.)$ *is characterized by*

$$\partial_\xi W_n(\omega,\xi) = \left\{ \frac{1}{n^N} \int_{nY} \sigma \, dy : \sigma \in X_n(\omega,\xi) \right\}$$

where $X_n(\omega,\xi)$ *is the subspace of* $L^2(nY,\mathbb{R}^N)$ *defined by*

$$X_n(\omega,\xi) = \left\{ \sigma : \; div\, \sigma = 0, \; \sigma(y) \in \partial_\xi W(\omega,y,q+\xi) \text{ a.e. in } nY, \; q \in \nabla(H^1_0(nY)) \right\};$$

(iii) assume that for a.e. $x \in \mathbb{R}^N$, $W(\omega,x,.)$ *is Gâteaux-differentiable, strictly convex, and that its Fenchel conjugate satisfies*

(D_3^*) *there exists* $\gamma^* > 0$ *such that* $\langle \xi_1^* - \xi_2^*, \xi^1 - \xi^2 \rangle \geq \gamma^* |\xi_1 - \xi_2|^2$ *for* $\mathbb{P}$*-a.e.* $\omega \in \Sigma$, *for a.e.* $x \in \mathbb{R}^N$, *for all* $(\xi_1,\xi_2) \in \mathbb{R}^N \times \mathbb{R}^N$ *and all* $(\xi_1^*,\xi_2^*) \in \partial_\xi W^*(\omega,x,\xi_1) \times \partial_\xi W^*(\omega,x,\xi_2)$.

Then $W_n(\omega,.)$ *and* $W^{hom}(\omega,.)$ *are Gâteaux-differentiable. More precisely, for all* $\xi \in \mathbb{R}^N$,

$$D_\xi W^{hom}(\omega,\xi) = \lim_{n\to+\infty} D_\xi W_n(\omega,\xi) \text{ for } \mathbb{P}\text{-a.e. } \omega \in \Sigma$$

where

$$D_\xi W_n(\omega,\xi) = \frac{1}{n^N} \int_{nY} D_\xi W(\omega,y,\nabla u_{\xi,n}(\omega)(y) + \xi) \, dy,$$

and $u_{\xi,n}(\omega)$ *is the unique solution of the random Dirichlet problem*

$$\begin{cases} div\,(D_\xi W(\omega,.,\xi + \nabla v(.))) = 0 \text{ a.e. in } nY; \\ \\ v = 0 \text{ on } \partial nY. \end{cases}$$

Proof. In all the proof, we fix ω in the set Σ' of full probability for which the limit $\lim_{n\to+\infty} W_n(\omega,\xi) = W^{hom}(\omega,\xi)$ exists. It is easy to prove that $W_n(\omega,.)$, $W^{hom}(\omega,.) : \mathbb{R}^N \to \mathbb{R}$ are convex and satisfy the growth conditions fulfilled by W, i.e.

$$\alpha|.|^2 \leq W_n(\omega,.), W^{hom}(\omega,.) \leq \beta(1+|.|^2).$$

Consequently they satisfy the standard local equi-Lipschitz condition 2.7. This implies that $W_n(\omega,.)$ almost surely Γ-converges to $W^{hom}(\omega,.)$ (this is the general feature of sequences of equi-lower semicontinuous functionals which pointwise converge (for a complete proof see (Attouch *et al.*, 2014, Proposition 17.4.6)). Assertion (i) follows since the Γ-convergence and the Mosco-convergence coincide in finite dimensional spaces.

Assertion *(ii)* follows from Theorem B.3. To express the subdifferential $\partial_\xi W_n(\omega,.)$, observe that $W_n(\omega,.)$ is a epigraphical sum and apply the standard Legendre-Fenchel calculus (for a complete proof we refer again to (Attouch *et al.*, 2014, Proposition 17.4.6)).

Let us prove (iii). The fact that $W_n(\omega, .)$ is Gâteaux-differentiable comes from the formula of its subdifferential operator expressed in (ii). Indeed, under the hypothesis of (iii), $\partial_\xi W_n(\omega, \xi)$ is reduced to

$$\frac{1}{n^N} \int_{nY} D_\xi W(\omega, y, \nabla u_{\xi,n}(y) + \xi)\, dy,$$

where $u_{\xi,n}$ is the unique minimizer in $H_0^1(nY)$ of (7.4), then satisfies the random Dirichlet problem

$$\begin{cases} \text{div}\,(D_\xi W(\omega, ., \xi + \nabla u_{\xi,n}(.))) = 0 \text{ a.e. in } nY; \\ u_{\xi,n} = 0 \text{ on } \partial nY. \end{cases}$$

In order to simplify the notation, we no longer indicate the dependence on ω and n for this minimizer and write it u_ξ.

Let $(\xi, \xi^*) \in \partial_\xi W^{hom}(\omega, .)$. Since $D_\xi W_n(\omega, .) \stackrel{G}{\to} \partial_\xi W^{hom}(\omega, .)$, according to Proposition B.7, there exists $\xi_n \in \mathbb{R}^N$ such that $\xi_n \to \xi$ and $D_\xi W_n(\omega, \xi_n) \to \xi^*$. We first claim that

$$|D_\xi W_n(\omega, \xi_n) - D_\xi W_n(\omega, \xi)| \leq \frac{1}{\gamma^*} |\xi_n - \xi|. \tag{7.9}$$

Indeed from Jensen's inequality we have

$$\begin{aligned} &|D_\xi W_n(\omega, \xi_n) - D_\xi W_n(\omega, \xi)|^2 \\ &\leq \frac{1}{n^N} \int_{nY} |D_\xi W(\omega, y, \nabla u_{\xi_n}(y) + \xi_n) - D_\xi W(\omega, y, \nabla u_\xi(y) + \xi)|^2\, dy. \end{aligned} \tag{7.10}$$

On the other hand, from (D_3^*)

$$\begin{aligned} &\gamma^* \int_{nY} |D_\xi W(\omega, y, \nabla u_{\xi_n}(y) + \xi_n) - D_\xi W(\omega, y, \nabla u_\xi(y) + \xi)|^2 dy \\ &\leq \int_{nY} (D_\xi W(\omega, y, \nabla u_{\xi_n}(y) + \xi_n) - D_\xi W(\omega, y, \nabla u_\xi(y) + \xi)) \cdot (\xi_n - \xi) dy \\ &\leq \left(\int_{nY} |D_\xi W(\omega, y, \nabla u_{\xi_n}(y) + \xi_n) - D_\xi W(\omega, y, \nabla u_\xi(y) + \xi)|^2 dy \right)^{\frac{N}{2}} n^{\frac{1}{2}} |\xi_n - \xi| \end{aligned}$$

from which we deduce

$$\begin{aligned} &\frac{1}{n^N} \int_{nY} |D_\xi W(\omega, y, \nabla u_{\xi_n}(y) + \xi_n) - D_\xi W(\omega, y, \nabla u_\xi(y) + \xi)|^2\, dy \\ &\qquad\qquad \leq \frac{1}{\gamma^{*2}} |\xi_n - \xi|^2. \end{aligned} \tag{7.11}$$

Combining (7.10) and (7.11) yields (7.9). From (7.9), we infer that $\xi^* = \lim_{n \to +\infty} DW_n(\omega, \xi)$. Thus $\partial_\xi W^{hom}(\omega, \cdot)$ is made up of a single point. This completes the proof. □

7.1.2 *The random reaction part*

We are given a random regular CP-structured reaction functional, i.e. a functional

$$F : \Sigma \times [0, +\infty) \times L^2(\Omega) \to \mathbb{R}^\Omega$$

defined by $F(\omega, t, v)(x) = f(\omega, t, x, v(x))$ where

$$f : \Sigma \times [0, +\infty) \times \mathbb{R}^N \times \mathbb{R} \to \mathbb{R}$$

is a $(\mathcal{A}\otimes\mathcal{B}(\mathbb{R})\otimes\mathcal{B}(\mathbb{R}^N)\otimes\mathcal{B}(\mathbb{R}), \mathcal{B}(\mathbb{R}))$-measurable function such that for $\mathbb{P}$-a.e. $\omega \in \Sigma$, $(t, x, \zeta) \mapsto f(\omega, t, x, \zeta)$ is a regular CP-structured reaction function associated with $(r(\omega, \cdot), g, q(\omega, \cdot))$. Furthermore, we make the following additional hypotheses on r and q:

(REA_1) for $\mathbb{P}$-a.e. $\omega \in \Sigma$, $r(\omega, \cdot, \cdot) \in W^{1,2}(0, T, L^2_{\text{loc}}(\mathbb{R}^N, \mathbb{R}^l))$, $r(\omega, \cdot, \cdot) \in W^{1,2}(0, T, L^2_{\text{loc}}(\mathbb{R}^N))$;

(REA_2) for all bounded Borel sets B of $\mathbb{R}^N$, the real valued functions

$$\omega \mapsto \|r(\omega, t, \cdot)\|^2_{L^2(B,\mathbb{R}^l)} \text{ for all } t \in [0, T], \tag{7.12}$$

$$\omega \mapsto \|q(\omega, t, \cdot)\|^2_{L^2(B)} \text{ for all } t \in [0, T], \tag{7.13}$$

$$\omega \mapsto \int_0^T \left\| \frac{dr}{dt}(\omega, \tau, \cdot) \right\|^2_{L^2(B,\mathbb{R}^l)} d\tau \tag{7.14}$$

$$\omega \mapsto \int_0^T \left\| \frac{dq}{dt}(\omega, \tau, \cdot) \right\|^2_{L^2(B)} d\tau \tag{7.15}$$

belong to $L^1_{\mathbb{P}}(\Sigma)$;

(REA_3) the functions r and q, satisfy the covariance property with respect to the probability dynamical system $(\Sigma, \mathcal{A}, \mathbb{P}, (T_z)_{z\in\mathbb{Z}^N})$, i.e. for all $z \in \mathbb{Z}^N$, all $t \in [0, +\infty)$, a.e. $x \in \mathbb{R}^N$ and $\mathbb{P}$-a.e. $\omega \in \Sigma$,

$$\begin{aligned} r(\omega, t, x + z) &= r(T_z\omega, t, x), \\ q(\omega, t, x + z) &= q(T_z\omega, t, x). \end{aligned} \tag{7.16}$$

For $\varepsilon > 0$ we set $f_\varepsilon(\omega, t, x, \zeta) := f(\omega, t, \frac{x}{\varepsilon}, \zeta)$ for all $(\omega, t, x, \zeta) \in \Sigma \times \mathbb{R}_+ \times \mathbb{R}$, and define the functional F_ε by $F_\varepsilon(\omega, t, v)(x) = f(\omega, t, \frac{x}{\varepsilon}, v(x))$ for all $v \in L^2(\Omega)$. Note that in the expression of the condition (CP), the functions $\underline{f}$, $\overline{f}$, $\underline{y}$, $\overline{y}$, and the numbers $\underline{\rho}$, $\overline{\rho}$ may depend on ω (we sometimes omit it to shorten the notation), and that $F_\varepsilon(\omega, \cdot, \cdot)$ is a CP-structured reaction functional whose condition (CP) is exactly that of $F(\omega, \cdot, \cdot)$, i.e. with the functions $\underline{f}$, $\overline{f}$, $\underline{y}$, $\overline{y}$, $\underline{\rho}$ and $\overline{\rho}$. Since $\underline{y}$ and $\overline{y}$ do not depend on ε, condition (2.27) is automatically satisfied. The following lemma will be useful for later.

Lemma 7.1. *There exists N_{lem} in $\mathcal{A}$ with $\mathbb{P}(N_{\text{lem}}) = 0$ such that for all $\omega \in \Sigma \backslash N_{\text{lem}}$ we have*

$$\lim_{\varepsilon \to 0} \int_0^T \left\| \frac{dr}{dt} \left(\omega, \tau, \frac{\cdot}{\varepsilon}\right) \right\|^2_{L^2(\Omega, \mathbb{R}^l)} d\tau = \mathcal{L}_N(\Omega) \mathbb{E}^{\mathcal{I}} \int_0^T \left\| \frac{dr}{dt}(\omega, \tau, \cdot) \right\|^2_{L^2(Y, \mathbb{R}^l)} d\tau; \quad (7.17)$$

$$\lim_{\varepsilon \to 0} \int_0^T \left\| \frac{dq}{dt} \left(\omega, \tau, \frac{\cdot}{\varepsilon}\right) \right\|^2_{L^2(\Omega)} d\tau = \mathcal{L}_N(\Omega) \mathbb{E}^{\mathcal{I}} \int_0^T \left\| \frac{dq}{dt}(\omega, \tau, \cdot) \right\|^2_{L^2(Y)} d\tau. \quad (7.18)$$

Proof. We only prove (7.18), the proof of (7.17) is similar. Consider the set function $\mathbb{A}$ from the class $\mathcal{B}_b(\mathbb{R}^N)$ of bounded Borel subsets of $\mathbb{R}^N$ into the space $L^1_{\mathbb{P}}(\Sigma)$ of $\mathbb{P}$-integrable real valued functions, defined by

$$\mathbb{A}(B)(\cdot) = \int_0^T \left\| \frac{dq}{d\tau}(\cdot, \tau, \cdot) \right\|^2_{L^2(B)} d\tau.$$

From (7.15), the process $\mathbb{A}$ is well defined. Then, for every $(A, B) \in \mathcal{B}_b(\mathbb{R}^N) \times \mathcal{B}_b(\mathbb{R}^N)$ with $A \cap B = \emptyset$, from additivity of the integral we have

$$\mathbb{A}(A \cup B) = \mathbb{A}(A) + \mathbb{A}(B).$$

Moreover, from (7.16) we infer that

$$\mathbb{A}(z + B) = \mathbb{A}(B) \circ T_z.$$

Furthermore, $\mathbb{A}$ fulfills the following domination property: for all Borel set A included in Y,

$$\mathbb{A}(A) \leq h := \int_0^T \left\| \frac{dq}{d\tau}(\cdot, \tau, \cdot) \right\|^2_{L^2(Y)} d\tau$$

with $h \in L^1_{\mathbb{P}}(\Sigma)$. Therefore, $\mathbb{A}$ is an additive process indexed by $\mathcal{B}_b(\mathbb{R}^N)$, covariant with respect to $(T_z)_{z \in \mathbb{Z}^N}$ (see Definition C.7 or (Attouch *et al.*, 2014, Definition 12.4.1) and references therein). According to the additive ergodic theorem, Theorem C.5 (see also (Attouch *et al.*, 2014, Theorem 12.4.1)), there exists $N_{\text{lem}} \in \mathcal{A}$ with $\mathbb{P}(N_{\text{lem}}) = 0$ such that for all[2] $\omega \in \Sigma \setminus N$,

$$\begin{aligned}
\lim_{\varepsilon \to 0} \frac{\mathbb{A}(\frac{1}{\varepsilon}\Omega)}{\mathcal{L}_N(\frac{1}{\varepsilon}\Omega)} &= \lim_{\varepsilon \to 0} \int_0^T \frac{\varepsilon^N}{\mathcal{L}_N(\Omega)} \left\| \frac{dq}{d\tau}(\omega, \tau, \cdot) \right\|^2_{L^2(\frac{1}{\varepsilon}\Omega)} d\tau \\
&= \mathbb{E}^{\mathcal{I}} \int_0^T \left\| \frac{dq}{d\tau}(\omega, \tau, \cdot) \right\|^2_{L^2(Y)} d\tau.
\end{aligned}$$

Hence, a change of scale gives

$$\lim_{\varepsilon \to 0} \int_0^T \left\| \frac{dq}{d\tau} \left(\omega, \tau, \frac{\cdot}{\varepsilon}\right) \right\|^2_{L^2(\Omega)} d\tau = \mathcal{L}_N(\Omega) \mathbb{E}^{\mathcal{I}} \int_0^T \left\| \frac{dq}{d\tau}(\omega, \tau, \cdot) \right\|^2_{L^2(Y)} d\tau.$$

This completes the proof □

[2]Strictly speaking the almost sure convergence holds when Ω is a convex set. Using approximation of Ω by finite union of convex subset, it is easy to show that the convergence holds for regular Ω of class C^1 (see (Chabi and Michaille, 1994, Remark 3.3)).

7.2 General homogenization theorems

Given a sequence $(u^0_\varepsilon)_\varepsilon$ of $(\mathcal{A}, \mathcal{B}(L^2(\Omega)))$-measurable functions $u^0_\varepsilon : \Sigma \to H^1(\Omega)$, by combining Theorem 2.6 of Chapter 2 together with the variational convergence of the sequence of random energies Φ_ε specified above, we intend to analyze when $\varepsilon \to 0$, the asymptotic behavior in $C(0, T, L^2(\Omega))$ of the solution $u_\varepsilon(\omega)$ of one of the two random reaction-diffusion problems associated with functionals (7.5) or (7.6):

$$(\mathcal{P}_\varepsilon(\omega)) \begin{cases} \dfrac{du_\varepsilon}{dt}(\omega,t) - \mathrm{div} D_\xi W\left(\omega, \dfrac{\cdot}{\varepsilon}, \nabla u_\varepsilon(\omega,t)\right) = F_\varepsilon(\omega, t, u_\varepsilon(\omega,t)) \text{ in } L^2(\Omega) \\ \text{for a.e. } t \in (0,T), \\ u_\varepsilon(\omega, 0) = u^0_\varepsilon(\omega) \in H^1(\Omega),\ \underline{\rho}(\omega) \le u^0_\varepsilon(\omega, \cdot) \le \overline{\rho}(\omega), \\ u_\varepsilon(\omega,t) \in H^1(\Omega), \mathrm{div} D_\xi W\left(\omega, \frac{\cdot}{\varepsilon}, \nabla u_\varepsilon(\omega,t)\right) \in L^2(\Omega) \text{ for all } t \in]0,T], \\ a_0 u_\varepsilon(\omega,t) + D_\xi W\left(\omega, \dfrac{\cdot}{\varepsilon}, \nabla u_\varepsilon(\omega,t)\right) \cdot \mathbf{n} = \phi \text{ on } \partial\Omega \text{ for all } t \in]0,T]. \end{cases}$$

where we assume $a_0 \underline{\rho}(\omega) \le \phi \le a_0 \overline{\rho}(\omega)$ for $\mathbb{P}$-a.e. $\omega \in \Sigma$;

$$(\mathcal{P}'_\varepsilon(\omega)) \begin{cases} \dfrac{du_\varepsilon}{dt}(\omega,t) - \mathrm{div} D_\xi W\left(\omega, \dfrac{\cdot}{\varepsilon}, \nabla u_\varepsilon(\omega,t)\right) = F_\varepsilon(\omega, t, u_\varepsilon(\omega,t)) \text{ in } L^2(\Omega) \\ \text{for a.e. } t \in (0,T), \\ u_\varepsilon(\omega, 0) = u^0_\varepsilon(\omega) \in H^1_\Gamma(\Omega),\ \underline{\rho}(\omega) \le u^0_\varepsilon(\omega, \cdot) \le \overline{\rho}(\omega), \\ u_\varepsilon(\omega,t) \in H^1_\Gamma(\Omega), \mathrm{div} D_\xi W(\omega, \frac{\cdot}{\varepsilon}, \nabla u_\varepsilon(\omega,t)) \in L^2(\Omega) \text{ for all } t \in]0,T], \\ D_\xi W\left(\omega, \dfrac{\cdot}{\varepsilon}, \nabla u_\varepsilon(\omega,t)\right) \cdot \mathbf{n} = 0 \text{ on } \partial\Omega \setminus \Gamma \text{ for all } t \in]0,T], \end{cases}$$

where we assume $\underline{\rho}(\omega) \le 0 \le \overline{\rho}(\omega)$ for $\mathbb{P}$-a.e. $\omega \in \Sigma$.

Theorem 7.1. *For each $\omega \in \Sigma$, let us denote by $u_\varepsilon(\omega, \cdot)$ the unique solution in $C([0,T], L^2(\Omega))$ of the reaction-diffusion problem $(\mathcal{P}_\varepsilon(\omega))$, respectively $(\mathcal{P}'_\varepsilon(\omega))$. Assume that for $\mathbb{P}$-a.e. $\omega \in \Sigma$,*

i) $\sup_{\varepsilon > 0} \Phi_\varepsilon(\omega, u^0_\varepsilon(\omega)) < +\infty$;
ii) $u^0_\varepsilon(\omega)$ *strongly converges to* $u^0(\omega)$ *in* $L^2(\Omega)$.

Then, for $\mathbb{P}$-a.e. $\omega \in \Sigma$, $u_\varepsilon(\omega, \cdot)$ uniformly converges in $C([0,T], L^2(\Omega))$ to the *unique solution of the reaction-diffusion problem*

$$(\mathcal{P}^{hom}(\omega))\begin{cases}\dfrac{du}{dt}(\omega,t)-\operatorname{div}\partial_\xi W^{hom}(\omega,\nabla u(\omega,t))\ni F^{hom}(\omega,t,u(\omega,t)) \text{ in } L^2(\Omega)\\ \text{for a.e. } t\in(0,T),\\ u(\omega,0)=u^0(\omega)\in H^1(\Omega),\ \underline{\rho}(\omega)\leq u^0(\omega,\cdot)\leq\overline{\rho}(\omega),\\ u(\omega,t)\in H^1(\Omega), \operatorname{div}\partial_\xi W^{hom}(\omega,\nabla u(\omega,t))\in L^2(\Omega) \text{ for a.e. } t\in(0,T),\\ a_0u(\omega,t)+\partial_\xi W^{hom}(\omega,\nabla u(\omega,t))\cdot\mathbf{n}\ni\phi, \text{ on } \partial\Omega \text{ for a.e. } t\in(0,T),\end{cases}$$

respectively

$$(\mathcal{P}'^{hom}(\omega))\begin{cases}\dfrac{du}{dt}(\omega,t)-\operatorname{div}\partial_\xi W^{hom}(\omega,\nabla u(\omega,t))\ni F^{hom}(\omega,t,u(\omega,t)) \text{ in } L^2(\Omega)\\ \text{for a.e. } t\in(0,T),\\ u(\omega,0)=u^0(\omega)\in H^1_\Gamma(\Omega),\ \underline{\rho}(\omega)\leq u^0(\omega,\cdot)\leq\overline{\rho}(\omega),\\ u(\omega,t)\in H^1_\Gamma(\Omega), \operatorname{div}\partial_\xi W^{hom}(\omega,\nabla u(\omega,t))\in L^2(\Omega) \text{ for a.e. } t\in(0,T),\\ \partial_\xi W^{hom}(\omega,\nabla u(\omega,t))\cdot\mathbf{n}\ni 0 \text{ on } \partial\Omega\setminus\Gamma, \text{ for a.e. } t\in(0,T),\end{cases}$$

where F^{hom} *is given by* $F^{hom}(\omega,t,v)(x)=f^{hom}(\omega,t,x,v(x))$ *with*

$$f^{hom}(\omega,t,x,\zeta)=\overline{r}(\omega,t)\cdot g(\zeta)+\overline{q}(\omega,t),$$

and

$$\overline{r}(\omega,t)=\mathbb{E}^{\mathcal{I}}\left(\int_{(0,1)^N} r(\omega,t,y)\,dy\right), \overline{q}(\omega,t)=\mathbb{E}^{\mathcal{I}}\left(\int_{(0,1)^N} q(\omega,t,y)\,dy\right).$$

Moreover, for $\mathbb{P}$*-a.e.* $\omega\in\Sigma$, $\frac{du_\varepsilon}{dt}(\omega,\cdot)\rightharpoonup\frac{du}{dt}(\omega,\cdot)$ *weakly in* $L^2(0,T,L^2(\Omega))$ *and* $\underline{y}(\omega,t)\leq u(\omega,t)\leq\overline{y}(\omega,t)$ *for all* $t\in[0,T]$.

When the dynamical system $(\Sigma,\mathcal{A},\mathbb{P},(T_z)_{z\in\mathbb{Z}^N})$ *is ergodic, the initial condition is deterministic, i.e.* $u^0_\varepsilon(\omega)=u^0_\varepsilon$ *for* $\mathbb{P}$*-a.e.* $\omega\in\Sigma$, *then* $(\mathcal{P}^{hom}(\omega)=\mathcal{P}^{hom})$ *and* $(\mathcal{P}'^{hom}(\omega)=\mathcal{P}'^{hom})$ *are deterministic, given by*

$$(\mathcal{P}^{hom})\begin{cases}\dfrac{du}{dt}(t)-\operatorname{div}\partial_\xi W^{hom}(\nabla u(t))\ni F^{hom}(t,u(t)) \text{ in } L^2(\Omega)\\ \text{for a.e. } t\in(0,T),\\ u(0)=u^0\in H^1(\Omega),\ \underline{\rho}(\omega)\leq u^0(\cdot)\leq\overline{\rho}(\omega),\\ u(t)\in H^1(\Omega), \operatorname{div}\partial_\xi W^{hom}(\nabla u(t))\in L^2(\Omega) \text{ for a.e. } t\in(0,T),\\ a_0u(t)+\partial_\xi W^{hom}(\nabla u(t))\cdot\mathbf{n}\ni\phi \text{ on } \partial\Omega \text{ for a.e. } t\in(0,T),\end{cases}$$

respectively

$$(\mathcal{P}'^{hom})\begin{cases} \dfrac{du}{dt}(t) - \text{div}\partial_\xi W^{hom}(\nabla u(t)) \ni F^{hom}(t, u(t)) \text{ in } L^2(\Omega) \\ \text{for a.e. } t \in (0, T), \\ u(0) = u^0 \in H^1_\Gamma(\Omega),\ \underline{\rho}(\omega) \le u^0(\cdot) \le \overline{\rho}(\omega), \\ u(t) \in H^1_\Gamma(\Omega), \text{div}\partial_\xi W^{hom}(\nabla u(t)) \in L^2(\Omega) \text{ for a.e. } t \in (0, T), \\ \partial_\xi W^{hom}(\nabla u(t)) \cdot \mathbf{n} \ni 0 \text{ on } \partial\Omega \setminus \Gamma, \text{ for a.e. } t \in (0, T), \end{cases}$$

where F^{hom} *is given by* $F^{hom}(t, v)(x) = f^{hom}(t, x, v(x))$ *with*

$$f^{hom}(t, x, \zeta) = \overline{r}(t) \cdot g(\zeta) + \overline{q}(t),$$

$$\overline{r}(t) = \mathbb{E}\left(\int_{(0,1)^N} r(\cdot, t, y)\, dy\right),$$

$$\overline{q}(t) = \mathbb{E}\left(\int_{(0,1)^N} q(\cdot, t, y)\, dy\right).$$

Moreover, for $\mathbb{P}$*-a.e.* $\omega \in \Sigma$, $\frac{du_\varepsilon}{dt}(\omega, \cdot) \rightharpoonup \frac{du}{dt}$ *weakly in* $L^2(0, T, L^2(\Omega))$ *and* $\underline{y}(\omega, t) \le u \le \overline{y}(\omega, t)$ *for all* $t \in [0, T]$.

If in addition the Legendre-Fenchel conjugate of W *satisfies* (D_3^*)*, then* $\partial_\xi W^{hom}(\omega, \nabla u(t))$ *or* $\partial_\xi W^{hom}(\nabla u(t))$ *are single valued, equal to* $D_\xi W^{hom}(\omega, \nabla u(t))$ *or* $D_\xi W^{hom}(\nabla u(t))$*, and differential inclusions are equalities.*

Proof. The proof is a straightforward consequence of Theorem 2.6, and consists in checking (H_1), (H_5) and (H_6). In the whole proof, we reason with the set of full probability $\Sigma' = \Sigma \setminus N_{\text{lem}}$ where N_{lem} is the $\mathbb{P}$-negligible set of Lemma 7.1.

Proof of (H_1). $\Phi_\varepsilon(\omega, \cdot) \overset{M}{\to} \Phi^{hom}(\omega, \cdot)$. According to (Attouch *et al.*, 2014, Theorem 12.4.7), we deduce that for $\mathbb{P}$-a.e. ω in Σ', the sequence of functional $\left(\widetilde{\Phi}_\varepsilon(\omega, \cdot)\right)_{\varepsilon > 0}$ defined in (7.1), Γ-converges to the random integral functional $\widetilde{\Phi}^{hom}(\omega, \cdot)$ defined in 7.2) when $L^2(\Omega)$ is equipped with its strong convergence. We conclude by using Proposition 2.7.

Proof of (H_5). Note that from (7.12)–(7.15), we can assert that for all bounded set of $\mathbb{R}^N$, the maps $\omega \mapsto \int_B r(\cdot, t, \frac{x}{\varepsilon})dx$, $\omega \mapsto \int_B r(\cdot, t, x)dx$, $\omega \mapsto \int_B q(\cdot, t, \frac{x}{\varepsilon})dx$, and $\omega \mapsto \int_B q(\cdot, t, x)dx$ belong to $L^1_{\mathbb{P}}(\Sigma)$. We have to establish that for $\mathbb{P}$-a.e. $\omega \in \Sigma'$, $\sup_{\varepsilon > 0} \|r(\omega, \cdot, \frac{\cdot}{\varepsilon})\|_{L^\infty([0,T] \times \mathbb{R}^N, \mathbb{R}^l)} < +\infty$ and $r(\omega, \cdot, \frac{\cdot}{\varepsilon}) \rightharpoonup \overline{r}(\omega, \cdot)$ in $L^2(0, T, L^2(\Omega, \mathbb{R}^l))$.

The first claim is obvious. To show that $r(\omega, \cdot, \frac{\cdot}{\varepsilon}) \rightharpoonup \overline{r}(\omega, \cdot)$ in $L^2(0, T, L^2(\Omega, \mathbb{R}^l))$ we need the following lemma.

Lemma 7.2. *There exists $N \in \mathcal{A}$ with $\mathbb{P}(N) = 0$, such that for all $t \in [0,T]$ and all $\omega \in \Sigma' \setminus N$,*

$$r\left(\omega, t, \frac{\cdot}{\varepsilon}\right) \rightharpoonup \overline{r}(\omega, \cdot) := \mathbb{E}^{\mathcal{I}}\left(\int_{(0,1)^N} r(\omega, t, y)\, dy\right)$$

weakly in $L^2(\Omega, \mathbb{R}^l)$.

Proof. Fix $t \in [0,T] \cap \mathbb{Q}$. From (7.12) we can apply (Chabi and Michaille, 1994, Theorem 4.2)), straightforward consequence of the additive ergodic theorem, Theorem C.5: there exists $N_t \in \mathcal{A}$ with $\mathbb{P}(N_t) = 0$ such that for every $\omega \in \Sigma' \setminus N_t$

$$r\left(\omega, t, \frac{\cdot}{\varepsilon}\right) \rightharpoonup \overline{r}(\omega, t)$$

weakly in $L^2(\Omega, \mathbb{R}^l)$. Set $N := \cup_{t \in [0,T] \cap \mathbb{Q}} N_t \cup N_{\text{lem}}$. We are going to show that for all $\omega \in \Sigma' \setminus N$, the weak convergence $r(\omega, t, \frac{\cdot}{\varepsilon}) \rightharpoonup \overline{r}(\omega, t)$, holds for all $t \in [0,T]$. Let $\omega \in \Sigma' \setminus N$, $\varphi \in L^2(\Omega, \mathbb{R}^l)$, $t \in [0,T]$ and $(t_n)_{n \in \mathbb{N}}$ be a sequence in $[0,T] \cap \mathbb{Q}$ converging to t with $t_n \leq t$. Let $n \in \mathbb{N}$. We have

$$\begin{aligned}
&\left\langle r\left(\omega, t, \frac{\cdot}{\varepsilon}\right), \varphi\right\rangle_{L^2(\Omega,\mathbb{R}^l)} \\
&= \left\langle r\left(\omega, t_n, \frac{\cdot}{\varepsilon}\right), \varphi\right\rangle_{L^2(\Omega,\mathbb{R}^l)} + \left\langle r\left(\omega, t, \frac{\cdot}{\varepsilon}\right) - r\left(\omega, t_n, \frac{\cdot}{\varepsilon}\right), \varphi\right\rangle_{L^2(\Omega,\mathbb{R}^l)},
\end{aligned} \tag{7.19}$$

with, from the weak convergence above,

$$\lim_{\varepsilon \to 0} \left\langle r\left(\omega, t_n, \frac{\cdot}{\varepsilon}\right), \varphi\right\rangle_{L^2(\Omega,\mathbb{R}^l)} = \langle \overline{r}(\omega, t_n), \varphi\rangle_{L^2(\Omega,\mathbb{R}^l)}.$$

Let us set

$$R_\varepsilon(\omega, t, t_n) = \left\langle r\left(\omega, t, \frac{\cdot}{\varepsilon}\right) - r\left(\omega, t_n, \frac{\cdot}{\varepsilon}\right), \varphi\right\rangle_{L^2(\Omega,\mathbb{R}^l)}.$$

Since $r(\omega, \cdot) \in W^{1,2}(0, T, L^2_{\text{loc}}(\mathbb{R}^N, \mathbb{R}^l))$, we infer that

$$\begin{aligned}
|R_\varepsilon(\omega, t, t_n)| &\leq \left\| r\left(\omega, t, \frac{\cdot}{\varepsilon}\right) - r\left(\omega, t_n, \frac{\cdot}{\varepsilon}\right)\right\|_{L^2(\Omega,\mathbb{R}^l)} \|\varphi\|_{L^2(\Omega,\mathbb{R}^l)} \\
&\leq \|\varphi\|_{L^2(\Omega,\mathbb{R}^l)} \int_{t_n}^t \left\| \frac{dr}{d\tau}\left(\omega, \tau, \frac{\cdot}{\varepsilon}\right)\right\|_{L^2(\Omega,\mathbb{R}^l)} d\tau \\
&\leq \|\varphi\|_{L^2(\Omega,\mathbb{R}^l)} (t_n - t)^{\frac{1}{2}} \left(\int_0^T \left\| \frac{dr}{d\tau}\left(\omega, \tau, \frac{\cdot}{\varepsilon}\right)\right\|^2_{L^2(\Omega,\mathbb{R}^l)} d\tau\right)^{\frac{1}{2}}.
\end{aligned}$$

Thus, from (7.17), we infer that

$$\begin{aligned}
&\limsup_{\varepsilon \to 0} |R_\varepsilon(\omega, t, t_n)| \\
&\leq \|\varphi\|_{L^2(\Omega,\mathbb{R}^l)} (t_n - t)^{\frac{1}{2}} \mathcal{L}_N(\Omega) \mathbb{E}^{\mathcal{I}} \int_0^T \left\| \frac{dr}{dt}(\omega, \tau, \cdot)\right\|^2_{L^2(Y,\mathbb{R}^l)} d\tau.
\end{aligned} \tag{7.20}$$

Passing to the limit $\varepsilon \to 0$, then $n \to +\infty$ in (7.19), from (7.20), we deduce that

$$\begin{aligned}\lim_{\varepsilon\to 0}\left\langle r\left(\omega,t,\frac{\cdot}{\varepsilon}\right),\varphi\right\rangle_{L^2(\Omega,\mathbb{R}^l)} &= \lim_{n\to+\infty}\lim_{\varepsilon\to 0}\left\langle r\left(\omega,t,\frac{\cdot}{\varepsilon}\right),\varphi\right\rangle_{L^2(\Omega,\mathbb{R}^l)}\\ &= \lim_{n\to+\infty}\langle\overline{r}(\omega,t_n),\varphi\rangle_{L^2(\Omega,\mathbb{R}^l)}\\ &= \langle\overline{r}(\omega,t),\varphi\rangle_{L^2(\Omega,\mathbb{R}^l)},\end{aligned}$$

which ends the proof of Lemma 7.2, provided that we justify the last equality

$$\lim_{n\to+\infty}\langle\overline{r}(\omega,t_n),\varphi\rangle_{L^2(\Omega,\mathbb{R}^l)} = \langle\overline{r}(\omega,t),\varphi\rangle_{L^2(\Omega,\mathbb{R}^l)}.$$

This convergence is a straightforward consequence of the continuity of the map $t \mapsto \int_{(0,1)^N} r(\omega,t,y)dy$ and the conditional Lebesgue dominated convergence theorem, Proposition C.9, which states that

$$\lim_{n\to+\infty}\mathbb{E}^{\mathcal{I}}\int_{(0,1)^N} r(\omega,t_n,y)dy = \mathbb{E}^{\mathcal{I}}\int_{(0,1)^N} r(\omega,t,y)dy,$$

provided that $\int_{(0,1)^N} r(\omega,t_n,y)dy$ be dominated by some function $h \in L^1_{\mathbb{P}}(\Sigma)$. An elementary calculation yields

$$\|r(\omega,t_n,\cdot)\|_{L^2(Y,\mathbb{R}^l)} \leq \|r(\omega,0,\cdot)\|_{L^2(Y,\mathbb{R}^l)} + \left(T\int_0^T\left\|\frac{dr}{dt}(\omega,\tau,\cdot)\right\|^2_{L^2(Y,\mathbb{R}^l)}dt\right)^{\frac{1}{2}}.$$

From (7.12) and (7.14), the function h, defined by

$$h(\omega) := \|r(\omega,0,\cdot)\|_{L^2(Y,\mathbb{R}^l)} + \left(T\int_0^T\left\|\frac{dr}{dt}(\omega,\tau,\cdot)\right\|^2_{L^2(Y,\mathbb{R}^l)}dt\right)^{\frac{1}{2}},$$

belongs to $L^1_{\mathbb{P}}(\Sigma)$, hence is suitable. □

Proof of* (H_5) *continued. Fix $\omega \in \Sigma' \setminus N$. Let $\varphi \in L^2(0,T,L^2(\Omega,\mathbb{R}^l))$. According to Lemma 7.2, for all $t \in [0,T]$ we have

$$\left\langle r\left(\omega,t,\frac{\cdot}{\varepsilon}\right),\varphi(t)\right\rangle_{L^2(\Omega,\mathbb{R}^l)} \to \langle\overline{r}(\omega,t),\varphi(t)\rangle_{L^2(\Omega,\mathbb{R}^l)}$$

and the conclusion follows from the Lebesgue dominated convergence theorem. The domination property is obtained as follows: we have

$$\begin{aligned}&\left|\left\langle r\left(\omega,t,\frac{\cdot}{\varepsilon}\right),\varphi(t)\right\rangle_{L^2(\Omega,\mathbb{R}^l)}\right|\\ &\leq \left\|r\left(\omega,t,\frac{\cdot}{\varepsilon}\right)\right\|_{L^2(\Omega,\mathbb{R}^l)}\|\varphi(t)\|_{L^2(\Omega,\mathbb{R}^l)}\\ &\leq \left(\left\|r\left(\omega,0,\frac{\cdot}{\varepsilon}\right)\right\|_{L^2(\Omega,\mathbb{R}^l)} + \left(T\int_0^T\left\|\frac{dr}{dt}\left(\omega,\tau,\frac{\cdot}{\varepsilon}\right)\right\|^2_{L^2(\Omega,\mathbb{R}^l)}d\tau\right)^{\frac{1}{2}}\right)\|\varphi(t)\|_{L^2(\Omega,\mathbb{R}^l)}\\ &\leq \sup_{\varepsilon>0}\left(\left\|r\left(\omega,0,\frac{\cdot}{\varepsilon}\right)\right\|_{L^2(\Omega,\mathbb{R}^l)} + \left(T\int_0^T\left\|\frac{dr}{dt}\left(\omega,\tau,\frac{\cdot}{\varepsilon}\right)\right\|^2_{L^2(\Omega,\mathbb{R}^l)}d\tau\right)^{\frac{1}{2}}\right)\|\varphi(t)\|_{L^2(\Omega,\mathbb{R}^l)}\end{aligned}$$

where, from Lemma 7.2,

$$\sup_{\varepsilon>0}\left\|r\left(\omega,0,\frac{\cdot}{\varepsilon}\right)\right\|_{L^2(\Omega,\mathbb{R}^l)}<+\infty,$$

and, from (7.17),

$$\sup_{\varepsilon>0}\int_0^T\left\|\frac{dr}{dt}\left(\omega,\tau,\frac{\cdot}{\varepsilon}\right)\right\|^2_{L^2(\Omega,\mathbb{R}^l)}d\tau<+\infty.$$

Proof of (H_6)**.** The proof is exactly that of condition (H_5), by substituting q for r, $L^2(\Omega)$ for $L^2(\Omega,\mathbb{R}^l)$, then using Lemma 7.3 below whose proof is an easy adaptation of that of Lemma 7.2.

Lemma 7.3. *There exists $N'\in\mathcal{A}$ with $\mathbb{P}(N')=0$, such that for all $t\in[0,T]$ and all $\omega\in\Sigma'\setminus N'$,*

$$q\left(\omega,t,\frac{\cdot}{\varepsilon}\right)\rightharpoonup\overline{q}(\omega,\cdot):=\mathbb{E}^{\mathcal{I}}\left(\int_{(0,1)^N}q(\omega,t,y)\,dy\right)$$

weakly in $L^2(\Omega,\mathbb{R})$.

The conclusion of Theorem 7.1 then holds for all $\omega\in\Sigma\setminus N_{lem}\cup N\cup N'$. □

Remark 7.3. Assume the domination property: there exists h_r and h_q in $L^1_{\mathbb{P}}(\Sigma)$ such that for a.e. $(t,x)\in(0,T)\times(0,1)^N$, and for $\mathbb{P}$-a.e. $\omega\in\Sigma$,

$$\left|\frac{\partial r}{\partial t}(\omega,t,x)\right|\le h_r(\omega),$$

$$\left|\frac{\partial q}{\partial t}(\omega,t,x)\right|\le h_q(\omega).$$

Then, noticing that $\frac{d}{dt}\mathbb{E}^{\mathcal{I}}\left(\int_{(0,1)^N}r(\cdot,t,y)\,dy\right)$ and $\frac{d}{dt}\mathbb{E}^{\mathcal{I}}\left(\int_{(0,1)^N}q(\cdot,t,y)\,dy\right)$ are $\mathcal{I}$-measurable (because invariant under the group $(T_z)_{z\in\mathbb{Z}^N}$), we easily infer that

$$\frac{d}{dt}\mathbb{E}^{\mathcal{I}}\left(\int_{(0,1)^N}r(\cdot,t,y)\,dy\right)=\mathbb{E}^{\mathcal{I}}\left(\int_{(0,1)^N}\frac{\partial r}{\partial t}(\cdot,t,y)\,dy\right),$$

$$\frac{d}{dt}\mathbb{E}^{\mathcal{I}}\left(\int_{(0,1)^N}q(\cdot,t,y)\,dy\right)=\mathbb{E}^{\mathcal{I}}\left(\int_{(0,1)^N}\frac{\partial q}{\partial t}(\cdot,t,y)\,dy\right).$$

If furthermore condition (REA_1) is reinforced by

$$r(\omega,\cdot,\cdot)\in W^{2,2}(0,T,L^2_{\mathrm{loc}}(\mathbb{R}^N,\mathbb{R}^l)),\quad q(\omega,\cdot,\cdot)\in W^{2,2}(0,T,L^2_{\mathrm{loc}}(\mathbb{R}^N)),$$

and condition (REA_2) by assuming that

$$\omega\mapsto\int_0^T\left\|\frac{d^2r}{dt^2}(\omega,\tau,\cdot)\right\|^2_{L^2(B,\mathbb{R}^l)}d\tau$$

$$\omega\mapsto\int_0^T\left\|\frac{d^2q}{dt^2}(\omega,\tau,\cdot)\right\|^2_{L^2(B)}d\tau,$$

belong to $L^1_{\mathbb{P}}(\Sigma)$ for all bounded Borel set of $\mathbb{R}^N$, then substituting $\frac{dr}{dt}(\omega,t,x)$ and $\frac{dq}{dt}(\omega,t,x)$ for $r(\omega,t,x)$ and $q(\omega,t,x)$ in the proofs of Lemmas 7.1, 7.2, 7.3, and the proofs of (H_5) and (H_6), we infer that for $\mathbb{P}$-a.e. $\omega\in\Sigma$, we have

$$\frac{dr}{dt}\left(\omega,\cdot,\frac{\cdot}{\varepsilon}\right) \rightharpoonup \mathbb{E}^{\mathcal{I}}\left(\int_{(0,1)^N}\frac{\partial r}{\partial t}(\omega,\cdot,y)\,dy\right) = \frac{d\overline{r}(\omega,\cdot)}{dt} \text{ in } L^2(0,T,L^2(\Omega,\mathbb{R}^l)),$$

$$\frac{dq}{dt}\left(\omega,\cdot,\frac{\cdot}{\varepsilon}\right) \rightharpoonup \mathbb{E}^{\mathcal{I}}\left(\int_{(0,1)^N}\frac{\partial q}{\partial t}(\omega,\cdot,y)\,dy\right) = \frac{d\overline{q}(\omega,\cdot)}{dt} \text{ in } L^2(0,T,L^2(\Omega,\mathbb{R})).$$

Hence, in the conclusion of Theorem 7.1, from Theorem 2.6 we can specify that for $\mathbb{P}$-a.e. $\omega\in\Sigma$, the solution $u(\omega,t)$ belongs to $\text{dom}(\partial\Phi^{hom}(\omega))$ for all $t\in]0,T]$.

Consider now random reaction-diffusion problems $(\mathcal{P}_\varepsilon(\omega))$ with the same diffusion term, and with a random reaction term fulfilling the two conditions of Theorem 2.7, i.e. $F_\varepsilon(\omega,t,v)(x) = f_\varepsilon(\omega,t,x,v(x))$ for all $v\in L^2(\Omega)$ with, for $\mathbb{P}$-a.e. $\omega\in\Sigma$,

(i) $\zeta\mapsto f_\varepsilon(\omega,t,x,\zeta)$ is locally Lipschitz continuous, uniformly with respect to $(t,x,n)\in\mathbb{R}\times\mathbb{R}^N$;
(ii) $f_\varepsilon(\omega,\cdot,\cdot,\cdot)$ satisfies (CP).

Set $G_\varepsilon(\omega,t) := F_\varepsilon(\omega,t,u_\varepsilon(\omega,t))$. As a straightforward corollary of Theorem 2.7 we obtain

Theorem 7.2. *For each $\omega\in\Sigma$, denote by $u_\varepsilon(\omega,\cdot)$ the unique solution in $C([0,T],L^2(\Omega))$ of the reaction-diffusion problem $(\mathcal{P}_\varepsilon(\omega))$, respectively $(\mathcal{P}'_\varepsilon(\omega))$. Assume that for $\mathbb{P}$-a.e. $\omega\in\Sigma$,*

i) $\sup_{\varepsilon>0}\Phi_\varepsilon(\omega,u^0_\varepsilon(\omega)) < +\infty$;
ii) $u^0_\varepsilon(\omega)$ *strongly converges to* $u^0(\omega)$ *in* $L^2(\Omega)$;
iii) for $\mathbb{P}$-a.e. $\omega\in\Sigma$, $G_\varepsilon(\omega,\cdot)\rightharpoonup G(\omega,\cdot)$ weakly in $L^2(0,T,L^2(\Omega))$.

Then, for $\mathbb{P}$-a.e. $\omega\in\Sigma$, $u_\varepsilon(\omega,\cdot)$ uniformly converges in $C([0,T],L^2(\Omega))$ to the unique solution of the reaction-diffusion problem

$$(\mathcal{P}^{hom}(\omega))\begin{cases}\dfrac{du}{dt}(\omega,t)-\text{div}\partial_\xi W^{hom}(\omega,\nabla u(\omega,t))\ni G(\omega,t) \text{ in } L^2(\Omega)\\ \text{for a.e. } t\in(0,T),\\ u(\omega,0)=u^0(\omega)\in H^1(\Omega),\ \underline{\rho}(\omega)\le u^0(\omega,\cdot)\le\overline{\rho}(\omega),\\ u(\omega,t)\in H^1(\Omega), \text{div}\partial_\xi W^{hom}(\omega,\nabla u(\omega,t))\in L^2(\Omega) \text{ for a.e. } t\in(0,T),\\ a_0u(\omega,t)+\partial_\xi W^{hom}(\omega,\nabla u(\omega,t))\cdot\mathbf{n}\ni\phi, \text{ on } \partial\Omega \text{ for a.e. } t\in(0,T),\end{cases}$$

respectively

$$(\mathcal{P}'^{hom}(\omega))\begin{cases}\frac{du}{dt}(\omega,t)-\operatorname{div}\partial_\xi W^{hom}(\omega,\nabla u(\omega,t))\ni G(\omega,t)\ in\ L^2(\Omega)\\ for\ a.e.\ t\in(0,T),\\ u(\omega,0)=u^0(\omega)\in H^1_\Gamma(\Omega),\ \underline{\rho}(\omega)\le u^0(\omega,\cdot)\le\overline{\rho}(\omega),\\ u(\omega,t)\in H^1_\Gamma(\Omega),\operatorname{div}\partial_\xi W^{hom}(\omega,\nabla u(\omega,t))\in L^2(\Omega)\ for\ a.e.\ t\in(0,T),\\ \partial_\xi W^{hom}(\omega,\nabla u(\omega,t))\cdot\mathbf{n}\ni 0\ on\ \partial\Omega\backslash\Gamma\ for\ a.e.\ t\in(0,T).\end{cases}$$

Moreover, for $\mathbb{P}$-a.e. $\omega\in\Sigma$, $\frac{du_\varepsilon}{dt}(\omega,\cdot)\rightharpoonup\frac{du}{dt}(\omega,\cdot)$ weakly in $L^2(0,T,L^2(\Omega))$ and $\underline{y}(\omega,t)\le u(\omega,t)\le\overline{y}(\omega,t)$ for all $t\in[0,T]$.

When the dynamical system $(\Sigma,\mathcal{A},\mathbb{P},(T_z)_{z\in\mathbb{Z}^N})$ is ergodic, the initial condition is deterministic, i.e. $u^0_\varepsilon(\omega)=u^0_\varepsilon$ for $\mathbb{P}$-a.e. $\omega\in\Sigma$, together with G, then $(\mathcal{P}^{hom}(\omega)=\mathcal{P}^{hom})$ and $(\mathcal{P}'^{hom}(\omega)=\mathcal{P}'^{hom})$ are deterministic, given by

$$(\mathcal{P}^{hom})\begin{cases}\frac{du}{dt}(t)-\operatorname{div}\partial_\xi W^{hom}(\nabla u(t))\ni G(t)\ in\ L^2(\Omega)\\ for\ a.e.\ t\in(0,T),\\ u(0)=u^0\in H^1(\Omega),\ \underline{\rho}(\omega)\le u^0(\cdot)\le\overline{\rho}(\omega),\\ u(t)\in H^1(\Omega),\operatorname{div}\partial_\xi W^{hom}(\nabla u(t))\in L^2(\Omega)\ for\ a.e.\ t\in(0,T),\\ a_0u(t)+\partial_\xi W^{hom}(\nabla u(t))\cdot\mathbf{n}\ni\phi\ on\ \partial\Omega\ for\ a.e.\ t\in(0,T),\end{cases}$$

respectively

$$(\mathcal{P}'^{hom})\begin{cases}\frac{du}{dt}(t)-\operatorname{div}\partial_\xi W^{hom}(\nabla u(t))\ni G(t)\ in\ L^2(\Omega)\\ for\ a.e.\ t\in(0,T),\\ u(0)=u^0\in H^1_\Gamma(\Omega),\ \underline{\rho}(\omega)\le u^0(\cdot)\le\overline{y}(\omega),\\ u(t)\in H^1_\Gamma(\Omega),\operatorname{div}\partial_\xi W^{hom}(\nabla u(t))\in L^2(\Omega)\ for\ a.e.\ t\in(0,T),\\ \partial_\xi W^{hom}(\nabla u(t))\cdot\mathbf{n}\ni 0\ on\ \partial\Omega\setminus\Gamma\ for\ a.e.\ t\in(0,T).\end{cases}$$

Moreover, for $\mathbb{P}$-a.e. $\omega\in\Sigma$, $\frac{du_\varepsilon}{dt}(\omega,\cdot)\rightharpoonup\frac{du}{dt}$ weakly in $L^2(0,T,L^2(\Omega))$ and $\underline{y}(\omega,t)\le u\le\overline{y}(\omega,t)$ for all $t\in[0,T]$.

If in addition the Legendre-Fenchel conjugate of W satisfies (D_3^*), then $\partial_\xi W^{hom}(\omega,\nabla u(t))$ or $\partial_\xi W^{hom}(\nabla u(t))$ are single valued, equal to $D_\xi W^{hom}(\omega,\nabla u(t))$ or $D_\xi W^{hom}(\nabla u(t))$, and differential inclusions are equalities.

7.3 Examples of stochastic homogenization of a diffusive Fisher food-limited population model with Allee effect

Applying Theorem 7.1, we deal with the stochastic homogenization of a reaction-diffusion problem which models the food-limited population dynamics. The reaction functional corresponds to that of the Fisher model with Allee effect described in Example 2.1, a). We assume that the growth rate r, along with the critical threshold a, below which the per-capita growth rate turns negative, are influenced by the heterogeneities of the spatial environment and change in each small habitats. In a first example, we assume that the heterogeneities are distributed following a regular random patch model, i.e. in the probabilistic setting, the dynamical system is that of a random checkerboard-like environment. In a second example, the heterogeneities are distributed following a Poisson point process. In the two examples, in order to simplify the model, we assume that r and a do not depend on the time variable t. Otherwise, it would be easy to make the appropriate assumptions concerning the absolute continuity of r and a with respect to the time variable, without changing the constructions below (see examples treated in Sections 8.4 and 9.5). We also assume that the carrying capacity K_{car} is constant. For a carrying capacity depending on the time-space variable in the case of a prey-predator model, see Section 9.5. In what follows $N = 2$ or 3.

7.3.1 *Random checkerboard-like environment*

Given two triples (r^-, a^-, W^-) and (r^+, a^+, W^+) in $[0, +\infty] \times [0, K_{car}] \times \mathrm{Conv}_{\alpha,\beta}$ where W^-, W^+ do not depend on x, and $\mathbf{p} \in [0,1]$, we consider the product $\Sigma = \{(r^-, a^-, W^-), (r^+, a^+, W^+)\}^{\mathbb{Z}^N}$ equipped with the σ-algebra $\mathcal{A}$, product of the trivial σ-algebra of subsets of $\{(r^-, a^-, W^-), (r^+, a^+, W^+)\}$. Each element of Σ is then of the form $(\omega_z)_{z\in\mathbb{Z}^N}$, with $\omega_z = (\omega_z^1, \omega_z^2, \omega_z^3)$, where $\omega_z^1 \in \{r^-, r^+\}$, $\omega_z^2 \in \{a^-, a^+\}$, and $\omega_z^3 \in \{W^-, W^+\}$.

We equip $(\Sigma, \mathcal{A})$ with the product probability measure $\mathbb{P}_{\mathbf{p}} = \otimes_{z\in\mathbb{Z}^N}\mu_z$ where for all $z \in \mathbb{Z}^N$, $\mu_z = \mathbf{p}\delta_{(r^-,a^-,W^-)} + (1-\mathbf{p})\delta_{(r^+,a^+,W^+)}$. By construction $\mathbb{P}_{\mathbf{p}}$ is invariant under the shift group $(T_z)_{z\in\mathbb{Z}^N}$ defined by $T_z(\omega_t)_{t\in\mathbb{Z}^N} = (\omega_{t+z})_{t\in\mathbb{Z}^N}$, i.e. $T_z\#\mathbb{P}_{\mathbf{p}} = \mathbb{P}_{\mathbf{p}}$ for all $z \in \mathbb{Z}^N$. We set for all $(\omega, t, x, \zeta) \in \Sigma \times \mathbb{R}_+ \times \mathbb{R}^N \times \mathbb{R}$,

$$r(\omega, x) := \omega_z^1, \quad a(\omega, x) = \omega_z^2, \quad W(\omega, x, \cdot) = \omega_z^3 \text{ whenever } x \in Y + z,$$

and $f(\omega, t, x, \zeta) = r(\omega, x)\zeta\left(1 - \frac{\zeta}{K_{car}}\right)\left(\frac{\zeta - a(\omega,x)}{K_{car}}\right)$ which define a random CP-structured reaction function provided that we write it

$$f(\omega, t, x, \zeta) = r(\omega, x)\frac{\zeta^2}{K_{car}}\left(1 - \frac{\zeta}{K_{car}}\right) - r(\omega, x)a(\omega, x)\frac{\zeta}{K_{car}}\left(1 - \frac{\zeta}{K_{car}}\right).$$

According to this definition it is straightforward to show that f is a random CP-structured reaction function, that $f(\omega, t, x+z, \zeta) = f(T_z\omega, t, x, \zeta)$ for all $(\omega, t, x, \zeta) \in \Sigma \times \mathbb{R}_+ \times \mathbb{R}^N \times \mathbb{R}$, and that conditions (7.12) and (7.13) hold. Regarding the random density W, one can easily show that it verify $W(\omega, x+z, \xi) = W(T_z\omega, x, \xi)$ for all $(\omega, t, x, z, \xi) \in \Sigma \times \mathbb{R}_+ \times \mathbb{R}^N \times \mathbb{Z}^N \times \mathbb{R}$.

Furthermore, it is easily seen that $(\Sigma, \mathcal{A}, \mathbb{P}_{\mathbf{p}}, (T_z)_{z\in\mathbb{Z}^N})$ is ergodic since its satisfies the mixing condition (C.1) (observe that (C.1) is satisfied with the cylinders which generate $\mathcal{A}$).

The random CP-structured reaction function f_ε defined by $f_\varepsilon(\omega,t,x,\zeta) = f(\omega,t,\frac{x}{\varepsilon},\zeta)$ may be seen as the Fisher reaction function defined in a ε-random checkerboard-like spatial environment: the growth rate and the threshold take two values at random on the lattice spanned by the cell $\varepsilon Y = (0,\varepsilon)^2$ modeling a mosaic of two kinds of small habitats. The diffusion is associated with a random density W_ε defined by $W_\varepsilon(\omega,x,\xi) = W(\omega,\frac{x}{\varepsilon},\xi)$ taking also two values at random on this lattice. The triples (r^-,a^-,W^-) and (r^+,a^+,W^+) represent a sample at scale 1 of two kinds of habitat whose probability of occurring is $\mathbf{p}$ and $1-\mathbf{p}$ respectively. Obviously we can easily generalize this model with r, a and W taking countable values. To shorten the notation, we assume that the Legendre-Fenchel conjugate of $W^\pm$ satisfies (D_3^*) so that, from Proposition 7.2, (iii), W^{hom} is Gâteaux-differentiable and is reduced to a pointwise limit.

The reaction-diffusion problem modeling the evolution of the density u of some specie during a time $T>0$, in a C^1-regular domain Ω, where Ω is included in a ε-random checkerboard-like environment, when no specie is located on the boundary, and when the density at time $t=0$ is deterministic and known equal to u_ε^0, is given by

$$\begin{cases} \dfrac{du_\varepsilon}{dt}(\omega,t) - \mathrm{div} D_\xi W\left(\omega,\dfrac{\cdot}{\varepsilon},\nabla u_\varepsilon(\omega,t)\right) \\ = r\left(\omega,\dfrac{\cdot}{\varepsilon}\right)u_\varepsilon(\omega,t)\left(1-\dfrac{u_\varepsilon(\omega,t)}{K_{car}}\right)\left(\dfrac{u_\varepsilon(\omega,t)-a(\omega,\frac{\cdot}{\varepsilon})}{K_{car}}\right) \quad \text{a.e. in } (0,T)\times\Omega \\ u_\varepsilon(\omega,0) = u_\varepsilon^0 \in H_0^1(\Omega),\ 0\le u_\varepsilon^0 \le K_{car}, \\ \\ u_\varepsilon(\omega,t)\in H_0^1(\Omega),\ \mathrm{div} D_\xi W(\omega,\frac{\cdot}{\varepsilon},\nabla u_\varepsilon(\omega,t)) \in L^2(\Omega) \text{ for all } t\in]0,T]. \end{cases}$$

We assume that the initial density u_ε^0 strongly converges to some u_0 in $L^2(\Omega)$, and that $\sup_{\varepsilon>0}\Phi_\varepsilon(\omega,u_\varepsilon) < +\infty$, where Φ_ε is the functional (7.6). According to Theorem 7.1, we can say that when ε is very small compared to the size of the domain Ω, a deterministic model which reflects the dynamic of the population, is given by

$$\begin{cases} \dfrac{du}{dt}(t) - \mathrm{div} D_\xi W^{hom}(\nabla u(t)) = \overline{r}u(t)\left(1-\dfrac{u(t)}{K_{car}}\right)\left(\dfrac{u(t)-\frac{\overline{ra}}{\overline{r}}}{K_{car}}\right) \\ \text{for a.e. } (t,x)\in(0,T)\times\Omega, \\ \\ u(0) = u_0 \in H_0^1(\Omega),\ 0\le u^0\le K_{car}, \\ \\ u(t)\in H_0^1(\Omega),\ \mathrm{div}(D_\xi W^{hom}(\nabla u(t))) \in L^2(\Omega) \text{ for a.e. } t\in(0,T), \end{cases} \tag{7.21}$$

where

$$W^{hom}(\xi) = \inf_{n\in\mathbb{N}^*}\mathbb{E}\inf\left\{\frac{1}{n^2}\int_{nY} W(\omega,y,\xi+\nabla u(y))\,dy : u\in H_0^1(Y)\right\},$$

$$\overline{r} = \mathbb{E}\left(\int_Y r(\omega,y)\,dy\right) = \mathbf{p}r^- + (1-\mathbf{p})r^+,$$

$$\overline{ra} = \mathbb{E}\left(\int_Y r(\omega,y)a(\omega,y)\,dy\right) = \mathbf{p}r^-a^- + (1-\mathbf{p})r^+a^+.$$

Actually, since the reaction functional F_ε is regular, and r and a do not depend on t, from Theorem 2.6, or from Remark 7.3, we can assert that $u(t) \in H_0^1(\Omega)$, and $\mathrm{div}(D_\xi W^{hom}(\nabla u(t))) \in L^2(\Omega)$ for all $t \in]0,T]$.

Everything happens as if the density evolution of the specie took place in a homogeneous environment following a Fisher diffusive model with Allee effect and constant coefficients. It is interesting to note that the growth rate is deterministic and constant in the environment, and that the critical density $\widetilde{a} = \frac{\overline{ra}}{\overline{r}}$ which still satisfies $0 \leq \widetilde{a} \leq K_{car}$, is now a function of the growth rate. This illustrates the interplay between the environment and the evolution. Observe that $\widetilde{a}$ is a monotone function of the probability $\mathbf{p}$. The diffusion operator is now governed by an homogeneous and deterministic operator obtained from an almost sure graph limit.

7.3.2 *Environment whose heterogeneities are independently randomly distributed with a frequency λ*

As a first step, we are going to define a probability dynamical system $(\Sigma, \mathcal{A}, \mathbb{P}_\lambda, (T_z)_{z\in\mathbb{Z}^N})$, which models the environment whose heterogeneities are made up of balls whose centers are randomly distributed with a frequency λ per unit volume. We assume that the number of centers is locally finite and that these numbers in two disjointed regions are two independent random variables. The growth rate and the threshold in the Fisher reaction function with Allee effect, together with the density associated with the random diffusion, take two different values outside or inside the heterogeneities.

Denote by $\mathcal{M}$ the set of countable and locally finite sums of Dirac measures in $\mathbb{R}^N$, equipped with the σ-algebra generated by all the evaluation maps $\mathcal{E}_B : m \mapsto m(B)$ from $\mathcal{M}$ into $\mathbb{N} \cup \{+\infty\}$ when B runs through $\mathcal{B}(\mathbb{R}^N)$. Then, given $\lambda > 0$, there exists a subset Σ of locally finite sequences $(\omega_i)_{i\in\mathbb{N}}$ in $\mathbb{R}^N$, a probability space $(\Sigma, \mathcal{A}, \mathbb{P})$ and a point process, called Poisson point process, $\mathcal{N} : \omega \mapsto \mathcal{N}(\omega, \cdot)$ from Σ into $\mathcal{M}$ satisfying

(i) $\mathcal{N}(\omega, \cdot) = \sum_{i\in\mathbb{N}} \delta_{\omega_i}$;

(ii) for every finite and pairwise disjoint family $(B_i)_{i\in I}$ of $\mathcal{B}(\mathbb{R}^N)$, the random variables $(\mathcal{N}(\cdot, B_i))_{i\in I}$ are independent;

(iii) for every bounded Borel set B and every $k \in \mathbb{N}$

$$\mathbb{P}_\lambda([\mathcal{N}(\cdot, B) = k]) = \lambda^k \mathcal{L}_N(B)^k \,\frac{\exp(-\lambda \mathcal{L}_N(B))}{k!}.$$

We denote by $\mathbb{E}_\lambda$ the expectation operator with respect to the probability $\mathbb{P}_\lambda$. Note that for every bounded Borel set B in $\mathbb{R}^N$, we have $\mathcal{N}(\omega, B) = \#(\Sigma \cap B)$, and that an easy calculation yields $\mathbb{E}_\lambda(\mathcal{N}(\cdot, B)) = \lambda \mathcal{L}_N(B)$. The parameter λ is called the intensity of the Poisson point process. For the existence of Poisson point processes and an explicit construction of the probability space $(\Sigma, \mathcal{A}, \mathbb{P}_\lambda)$, we refer the reader to Bouleau (2000). We define the group $(T_z)_{z\in\mathbb{Z}^N}$ of $\mathbb{P}_\lambda$-preserving transformation on $(\Sigma,\mathcal{A},\mathbb{P}_\lambda)$, by $T_z\omega=\omega-z$. From (ii), and using the mixing condition (C.1), we can easily show that $(\Sigma,\mathcal{A},\mathbb{P}_\lambda,(T_z)_{z\in\mathbb{Z}^N})$ is ergodic. As we will see below, the probability dynamical system $(\Sigma,\mathcal{A},\mathbb{P}_\lambda,(T_z)_{z\in\mathbb{Z}^N})$ is a good description of the heterogeneous environment described above.

We now define the random diffusion and the random reaction part. Given $R>0$, (r^-,a^-,W^-) and (r^+,a^+,W^+) in $[0,+\infty]\times[0,K_{car}]\times \mathrm{Conv}_{\alpha,\beta}$, where $W^\pm$ does not depend on x and whose Legendre-Fenchel conjugate satisfies (D_3^*), we define the random density W associated with the random diffusion part, by

$$W(\omega,x,\xi)=\begin{cases}W^-(\xi) \text{ if } x\in \underset{i\in\mathbb{N}}{\cup} B_R(\omega_i),\\ W^+(\xi) \text{ otherwise.}\end{cases}$$

It is easy to show that we have the following convenient expression of W:

$$W(\omega,x,\xi)=W^+(\xi)+(W^-(\xi)-W^+(\xi))\min(1,\mathcal{N}(\omega,B_R(x))).$$

Similarly we define the random growth rate and the random threshold by

$$r(\omega,x)=r^+ +(r^- -r^+)\min(1,\mathcal{N}(\omega,B_R(x))),$$
$$a(\omega,x)=a^+ +(a^- -a^+)\min(1,\mathcal{N}(\omega,B_R(x))).$$

The random CP-structured reaction function is given by

$$f(\omega,t,x,\zeta)=r(\omega,x)\frac{\zeta^2}{K}\left(1-\frac{\zeta}{K}\right)-r(\omega,x)a(\omega,x)\frac{\zeta}{K}\left(1-\frac{\zeta}{K}\right)$$

which is a Fisher reaction function with Allee effect whose growth rate and threshold are (r^-,a^-) when $x\in\cup_{i\in\mathbb{N}}B_R(\omega_i)$, and (r^+,a^+) otherwise. We set $W_\varepsilon(\omega,x,\xi)=W(\omega,\frac{x}{\varepsilon},\xi)$, and $f_\varepsilon(\omega,t,x,\zeta)=f(\omega,t,\frac{x}{\varepsilon},\zeta)$. It is easy to check that conditions (7.12) and (7.13) hold so that the homogenized problem is still given by (7.21) where this time (when $N=2$)

$$\overline{r}=r^- +(r^+ -r^-)\exp(-\lambda\pi R^2),$$
$$\overline{ra}=r^-a^- +(r^+a^+ -r^-a^-)\exp(-\lambda\pi R^2).$$

Indeed, noticing that $\#(\Sigma\cap B_R(y))\geq 1$ if and only if there exists $\omega\in\Sigma$ such that $y\in\bigcup_{i\in\mathbb{N}}B_R(\omega_i)$, and using Fubini's theorem, we infer that

$$\begin{aligned}\overline{r}&=\mathbb{E}_\lambda\left(\int_Y r(\omega,y)\,dy\right)\\&=r^+\int_\Sigma\int_{(0,1)^N}\mathbb{1}_{[\#(\Sigma\cap B_R(y))=0]}(\omega,y)dy\,d\mathbb{P}_\lambda(\omega)\\&\quad+r^-\int_\Sigma\int_{(0,1)^N}\mathbb{1}_{[\#(\Sigma\cap B_R(y))\geq 1]}(\omega,y)dy\,d\mathbb{P}_\lambda(\omega)\\&=r^+\int_{(0,1)^N}\int_\Sigma\mathbb{1}_{[\#(\Sigma\cap B_R(y))=0]}(\omega,y)d\mathbb{P}_\lambda(\omega)dy\\&\quad+r^-\int_{(0,1)^N}\int_\Sigma\mathbb{1}_{[\#(\Sigma\cap B_R(y))\geq 1]}(\omega,y)d\mathbb{P}_\lambda(\omega)\,dy\\&=r^+\exp(-\lambda\pi R^2)+r^-(1-exp(-\lambda\pi R^2)).\end{aligned}$$

A similar calculation holds for $\overline{ra}$.

Remark 7.4. The distribution of $\omega \mapsto f(\omega, t, x, \zeta)$ is unchanged whenever λ and R are replaced by $r^{-N}\lambda$ and rR respectively, with $r \in (0, +\infty)$. Therefore we can assume $R = 1$ without loss of generality.

Remark 7.5. The group of $\mathbb{P}_\lambda$-preserving transformation on $(\Sigma, \mathcal{A}, \mathbb{P}_\lambda)$ is actually $(T_x)_{x\in\mathbb{R}^N}$. We consider its restriction $(T_z)_{z\in\mathbb{Z}^N}$ which is sufficient for applying the additive and subadditive ergodic theorems.

7.4 Stochastic homogenization of a reaction-diffusion problem stemming from a hydrogeological model

Consider the random reaction-diffusion Cauchy problems corresponding to Examples 2.5 and 2.6: for every $\omega \in \Sigma$

$$(\mathcal{P}'_\varepsilon(\omega)) \begin{cases} \dfrac{du_\varepsilon}{dt}(\omega,t) - \mathrm{div} D_\xi W\left(\omega, \dfrac{\cdot}{\varepsilon}, \nabla u_\varepsilon(\omega,t)\right) = F_\varepsilon(\omega,t,u_\varepsilon(\omega,t)) \text{ in } L^2(\Omega) \\ \text{for a.e. } t \in (0,T), \\ \\ u_\varepsilon(\omega,0) = u^0_\varepsilon(\omega) \in H^1_\Gamma(\Omega),\ 0 \le u^0_\varepsilon(\omega,\cdot) \le \overline{\rho}(\omega), \\ \\ u_\varepsilon(\omega,t) \in H^1_\Gamma(\Omega), \mathrm{div} D_\xi W(\omega, \frac{\cdot}{\varepsilon}, \nabla u_\varepsilon(\omega,t)) \in L^2(\Omega) \\ \text{for a.e. } t \in (0,T), \\ \\ D_\xi W\left(\omega, \dfrac{\cdot}{\varepsilon}, \nabla u_\varepsilon(\omega,t)\right) \cdot \mathbf{n} = 0 \text{ on } \partial\Omega \setminus \Gamma, \quad u_\varepsilon(\omega,t) = 0 \text{ on } \Gamma \\ \text{for a.e. } t \in (0,T). \end{cases}$$

The random reaction functional is given by

$$F_\varepsilon(\omega,t,u_\varepsilon(\omega,t))(x) = a\left(\omega, \frac{x}{\varepsilon}\right) R\left(t - g\left(\frac{L - u_\varepsilon(\omega,t,x)}{K}\right)\right)$$

where

- $R : \mathbb{R} \to \mathbb{R}_+$ is Lipschitz continuous, i.e. there exists $L_R > 0$ such that $\forall (t,t') \in \mathbb{R}^2$, $|R(t) - R(t')| \le L_R|t - t'|$ for all $(t,t') \in \mathbb{R}^2_+$;
- $g : \mathbb{R} \to \mathbb{R}$ is locally Lipschitz continuous: for all interval $I \subset \mathbb{R}$, there exists $L_I > 0$ such that $|g(\zeta) - g(\zeta')| \le L_I|\zeta - \zeta'|$, for all $(\zeta, \zeta') \in I \times I$;
- $a : \Sigma\times\mathbb{R}^N \to \mathbb{R}$ is $(\mathcal{A}\otimes\mathcal{B}(\mathbb{R}^N), \mathcal{B}(\mathbb{R}))$-measurable, covariant i.e. $a(\omega, x+z) = a(T_z\omega, x)$, $a(\omega,\cdot) \ge 0$, and $a(\omega,\cdot) \in L^\infty(\mathbb{R})$ for all $\omega \in \Sigma$, and all $x \in \mathbb{R}^N$.

Recall that for the hydrogeological model succinctly described in Example 2.5, the state variable $u(\omega,t,x)$ may denotes the piezometric load, or the height at time t and position $x \in \Omega \subset \mathbb{R}^N$, $(N = 2)$, of a water table. The transmissivity, i.e. the speed at which groundwater flows horizontally, is assumed to randomly fluctuate. This is due to the particle size of the porous media that stores water, of which we only know a statistical distribution. This explains the randomness of the density W. The randomness of the source is due to the statistical knowledge of the media between the soil surface and the phreatic water table. As in the previous examples, the spatial environment is assumed to be statistically homogeneous. From a strictly mathematical point of view, this means that a probability dynamical system $(\Sigma, \mathcal{A}, \mathbb{P}, (T_z)_{z\in\mathbb{Z}^N})$ models the spatial environment in $\mathbb{R}^N$. The coefficient $\varepsilon > 0$ accounts for the size of the porosities.

Applying (Chabi and Michaille, 1994, Theorem 4.2), straightforward consequence of the additive ergodic theorem, Theorem C.5: there exists $N \in \mathcal{A}$ with $\mathbb{P}(N) = 0$ such that for every $\omega \in \Sigma' \setminus N$

$$a\left(\omega, \frac{\cdot}{\varepsilon}\right) \rightharpoonup \overline{a}(\omega)$$

where $\overline{a}(\cdot) = \mathbb{E}^{\mathcal{I}} \int_{(0,1)^N} a(\cdot, y)dy$. On the other hand, it is easy to establish from the local Lipschitz continuity of g, and the boundedness of $u_\varepsilon(\omega, \cdot)$, that $t \mapsto R\left(t - g\left(\frac{L-u_\varepsilon(\omega,t,\cdot)}{K}\right)\right)$ strongly converges to $t \mapsto R\left(t - g\left(\frac{L-u(\omega,t,\cdot)}{K}\right)\right)$ in $L^2(0, T, L^2(\Omega))$, where $u(\omega, \cdot)$ is the limit of $u_\varepsilon(\omega, \cdot)$ in $C([0, T], L^2(\Omega))$.

Assume that the initial density $u^0_\varepsilon(\omega, \cdot)$ strongly converges to some $u_0(\omega, \cdot)$ in $L^2(\Omega)$ for $\mathbb{P}$-a.e. $\omega \in \Sigma$, and that $\sup_{\varepsilon>0} \Phi_\varepsilon(\omega, u_\varepsilon(\omega, \cdot)) < +\infty$, where Φ_ε is the functional (7.6). According to Theorem 2.7, we can say that when $\varepsilon > 0$ is very small compared to the size of the domain Ω, a model which reflects the dynamic of the phreatic water table, is given by

$$(\mathcal{P}'^{hom}(\omega))\begin{cases} \dfrac{du}{dt}(t) - \mathrm{div}\partial_\xi W^{hom}(\omega, \nabla u(t)) \ni G(t) \text{ in } L^2(\Omega) \text{ for a.e. } t \in (0, T), \\ u(\omega, 0) = u^0(\omega) \in H^1_\Gamma(\Omega),\ 0 \le u^0(\omega) \le \overline{p}, \\ u(t) \in H^1_\Gamma(\Omega), \mathrm{div}\partial_\xi W^{hom}(\nabla u(t)) \in L^2(\Omega) \text{ for a.e. } t \in (0, T), \\ \partial_\xi W^{hom}(\nabla u(t)) \cdot \mathbf{n} \ni 0 \text{ on } \partial\Omega \setminus \Gamma \text{ for a.e. } t \in (0, T), \end{cases}$$

where $G(\omega, t) := \overline{a}(\omega) R\left(t - g\left(\frac{L-u_\varepsilon(\omega,t,x)}{K}\right)\right)$. Moreover, for $\mathbb{P}$-a.e. $\omega \in \Sigma$, $\frac{du_\varepsilon}{dt}(\omega, \cdot) \rightharpoonup \frac{du}{dt}$ weakly in $L^2(0, T, L^2(\Omega))$ and $0 \le u(\omega, t) \le \overline{y}(t)$ for all $t \in [0, T]$, where $\overline{y}(\omega, \cdot)$ is the solution (random) of o.d.e.

$$\overline{y}' = \|a(\omega, \cdot)\| R\left(t - g\left(\frac{L - y}{K}\right)\right),\ y(0) = \overline{p}.$$

Chapter 8

Stochastic homogenization of nonlinear distributed time delays reaction-diffusion equations

We place this chapter within the framework of stochastic homogenization introduced in Chapter 7. We especially seek to identify the effective growth rates, the various effectives thresholds and effective time delays relating to time delays reaction-diffusion problems with rapidly random oscillating coefficients. We recall that the triple $(\Sigma, \mathcal{A}, \mathbb{P}, (T_z)_{z\in\mathbb{Z}^N})$ is a probability dynamical system, $\mathcal{I}$ denotes the σ-algebra of invariant sets of $\mathcal{A}$ by the group $(T_z)_{z\in\mathbb{Z}^N}$ and, for every $\mathbf{h}$ in the space $L^1_{\mathbb{P}}(\Sigma)$ of $\mathbb{P}$-integrable functions, $\mathbb{E}^{\mathcal{I}}\mathbf{h}$ denotes the conditional expectation of $\mathbf{h}$ with respect to $\mathcal{I}$. Theorem 8.1, states that a mixing phenomena appears at the limit between the various growths rate and time delays. This result is illustrated through the homogenization of a vector disease model and a delay logistic equation with immigration. In the first example, and in the case of random multiple delays, the growth rate of the uninfected population and the time delays coefficients of the homogenized problem are mixed in such a way that the effective distributed delays are associated with the measure of multiple delays $\sum_{k\in\mathbb{N}} \mathbb{E}\left[\int_Y a(\cdot, y)\mathbf{d}_k(\cdot, y)\, dy\right] \delta_{t-\tau_k}$ incorporating the growth rate. In the second example, we can interpret the homogenized problem as a diffuse delay logistic equation modeling the evolution of a density population spreading in an homogeneous environment, whose carrying capacity is function of the growth rate, time dependent, and bigger than the carrying capacity of the heterogeneous spatial domain at scale $\varepsilon > 0$. In Section 8.5 we take advantage of the vector disease model to introduce some basic concepts in percolation theory.

8.1 The random diffusion part

The diffusion part, is that of Chapter 7, Section 7.1.1 to which we refer. To facilitate reading, we briefly recall that the diffusion operator is the subdifferential of the random energy functionals (7.5) or (7.6) below

$$\Phi_\varepsilon(\omega, u) = \begin{cases} \widetilde{\Phi}_\varepsilon(\omega, u) + \dfrac{1}{2}\displaystyle\int_{\partial\Omega} a_0 u \, d\mathcal{H}_{N-1} - \int_{\partial\Omega} \phi u \, d\mathcal{H}_{N-1} & \text{if } u \in H^1(\Omega) \\ +\infty & \text{otherwise;} \end{cases}$$

or

$$\Phi_\varepsilon(\omega,u)=\begin{cases}\widetilde{\Phi}_\varepsilon(\omega,u) \text{ if } u\in H^1_\Gamma(\Omega)\\ \\ +\infty \quad \text{otherwise,}\end{cases}$$

where

$$\widetilde{\Phi}_\varepsilon(\omega,u)=\begin{cases}\displaystyle\int_\Omega W\left(\omega,\frac{x}{\varepsilon},\nabla u\right)dx \text{ if } u\in H^1(\Omega)\\ \\ +\infty \qquad \text{otherwise.}\end{cases}$$

We also recall that for $\mathbb{P}$-a.e. $\omega\in\Sigma$, these functionals respectively Mosco-converges to the functionals

$$\Phi^{hom}(\omega,u)=\begin{cases}\widetilde{\Phi}^{hom}(\omega,u)+\dfrac{1}{2}\displaystyle\int_{\partial\Omega}a_0u\, d\mathcal{H}_{N-1}-\int_{\partial\Omega}\phi u\, d\mathcal{H}_{N-1} \text{ if } u\in H^1(\Omega)\\ \\ +\infty \qquad \text{otherwise;}\end{cases}$$

or

$$\Phi^{hom}(\omega,u)=\begin{cases}\widetilde{\Phi}^{hom}(\omega,u) \text{ if } u\in H^1_\Gamma(\Omega)\\ \\ +\infty \quad \text{otherwise.}\end{cases}$$

where

$$\widetilde{\Phi}^{hom}(\omega,u)=\begin{cases}\displaystyle\int_\Omega W^{hom}(\omega,\nabla u)dx \text{ if } u\in H^1(\Omega)\\ \\ +\infty \qquad \text{otherwise,}\end{cases}$$

and W^{hom} is given by (7.3).

8.2 The random reaction part

We are given a random DCP-structured reaction functional, namely, a functional

$$F:\Sigma\times[0,+\infty)\times L^2(\Omega)\times L^2(\Omega)\to\mathbb{R}^\Omega$$

defined by $F(\omega,t,u,v)(x)=f(\omega,t,x,u(x),v(x))$ where

$$f:\Sigma\times[0,+\infty)\times\mathbb{R}^N\times\mathbb{R}\times\mathbb{R}\to\mathbb{R}$$

is a $(\mathcal{A}\otimes\mathcal{B}(\mathbb{R})\otimes\mathcal{B}(\mathbb{R}^N)\otimes\mathcal{B}(\mathbb{R}),\mathcal{B}(\mathbb{R}))$ measurable function such that for $\mathbb{P}$-a.e. $\omega\in\Sigma$, the function $f(\omega,\cdot,\cdot,\cdot,\cdot)$ is a DCP-structured reaction function associated with $(r(\omega,\cdot),g,h,q(\omega,\cdot))$. As in Chapter 7, we assume that r and q fulfills conditions (REA$_1$), (REA$_2$), and (REA$_3$).

We set $f_\varepsilon(\omega,t,x,\zeta,\zeta'):=f(\omega,t,\frac{x}{\varepsilon},\zeta,\zeta')$, and define the functional F_ε for all $(u,v)\in L^2(\Omega)\times L^2(\Omega)$ by

$$F_\varepsilon(\omega,t,u,v)(x)=f\left(\omega,t,\frac{x}{\varepsilon},u(x),v(x)\right).$$

Note that in the expression of (DCP), the functions $\overline{f}$, $\overline{y}$, and $\overline{\rho}$ may depend on ω but we sometimes omit it to shorten the notation. Observe that $F_\varepsilon(\omega,\cdot,\cdot,\cdot)$ is a DCP-structured reaction functional whose (DCP) condition is exactly that of $F(\omega,\cdot,\cdot,\cdot)$, i.e. with $\overline{f}$, $\overline{y}$, and $\overline{\rho}$. Since $\overline{y}$ does not depend on ε, condition (3.18) is automatically satisfied. Recall that, according to Lemma 7.1, for $\mathbb{P}$-a.e. $\omega \in \Sigma$, we have

$$\lim_{\varepsilon\to 0}\int_0^T \left\| \frac{dr}{dt}\left(\omega,\tau,\frac{\cdot}{\varepsilon}\right)\right\|^2_{L^2(\Omega,\mathbb{R}^l)} d\tau = \mathcal{L}_N(\Omega)\mathbb{E}^{\mathcal{I}}\int_0^T \left\|\frac{dr}{dt}(\omega,\tau,\cdot)\right\|^2_{L^2(Y,\mathbb{R}^l)} d\tau;$$

$$\lim_{\varepsilon\to 0}\int_0^T \left\| \frac{dq}{dt}\left(\omega,\tau,\frac{\cdot}{\varepsilon}\right)\right\|^2_{L^2(\Omega)} d\tau = \mathcal{L}_N(\Omega)\mathbb{E}^{\mathcal{I}}\int_0^T \left\|\frac{dq}{dt}(\omega,\tau,\cdot)\right\|^2_{L^2(Y)} d\tau.$$

The sequence of random history function, is defined by considering a sequence $(\eta_\varepsilon)_{\varepsilon>0}$ of measurable maps $\omega \mapsto \eta_\varepsilon(\omega)$ from Σ into $C_c((-\infty,0],L^2(\Omega))$ with support$(\eta_\varepsilon(\omega))$ included of a fixed compact set $[-M(\omega),0]$, $M(\omega)>0$, satisfying $\eta_\varepsilon(\omega)(0)\in \operatorname{dom}(\Phi_\varepsilon(\omega,\cdot))$, $0\le \eta_\varepsilon(\omega)\le \overline{\rho}(\omega)$, and

$$\sup_{\varepsilon>0}\int_{-\infty}^0 \left\|\frac{d\eta_\varepsilon(\omega)}{dt}(t,\cdot)\right\|_{L^2(\Omega)} dt < +\infty$$

for $\mathbb{P}$-a.e. $\omega\in\Sigma$. We sometimes write $\eta_\varepsilon(\omega,\cdot)$ for $\eta_\varepsilon(\omega)$.

Finally, we assume that condition (3.19) is satisfied, more precisely, for $\mathbb{P}$-a.e. $\omega\in\Sigma$,

$$0\le\phi\le a_0\overline{\rho}(\omega) \text{ on } \partial\Omega.$$

We restrict our analysis to sequences $(\mathbf{m}_t^\varepsilon)_{t\ge 0,\varepsilon}$ of random vector measures i.e., measurable maps $\omega\mapsto \mathbf{m}_t^\varepsilon(\omega)$ from Σ into $\mathbf{M}_1^+(\mathbb{R},L^\infty(\Omega))$, satisfying one of the two special structures described below.

8.2.1 *First structure*

Let $\tau:\Sigma\to[0,+\infty)^{\mathbb{R}^N}$ be a measurable map such that for $\mathbb{P}$-a.e. $\omega\in\Sigma$, $\#(\tau(\omega,\cdot)(\mathbb{R}^N))$ is a constant denoted by $\#\tau$ (recall that $\#(E)$ denotes the cardinal of the set E). Furthermore, we assume that τ satisfies the covariance property $\tau(\omega)(x+z)=\tau(T_z\omega)(x)$ for all $z\in\mathbb{Z}^N$, a.e. $x\in\mathbb{R}^N$, and $\mathbb{P}$-a.e. $\omega\in\Sigma$. According to Example 3.2, for each $\omega\in\Sigma$, $t\ge 0$ and $\varepsilon>0$, we consider the vector measure $\mathbf{m}_t^\varepsilon(\omega)\in\mathbf{M}_1^+(\mathbb{R},L^\infty(\Omega))$ defined by

$$\mathbf{m}_t^\varepsilon(\omega)=\frac{1}{\#\tau}\delta_{t-\tau(\omega,\frac{\cdot}{\varepsilon})}.$$

For each $\omega\in\Sigma$ and each $\varepsilon>0$, $(\mathbf{m}_t^\varepsilon(\omega))_{t\ge0}$ is clearly a family of vector measures in $\mathbf{M}_1^+(\mathbb{R},L^\infty(\Omega))$ which satisfies conditions (M_1), (M_2), and (M_3) of Chapter 3. For all $v\in C_c((-\infty,T],L^2(\Omega))$, all $\varepsilon>0$ and all $\omega\in\Sigma$, we set

$$\mathcal{T}_\varepsilon(\omega)v(t)=\int_{-\infty}^t v(s)d\mathbf{m}_t^\varepsilon(\omega)(s)=\frac{1}{\#\tau}v\left(t-\tau\left(\omega,\frac{\cdot}{\varepsilon}\right)\right),$$

which defines a linear continuous operator $\mathcal{T}_\varepsilon(\omega) : C_c((-\infty,T],L^2(\Omega)) \to L^2(0,T,L^2(\Omega))$. Condition (M$'_4$) is almost surely satisfied. Indeed, from Example 3.2 and Remark 3.1, the probability measure

$$\mu_\varepsilon(\omega) = \frac{1}{\#(\mathbb{R}^N)} \sum_{\tau_i \in \tau(\omega,\cdot)(\mathbb{R}^N)} \delta_{\tau_i}$$

satisfies for all $v \in C_{\eta_\varepsilon(\omega)}((-\infty,T],L^2(\Omega))$ and all $(s,t) \in [0,T]^2$:

$$\forall (s,t) \in [0,T]^2, \quad s < t$$
$$\Longrightarrow \|\mathcal{T}_\varepsilon(\omega)v(t) - \mathcal{T}_\varepsilon(\omega)v(s)\|_{L^2(\Omega)} \leq \int_s^t \left(\left\| \frac{dv}{d\sigma}(\cdot) \right\|_{L^2(\Omega)} \star \mu_\varepsilon(\omega) \right)(\sigma) d\sigma$$

for all $\omega \in \Sigma$.

8.2.2 *Second structure*

For each $k \in \mathbb{N}$, we are given a $(\mathcal{A},\mathcal{B}(L^\infty(\mathbb{R}^N)))$-measurable map $\omega \mapsto \mathbf{d}_k(\omega,\cdot)$, $\mathbf{d}_k \geq 0$, such that for $\mathbb{P}$-a.e. $\omega \in \Sigma$, $\sum_{k\in\mathbb{N}} \|\mathbf{d}_k(\omega,\cdot)\|_{L^\infty(\mathbb{R}^N)} = 1$. We set $\mathbf{d}(\omega) = (\mathbf{d}_k(\omega))_{k\in\mathbb{N}}$, and $\mathbf{d}^\varepsilon_k(\omega,x) := \mathbf{d}_k(\omega,\frac{x}{\varepsilon})$, $\mathbf{d}^\varepsilon(\omega) = (\mathbf{d}^\varepsilon_k(\omega))_{k\in\mathbb{N}}$ for all $\omega \in \Sigma$ and all $\varepsilon > 0$. On the other hand, we consider a family $(\nu_t)_{t\geq 0}$ where $\nu_t = (\nu_{k,t})_{k\in\mathbb{N}}$ and $\nu_{k,t}$ is a Borel probability measure in $\mathcal{P}_+(\mathbb{R})$. We assume that the maps $t \mapsto \mathbb{1}_{(-\infty,t]}\nu_{k,t}$ are measurable, and define the sequence $(\mathbf{m}^\varepsilon_t)_{t\geq 0}$ of random measures by

$$\omega \mapsto \mathbf{m}^\varepsilon_t(\omega) = \mathbf{d}^\varepsilon(\omega,\cdot) \cdot \nu_t.$$

It is easily seen that for each $\omega \in \Sigma$ and $\varepsilon > 0$, $(\mathbf{m}^\varepsilon_t(\omega))_{t\geq 0}$ is a family of vector measures in $\mathbf{M}^+_1(\mathbb{R},L^\infty(\Omega))$ satisfying (M$_1$), (M$_2$), and (M$_3$). Note that the vector measures considered in Example 3.1, and Example 3.3 with the Γ_l-distribution delays, fulfill this particular structure. Indeed, take $\nu_{k,t} = \delta_{t-\tau_k}$ for Example 3.1, and, for Example 3.3, $\nu_{k,t} = K(t-\tau)\,d\tau$, $\mathbf{d}^\varepsilon_k(\omega,x) = \mathbf{d}(\omega,\frac{x}{\varepsilon})$ if $k = 0$, $\mathbf{d}^\varepsilon_k(\omega,x) = 0$ if $k \in \mathbb{N}^*$, for all $\omega \in \Sigma$, all $x \in \mathbb{R}^N$ and all $\varepsilon > 0$.

For all $\omega \in \Sigma$, and all v in $C_c((-\infty,T],L^2(\Omega))$, we set

$$\mathcal{T}_\varepsilon(\omega)v(t) = \int_{-\infty}^t v(s) d\mathbf{m}^\varepsilon_t(\omega)(s),$$

which defines for all $\varepsilon > 0$ a linear continuous operator $\mathcal{T}_\varepsilon(\omega) : C_c((-\infty,T], L^2(\Omega)) \to L^2(0,T,L^2(\Omega))$. We assume that $(\mathbf{m}^\varepsilon_t)_{t\geq 0}$ is such that each operator $\mathcal{T}_\varepsilon$ satisfies almost surely (M$'_4$). This is the case for the previous two examples, with $\mu_\varepsilon(\omega) = \sum_{k\in\mathbb{N}} \|\mathbf{d}_k(\omega,\cdot)\|_{L^\infty(\mathbb{R}^N)}\delta_{\tau_k}$ for Example 3.1, and $\mu_\varepsilon(\omega) = K(\sigma)d\sigma$ for Example 3.3.

8.3 Almost sure convergence to the homogenized reaction-diffusion problem

Under above conditions, by combining Theorem 3.4 together with the variational convergence of the sequence of random energies Φ_ε, we intend to analyze when

$\varepsilon \to 0$, the asymptotic behavior in $C(0,T,L^2(\Omega))$ of the sequence $(u_\varepsilon(\omega,\cdot))_{\varepsilon>0}$, where $u_\varepsilon(\omega,\cdot)$ solves the random delays reaction-diffusion problems associated with (7.5) or (7.6)

$$(\mathcal{P}_\varepsilon(\omega)) \begin{cases} \dfrac{du_\varepsilon}{dt}(\omega,t) - \mathrm{div} D_\xi W\left(\omega, \dfrac{\cdot}{\varepsilon}, \nabla u_\varepsilon(\omega,t)\right) \\ = F_\varepsilon(\omega,t,u_\varepsilon(\omega,t), \mathcal{T}_\varepsilon(\omega) u_\varepsilon(\omega,\cdot)(t)) \text{ for a.e. } t \in (0,T), \\ \\ u_\varepsilon(\omega,t) = \eta_\varepsilon(\omega,t),\ 0 \le \eta_\varepsilon(\omega,t) \le \overline{\rho}(\omega) \\ \text{for all } t \in (-\infty, 0], \\ \\ u_\varepsilon(\omega,t) \in H^1(\Omega),\ \mathrm{div} D_\xi W\left(\omega, \frac{\cdot}{\varepsilon}, \nabla u_\varepsilon(\omega,t)\right) \in L^2(\Omega) \\ \text{for all } t \in]0,T], \\ \\ a_0 u_\varepsilon(\omega,t) + D_\xi W\left(\omega, \frac{\cdot}{\varepsilon}, \nabla u_\varepsilon(\omega,t)\right) \cdot \mathbf{n} = \phi \text{ on } \partial\Omega \\ \text{for all } t \in [0,T], \end{cases}$$

or

$$(\mathcal{P}'_\varepsilon(\omega)) \begin{cases} \dfrac{du_\varepsilon}{dt}(\omega,t) - \mathrm{div} D_\xi W\left(\omega, \dfrac{\cdot}{\varepsilon}, \nabla u_\varepsilon(\omega,t)\right) \\ = F_\varepsilon(\omega,t,u_\varepsilon(\omega,t), \mathcal{T}_\varepsilon(\omega) u_\varepsilon(\omega,\cdot)(t)) \text{ for a.e. } t \in (0,T), \\ \\ u_\varepsilon(\omega,t) = \eta_\varepsilon(\omega,t),\ 0 \le \eta_\varepsilon(\omega,t) \le \overline{\rho}(\omega) \\ \text{for all } t \in (-\infty, 0], \\ \\ u_\varepsilon(\omega,t) \in H^1_\Gamma(\Omega),\ \mathrm{div} D_\xi W\left(\omega, \frac{\cdot}{\varepsilon}, \nabla u_\varepsilon(\omega,t)\right) \in L^2(\Omega) \text{ for all } t \in]0,T], \\ \\ D_\xi W\left(\omega, \frac{\cdot}{\varepsilon}, \nabla u_\varepsilon(\omega,t)\right) \cdot \mathbf{n} = 0 \text{ on } \partial\Omega \setminus \Gamma, \text{ for all } t \in [0,T]. \end{cases}$$

From Theorem 3.1, we know that each two problems admits a unique solution. We restrict our analysis to the case when $h = (h_j)_{j=1,\dots,l}$ is given by $h_j(\zeta') = \zeta'^{\alpha_j}$ where $\alpha_j \in \{0,1\}$ for $j = 1,\dots,l$. This is the case of Examples 3.4 and 3.5. To shorten the notation, for any measurable map $\omega \mapsto \mathbf{v}(\omega,\cdot)$ from Σ into $L^2(\Omega)$ such that $\mathbf{v}(\cdot,x)$ belongs to $L^1_{\mathbb{P}}(\Sigma)$, we write $\mathbb{E}^{\mathcal{I}}\mathbf{v}$ for the function $x \mapsto \mathbb{E}^{\mathcal{I}}\mathbf{v}(\cdot,x)$. Recall that we write indifferently $\partial_\xi W^{hom}(\omega,\cdot)$ to denote the subdifferential of $W^{hom}(\omega,\cdot)$ or any of its elements.

Theorem 8.1. *For each $\omega \in \Sigma$, let denote by $u_\varepsilon(\omega,\cdot)$ the unique solution in $C([0,T],L^2(\Omega))$ of the random delays reaction-diffusion problem $(\mathcal{P}_\varepsilon(\omega))$, respectively $(\mathcal{P}'_\varepsilon(\omega))$. Assume that for $\mathbb{P}$-a.e. $\omega \in \Sigma$,*

i) $\sup_{\varepsilon>0} \Phi_\varepsilon(\omega, \eta_\varepsilon(\omega,0)) < +\infty$;

ii) there exists $\eta(\omega) \in C_c((-\infty,0],L^2(\Omega))$ such that $\eta_\varepsilon(\omega) \to \eta(\omega)$ in $C((-\infty,0],L^2(\Omega))$. In the case when F_ε satisfies the first structure, we assume that the support of $\eta_\varepsilon(\omega)$ is include in a fixed compact set $[-M,0]$, $M>0$ independent of ω.

Then, for $\mathbb{P}$-a.e. $\omega \in \Sigma$, $u_\varepsilon(\omega,\cdot)$ uniformly converges in $C([0,T],L^2(\Omega))$ to the a function $u(\omega)$ whose extension by $\eta(\omega)$ in $(-\infty,0]$ is the unique solution of the reaction-diffusion problem

$$(\mathcal{P}^{hom}(\omega))\begin{cases}\dfrac{du}{dt}(\omega,t)-\mathrm{div}\partial_\xi W^{hom}(\omega,\nabla u(\omega,t))\ni r_{\mathcal{T}}(\omega,u)(t)\cdot g(u(\omega,t))\\ +\overline{q}(\omega,t) \text{ for a.e. } t\in(0,T),\\ \\ u(\omega,t)=\eta(\omega,t),\ 0\le\eta(\omega,t)\le\overline{y}(\omega)\\ \text{for all } t\in(-\infty,0],\\ \\ u(\omega,t)\in H^1(\Omega),\ \mathrm{div}\partial_\xi W^{hom}(\omega,\nabla u(\omega,t))\in L^2(\Omega)\\ \text{for a.e. } t\in(0,T),\\ \\ a_0u(\omega,t)+\partial_\xi W^{hom}(\omega,\nabla u(\omega,t))\cdot\mathbf{n}\ni\phi \text{ on } \partial\Omega\\ \text{for a.e. } t\in(0,T),\end{cases}$$

respectively

$$(\mathcal{P}'^{hom}(\omega))\begin{cases}\dfrac{du}{dt}(\omega,t)-\mathrm{div}\partial_\xi W^{hom}(\omega,\nabla u(\omega,t))\ni r_{\mathcal{T}}(\omega,u)(t)\cdot g(u(\omega,t))\\ +\overline{q}(\omega,t) \text{ for a.e. } t\in(0,T),\\ \\ u(\omega,t)=\eta(\omega,t),\ 0\le\eta(\omega,t)\le\overline{y}(\omega) \text{ for all } t\in(-\infty,0]\\ \\ u(\omega,t)\in H^1_\Gamma(\Omega),\ \mathrm{div}\partial_\xi W^{hom}(\omega,\nabla u(\omega,t))\in L^2(\Omega)\\ \text{for a.e. } t\in(0,T),\\ \\ \partial_\xi W^{hom}(\omega,\nabla u(\omega,t))\cdot\mathbf{n}\ni 0 \text{ on } \partial\Omega\setminus\Gamma \text{ for a.e. } t\in(0,T),\end{cases}$$

where, for all $\psi\in C_c((-\infty,T],L^2(\Omega))$

- *First structure case:*

$$r_{\mathcal{T}}(\omega,\psi)(t)=\mathbb{E}^{\mathcal{I}}\left[\int_Y r(\cdot,t,y)\odot h\left(\frac{1}{\#\tau}\psi(t-\tau(\cdot,y))\right)dy\right](\omega),$$
$$\overline{q}(\omega,t)=\mathbb{E}^{\mathcal{I}}\left[\int_Y q(\cdot,t,y)dy\right](\omega).$$

- *Second structure case:*

$$r_{\mathcal{T}}(\omega,\psi)(t)=\mathbb{E}^{\mathcal{I}}\left[\int_Y r(\cdot,t,y)h\left(\mathbf{d}(\cdot,y)\cdot\int_{-\infty}^t\psi(s)d\nu_t(s)\right)dy\right](\omega),$$
$$\overline{q}(\omega,t)=\mathbb{E}^{\mathcal{I}}\left[\int_Y q(\cdot,t,y)dy\right](\omega).$$

Moreover, for $\mathbb{P}$-a.e. $\omega\in\Sigma$, we have $\dfrac{du_\varepsilon(\omega,\cdot)}{dt}\rightharpoonup\dfrac{du(\omega,\cdot)}{dt}$ weakly in $L^2(0,T,L^2(\Omega))$ and $0\le u(\omega,t)\le\overline{y}(\omega,T)$ for all $t\in[0,T]$.

When the dynamical system $(\Sigma, \mathcal{A}, \mathbb{P}, (T_z)_{z\in\mathbb{Z}^N})$ is ergodic, the history function η_ε is deterministic, i.e. $\eta_\varepsilon(\omega) = \eta_\varepsilon$ for $\mathbb{P}$-a.e. $\omega \in \Sigma$, together with $\overline{\rho}$, and $\overline{f}$, then $(\mathcal{P}^{hom}(\omega))$ and $(\mathcal{P}'^{hom}(\omega))$ are deterministic: the diffusion operator becomes $-\mathrm{div}\partial_\xi W^{hom}(\nabla u(t))$, and, for all $\psi \in C_c((-\infty, T], L^2(\Omega))$, $r_\mathcal{T}(\omega, \psi)$ is defined as above with the conditional expectation operator replaced by the mathematical expectation operator.

If the Legendre-Fenchel conjugate of W satisfies (D_3^*), then the subdifferentials $\partial_\xi W^{hom}(\omega, \nabla u(t))$ or $\partial_\xi W^{hom}(\nabla u(t))$ are single valued equal to $D_\xi W^{hom}(\omega, \nabla u(t))$ or $D_\xi W^{hom}(\nabla u(t))$, and differential inclusions are equalities.

Proof. All conditions of Theorem 3.4 are easy to check for $\mathbb{P}$-a.e. $\omega \in \Sigma$, except condition (Hd_4') (for the proof of (Hd_7) see [Proof of (Hd_6)] in Theorem 7.1). To conclude it remains to establish that for $\mathbb{P}$-a.e. ω in Σ, and all $\psi \in C_c([-M(\omega), T], L^2(\Omega))$, the following convergence holds weakly in $L^2(0, T, L^2(\Omega, \mathbb{R}^l))$:

$$r\left(\omega, \cdot, \frac{\cdot}{\varepsilon}\right) \odot h(\mathcal{T}_\varepsilon(\omega)\psi) \rightharpoonup r_\mathcal{T}(\omega, \psi). \tag{8.1}$$

Proof for the first structure. The proof proceeds in two steps.

Step 1. Let denote by $\mathcal{E}([-M, T], L^2(\Omega))$ the space of step functions $v : [-M, T] \to L^2(\Omega)$. Since $L^2(\Omega)$ is separable, it is easy to show that there exists a countable subset $\mathcal{D} \subset \mathcal{E}([-M, T], L^2(\Omega))$ which is dense in $C([-M, T], L^2(\Omega))$ for the uniform convergence. We establish the claim for any $\psi \in \mathcal{D}$.

Let $\psi \in \mathcal{D}$ then for all $s \in [-M, T]$, $\psi(s) = \sum_{i\in I} \mathbb{1}_{B_i}(s)\psi_i$ where $B_i \in \mathcal{B}(\mathbb{R})$, are mutually disjoint and $\psi_i \in L^2(\Omega)$. For every $j \in \{1, \dots, l\}$ and every $t \in [0, T]$, we have (recall that $h = (h_j)_{j=1,\dots,l}$ with $h_j(\zeta') = \zeta'^{\alpha_j}$, $\alpha_j \in \{0, 1\}$),

$$\left(r\left(\omega, t, \frac{\cdot}{\varepsilon}\right) \odot h(\mathcal{T}_\varepsilon\psi)(t)\right)_j = \left(\frac{1}{\#\tau}\right)^{\alpha_j} \sum_{i\in I} \mathbb{1}_{B_i}\left(t - \tau\left(\omega, \frac{\cdot}{\varepsilon}\right)\right) r_j\left(\omega, t, \frac{\cdot}{\varepsilon}\right) \psi_i^{\alpha_j}.$$

As a consequence of the additive ergodic theorem, Theorem C.5 (see also (Attouch *et al.*, 2014, Theorem 12.4.1)), from an easy adaptation of Lemma 7.2, there exists $\Sigma_\psi \in \mathcal{A}$ which does not depend on t, with $\mathbb{P}(\Sigma_\psi) = 1$, such that, for all $\omega \in \Sigma_\psi$, all t, ε, j,

$$\left(\frac{1}{\#\tau}\right)^{\alpha_j} \mathbb{1}_{B_i}\left(t - \tau\left(\omega, \frac{\cdot}{\varepsilon}\right)\right) r_j\left(\omega, t, \frac{\cdot}{\varepsilon}\right)$$

converges as $\varepsilon \to 0$ for the weak $\sigma(L^\infty(\Omega), L^1(\Omega))$ topology toward

$$\left(\frac{1}{\#\tau}\right)^{\alpha_j} \mathbb{E}^\mathcal{I}\left[\int_Y \mathbb{1}_{B_i}(t - \tau(\cdot, y)) r_j(\cdot, t, y)dy\right](\omega).$$

Since for any $\varphi \in L^2(0, T, L^2(\Omega, \mathbb{R}^l))$ the function $\psi_i^{\alpha_j}\varphi_j(t, \cdot)$ belongs to $L^1(\Omega)$, we infer that for all $\omega \in \Sigma_\psi$, and for a.e. $t \in [0, T]$

$$\left\langle \left(r\left(\omega, \cdot, \frac{\cdot}{\varepsilon}\right) \odot h(\mathcal{T}_\varepsilon\psi)(t)\right)_j, \varphi_j(t, \cdot)\right\rangle$$

converges to

$$\int_\Omega \left(\frac{1}{\#\mathcal{T}}\right)^{\alpha_j} \mathbb{E}^{\mathcal{I}}\left[\int_Y \left(\sum_{i\in I} \mathbb{1}_{B_i}(t-\tau(\cdot,y))r_j(\cdot,t,y)\right) dy\right](\omega)\psi_i^{\alpha_j}\varphi_j(t,\cdot)\, dx$$

which is nothing but

$$\left\langle \left(\frac{1}{\#\mathcal{T}}\right)^{\alpha_j} \mathbb{E}^{\mathcal{I}}\left[\int_Y r(\cdot,t,y)\odot h(\psi(t-\tau(\cdot,y)))dy\right]_j(\omega), \varphi_j(t,\cdot)\right\rangle$$

(recall that $\langle\cdot,\cdot\rangle$ denotes the duality bracket in $L^2(\Omega)$). Therefore, for all $\varphi \in L^2(0,T,L^2(\Omega,\mathbb{R}^l))$, and for almost all $t\in[0,T]$, we have for all $\omega\in\Sigma' = \bigcap_{\psi\in\mathcal{D}}\Sigma_\psi$,

$$\left\langle r\left(\omega,t,\frac{\cdot}{\varepsilon}\right)\odot h(\mathcal{T}_\varepsilon(\omega)\psi), \varphi(t,\cdot)\right\rangle \to \langle r_{\mathcal{T}}(\omega,\psi),\varphi(t,\cdot)\rangle$$

as $\varepsilon\to 0$. The claim follows by applying the Lebesgue dominated convergence theorem.

Step 2. ***(General case)*** Let $\psi\in C([-M,T],L^2(\Omega))$. Then there exists a sequence $(\psi_n)_{n\in\mathbb{N}}$ of $\mathcal{D}$ such that $\psi_n\to\psi$ uniformly in $C([-M,T],L^2(\Omega))$. From ***Step 1***, there exists a set $\Sigma'\in\mathcal{A}$ of full probability such that for all $\omega\in\Sigma'$,

$$r\left(\omega,\cdot,\frac{\cdot}{\varepsilon}\right)\odot h(\mathcal{T}_\varepsilon(\omega)\psi_n) \rightharpoonup r_{\mathcal{T}}(\omega,\psi_n) \tag{8.2}$$

weakly in $L^2(0,T,L^2(\Omega,\mathbb{R}^l))$ when $\varepsilon\to 0$. On the other hand, by using the conditional Lebesgue dominated convergence theorem, Proposition C.9, it is easily seen that for all $\omega\in\Sigma'$, $r_{\mathcal{T}}(\omega,\psi_n)$ strongly converges to $r_{\mathcal{T}}(\omega,\psi)$ in $L^2(0,T,L^2(\Omega))$ when $n\to+\infty$. Writing for every $\omega\in\Sigma$

$$\begin{aligned} & r\left(\omega,\cdot,\frac{\cdot}{\varepsilon}\right)\odot h(\mathcal{T}_\varepsilon(\omega)\psi) \\ &= \left(r\left(\omega,\cdot,\frac{\cdot}{\varepsilon}\right)\odot h(\mathcal{T}_\varepsilon(\omega)\psi) - r\left(\omega,\cdot,\frac{\cdot}{\varepsilon}\right)\odot h(\mathcal{T}_\varepsilon(\omega)\psi_n)\right) \\ &\quad + r\left(\omega,\cdot,\frac{\cdot}{\varepsilon}\right)\odot h(\mathcal{T}_\varepsilon(\omega)\psi_n), \end{aligned} \tag{8.3}$$

the claim follows easily from (8.2), the following estimate,

$$\begin{aligned} & \left\| r\left(\omega,t,\frac{\cdot}{\varepsilon}\right)\odot h(\mathcal{T}_\varepsilon(\omega)\psi(t)) - r\left(\omega,t,\frac{\cdot}{\varepsilon}\right)\odot h(\mathcal{T}_\varepsilon(\omega)\psi_n(t))\right\|_{L^2(\Omega;\mathbb{R}^l)} \\ &\le \|r(\omega)\|_{L^\infty((0,+\infty)\times\mathbb{R}^N,\mathbb{R}^l)} \left\| h\left(\psi\left(t-\tau\left(\omega,\frac{\cdot}{\varepsilon}\right)\right)\right) - h\left(\psi_n\left(t-\tau\left(\omega,\frac{\cdot}{\varepsilon}\right)\right)\right)\right\|_{L^2(\Omega,\mathbb{R}^l)} \\ &\le l\,\|r(\omega)\|_{L^\infty((0,+\infty)\times\mathbb{R}^N,\mathbb{R}^l)} \left\|\psi\left(t-\tau\left(\omega,\frac{\cdot}{\varepsilon}\right)\right) - \psi_n\left(t-\tau\left(\omega,\frac{\cdot}{\varepsilon}\right)\right)\right\|_{L^2(\Omega)} \\ &\le l\,\|r(\omega)\|_{L^\infty((0,+\infty)\times\mathbb{R}^N,\mathbb{R}^l)} \|\psi_n-\psi\|_{C_c([-M,T],L^2(\Omega))}, \end{aligned}$$

and by letting first $\varepsilon\to 0$, then $n\to+\infty$ in (8.3).

Proof for the second structure. Let $t \in [0,T]$ and $i \in \{1,\ldots,l\}$ be such that $\alpha_i = 1$. Let $\omega \in \Sigma$, then

$$
\begin{aligned}
r\left(\omega,t,\frac{\cdot}{\varepsilon}\right) \odot h(\mathcal{T}_\varepsilon\psi(t))_i &= r_i\left(\omega,t,\frac{\cdot}{\varepsilon}\right)\int_{-\infty}^{t}\psi(s)d\left(\mathbf{d}\left(\omega,\frac{\cdot}{\varepsilon}\right)\cdot\nu_t\right)(s)\\
&= r_i\left(\omega,t,\frac{\cdot}{\varepsilon}\right)\mathbf{d}\left(\omega,\frac{\cdot}{\varepsilon}\right)\cdot\int_{-\infty}^{t}\psi(s)d\nu_t(s)\\
&= \sum_{k\in\mathbb{N}} r_i\left(\omega,t,\frac{\cdot}{\varepsilon}\right)\mathbf{d}_k\left(\omega,\frac{\cdot}{\varepsilon}\right)\int_{-\infty}^{t}\psi(s)d\nu_{k,t}(s).
\end{aligned}
$$

Hence, for all $\varphi \in L^2(0,T,L^2(\Omega))$,

$$
\begin{aligned}
&\left\langle r\left(\omega,t,\frac{\cdot}{\varepsilon}\right)\odot h(\mathcal{T}_\varepsilon\psi(t))_i,\ \varphi(t,\cdot)\right\rangle\\
&= \left\langle \sum_{k\in\mathbb{N}} r_i\left(\omega,t,\frac{\cdot}{\varepsilon}\right)\mathbf{d}_k\left(\omega,\frac{\cdot}{\varepsilon}\right)\int_{-\infty}^{t}\psi(s)d\nu_{k,t}(s),\ \varphi(t,\cdot)\right\rangle\\
&= \sum_{k\in\mathbb{N}}\left\langle r_i\left(\omega,t,\frac{\cdot}{\varepsilon}\right)\mathbf{d}_k\left(\omega,\frac{\cdot}{\varepsilon}\right)\int_{-\infty}^{t}\psi(s)d\nu_{k,t}(s),\ \varphi(t,\cdot)\right\rangle\\
&= \sum_{k\in\mathbb{N}}\left\langle r_i\left(\omega,t,\frac{\cdot}{\varepsilon}\right)\mathbf{d}_k\left(\omega,\frac{\cdot}{\varepsilon}\right),\varphi(t,\cdot)\int_{-\infty}^{t}\psi(s)d\nu_{k,t}(s)\right\rangle. \qquad (8.4)
\end{aligned}
$$

We have used the domination inequality

$$
\begin{aligned}
&\left\| r_i\left(\omega,t,\frac{\cdot}{\varepsilon}\right)\mathbf{d}_k\left(\omega,\frac{\cdot}{\varepsilon}\right)\int_{-\infty}^{t}\psi(s)d\nu_{k,t}(s)\right\|_{L^2(\Omega)}\\
&\le \|r\|_{L^\infty(\Omega,\mathbb{R}^l)}\|\mathbf{d}_k(\omega,\cdot)\|_{L^\infty(\mathbb{R}^N)}\left\|\int_{-\infty}^{t}\psi(s)d\nu_{k,t}(s)\right\|_{L^2(\Omega)} \qquad (8.5)
\end{aligned}
$$

together with $\sum_{k\in\mathbb{N}}\|\mathbf{d}_k(\omega,\cdot)\|_{L^\infty(\mathbb{R}^N)} = 1$ to interchange the sum and the scalar product in the second equality. As a consequence of the additive ergodic theorem, Theorem C.5 and Lemma 7.2, there exists $\Sigma' \in \mathcal{A}$ which does not depend on t, with $\mathbb{P}(\Sigma') = 1$ such that, for all $\omega \in \Sigma'$,

$$
r_i\left(\omega,t,\frac{\cdot}{\varepsilon}\right)d_k\left(\omega,\frac{\cdot}{\varepsilon}\right) \rightharpoonup \mathbb{E}^{\mathcal{I}}\left[\int_Y r_i(\cdot,t,y)\odot d_k(y)dy\right](\omega) \qquad (8.6)
$$

weakly in $L^2(\Omega)$. Combining (8.4), (8.6), and the domination property (8.5), we deduce that for all $\omega \in \Sigma'$,

$$
\begin{aligned}
&\left\langle r\left(\omega,t,\frac{\cdot}{\varepsilon}\right)\odot h(\mathcal{T}_\varepsilon\psi(t))_i,\varphi(t,\cdot)\right\rangle\\
&\rightharpoonup \left\langle \mathbb{E}^{\mathcal{I}}\left[\int_Y r_i(\cdot,t,y)\mathbf{d}(y)\ dy\right](\omega)\cdot\int_{-\infty}^{t}\psi(s)\ d\nu_t(s),\varphi(t,\cdot)\right\rangle.
\end{aligned}
$$

According to the Lebesgue dominated convergence theorem, we infer that

$$
\begin{aligned}
&\left\{t\mapsto r\left(\omega,t,\frac{\cdot}{\varepsilon}\right)\odot h(\mathcal{T}_\varepsilon\psi(t))_i\right\}\\
&\rightharpoonup\left\{t\mapsto \mathbb{E}^{\mathcal{I}}\left[\int_Y r_i(\cdot,t,y)\mathbf{d}(y)\ dy\right](\omega)\cdot\int_{-\infty}^{t}\psi(s)\ d\nu_t(s)\right\}
\end{aligned}
$$

weakly in $L^2(0,T,L^2(\Omega))$ as $\varepsilon \to 0$. By using the same arguments with $\alpha_i = 0$, we obtain that there exists $\Sigma'' \in \mathcal{A}$ with $\mathbb{P}(\Sigma'') = 1$, such that for all $\omega \in \Sigma'$,

$$\left\{t \mapsto r\left(\omega, t, \frac{\cdot}{\varepsilon}\right) \odot h(\mathcal{T}_\varepsilon \psi(t))_i\right\} \rightharpoonup \left\{t \mapsto \mathbb{E}^{\mathcal{I}}\left[\int_Y r_i(\cdot, t, y)dy\right](\omega)\right\}$$

weakly in $L^2(0,T,L^2(\Omega))$ as $\varepsilon \to 0$. □

8.4 Application to some examples

In examples below, $N = 2$ or 3.

8.4.1 *Homogenization of vector disease models*

Against the background of vector disease models illustrated in Example 3.4, we consider the delays reaction-diffusion problem modeling the evolution of some infected population during a time $T > 0$, in a C^1-regular domain Ω. We assume that Ω is included in an ε-random checkerboard-like environment, or in an environment whose spherical heterogeneities of size $\varepsilon > 0$ are independently randomly distributed, with a given frequency, following a Poisson point process (see Section 7.3.2 for a complete description). Recall that the probability dynamical system $(\Sigma, \mathcal{A}, \mathbb{P}, (T_z)_{z\in\mathbb{Z}^N})$ modeling these two situations is ergodic. We assume that each two functions a and b belong to $W^{1,2}(0,T,L^2_{loc}(\mathbb{R}^N)) \cap L^\infty([0,T]\times\mathbb{R}^N)$. When no infected population is located at the boundary, and the history function η_ε is deterministic, satisfies $0 \le \eta_\varepsilon \le \overline{\rho}$ for some $\overline{\rho} \ge 1$, and $\eta_\varepsilon(0) \in H^1_0(\Omega)$, then the density $u_\varepsilon(\omega,\cdot)$ of the infected population is the unique solution of

$$\begin{cases} \dfrac{du_\varepsilon}{dt}(\omega,t) - \mathrm{div} D_\xi W\left(\omega, \dfrac{x}{\varepsilon}, \nabla u_\varepsilon(\omega,t)\right) \\ = a\left(\omega, t, \dfrac{\cdot}{\varepsilon}\right)(1 - u_\varepsilon(\omega,t)) \displaystyle\int_{-\infty}^t u_\varepsilon(\omega,s)d\mathbf{m}_t^\varepsilon(s) - b\left(\omega, t, \dfrac{\cdot}{\varepsilon}\right) u_\varepsilon(\omega,t) \\ \text{for a.e. } t \in (0,T), \\ \\ u_\varepsilon(\omega,t) = \eta_\varepsilon(t),\ 0 \le \eta_\varepsilon(t) \le \overline{\rho} \text{ for all } t \in (-\infty, 0], \\ \\ 0 \le u_\varepsilon(\omega) \le \overline{\rho} \text{ for all } t \in [0,T], \\ \\ u_\varepsilon(\omega,t) \in H^1_0(\Omega),\ \mathrm{div} D_\xi W(\omega, \frac{\cdot}{\varepsilon}, \nabla u_\varepsilon(\omega,t)) \in L^2(\Omega) \text{ for all } t \in]0,T]. \end{cases}$$

We are in the conditions of Theorem 8.1, with $l = 2$ and

$$r(t,x) = (a(t,x), -b(t,x)),\ h(\zeta') = (\zeta', 1),\ g(\zeta) = (1-\zeta, \zeta).$$

To shorten the notation in all the situations below, we assume that the Legendre-Fenchel conjugate of $W^\pm$ satisfy (D_3^*) so that, from Proposition 7.2, (iii), W^{hom} is Gâteaux-differentiable.

a) The case of random single delay. Concerning the distributed delays, we first consider the family of random measures associated with the random single delay of the first structure described in Section 8.2.1. These measures models a disease whose incubation period varies according to the spatial randomly distributed heterogeneities: we have $\mathbf{m}_t^\varepsilon(\omega) = \frac{1}{\#\tau}\delta_{t-\tau(\omega,\frac{\cdot}{\varepsilon})}$, and the reaction functional is given by

$$a\left(\omega, t, \frac{\cdot}{\varepsilon}\right)(1 - u_\varepsilon(\omega,t))\frac{1}{\#\tau}u_\varepsilon\left(\omega, t - \tau\left(\omega, \frac{\cdot}{\varepsilon}\right)\right) - b\left(\omega, t, \frac{\cdot}{\varepsilon}\right)u_\varepsilon(\omega, t).$$

Then, assuming that for $\mathbb{P}$-a.e. $\omega \in \Sigma$ it hold $\sup_\varepsilon \Phi_\varepsilon(\eta_\varepsilon(0)) < +\infty$ and that η_ε converges to η in $C_c((-\infty,0], L^2(\Omega))$ whose support is included in $[-M, 0]$ for some $M > 0$, according to Theorem 8.1, we can say that when $\varepsilon > 0$ is very small compared to the size of the domain, a deterministic model, well aware with the evolution of the infected population, is given by the homogenized problem

$$(\mathcal{P}^{hom})\begin{cases} \frac{du}{dt}(t) - \mathrm{div} D_\xi W^{hom}(\nabla u(t)) = F^{hom}(t, u(t)) \text{ for a.e. } t \in (0,T) \\ u(t) = \eta(t), \quad 0 \le \eta(t) \le \overline{\rho} \text{ for all } t \in (-\infty, 0], \\ 0 \le u(t) \le \overline{\rho} \text{ for all } t \in [0,T], \\ u(t) \in H_0^1(\Omega), \ \mathrm{div} D_\xi W^{hom}(\nabla u(t)) \in L^2(\Omega) \text{ for a.e. } t \in (0,T), \end{cases}$$

where,

$$\begin{aligned} &F^{hom}(t, u(t)) \\ &= \frac{1}{\#\tau}\mathbb{E}\left[\int_Y a(\cdot,t,y)u(t - \tau(\cdot,y))dy\right](1 - u(t)) - \mathbb{E}\left[\int_Y b(\cdot,t,y)\,dy\right]u(t). \end{aligned}$$

We see that the growth rate of the uninfected population and the time delay have been mixed and averaged. Let us look at two types of spatial environments.

The ε-random checkerboard-like environment. Assume that the spread of the disease occurs in a ε-random checkerboard-like environment modeling a mosaic of two kinds of small habitats, with two virus strains: the growth rate of the uninfected and infected population, the diffusion density, together with the time delay (incubation period), take two values at random on the lattice spanned by the cell $Y = (0,1)^N$ at scale 1, say a^-, a^+, b^-, b^+, W^-, W^+, and τ^-, τ^+ respectively, with a probability $\mathbf{p} > 0$ and $1 - \mathbf{p}$ in each cell. The ε-random checkerboard-like environment is then modeled by a ergodic dynamical system $(\Sigma, \mathcal{A}, \mathbb{P}_\mathbf{p}, (T_z)_{z\in\mathbb{Z}^N})$ where $(\Sigma, \mathcal{A}, \mathbb{P}_\mathbf{p})$ is a Bernoulli product probability space, and T_z the shift operator (see Section 7.3.1 for a precise mathematical explanation). In this specific case we have for all $t \ge 0$

$$\begin{aligned} &F^{hom}(t, u(t)) \\ &= \frac{1}{2}\left[\mathbf{p}a^-u(t - \tau^-) + (1 - \mathbf{p})a^+u(t - \tau^+)\right](1 - u(t)) - \left[\mathbf{p}b^- + (1 - \mathbf{p})b^+\right]u(t). \end{aligned}$$

For a formula giving W^{hom}, A^{hom} and their properties, we refer the reader to Section 7.1.1.

The Poisson point process environment. The ε-random checkerboard-like environment model being somewhat academic, we now consider the environment described in Section 7.3.2 with $N = 2$, a little more realistic, where spherical heterogeneities with radius ε, are independently randomly distributed with a given frequency λ following a Poisson point process with intensity λ. Recall that this random environment is modeled by an ergodic dynamical system $(\Sigma, \mathcal{A}, \mathbb{P}_\lambda, (T_z)_{z\in\mathbb{R}^2})$ where $T_z\omega = \omega - z$ for all $\omega \in \Sigma$ and all $z \in \mathbb{Z}^2$,[1] and, for every bounded Borel set B, and every $k \in \mathbb{N}$,

$$\mathbb{P}_\lambda(\#(\Sigma \cap B) = k) = \lambda^k \mathcal{L}_2(B)^k \,\frac{\exp(-\lambda\mathcal{L}_2(B))}{k!}$$

so that $\mathbb{E}_\lambda\left[\#(\Sigma \cap B)\right] = \lambda\mathcal{L}_2(B)$.

Assume that the disease consists of two virus strains, one of these being concentrated in the environment, at scale 1, made up of all the balls with radius R centered at the points of the Poisson process. Then, given $R > 0$, we define the random density W associated with the random diffusion part, for all ω, x, ξ, t, by

$$W(\omega, x, \xi) = \begin{cases} W^-(\xi) \text{ if } x \in \displaystyle\bigcup_{i\in\mathbb{N}} B_R(\omega_i), \\ W^+(\xi) \text{ otherwise.} \end{cases}$$

Similarly we set

$$a(\omega, t, x) = \begin{cases} a^-(t) \text{ if } x \in \displaystyle\bigcup_{i\in\mathbb{N}} B_R(\omega_i), \\ a^+(t) \text{ otherwise.} \end{cases}$$

$$b(\omega, t, x) = \begin{cases} b^-(t) \text{ if } x \in \displaystyle\bigcup_{i\in\mathbb{N}} B_R(\omega_i), \\ b^+(t) \text{ otherwise.} \end{cases}$$

$$\tau(\omega, t, x) = \begin{cases} \tau^- \text{ if } x \in \displaystyle\bigcup_{i\in\mathbb{N}} B_R(\omega_i), \\ \tau^+ \text{ otherwise.} \end{cases}$$

Noticing that

$$\left(\exists\omega \in \Sigma, \quad y \in \bigcup_{i\in\mathbb{N}} B_R(\omega_i)\right) \iff \#(\Sigma \cap B_R(y)) \geq 1,$$

and using Fubini's theorem, we can express for all $t \geq 0$,

$$F^{hom}(t, u(t))$$
$$= \frac{1}{\#\tau}\mathbb{E}_\lambda\left[\int_Y a(\cdot, t, y)u(t - \tau(\cdot, y))dy\right](1 - u(t)) - \mathbb{E}_\lambda\left[\int_Y b(\cdot, t, y)\, dy\right] u(t).$$

[1] Here we could take the group $(T_x)_{x\in\mathbb{R}^2}$, $T_x\omega = \omega - x$, without changing the results.

Indeed,

$$\begin{aligned}
&\mathbb{E}_\lambda \left[\int_Y a(\cdot,t,y)u(t-\tau(\cdot,y))dy \right] (1-u(t)) \\
&= a^+(t)u(t-\tau^+)(1-u(t)) \int_\Sigma \int_{(0,1)^2} \mathbb{1}_{[\#(\Sigma\cap B_R(y))=0]}(\omega,y)dy\, d\mathbb{P}(\omega) \\
&\quad + a^-(t)u(t-\tau^-)(1-u(t)) \int_\Sigma \int_{(0,1)^2} \mathbb{1}_{[\#(\Sigma\cap B_R(y))\geq 1]}(\omega,y)dy\, d\mathbb{P}(\omega) \\
&= a^+(t)u(t-\tau^+)(1-u(t)) \int_{(0,1)^2} \int_\Sigma \mathbb{1}_{[\#(\Sigma\cap B_R(y))=0]}(\omega,y)d\mathbb{P}(\omega)\, dy \\
&\quad + a^-(t)u(t-\tau^-)(1-u(t)) \int_{(0,1)^2} \int_\Sigma \mathbb{1}_{[\#(\Sigma\cap B_R(y))\geq 1]}(\omega,y)d\mathbb{P}(\omega)\, dy \\
&= a^+(t)u(t-\tau^+)(1-u(t))\exp(-\lambda\pi R^2) \\
&\quad + a^-(t)u(t-\tau^-)(1-u(t))(1-\exp(-\lambda\pi R^2)) \\
&= \left[a^-(t)u(t-\tau^-) + (a^+(t)u(t-\tau^+) - a^-(t)u(t-\tau^-))\exp(-\lambda\pi R^2)\right](1-u(t))
\end{aligned}$$

and, from a similar calculation

$$\mathbb{E}_\lambda \left[\int_Y b(\cdot,t,y)\, dy \right] u(t) = \left[b^-(t) + (b^+(t)-b^-(t))\exp(-\lambda\pi R^2)\right] u(t).$$

Consequently, the homogenized reaction functional is given by

$$\begin{aligned}
&F^{hom}(t,u(t)) \\
&= \frac{1}{2}\left[a^-(t)u(t-\tau^-) + (a^+(t)u(t-\tau^+) - a^-(t)u(t-\tau^-))\exp(-\lambda\pi R^2)\right](1-u(t)) \\
&\quad + \frac{1}{2}\left[b^-(t) + (b^+(t)-b^-(t))\exp(-\lambda\pi R^2)\right] u(t) \qquad (8.7)
\end{aligned}$$

for all $t \geq 0$.

b) The case of a random multiple delays. We consider now the case when the family of random measures corresponds to the second structure of Section 8.2.2 derived from Example 3.1: $\mathbf{m}^\varepsilon_t(\omega) = \sum_{i\in\mathbb{N}} \mathbf{d}_k(\omega, \frac{\cdot}{\varepsilon})\delta_{t-\tau_k}$ for all $\varepsilon > 0$, $t \geq 0$ and $\omega \in \Sigma$. Then, under the same hypotheses, the limit second member becomes

$$\begin{aligned}
&F^{hom}(t,u(t)) \\
&= (1-u(t)) \sum_{k\in\mathbb{N}} \mathbb{E}\left[\int_Y a(\cdot,t,y)\mathbf{d}_k(\cdot,y)\, dy \right] u(t-\tau_k) - u(t)\mathbb{E}\left[\int_Y b(\cdot,t,y)\, dy \right]
\end{aligned}$$

which can be expressed in the two previous probabilistic cases by reproducing the calculations of a). We see that the growth rate of the uninfected population and the time delays coefficients have been mixed and averaged in such a way that the limit delays reaction-diffusion problem is a reaction-diffusion equation with a multiple delays associated with the measure $\sum_{k\in\mathbb{N}} \mathbb{E}\left[\int_Y a(\cdot,y)\mathbf{d}_k(\cdot,y)\, dy\right] \delta_{t-\tau_k}$ incorporating the growth rate for all $t \geq 0$.

8.4.2 *Homogenization of delays logistic equations with immigration*

We consider the food limited model with time delays corresponding to Example 3.5 which describes the evolution of the population density of some specie at time t located at x, during a time $T > 0$, in a C^1-regular domain Ω. We assume, as in the previous example, that Ω is included in a ε-random checkerboard-like environment, or in an environment whose spherical heterogeneities of size ε, are independently randomly distributed with a given frequency, following a Poisson point process. We assume that no population is located at the boundary. When the history function η_ε is deterministic, satisfies $0 \le \eta_\varepsilon \le \overline{\rho}$ for some $\overline{\rho} > 0$, $\eta_\varepsilon(0) \in H_0^1(\Omega)$, and each two functions a and q belongs to $W^{1,2}(0, T, L^\infty_{loc}(\mathbb{R}^N)) \cap L^\infty([0, T] \times \mathbb{R}^N)$, with $0 \le a(\omega, t, x) \le \overline{a}$, and $0 \le q(\omega, t, x) \le \overline{q}$ for all (ω, t, x), then, the density $u_\varepsilon(\omega, \cdot)$ is the unique solution of

$$
\begin{cases}
\dfrac{du_\varepsilon}{dt}(\omega, t) - \operatorname{div} D_\xi W\left(\omega, \dfrac{x}{\varepsilon}, \nabla u_\varepsilon(\omega, t)\right) \\
= a\left(\omega, t, \dfrac{\cdot}{\varepsilon}\right) u_\varepsilon(\omega, t)\left(1 - \dfrac{1}{K_{car}} \displaystyle\int_{-\infty}^t u_\varepsilon(\omega, s) d\mathbf{m}_t^\varepsilon(s)\right) - q(\omega, t) \\
\text{for a.e. } t \in (0, T), \\
u_\varepsilon(\omega, t) = \eta_\varepsilon(t), \quad 0 \le \eta_\varepsilon(t) \le \overline{\rho} \text{ for all } t \in (-\infty, 0], \\
0 \le u(t) \le \overline{\rho} \text{ for all } t \in [0, T], \\
u_\varepsilon(\omega, t) \in H_0^1(\Omega), \ \operatorname{div} D_\xi W\left(\omega, \dfrac{\cdot}{\varepsilon}, \nabla u_\varepsilon(\omega, t)\right) \in L^2(\Omega) \text{ for all } t \in]0, T].
\end{cases}
$$

Moreover $0 \le u_\varepsilon(\omega, t) \le \left(\overline{\rho} + \dfrac{\overline{q}}{\overline{a}}\right) \exp(\overline{a}T) - \dfrac{\overline{q}}{\overline{a}} = \overline{y}(T)$ for all $t \in [0, T]$. Recall that $a(\omega, t, x)$ is the growth rate of the population at time t, located at x in the $\frac{1}{\varepsilon}\Omega$ domain, and $K_{car} > 0$ is the carrying capacity, i.e. the capacity of the environment to sustain the population. To shorten the calculations, we assume that K_{car} is independent on the time variable. Writing the reaction functional as

$$
a\left(t, \frac{\cdot}{\varepsilon}\right) u_\varepsilon(\omega, t) - \frac{1}{K_{car}} a\left(t, \frac{\cdot}{\varepsilon}\right) u_\varepsilon(\omega, t) \int_{-\infty}^t u_\varepsilon(\omega, s) d\mathbf{m}_t^\varepsilon(s),
$$

we see that we are in the context of Theorem 8.1, with $l = 2$ and

$$
r(t, x) = \left(a(t, x), -\frac{1}{K_{car}} a(t, x)\right), \ h(\zeta') = (1, \zeta'), \ g(\zeta) = (\zeta, \zeta)
$$

for all $(t, x, \zeta, \zeta') \in \mathbb{R}_+ \times \mathbb{R}^N \times \mathbb{R}^2$. To model the time delays, let us take the random measure derived from Example 3.3, associated with a diffuse distributed delays $\mathbf{m}_t^\varepsilon(\omega) = \mathbf{d}(\omega, \frac{\cdot}{\varepsilon}) \mathcal{K}(t - \cdot) ds$ where $\mathbf{d} > 0$, $\|\mathbf{d}(\omega, \cdot)\|_{L^\infty(\mathbb{R}^N)} = 1$, and $\mathcal{K}$ is the Γ_l-distributed delays. Observe that everything happens as if the carrying capacity is random, depends on the spatial variable, and given by $K_{car,\varepsilon}(\omega, x) = \frac{K_{car}}{d(\omega, \frac{x}{\varepsilon})}$.

Then, assuming that for $\mathbb{P}$-a.e. $\omega \in \Sigma$, $\sup_\varepsilon \Phi_\varepsilon(\eta_\varepsilon(0)) < +\infty$ and that η_ε converges to η in $C_c((-\infty,0], L^2(\Omega))$ whose support is included in $[-M,0]$ for some $M > 0$, according to Theorem 8.1 as $\varepsilon \to 0$, we can say that when ε is very small compared to the size of the domain, a deterministic model, well aware with the evolution of the density population, is given by

$$(\mathcal{P}^{hom}) \begin{cases} \dfrac{du}{dt}(t) - \mathrm{div} D_\xi W^{hom}(u(t)) = F^{hom}(t,u(t)) \text{ for a.e. } t \in (0,T) \\ u(t) = \eta(t),\ 0 \leq \eta(t) \leq \overline{\rho} \text{ for all } t \in (-\infty, 0], \\ 0 \leq u(t) \leq \left(\overline{\rho} + \dfrac{\overline{q}}{\overline{a}}\right) \exp(\overline{a}T) - \dfrac{\overline{q}}{\overline{a}} \text{ for all } t \in [0,T], \\ u(t) \in H_0^1(\Omega),\ \mathrm{div} D_\xi W^{hom}(\nabla u(t)) \in L^2(\Omega) \text{ for a.e. } t \in (0,T), \end{cases}$$

where, for every $t \geq 0$,

$$\begin{aligned} &F^{hom}(t,u(t)) \\ &= \mathbb{E}\left[\int_Y a(\cdot,t,y)\, dy\right] u(t) - u(t)\frac{1}{K_{car}}\mathbb{E}\left[\int_Y a(\cdot,t,y)\mathbf{d}(\cdot,y)dy\right]\int_{-\infty}^t \mathcal{K}(t-s)u(s)ds \\ &\quad - \mathbb{E}\left[\int_Y q(\cdot,t,y)dy\right]. \end{aligned}$$

By reproducing the calculations of Section 8.4.1, it is easy to express F^{hom} for each of the two probabilistic environments. Let $t \geq 0$. Writing F^{hom} of the form

$$\begin{aligned} &F^{hom}(t,u(t)) \\ &= \mathbb{E}\left[\int_Y a(\cdot,t,y)\, dy\right] u(t) \left(1 - \frac{\mathbb{E}\left[\int_Y a(\cdot,t,y)\mathbf{d}(\cdot,y)dy\right]}{K_{car}\mathbb{E}\left[\int_Y a(\cdot,t,y)\, dy\right]}\int_{-\infty}^t \mathcal{K}(t-s)u(s)ds\right) \\ &\quad - \mathbb{E}\left[\int_Y q(\cdot,t,y)dy\right], \end{aligned}$$

we can interpret the homogenized problem as a diffuse delays logistic equation, modeling the evolution of a density population spreading in an homogeneous environment, with a Γ_l-distributed delays. The growth and the immigration rates are given by $\mathbb{E}\left[\int_Y a(\cdot,t,y)\, dy\right]$ and $\mathbb{E}\left[\int_Y q(\cdot,t,y)dy\right]$ respectively. The effective carrying capacity, given by

$$\overline{K}_{car}(t) = \frac{K_{car}\mathbb{E}\left[\int_Y a(\cdot,t,y)\, dy\right]}{\mathbb{E}\left[\int_Y a(\cdot,t,y)\mathbf{d}(\cdot,y)dy\right]},$$

is now deterministic, constant with respect to the spatial variable, but time dependent and bigger than the carrying capacity K_{car}. Note that, if the growth rate

$a(\cdot,\cdot,\cdot)$ is constant, then the carrying capacity is time-space constant and given by

$$\overline{K}_{car} = K_{car} \frac{1}{\mathbb{E}\left[\int_Y \mathbf{d}(\cdot, y)dy\right]},$$

which is not the almost sure weak limit of $K_{car,\varepsilon}(\omega, x)$ obtained by applying the additive ergodic theorem, Theorem C.5, given by

$$K_{car}\, \mathbb{E}\left[\int_Y \frac{1}{\mathbf{d}(\cdot, y)} dy\right].$$

8.5 A short digression around percolation for the disease model

From the vector disease model of Section 8.4.1, we are going to briefly illustrate some basic concepts of percolation theory. We restrict our analysis to the case of a random single delay.

8.5.1 *Percolation in the ε-random checkerboard-like environment*

Recall that the disease is modeled by the reaction-diffusion problem

$$(\mathcal{P}^{hom}) \begin{cases} \dfrac{du}{dt}(t) - \mathrm{div} D_\xi W^{hom}(u(t)) = F^{hom}(t, u(t)) \text{ for a.e. } t \in (0,T) \\ u(t) = \eta(t),\ 0 \le \eta(t) \le \overline{p} \text{ for all } t \in (-\infty, 0], \\ 0 \le u(t) \le \overline{p} \text{ for all } t \in (-\infty, 0], \\ u(t) \in H_0^1(\Omega),\ \mathrm{div} D_\xi W^{hom}(\nabla u(t)) \in L^2(\Omega) \text{ for a.e. } t \in (0,T), \end{cases}$$

where the reaction functional is the function of the probability $\mathbf{p}$ given by

$$\begin{aligned} F^{hom}(t, u(t)) = &\frac{1}{2}\left[\mathbf{p} a^- u(t-\tau^-) + (1-\mathbf{p}) a^+ u(t-\tau^+)\right](1 - u(t)) \\ &- \left[\mathbf{p} b^- + (1-\mathbf{p}) b^+\right] u(t) \end{aligned} \tag{8.8}$$

for all $t \ge 0$. A cluster of cells relating to (a^-, b^-, τ^-, W^-) is defined as a union of all $\mathbb{Z}^N$-translated cells of Y connected via their edges where the disease has the same characteristics (a^-, b^-, τ^-, W^-). For each $z \in \mathbb{Z}^N$, and each $\omega \in \Sigma$, let denote by $C_z(\omega)$ the cluster of cells relating to (a^-, b^-, τ^-, W^-), which are connected to $z + Y$. We define the percolation probability relating to (a^-, b^-, τ^-, W^-) by

$$\theta(\mathbf{p}) = \mathbb{P}_{\mathbf{p}}(\{\omega \in \Sigma : \mathcal{L}_N(C_0(\omega)) = +\infty\}).$$

Since $\mathbb{P}_{\mathbf{p}}$ is invariant under $(T_z)_{z \in \mathbb{Z}^N}$ we also have

$$\theta(\mathbf{p}) = \mathbb{P}_{\mathbf{p}}(\{\omega \in \Sigma : \mathcal{L}_N(C_z(\omega)) = +\infty\})$$

for any $z \in \mathbb{Z}^N$. It is known that there exists a critical probability of percolation $\mathbf{p}_c$, defined by

$$\mathbf{p}_c = \sup\{\mathbf{p} \in [0,1] : \theta(\mathbf{p}) = 0\},$$

and if $N \geq 2$, then $0 < \mathbf{p}_c < 1$, if $N = 2$, $\mathbf{p}_c = \frac{1}{2}$, and $\theta(\frac{1}{2}) = 0$. For a proof we refer the reader to (Grimmett, 1999, Theorem 1.10 and Theorem 1.11).

Now, let us consider the event

$$E = \left\{ \omega \in \Sigma : \exists z \in \mathbb{Z}^N \mathcal{L}_N(C_z(\omega)) = +\infty \right\}.$$

Clearly, E is invariant under the group $(T_z)_{z\in\mathbb{Z}^N}$. Then, since the dynamical system $(\Sigma, \mathcal{A}, \mathbb{P}_\mathbf{p}, (T_z)_{z\in\mathbb{Z}^N})$ is ergodic, we have $\mathbb{P}_\mathbf{p}(E) \in \{0,1\}$. On the other hand, by definition of $\theta(\mathbf{p})$, we have $\mathbb{P}_\mathbf{p}(E) \geq \theta(\mathbf{p})$. Consequently $\mathbb{P}_\mathbf{p}(E) = 1$ if $\mathbf{p} > \mathbf{p}_c$. Therefore one can say that for $\mathbf{p} > \mathbf{p}_c$, almost surely, there exists a cluster made up of an infinity of cells in which the disease takes the same characteristics, thus which is a crossing of $\frac{1}{\varepsilon}\Omega$. Furthermore, there exists almost surely exactly one such cluster (see (Grimmett, 1999, Theorem 8.1)) which splits the domain $\frac{1}{\varepsilon}\Omega$ into two subsets. One can say that the solution u of $(\mathcal{P}^{hom})$ whose reaction functional is given by (8.8) with $\mathbf{p} > \mathbf{p}_c$, accounts for the density of the infected population for which there is percolation for the incubation period τ^- of the disease.

8.5.2 *Percolation in the Poisson point process environment*

From Section 8.4.1 a), recall that the disease is modeled by the reaction-diffusion problem $(\mathcal{P}^{hom})$ where the reaction functional is the function of the intensity λ given by (8.7). A cluster of balls is defined as a union of all connected overlapping balls of radius R centered at points of the Poisson process. Let denote by $C_x(\omega)$ the cluster containing the point $x \in \mathbb{R}^2$ whose overlapping balls are centered at $\omega = (\omega_i)_{i\in\mathbb{N}}$. As in the previous random model, we denote by $\theta(\lambda)$ the percolation probability related to (a^-, b^-, τ^-, W^-), defined by

$$\theta(\lambda) = \mathbb{P}_\lambda(\{\omega \in \Sigma : \mathcal{L}_2(C_0(\omega)) = +\infty\}).$$

Since $\mathbb{P}_\lambda$ is invariant under the translations of $\mathbb{R}^2$, we have

$$\theta(\lambda) = \mathbb{P}_\lambda(\{\omega \in \Sigma : \mathcal{L}_2(C_x(\omega)) = +\infty\})$$

for all $x \in \mathbb{R}^2$. It is known that there exists a critical intensity λ_c, defined by

$$\lambda_c = \sup\{\lambda : \theta(\lambda) = 0\},$$

which satisfies $0 < \lambda_c < +\infty$. For a proof, we refer the reader to (Grimmett, 1999, Theorem 12.35). The percolation threshold λ_c is usually estimated using numerical simulations (see Zuyev and Quintanilla (2003)) and, to our knowledge, we can say nothing if $\lambda = \lambda_c$. The event

$$E = \left\{ \omega \in \Sigma : \exists x \in \mathbb{R}^2 \quad \mathcal{L}_2(C_x(\omega)) = +\infty \right\}$$

is invariant under the group $(T_x)_{x\in\mathbb{R}^2}$. Then, since the dynamical system $(\Sigma, \mathcal{A}, \mathbb{P}_\lambda, (T_x)_{x\in\mathbb{R}^2})$ is ergodic, we have $\mathbb{P}_\lambda(E) \in \{0,1\}$. Consequently, from $\mathbb{P}_\lambda(E) \geq \theta(\lambda)$ we infer that $\mathbb{P}_\lambda(E) = 1$ if $\lambda > \lambda_c$. Furthermore, there exists almost surely exactly one such infinite cluster of overlapping balls (see Zuyev and Sidorenko (1985)). Then, as in the previous model, this cluster splits the domain $\frac{1}{\varepsilon}\Omega$ into two subdomains. One can say that for $\lambda > \lambda_c$, the unique solution u of $(\mathcal{P}^{hom})$ whose reaction functional is given by (8.7), accounts for the density of the infected population for which there is percolation for the incubation period τ^- of the disease.

Chapter 9

Stochastic homogenization of two components nonlinear reaction-diffusion systems

As far as we know, the homogenization of reaction-diffusion systems was first addressed in Mielke *et al.* (2014), Reichelt (2015) by means of the two scale convergence; see also Peter (2009) where the homogenization with evolving microstructure is performed using the method of transformation to a periodic reference domain. In this chapter we contribute to this study in the framework of stochastic homogenization. The main results, which are direct consequences of Theorems 4.3, 4.4, are stated in Theorems 9.1, 9.2. They are illustrated in Section 9.5 through the homogenization of a prey-predator model with a saturation effect. The model brings into play two species spreading in an heterogeneous environment whose small spatial heterogeneities are distributed at random following a Poisson point process. The homogenized problem illustrates the interplay between the growth rate of the prey and the maximum carrying capacity of the environment when the size of the spatial heterogeneities is very small. We place this chapter within the framework of stochastic homogenization introduced in Chapter 7.

9.1 The random diffusion parts

We briefly specify the random diffusion parts by recalling some results and the notation of Chapter 7. For each $i = 1, 2$, we are given a random convex integrand $W_i : \Sigma \times \mathbb{R}^N \times \mathbb{R}^N \to \mathbb{R}$, that is to say, a $(\mathcal{A} \otimes \mathcal{B}(\mathbb{R}^N) \otimes \mathcal{B}(\mathbb{R}^N), \mathcal{B}(\mathbb{R}))$ measurable function such that for every $\omega \in \Sigma$, the function $W_i(\omega, \cdot, \cdot)$, belongs to the class $\mathrm{Conv}_{\alpha,\beta}$. We assume that W_i fulfills the *covariance* property with respect to the dynamical system $(\Sigma, \mathcal{A}, \mathbb{P}, (T_z)_{z \in \mathbb{Z}^N})$. For each $i = 1, 2$ consider $\Gamma_i \subset \partial\Omega$ with $\mathcal{H}^{N-1}(\Gamma_i) > 0$, and the functionals $\Phi_{i,\varepsilon} : \Sigma \times L^2(\Omega) \longrightarrow \mathbb{R}_+ \cup \{+\infty\}$ defined of the one of two following forms:

$$\Phi_{i,\varepsilon}(\omega, u) = \begin{cases} \displaystyle\int_\Omega W_i\left(\omega, \frac{x}{\varepsilon}, \nabla u\right) dx + \frac{1}{2}\int_{\partial\Omega} a_i u^2 d\mathcal{H}_{N-1} - \int_{\partial\Omega} \phi_i u d\mathcal{H}_{N-1} & \text{if } u \in H^1(\Omega) \\ +\infty & \text{otherwise,} \end{cases} \tag{9.1}$$

where $\phi_i \in L^2_{\mathcal{H}_{N-1}}(\partial\Omega)$, $a_i \in L^\infty_{\mathcal{H}_{N-1}}(\partial\Omega)$ with $a_i \geq 0$ $\mathcal{H}_{N-1}$ a.e. in $\partial\Omega$, and $a_i \geq \sigma$ on $\Gamma_i \subset \partial\Omega$ with $\mathcal{H}_{N-1}(\Gamma_i) > 0$, for some $\sigma > 0$; or

$$\Phi_{i,\varepsilon}(\omega, u) = \begin{cases} \displaystyle\int_\Omega W_i\left(\omega, \frac{x}{\varepsilon}, \nabla u\right) dx & \text{if } u \in H^1_{\Gamma_i}(\Omega) \\ +\infty & \text{otherwise.} \end{cases} \tag{9.2}$$

These functionals model random energies concerning various steady-states situations, where the small parameter ε accounts for the size of small and randomly distributed heterogeneities in the context of a statistically homogeneous media.

Let $i = 1, 2$. Under above hypotheses with respect to the probability dynamical system $(\Sigma, \mathcal{A}, \mathbb{P}, (T_z)_{z\in\mathbb{Z}^N})$, for $\mathbb{P}$-a.e. $\omega \in \Sigma$, the sequence of functional $\Phi_{i,\varepsilon}(\omega, \cdot)_{\varepsilon>0}$ Mosco-converges respectively to the integral functional $\Phi_i^{hom}(\omega, \cdot)$, where Φ_i^{hom} : $\Sigma \times L^2(\Omega) \longrightarrow \mathbb{R}^+ \cup \{+\infty\}$ is defined respectively by

$$\Phi_i^{hom}(\omega, u) = \begin{cases} \displaystyle\int_\Omega W_i^{hom}(\omega, \nabla u)\, dx + \frac{1}{2}\int_{\partial\Omega} a_i u^2 d\mathcal{H}_{N-1} - \int_{\partial\Omega} \phi_i u d\mathcal{H}_{N-1} & \\ & \text{if } u \in H^1(\Omega), \\ +\infty & \text{otherwise,} \end{cases}$$

or

$$\Phi_i^{hom}(\omega, u) = \begin{cases} \displaystyle\int_\Omega W_i^{hom}(\omega, \nabla u) dx & \text{if } u \in H^1_{\Gamma_i}(\Omega), \\ +\infty & \text{otherwise.} \end{cases}$$

For a proof, we refer the reader to the proof of Theorem 7.1, proof of ***Proof of*** (H_1).

9.2 The random reaction parts

We are given a random TCCP-structured reaction functional, i.e. a pair (F_1, F_2) with $F_i : \Sigma \times [0, +\infty) \times L^2(\Omega) \times L^2(\Omega) \to \mathbb{R}^\Omega$, $i = 1, 2$ defined by $F_i(\omega, t, u, v)(x) = f_i(\omega, t, x, u(x), v(x))$ where

$$f_i : \Sigma \times [0, +\infty) \times \mathbb{R}^N \times \mathbb{R} \times \mathbb{R} \to \mathbb{R}$$

is a $(\mathcal{A} \otimes \mathcal{B}(\mathbb{R}) \otimes \mathcal{B}(\mathbb{R}^N) \otimes \mathcal{B}(\mathbb{R}) \otimes \mathcal{B}(\mathbb{R}), \mathcal{B}(\mathbb{R}))$ measurable function such that for $\mathbb{P}$ a.s. $\omega \in \Sigma$, $(f_1(\omega, \cdot, \cdot, \cdot, \cdot), f_2(\omega, \cdot, \cdot, \cdot, \cdot))$ is a TCCP-structured reaction function associated with $(r_i(\omega, \cdot), g_i, h_i, q_i(\omega, \cdot))$. We assume that for $\mathbb{P}$-a.e. $\omega \in \Sigma$, r_i and q_i satisfy (REA_1), (REA_2), and (REA_3), and set $f_{i,\varepsilon}(\omega, t, x, \zeta, \zeta') := f_i(\omega, t, \frac{x}{\varepsilon}, \zeta, \zeta')$ for all $(\omega, t, x, \zeta, \zeta')$. Each reaction functional $F_{i,\varepsilon}$ is the defined for all u, v in $L^2(\Omega)$ by

$$F_{i,\varepsilon}(\omega, t, u, v)(x) = f_i\left(\omega, t, \frac{x}{\varepsilon}, u(x), v(x)\right)$$

for all $\varepsilon > 0$. Note that in condition (TCCP) satisfied by $(F_1(\omega,\cdot,\cdot,\cdot,\cdot), F_2(\omega,\cdot,\cdot,\cdot,\cdot))$, the functions $\overline{f}_i$, $\overline{y}_i$, $\overline{\rho}_i$, and $\underline{f}_i$, $\underline{y}_i$, $\underline{\rho}_i$ may depend on ω (we sometimes omit it to shorten the notation), and $(F_{1,\varepsilon}(\omega,\cdot,\cdot,\cdot), F_{2,\varepsilon}(\omega,\cdot,\cdot,\cdot))$ is a TCCP-structured functional whose condition (TCCP) is exactly that of $(F_1(\omega,\cdot,\cdot,\cdot), F_2(\omega,\cdot,\cdot,\cdot))$, i.e. with $\overline{f}_i$, $\overline{y}_i$, $\overline{\rho}_i$, and $\underline{f}_i$, $\underline{y}_i$, $\underline{\rho}_i$. Since $\underline{y}_i$ and $\overline{y}_i$ do not depend on ε, condition (4.18) is automatically satisfied. In Lemma 7.1 we obtained the following limits for $\mathbb{P}$-a.s. $\omega \in \Sigma$, and for $i = 1, 2$:

$$\lim_{\varepsilon\to 0}\int_0^T \left\| \frac{dr_i}{dt}\left(\omega,\tau,\frac{\cdot}{\varepsilon}\right)\right\|^2_{L^2(\Omega,\mathbb{R}^l)} d\tau = \mathcal{L}_N(\Omega)\mathbb{E}^{\mathcal{I}}\int_0^T \left\|\frac{dr_i}{dt}(\omega,\tau,\cdot)\right\|^2_{L^2(Y,\mathbb{R}^l)} d\tau;$$

$$\lim_{\varepsilon\to 0}\int_0^T \left\| \frac{dq_i}{dt}\left(\omega,\tau,\frac{\cdot}{\varepsilon}\right)\right\|^2_{L^2(\Omega)} d\tau = \mathcal{L}_N(\Omega)\mathbb{E}^{\mathcal{I}}\int_0^T \left\|\frac{dq_i}{dt}(\omega,\tau,\cdot)\right\|^2_{L^2(Y)} d\tau.$$

Finally we assume that for $\mathbb{P}$-a.e. $\omega \in \Sigma$,

$$a_i\underline{\rho}_i(\omega) \le \phi_i \le a_i\overline{\rho}_i(\omega) \text{ if } \Phi_{i,\varepsilon} \text{ is of the form (9.1)},$$

or

$$\underline{\rho}_i(\omega) \le 0 \le \overline{\rho}_i(\omega) \text{ if } \Phi_{i,\varepsilon} \text{ is of the form (9.2)}.$$

9.3 Almost sure convergence to the homogenized system

Under above conditions, by combining Theorem 4.3 with the variational convergence of the sequence of random energies $\Phi_{i,\varepsilon}$ specified above, we intend to analyze the asymptotic behavior in $C(0,T,L^2(\Omega)) \times C(0,T,L^2(\Omega))$ of the sequence $((u_\varepsilon(\omega), v_\varepsilon(\omega)))_{\varepsilon>0}$ solving the random reaction-diffusion system whose diffusions terms are the subdifferentials of the functionals of the form (9.1) or (9.2):

$$(\mathcal{S}_\varepsilon(\omega))\begin{cases} \dfrac{du_\varepsilon(\omega)}{dt}(t) + D\Phi_{1,e}(\omega)(u_\varepsilon(\omega,t)) = F_{1,\varepsilon}(\omega,t,u_\varepsilon(\omega,t),v_\varepsilon(\omega,t)) \\ \text{for a.e. } t \in (0,T), \\ \\ \dfrac{dv_\varepsilon(\omega)}{dt}(t) + D\Phi_{2,e}(\omega)(u_\varepsilon(\omega,t)) = F_{2,\varepsilon}(\omega,t,u_\varepsilon(\omega,t),v_\varepsilon(\omega,t)) \\ \text{for a.e. } t \in (0,T), \\ \\ (u_\varepsilon(\omega,0),v_\varepsilon(\omega,0)) = (u^0_\varepsilon(\omega),v^0_\varepsilon(\omega)) \in \mathrm{dom}(\Phi_{1,\varepsilon}) \times \mathrm{dom}(\Phi_{2,\varepsilon}), \\ \\ \underline{\rho}_1(\omega) \le u_\varepsilon(\omega,\cdot) \le \overline{\rho}_1(\omega),\ \underline{\rho}_2(\omega) \le v_\varepsilon(\omega,\cdot) \le \overline{\rho}_2(\omega), \\ \\ (u_\varepsilon(\omega,\cdot),v_\varepsilon(\omega,\cdot)) \in \mathrm{dom}(D\Phi_{1,\varepsilon}) \times \mathrm{dom}(D\Phi_{2,\varepsilon}) \text{ for all } t \in]0,T]. \end{cases}$$

Theorem 9.1. *For each $\omega \in \Sigma$, let denote by $(u_\varepsilon(\omega), v_\varepsilon(\omega))$ the unique solution in $C([0,T],L^2(\Omega)) \times C([0,T],L^2(\Omega))$ of the reaction-diffusion system $(\mathcal{S}_\varepsilon(\omega))$. Assume that for $\mathbb{P}$-a.s. $\omega \in \Sigma$,* *i)* $\sup_{\varepsilon>0} \Phi_{1,\varepsilon}(\omega, u^0_\varepsilon(\omega)) < +\infty$ *and* $\sup_{\varepsilon>0} \Phi_{2,\varepsilon}(\omega, v^0_\varepsilon(\omega)) < +\infty$*;*

ii) $(u^0_\varepsilon(\omega), v^0_\varepsilon(\omega))$ *strongly converges to* $(u^0(\omega), v^0(\omega))$ *in* $L^2(\Omega) \times L^2(\Omega)$.

Then, for $\mathbb{P}$*-a.e.* $\omega \in \Sigma$, $(u_\varepsilon(\omega,\cdot), v_\varepsilon(\omega,\cdot))$ *uniformly converges in* $C([0,T], L^2(\Omega)) \times C([0,T], L^2(\Omega))$ *to the unique solution of the reaction-diffusion system*

$$(\mathcal{S}^{hom}(\omega)) \begin{cases} \dfrac{du(\omega)}{dt}(t) + \partial\Phi_1^{hom}(\omega)(u(\omega,t)) \ni F_1^{hom}(\omega,t,u(\omega,t),v(\omega,t)) \\ \text{for a.e. } t \in (0,T) \\ \\ \dfrac{dv(\omega)}{dt}(t) + \partial\Phi_2^{hom}(\omega)(v(\omega,t)) \ni F_2^{hom}(\omega,t,u(\omega,t),v(\omega,t)) \\ \text{for a.e. } t \in (0,T) \\ \\ (u(\omega,0), v(\omega,0)) = (u^0(\omega), v^0(\omega)) \in \mathrm{dom}(\Phi_1) \times \mathrm{dom}(\Phi_2), \\ \\ \underline{\rho}_1(\omega) \le u(\omega,\cdot) \le \overline{\rho}_1(\omega),\ \underline{\rho}_2(\omega) \le v(\omega,\cdot) \le \overline{\rho}_2(\omega), \\ \\ (u(\omega,t), v(\omega,t)) \in \mathrm{dom}(D\Phi_1^{hom}) \times \mathrm{dom}(D\Phi_2^{hom}) \text{ for a.e. } t \in (0,T), \end{cases}$$

where for $i = 1,2$, F_i^{hom} *is given for all* $(\omega,t) \in \Sigma \times [0,T]$ *and all* $(u,v) \in L^2(\Omega)^2$ *by* $F_i^{hom}(\omega,t,u,v)(x) = f_i^{hom}(\omega,t,x,u(x),v(x))$*, with*

$$f_1^{hom}(\omega,t,\zeta,\zeta') = r_1^{hom}(\omega,t) \odot h_1(\zeta') \cdot g_1(\zeta) + q_1^{hom}(\omega,t),$$

$$f_2^{hom}(\omega,t,\zeta,\zeta') = r_2^{hom}(\omega,t) \odot h_2(\zeta) \cdot g_2(\zeta') + q_2^{hom}(\omega,t)$$

$$r_i^{hom}(\omega,t) = \mathbb{E}^{\mathcal{I}}\left(\int_{(0,1)^N} r_i(\omega,t,y)\, dy\right),$$

$$q_i^{hom}(\omega,t) = \mathbb{E}^{\mathcal{I}}\left(\int_{(0,1)^N} q_i(\omega,t,y)\, dy\right).$$

Moreover, for $\mathbb{P}$*-a.e.* $\omega \in \Sigma$,

$$\left(\frac{du_\varepsilon(\omega)}{dt}, \frac{dv_\varepsilon(\omega)}{dt}\right) \rightharpoonup \left(\frac{du(\omega)}{dt}, \frac{du(\omega)}{dt}\right)$$

weakly in $L^2(0,T,L^2(\Omega)) \times L^2(0,T,L^2(\Omega))$*, and for all* $t \in [0,T]$,

$$\underline{y}_1(\omega,t) \le u(\omega,t) \le \overline{y}_1(\omega,t),\ \underline{y}_2(\omega,t) \le v(\omega,t) \le \overline{y}_2(\omega,t).$$

When the dynamical system $(\Sigma, \mathcal{A}, \mathbb{P}, (T_z)_{z\in\mathbb{Z}^N})$ *is ergodic, the initial conditions are deterministic, then* $(\mathcal{S}^{hom})$ *is deterministic and the expectation operator must be replaced by the mathematical expectation operator in formulas expressing* $\overline{r}_i$ *and* $\overline{q}_i$.

If in addition W_i *satisfies* (D_3^*)*, then* $\partial\Phi_1^{hom}(\omega)$ *or* $\partial\Phi_1^{hom}$ *are univalent and the differential inclusions are equalities.*

Proof. The proof is a straightforward consequence of Theorem 4.4, and Lemmas 7.1, 7.2. ∎

9.4 The case of a coupling between a random r.d.e. and a random n.d.r.e.

We place this section within the framework of Section 4.3. We assume that the random reaction functionals F_1 and F_2 fulfill the conditions of Section 9.2, that f_2 does not depend on the space variable, and that g_2 and h_2 belong to $C^1_{\text{loc}}(\mathbb{R}, \mathbb{R}^l)$. Note that under these conditions, $F_{2,\varepsilon}(\omega, \cdot, \cdot, \cdot) = F_2(\omega, \cdot, \cdot, \cdot)$ for all $\omega \in \Sigma$. Theorem 9.2 below whose proof is a direct consequence of Theorem 4.4 and Lemmas 7.1, 7.2, expresses the homogenized problem of the following random system:

$$(\mathcal{S}_\varepsilon(\omega)) \begin{cases} \dfrac{du_\varepsilon(\omega)}{dt}(t) + D\Phi_e(\omega)(u_\varepsilon(\omega,t)) = F_{1,\varepsilon}(\omega, t, u_\varepsilon(\omega,t), v_\varepsilon(\omega,t)) \\ \text{for a.e. } t \in (0,T), \\ \dfrac{dv_\varepsilon(\omega)}{dt}(t) = F_2(\omega, t, u_\varepsilon(\omega,t), v_\varepsilon(\omega,t)) \text{ for a.e. } t \in (0,T), \\ (u_\varepsilon(\omega,0), v_\varepsilon(\omega,0)) = (u^0_\varepsilon(\omega), v^0_\varepsilon(\omega)) \in \operatorname{dom}(\Phi_\varepsilon) \times H^1(\Omega) \\ \underline{\rho}_1(\omega) \le u_\varepsilon(\omega) \le \overline{\rho}_1(\omega),\ \underline{\rho}_2(\omega) \le v_\varepsilon(\omega) \le \overline{\rho}_2(\omega), \\ u_\varepsilon(\omega,t) \in \operatorname{dom}(D\Phi_\varepsilon(\omega)) \text{ for all } t \in]0,T]. \end{cases}$$

Theorem 9.2. *For each $\omega \in \Sigma$, let denote by $(u_\varepsilon(\omega), v_\varepsilon(\omega))$ the unique solution in $C([0,T], L^2(\Omega)) \times C([0,T], L^2(\Omega))$ of the system $(\mathcal{S}_\varepsilon(\omega))$. Assume that for $\mathbb{P}$-a.e. $\omega \in \Sigma$,*

i) $\sup_{\varepsilon>0} \Phi_\varepsilon(\omega, u^0_\varepsilon(\omega)) < +\infty$;

ii) *$u^0_\varepsilon(\omega)$ strongly converges to $u^0(\omega)$ in $L^2(\Omega)$, and $v^0_\varepsilon(\omega) \rightharpoonup v^0(\omega)$ weakly in $H^1(\Omega)$.*

Then, for $\mathbb{P}$-a.e. $\omega \in \Sigma$, $(u_\varepsilon(\omega), v_\varepsilon(\omega))$ uniformly converges in $C([0,T], L^2(\Omega)) \times C([0,T], L^2(\Omega))$ to the unique solution of the system

$$(\mathcal{S}^{hom}(\omega)) \begin{cases} \dfrac{du(\omega)}{dt}(t) + D\Phi^{hom}(\omega)(u(\omega,t)) \ni F_1^{hom}(\omega, t, u(\omega,t), v(\omega,t)) \\ \text{for a.e. } t \in (0,T) \\ \dfrac{dv(\omega)}{dt}(t) = F_2(\omega, t, u(\omega,t), v(\omega,t)) \text{ for a.e. } t \in (0,T) \\ (u(\omega,0), v(\omega,0)) = (u^0(\omega), v^0(\omega)) \in \operatorname{dom}(\Phi^{hom}) \times H^1(\Omega), \\ \underline{\rho}_1(\omega) \le u(\omega) \le \overline{\rho}_1(\omega),\ \underline{\rho}_2(\omega) \le v(\omega) \le \overline{\rho}_2(\omega), \\ u(\omega,t) \in \operatorname{dom}(D\Phi^{hom}) \text{ for a.e. } t \in (0,T), \end{cases}$$

where F_1^{hom} *is given by* $F_1^{hom}(\omega,t,u,v)(x) = f_i^{hom}(\omega,t,x,u(x),v(x))$ *with*

$$f_1^{hom}(\omega,t,\zeta,\zeta') = r_1^{hom}(\omega,t) \odot h_1(\zeta') \cdot g_1(\zeta) + q_1^{hom}(\omega,t),$$

$$r_1^{hom}(\omega,t) = \mathbb{E}^{\mathcal{I}}\left(\int_{(0,1)^N} r_1(\omega,t,y)\, dy\right),$$

$$q_1^{hom}(\omega,t) = \mathbb{E}^{\mathcal{I}}\left(\int_{(0,1)^N} q_1(\omega,t,y)\, dy\right).$$

Moreover, for $\mathbb{P}$*-a.e.* $\omega \in \Sigma$,

$$\left(\frac{du_\varepsilon(\omega)}{dt}, \frac{dv_\varepsilon(\omega)}{dt}\right) \rightharpoonup \left(\frac{du(\omega)}{dt}, \frac{du(\omega)}{dt}\right)$$

weakly in $L^2(0,T,L^2(\Omega)) \times L^2(0,T,L^2(\Omega))$ *and*

$$\underline{y}_1(\omega,t) \le u(\omega,t) \le \overline{y}_1(\omega,t),\ \underline{y}_2(\omega,t) \le v(\omega,t) \le \overline{y}_2(\omega,t)$$

for all $t \in [0,T]$.

When the dynamical system $(\Sigma, \mathcal{A}, \mathbb{P}, (T_z)_{z\in\mathbb{Z}^N})$ *is ergodic, the initial conditions are deterministic,* F_2 *is deterministic, then* $(\mathcal{S}^{hom})$ *is deterministic and the expectation operator must be replaced by the mathematical expectation operator in formulas expressing* $\overline{r}_1^{hom}$ *and* $\overline{q}_1^{hom}$.

If in addition W *satisfies* (D_3^*)*, then* $\partial\Phi^{hom}(\omega)$ *or* $\partial\Phi^{hom}$ *are univalent and the differential inclusions are equalities.*

Remark 9.1. By applying Theorem 4.5 and Lemmas 7.1, 7.2, one can easily express the homogenized problem of a random system of two non diffusive reaction equations with obvious adaptations.

9.5 Application to stochastic homogenization of a prey-predator random model with saturation effect

For the notation we refer the reader to Example 4.3. For each $i = 1,2$, we are given two functions $W_i^{\pm}$ in $\mathrm{Conv}_{\alpha,\beta}$, where $W_i^{\pm}$ do not depend on x, and satisfy (D_3^*); and two functions $\alpha_i^{\pm} : [0,T] \to (0,+\infty)$ in $W^{1,2}(0,T)$ for which there exist positive real numbers $\underline{\alpha}_1^{\pm}$, $\underline{\alpha}_2^{\pm}$, and $\overline{\alpha}_2^{\pm}$ such that

$$\begin{cases} 0 < \underline{\alpha}_1^{\pm} \le \alpha_1^{\pm}(t); \\ 0 < \underline{\alpha}_2^{\pm} \le \alpha_2^{\pm}(t) \le \overline{\alpha}_2^{\pm}. \end{cases}$$

We also consider two functions $K^{\pm} : [0,T] \to (0,+\infty)$ in $W^{1,2}(0,T)$ satisfying $0 < \underline{K}^{\pm} \le K(t)^{\pm}$ for some positive real numbers $\underline{K}^{\pm}$, and two functions $a^{\pm} : [0,T] \to (0,+\infty)$ in $W^{1,2}(0,T)$ for which there exist a constant $\overline{a}^{\pm} > 0$ such that $0 < a^{\pm}(t) \le \overline{a}^{\pm}$.

The probabilistic framework is that of the random environment described in Section 7.3.2 with $N = 2$. Recall that the spherical heterogeneities of size of order ε, have centers independently randomly distributed with a given frequency λ, following a Poisson point process with intensity λ. This random environment is modeled by an ergodic dynamical system $(\Sigma, \mathcal{A}, \mathbb{P}_\lambda, (T_x)_{x\in\mathbb{R}^2})$ where $T_x\omega = \omega - x$ for all $(\omega, x) \in \Sigma \times \mathbb{R}^2$, and for every bounded Borel set B, and every $k \in \mathbb{N}$,

$$\mathbb{P}_\lambda(\#(\Sigma \cap B) = k) = \lambda^k \mathcal{L}_2(B)^k \frac{\exp(-\lambda \mathcal{L}_N(B))}{k!}$$

so that $\mathbb{E}_\lambda\left[\#(\Sigma \cap B)\right] = \lambda \mathcal{L}_N(B)$. Given $R > 0$, for $i = 1, 2$, we define the random density W_i associated with the random diffusion part by

$$W_i(\omega, x, \xi) = \begin{cases} W_i^-(\xi) \text{ if } x \in \bigcup_{i\in\mathbb{N}} B_R(\omega_i), \\ W_i^+(\xi) \text{ otherwise} \end{cases}$$

for all $(\omega, x, \xi) \in \Sigma \times \mathbb{R}^2 \times \mathbb{R}^2$. Similarly we set

$$\alpha_i(\omega, t, x) = \begin{cases} \alpha_i^-(t) \text{ if } x \in \bigcup_{j\in\mathbb{N}} B_R(\omega_j), \\ \alpha_i^+(t) \text{ otherwise.} \end{cases}$$

$$K_{car}(\omega, t, x) = \begin{cases} K^-(t) \text{ if } x \in \bigcup_{j\in\mathbb{N}} B_R(\omega_j), \\ K^+(t) \text{ otherwise.} \end{cases}$$

$$a(\omega, t, x) = \begin{cases} a^-(t) \text{ if } x \in \bigcup_{j\in\mathbb{N}} B_R(\omega_j), \\ a^+(t) \text{ otherwise.} \end{cases}$$

We define the following constants

$$\begin{aligned}
\underline{\alpha}_i &= \inf_{t\in[0,+\infty)} \min(\alpha_i^-(t), \alpha_i^+(t)),\ i = 1, 2; \\
\overline{\alpha}_2 &= \sup_{t\in[0,+\infty)} \max(\alpha_2^-(t), \alpha_2^+(t)); \\
\overline{a} &= \sup_{t\in[0,+\infty)} \max(a^-(t), a^+(t)); \\
\underline{K} &= \inf_{t\in[0,+\infty)} \min(K^-(t), K^+(t)).
\end{aligned}$$

Let b and c be two positive constants. Assume that *the extinction threshold* satisfies $\mu_{ext} := c\frac{\underline{\alpha}_1\underline{\alpha}_2}{\overline{a}\overline{\alpha}_2} \geq 4$, then, choosing $\underline{\rho}_1$, $\overline{\rho}_1$ and $\overline{\rho}_2$ fulfilling (4.5), we consider the following system stemming from Example 4.3:

$$(\mathcal{S}_\varepsilon(\omega))\begin{cases} \dfrac{du_\varepsilon}{dt}(\omega,t) - \mathrm{div} D_\xi W_1\left(\omega, \dfrac{\cdot}{\varepsilon}, \nabla u_\varepsilon(\omega,t)\right) = F_1\left(\omega, t, \dfrac{\cdot}{\varepsilon}, u_\varepsilon(\omega,t), v_\varepsilon(\omega,t)\right) \\ \text{for a.e. } t \in (0,T), \\ \dfrac{dv_\varepsilon(\omega)}{dt}(t) - \mathrm{div} D_\xi W_2\left(\omega, \dfrac{\cdot}{\varepsilon}, \nabla v_\varepsilon(\omega,t)\right) = F_2\left(\omega, t, \dfrac{\cdot}{\varepsilon}, u_\varepsilon(\omega,t), v_\varepsilon(\omega,t)\right) \\ \text{for a.e. } t \in (0,T), \\ (u_\varepsilon(\omega,0), v_\varepsilon(\omega,0)) = (u_\varepsilon^0(\omega), v_\varepsilon^0(\omega)) \in H^1(\Omega) \times H^1(\Omega), \\ \underline{\rho}_1 \le u_\varepsilon^0(\omega) \le \overline{\rho}_1,\ 0 \le v_\varepsilon^0(\omega) \le \overline{\rho}_2, \\ a_1 u_\varepsilon(\omega,t) + D_\xi W_1\left(\omega, \frac{\cdot}{\varepsilon}, \nabla u_\varepsilon(\omega,t)\right) \cdot \mathbf{n} = \phi_1 \text{ on } \partial\Omega \text{ for all } t \in]0,T], \\ a_2 v_\varepsilon(\omega,t) + D_\xi W_2\left(\omega, \frac{\cdot}{\varepsilon}, \nabla v_\varepsilon(\omega,t)\right) \cdot \mathbf{n} = \phi_2 \text{ on } \partial\Omega \text{ for all } t \in]0,T], \end{cases}$$

where

$$\begin{aligned} F_1\left(\omega, t, \frac{\cdot}{\varepsilon}, u_\varepsilon(\omega,t), v_\varepsilon(\omega,t)\right) &= \alpha_1\left(\omega, t, \frac{\cdot}{\varepsilon}\right) u_\varepsilon(\omega,t)\left(1 - \frac{u_\varepsilon(\omega,t)}{K_{car}\left(\omega, t, \frac{\cdot}{\varepsilon}\right)}\right) \\ &\quad - a\left(\omega, t, \frac{\cdot}{\varepsilon}\right) v_\varepsilon(\omega,t)\left(1 - \exp(-b u_\varepsilon(\omega,t))\right) \end{aligned}$$

and

$$F_2\left(\omega, t, \frac{\cdot}{\varepsilon}, u_\varepsilon(\omega,t), v_\varepsilon(\omega,t)\right) = \alpha_2\left(\omega, t, \frac{\cdot}{\varepsilon}\right) v_\varepsilon(\omega,t)\left(1 - c\frac{v_\varepsilon(\omega,t)}{u_\varepsilon(\omega,t)}\right)$$

for all $\omega \in \Sigma$, $t \in \mathbb{R}_+$, $\varepsilon > 0$. The functionals $\Phi_{i,\varepsilon}$ are of type (9.1) for $i = 1, 2$. According to Proposition 4.3 and Theorem 4.1, $(\mathcal{S}_\varepsilon(\omega))$ admits a unique solution $(u_\varepsilon(\omega,\cdot), v_\varepsilon(\omega,\cdot))$ in the space $C([0,T], L^2(\Omega)) \times C([0,T], L^2(\Omega))$, which satisfies $\underline{\rho}_1 \le u_\varepsilon(\omega,t) \le \overline{\rho}_1 \exp(\overline{\alpha}_1 t)$ and $0 \le v_\varepsilon(\omega,t) \le \overline{\rho}_2$ for all $t \in [0,T]$. Furthermore, $u_\varepsilon(\omega,\cdot)$ and $v_\varepsilon(\omega,\cdot)$ admit a right derivative at each $t \in]0,T[$. The system models the evolution of two species with density $u_\varepsilon(\omega,\cdot)$ and $v_\varepsilon(\omega,\cdot)$ of a prey and a predator respectively, whose birth growth rate, maximum carrying capacity, and saturation effect, take two values at random depending on whether the species reside in the environment made up of the union of small balls of size $\varepsilon > 0$ or not (refer to the comments of Example 4.3). The system is subject to Robin boundary conditions for both the prey and the predator. Assuming $\Gamma_i = \partial\Omega$ for $i = 1, 2$, this means that at each time $t > 0$, the flux across the boundary is proportional to the difference between the outside surrounding density and the density inside Ω: the boundary condition can be written as $D_\xi W_1(\omega, \frac{\cdot}{\varepsilon}, \nabla u_\varepsilon(\omega,t)) \cdot \mathbf{n} = \frac{1}{a_1}(\phi - u_\varepsilon(\omega,t))$, idem for the prey. We can also choose a Dirichlet boundary condition for the evolution of the predator (take Φ_2 of the form (9.2)). But we must underline the fact that a Dirichlet boundary condition for the evolution of the prey is incompatible with the initial condition. Indeed condition $\underline{\rho}_1 \leq 0 \leq \overline{\rho}_1$ is not fulfilled since, from (4.5), we must have $0 < \underline{\rho}_1 < \overline{\rho}_1$.

The homogenized system is expressed in the Proposition below. It is interesting to note that the effective growth rate α_i^{hom} of each two species is the mean value of α_i with respect to the product probability measure $\mathcal{L}_2\lfloor(0,1)^2 \otimes \mathbb{P}_\lambda$, while the effective maximum carrying capacity K_{car}^{hom} is now a function of the growth rate α_1 and K_{car}. This illustrates the interplay between the growth rate of the prey and the maximum carrying capacity of the environment when the size of the spatial heterogeneities, with a constant frequency λ, is very small compared with the size of the domain.[1]

Proposition 9.1. *Assume that the initial conditions are deterministic, that $(u_\varepsilon^0, v_\varepsilon^0)$ strongly converges to (u^0, v^0) in $L^2(\Omega) \times L^2(\Omega)$ and that $\sup_\varepsilon \Phi_{1,\varepsilon}(u_\varepsilon^0) < +\infty$, $\sup_\varepsilon \Phi_{2,\varepsilon}(v_\varepsilon^0) < +\infty$. Then, for $\mathbb{P}$-a.e. $\omega \in \Sigma$, $(u_\varepsilon, v_\varepsilon)$ uniformly converges in $C([0,T], L^2(\Omega)) \times C([0,T], L^2(\Omega))$ to the unique solution (u,v) of the deterministic reaction-diffusion system*

$$(\mathcal{S}) \begin{cases} \dfrac{du}{dt}(t) - \mathrm{div} D_\xi W_1^{hom}(\nabla u(t)) = \alpha_1^{hom}(t) u(t) \left(1 - \dfrac{u(t)}{K_{car}^{hom}(t)}\right) \\ -a^{hom}(t) v(t)(1 - \exp(-bu(t))) \\ \textit{for a.e. } t \in (0,T), \\ \\ \dfrac{dv}{dt}(t) - \mathrm{div} D_\xi W_2^{hom}(\nabla v(t)) = \alpha_2^{hom}(t) v(t) \left(1 - c\dfrac{v(t)}{u(t)}\right) \\ \textit{for a.e. } t \in (0,T), \\ \\ u(0) = u^0 \in H^1(\Omega),\ v(0) = v^0 \in H^1(\Omega), \\ \\ \underline{\rho}_1 \leq u^0 \leq \overline{\rho}_1,\ 0 \leq v^0 \leq \overline{\rho}_2, \\ \\ a_1 u(t) + D_\xi W_1^{hom}(\nabla u(t)) \cdot \mathbf{n} = \phi_1 \textit{ on } \partial\Omega \textit{ for a.e. } t \in (0,T), \\ \\ a_2 v(t) + D_\xi W_1^{hom}(\nabla v(t)) \cdot \mathbf{n} = \phi_2 \textit{ on } \partial\Omega \textit{ for a.e. } t \in (0,T), \end{cases}$$

[1]To shorten the notation, for $i = 1,2$, we assume that $W_i^\pm$ satisfy (D_3^*) so that, from Proposition 7.2, (iii), W_i^{hom} is Gâteaux-differentiable.

where for every $t \geq 0$ and $i = 1, 2$,

$$\alpha_i^{hom}(t) = \mathbb{E}_\lambda\left(\int_{(0,1)^2} \alpha_i(\cdot, t, y)\, dy\right)$$
$$= \alpha_i^-(t) + (\alpha_i^+(t) - \alpha_i^-(t)) \exp(-\lambda\pi R^2);$$
$$a^{hom}(t) = \mathbb{E}_\lambda\left(\int_{(0,1)^2} a(\cdot, t, y)\, dy\right)$$
$$= a^-(t) + (a^+(t) - a^-(t)) \exp(-\lambda\pi R^2);$$
$$K_{car}^{hom}(t) = \frac{\mathbb{E}_\lambda\left(\int_{(0,1)^2} \alpha_1(\cdot, t, y)\, dy\right)}{\mathbb{E}_\lambda\left(\int_{(0,1)^2} \frac{\alpha_1(\cdot, t, y)}{K_{car}(\cdot, t, y)}\, dy\right)}$$
$$= \frac{\alpha_1^-(t) + (\alpha_1^+(t) - \alpha_1^-(t)) \exp(-\lambda\pi R^2)}{\frac{\alpha_1^-(t)}{K^-(t)} + (\frac{\alpha_1^+(t)}{K^+(t)} - \frac{\alpha_1^-(t)}{K^-(t)}) \exp(-\lambda\pi R^2)}.$$

Furthermore $\underline{\rho}_1 \leq u \leq \overline{\rho}_1 \exp(\overline{\alpha}_1 T)$ *and* $0 \leq v \leq \overline{\rho}_2$.

Proof. Let $(\omega, t, x, \zeta, \zeta') \in \Sigma \times \mathbb{R}_+ \times \mathbb{R}^N \times \mathbb{R}^2)$. Apply Theorem 9.1 with $\Phi_{1,\varepsilon}$ and $\Phi_{2,\varepsilon}$ of the form (9.1), and

$$f_1(\omega, t, x, \zeta, \zeta') = \alpha_1(\omega, t, x)\zeta\left(1 - \frac{\zeta}{K_{car}(\omega, t, x)}\right) - a(\omega, t, x)\zeta'(1 - \exp(-b\zeta))$$
$$= \alpha_1(\omega, t, x)\zeta - \frac{\alpha_1(\omega, t, x)}{K_{car}(\omega, t, x)}\zeta^2 - a(\omega, x, t)\zeta'(1 - \exp(-b\zeta)),$$
$$f_2(\omega, t, x, \zeta, \zeta') = \alpha_2(\omega, t, x)\zeta' - c\alpha_2(\omega, t, x)\frac{\zeta'^2}{\zeta}.$$

Then

$$f_1^{hom}(t, \zeta, \zeta') = \mathbb{E}_\lambda\left(\int_{(0,1)^2} \alpha_1(\cdot, t, y)\, dy\right)\zeta - \mathbb{E}_\lambda\left(\int_{(0,1)^2} \frac{\alpha_1(\cdot, t, y)}{K_{car}(\cdot, t, y)}\, dy\right)\zeta^2$$
$$- \mathbb{E}_\lambda\left(\int_{(0,1)^2} a(\cdot, t, y)\, dy\right)\zeta'(1 - \exp(-b\zeta))$$

which can be written as,

$$f_1^{hom}(t, \zeta, \zeta') = \mathbb{E}_\lambda\left(\int_{(0,1)^2} \alpha_1(\cdot, t, y)\, dy\right)\zeta\left(1 - \frac{\mathbb{E}_\lambda\left(\int_{(0,1)^2} \frac{\alpha_1(\cdot, t, y)}{K_{car}(\cdot, t, y)}\, dy\right)}{\mathbb{E}_\lambda\left(\int_{(0,1)^2} \alpha_1(\cdot, t, y)\, dy\right)}\zeta\right)$$
$$- \mathbb{E}_\lambda\left(\int_{(0,1)^2} a(\cdot, t, y)\, dy\right)\zeta'(1 - \exp(-b\zeta))$$
$$= \alpha_1^{hom}(t)\zeta\left(1 - \frac{\zeta}{K_{car}^{hom}(t)}\right) - a^{hom}(t)\zeta'(1 - \exp(-b\zeta)).$$

Similarly we have

$$f_2^{hom}(t,\zeta,\zeta') = \alpha_2^{hom}(t)\zeta' - c\alpha_2^{hom}(t)\frac{\zeta'^2}{\zeta} = \alpha_2^{hom}(t)\zeta'\left(1 - c\frac{\zeta'}{\zeta}\right).$$

It remains to compute α_i^{hom}, a^{hom} and K_{car}^{hom}. Observe the equivalence

$$\left(\exists\omega \in \Sigma \quad y \in \bigcup_{i\in\mathbb{N}} B_R(\omega_i)\right) \iff \#(\Sigma \cap B_R(y)) \geq 1.$$

Then using Fubini's theorem, we infer that

$$\begin{aligned}
\alpha_i^{hom} &= \alpha_i^+(t)\int_\Sigma\int_{(0,1)^2} \mathbb{1}_{[\#(\Sigma\cap B_R(y))=0]}(\omega,y)dy\; d\mathbb{P}(\omega) \\
&\quad + \alpha_i^-(t)\int_\Sigma\int_{(0,1)^2} \mathbb{1}_{[\#(\Sigma\cap B_R(y))\geq 1]}(\omega,y)dy\; d\mathbb{P}(\omega) \\
&= \alpha_i^+(t)\int_{(0,1)^2}\int_\Sigma \mathbb{1}_{[\#(\Sigma\cap B_R(y))=0]}(\omega,y)d\mathbb{P}(\omega)dy \\
&\quad + \alpha_i^-(t)\int_{(0,1)^2}\int_\Sigma \mathbb{1}_{[\#(\Sigma\cap B_R(y))\geq 1]}(\omega,y)d\mathbb{P}(\omega)dy \\
&= \alpha_i^+(t)\exp(-\lambda\pi R^2) + \alpha_i^-(t)((1-\exp(-\lambda\pi R^2))) \\
&= \alpha_i^-(t) + (\alpha_i^+(t) - \alpha_i^-(t))\exp(-\lambda\pi R^2).
\end{aligned}$$

We express a^{hom} and K_{car}^{hom} by a similar calculation. □

Chapter 10

Stochastic homogenization of integrodifferential reaction-diffusion equations

We apply the convergence theorem of Chapter 5 to the stochastic homogenization analysis of equations

$$(\mathcal{P}_\varepsilon(\omega)) \begin{cases} \dfrac{du_\varepsilon(\omega)}{dt}(t) + \partial\Phi(\omega, u_\varepsilon(\omega,t)) + \displaystyle\int_0^t K(t-s)\partial\Psi_\varepsilon(\omega, u_\varepsilon(\omega,s))\, ds \\ = F_\varepsilon(\omega, t, u(t)) \text{ for a.e. } t \in (0,T) \\ \\ u_\varepsilon(\omega,0) = u^0(\omega),\ u^0(\omega) \in \mathrm{dom}(\partial\Phi_\varepsilon(\omega,\cdot)), \end{cases}$$

when Φ_ε, Ψ_ε are random functionals of the Calculus of Variations, and F_ε is a random reaction functional. In Section 10.1 we address the stochastic homogenization of a random problems modeled from a Fick's law with delay: the non Fickian flux is superimposed on the first flow at each time t. Depending on the model, the delay reflects a maturation period, a resource regeneration time, a mating process, or an incubation period. In Section 10.2, we treat the stochastic homogenization of general nonlinear integrodifferential reaction-diffusion equations in one dimension space, in the setting of a Poisson point process. For recent developments in periodic homogenization of parabolic problems in $L^2(0,T,L^2(\mathbb{R}^d))$ with a convolution type operator, we refer the reader to Piatnitski and Zhizhina (2017, 2019).

10.1 Stochastic homogenization of a random problem modeled from a Fick's law with delay

With the notation of Chapter 5, Ω is a C^1-domain of $\mathbb{R}^N$, $X = L^2(\Omega)$ and $V = H^1_0(\Omega)$. To model the spatial environment, we consider a probability dynamical system $(\Sigma, \mathcal{A}, \mathbb{P}, (T_z)_{z\in\mathbb{Z}^N})$. As in the previous chapters, $\mathcal{I}$ denotes the σ-algebra of invariant sets of $\mathcal{A}$ by the group of $\mathbb{P}$-preserving transformations $(T_z)_{z\in\mathbb{Z}^N}$. We denote by $u(\omega,t,x)$ a scalar state variable of a physical, biological or ecological model at position x and time t, subjected to an alea $\omega \in \Sigma$; according to the cases $u(\omega,\cdot,\cdot)$ is a concentration, or a density. For the model considered, we assume that the diffusion flux in Ω related to $u(\omega,t,x)$ has two contributions:

- the Fickian flux which locally has at each time t the direction of the negative spatial gradient of the state variable, given by $J_F(\omega,t,x) = -D(\omega,x)\nabla u(\omega,t,x)$,

- the non Fickian flux which locally has the direction of the negative spatial gradient of the state variable at some past time $\tau > 0$, given by $J_{NF}(\omega,t,x) = -D(\omega,x)\nabla u(\omega,t-\tau,x)$.

For example, in population dynamics, the non Fickian flux models a maturation period, a resource regeneration time, a mating processes, or an incubation period; it is superimposed on the first flow at each time t. The coefficient D accounts for the rate of movement in the heterogeneous spatial environment modeled by $(\Sigma,\mathcal{A},\mathbb{P},(T_z)_{z\in\mathbb{Z}^N})$ in $\mathbb{R}^N$. From the mass conservation principle, for a given source $F(\omega,t,x)$, the variable u satisfies the equation

$$\frac{du}{dt}(\omega,t) + \operatorname{div}(J_F(\omega,t,x)) + \operatorname{div}(J_{NF}(\omega,t,x)) = F(\omega,t,x). \tag{10.1}$$

Assume τ small. Then we can express $\operatorname{div}(J_{NF}(\omega,t,x))$ as a divergence of the gradient field distributed following a suitable time kernel. Indeed, from $J_{NF}(\omega,t+\tau,x) = -D(\omega,x)\nabla u(t)$, using the first order time approximation, we infer that

$$J_{NF}(\omega,t+\tau,x) \sim J_{NF}(\omega,t) + \tau\frac{\partial J_{NF}}{\partial t}(\omega,t,x)$$

so that J_{NF} satisfies the first order differential equation

$$\tau\frac{\partial J_{NF}}{\partial t}(\omega,t,x) + J_{NF}(\omega,t) = -D(\omega,x)\nabla u(\omega,t).$$

Assume that for $\mathbb{P}$-a.e. ω, $J_{NF}(\omega,0,\cdot) = 0$. By an elementary computation using the variation of constants method, we see that J_{NF} is given by

$$J_{NF}(\omega,t,x) = -\frac{1}{\tau}\int_0^t \exp\left(\frac{s-t}{\tau}\right) D(\omega,x)\nabla u(\omega,s,x)ds.$$

Therefore, (10.1) becomes

$$\frac{du}{dt}(\omega,t,\cdot) - \operatorname{div}(D(\omega,\cdot)\nabla u(\omega,t),\cdot) - \operatorname{div}\left(\frac{1}{\tau}\int_0^t \exp\left(\frac{s-t}{\tau}\right) D(\omega,\cdot)\nabla u(\omega,s,\cdot)ds\right)$$
$$= F(\omega,t,\cdot)$$

and can be written as

$$\frac{du}{dt}(\omega,t,\cdot) - \operatorname{div}(D(\omega,\cdot)\nabla u(\omega,t,\cdot)) - \frac{1}{\tau}\int_0^t \exp\left(-\frac{t-s}{\tau}\right)\operatorname{div}(D(\omega,\cdot)\nabla u(\omega,s,\cdot))ds$$
$$= F(\omega,t,\cdot),$$

which is an integrodifferential diffusion equation as treated in Section 5.2, with the kernel K defined by $K(t) = \frac{1}{\tau}\exp\left(-\frac{t}{\tau}\right)$. To take into account the size of order $\varepsilon > 0$ of the spatial heterogeneities, the last integrodifferential diffusion equation becomes

$$\frac{du_\varepsilon}{dt}(\omega,t,\cdot) - \operatorname{div}\left(D\left(\omega,\frac{\cdot}{\varepsilon}\right)\nabla u_\varepsilon(\omega,t,\cdot)\right)$$
$$-\frac{1}{\tau}\int_0^t \exp\left(\frac{s-t}{\tau}\right)\operatorname{div}\left(D\left(\omega,\frac{\cdot}{\varepsilon}\right)\nabla u_\varepsilon(\omega,s,\cdot)\right)ds = F_\varepsilon(\omega,t,\cdot).$$

Therefore, we are led to consider the following more general problem, written in a mathematically rigorous formulation as follows: for $\varepsilon > 0$, $u_\varepsilon(\omega, \cdot) \in L^2(0, T, L^2(\Omega))$ solves

$$(\mathcal{P}_\varepsilon(\omega)) \begin{cases} \dfrac{du_\varepsilon(\omega)}{dt}(t) + \partial\Phi_\varepsilon(\omega, u_\varepsilon(\omega, t)) + \displaystyle\int_0^t K(t-s)\partial\Psi_\varepsilon(\omega, u_\varepsilon(\omega, s))ds \\ \qquad = F_\varepsilon(\omega, t, u_\varepsilon(\omega, t)) \text{ for a.e. } t \in (0, T), \\ u_\varepsilon(\omega, 0) = u_\varepsilon^0(\omega) \in \operatorname{dom}(\partial\Phi_\varepsilon(\omega)). \end{cases}$$

The kernel K is given in Section 5.1. We restrict the study to the functionals $\Phi_\varepsilon, \Psi_\varepsilon : L^2(\Omega) \to]-\infty, +\infty]$ defined by: $a > 0$ and $b \geq 0$ are two given constants, and

$$\Phi_\varepsilon(\omega, u) = \begin{cases} a\displaystyle\int_\Omega D\left(\omega, \frac{x}{\varepsilon}\right) \nabla u \cdot \nabla u \, dx + \frac{b}{2}\int_\Omega u^2 dx & \text{if } u \in H_0^1(\Omega), \\ +\infty & \text{otherwise,} \end{cases}$$

$$\Psi_\varepsilon(\omega, u) = \begin{cases} \displaystyle\int_\Omega D\left(\omega, \frac{x}{\varepsilon}\right) \nabla u \cdot \nabla u \, dx & \text{if } u \in H_0^1(\Omega), \\ +\infty & \text{otherwise,} \end{cases}$$

where the random matrix valued map

$$D = (d_{i,j})_{i,j=1\dots N} : \Sigma \times \mathbb{R}^N \to \mathbb{M}_N$$

is $(\mathcal{A} \otimes \mathcal{B}(\mathbb{R}^N), \mathcal{B}(\mathbb{M}_N))$-measurable and covariant with respect to the group $(T_z)_{z\in\mathbb{Z}^N}$, that is, for $\mathbb{P}$-a.e. $\omega \in \Sigma$

$$D(T_z\omega, x) = D(\omega, x + z)$$

for all $x \in \mathbb{R}^N$ and all $z \in \mathbb{Z}^N$. We also assume that D satisfies the two borns: there exist $\alpha > 0$ and $\beta > 0$ such that

$$\alpha|\xi|^2 \leq \sum_{i,j=1}^N d_{i,j}(\omega, x)\xi_i\xi_j \leq \beta|\xi|^2$$

for all $\omega \in \Sigma$, all $x \in \mathbb{R}^N$ and all $\xi \in \mathbb{R}^N$.

The random reaction functional F_ε is structured as follows: for all $u \in L^2(\Omega)$, all $t \in [0, T]$, and all $x \in \Omega$,

$$F_\varepsilon(\omega, t, u)(x) = r\left(\omega, t, \frac{x}{\varepsilon}\right) \cdot g(u(x)) + q\left(\omega, t, \frac{x}{\varepsilon}\right)$$

where

- $g : \mathbb{R} \to \mathbb{R}^l$ is a bounded L_g-Lipschitz continuous function;

- $r : \Sigma \times [0,T] \times \mathbb{R}^N \to \mathbb{R}^l$ is $(\mathcal{A} \otimes \mathcal{B}([0,T]) \otimes \mathcal{B}(\mathbb{R}^N), \mathcal{B}(\mathbb{R}^l))$-measurable, r is covariant with respect to the group $(T_z)_{z\in\mathbb{Z}^N}$, i.e. for $\mathbb{P}$-a.e. $\omega \in \Sigma$ $r(T_z\omega,\cdot,\cdot) = r(\omega,\cdot,\cdot+z)$ for all $x \in \mathbb{R}^N$ and all $z \in \mathbb{Z}^N$, for $\mathbb{P}$-a.e. $\omega \in \Sigma$ we have $r \in L^\infty([0,T]\times\mathbb{R}^N,\mathbb{R}^l)\cap W^{1,1}(0,T,L^2_{\text{loc}}(\mathbb{R}^N,\mathbb{R}^l))$, for all bounded Borel subsets B of $\mathbb{R}^N$, the real valued functions

$$\omega \mapsto \|r(\omega,t,\cdot)\|_{L^2(B,\mathbb{R}^l)} \text{ for all } t \in [0,T],$$

$$\omega \mapsto \int_0^T \left\| \frac{dr}{dt}(\omega,\tau,\cdot) \right\|_{L^2(B,\mathbb{R}^l)} d\tau$$

  belong to $L_{\mathbb{P}}(\Sigma)$;
- $q : \Sigma \times [0,T] \times \mathbb{R}^N \to \mathbb{R}$ is $(\mathcal{A} \otimes \mathcal{B}([0,T]) \otimes \mathcal{B}(\mathbb{R}^N), \mathcal{B}(\mathbb{R}))$-measurable, q is covariant with respect to the group $(T_z)_{z\in\mathbb{Z}^N}$, for $\mathbb{P}$-a.e. $\omega \in \Sigma$, we have $q \in W^{1,2}(0,T,L^2_{\text{loc}}(\mathbb{R}^N))$, for all bounded Borel sets B of $\mathbb{R}^N$, the real valued functions

$$\omega \mapsto \|q(\omega,t,\cdot)\|^2_{L^2(B)} \text{ for all } t \in [0,T], \tag{10.2}$$

$$\omega \mapsto \int_0^T \left\| \frac{dq}{dt}(\omega,\tau,\cdot) \right\|^2_{L^2(B)} d\tau \tag{10.3}$$

  belong to $L_{\mathbb{P}}(\Sigma)$.

Taking into account the expression of each two subdifferentials $\partial\Phi_\varepsilon(\omega,\cdot)$ and $\partial\Psi_\varepsilon(\omega,\cdot)$, the problem $(\mathcal{P}_\varepsilon(\omega))$ can be written as

$$(\mathcal{P}_\varepsilon(\omega)) \begin{cases} \dfrac{du_\varepsilon(\omega,\cdot)}{dt}(t) - a \operatorname{div}\left(D\left(\omega,\dfrac{\cdot}{\varepsilon}\right)\nabla u_\varepsilon(\omega,t)\right) + b\, u_\varepsilon(\omega,t) \\ \qquad - \displaystyle\int_0^t K(t-s)\operatorname{div}\left(D\left(\omega,\frac{\cdot}{\varepsilon}\right)\nabla u_\varepsilon(\omega,s)\right)ds \\ \qquad\qquad = F_\varepsilon(\omega,t,u_\varepsilon(\omega,t)) \text{ for a.e. } t \in (0,T), \\ u_\varepsilon(\omega,0) = u^0_\varepsilon(\omega) \in \operatorname{dom}(\partial\Phi_\varepsilon(\omega)), \end{cases}$$

where

$$\operatorname{dom}(\partial\Phi_\varepsilon(\omega,\cdot)) = \operatorname{dom}(\partial\Psi_\varepsilon(\omega,\cdot)) = \left\{v \in H^1_0(\Omega) : \operatorname{div}(D(\omega,\cdot)\nabla v) \in L^2(\Omega)\right\}.$$

Condition (5.1) is clearly uniformly satisfied: take $\alpha_{\Psi_\varepsilon} = \alpha$. According to Example 5.1, condition (5.2) is uniformly satisfied. Moreover, since $\Psi_\varepsilon(\omega,\cdot)$ is quadratic, from Proposition 5.2 $(\mathcal{P}_\varepsilon(\omega))$ admits a unique solution $u_\varepsilon(\omega,\cdot)$.

For all $\omega \in \Sigma$ and all $(x,\xi) \in \mathbb{R}^N \times \mathbb{R}^N$ set

$$W(\omega,x,\xi) := D(\omega,x)\xi.\xi$$

and define W^{hom} for $\mathbb{P}$-a.e. $\omega \in \Sigma$ and all $\xi \in \mathbb{M}_N$ by

$$\begin{aligned} W^{hom}(\omega,\xi) &= \lim_{n\to+\infty} \inf\left\{\frac{1}{n^N}\int_{nY} W(\omega,y,\xi+\nabla u(y))dy : u \in H^1_0(nY)\right\} \\ &= \inf_{n\in\mathbb{N}^*} \mathbb{E}^{\mathcal{I}} \inf\left\{\frac{1}{n^N}\int_{nY} W(\omega,y,\xi+\nabla u(y))dy : u \in H^1_0(nY)\right\}. \end{aligned}$$

According to Chapter 7, Section 7.1.1, this limit exists for $\mathbb{P}$-a.e. $\omega \in \Sigma$ and is given by the formula above. Note that if $(\Sigma, \mathcal{A}, \mathbb{P}, (T_z)_{z\in\mathbb{Z}^N})$ is ergodic, then W^{hom} is deterministic and given for $\mathbb{P}$-a.e. $\omega \in \Sigma$ by

$$\begin{aligned} W^{hom}(\xi) &= \lim_{n\to+\infty} \inf \left\{ \frac{1}{n^N} \int_{nY} W(\omega, y, \xi + \nabla u(y))dy : u \in H_0^1(nY) \right\} \\ &= \inf_{n\in\mathbb{N}^*} \mathbb{E} \inf \left\{ \frac{1}{n^N} \int_{nY} W(\cdot, y, \xi + \nabla u(y))dy : u \in H_0^1(nY) \right\}. \end{aligned}$$

As a consequence of Theorem 5.5 we obtain

Corollary 10.1. *Assume that for $\mathbb{P}$-a.e. $\omega \in \Sigma$*

i) $\sup_{\varepsilon>0} \Phi_\varepsilon(u_\varepsilon^0(\omega)) < +\infty$;

ii) $u_\varepsilon^0(\omega) \to u^0(\omega)$ *strongly in* $L^2(\Omega)$.

Then for $\mathbb{P}$-a.e. $\omega \in \Sigma$, the solution $u_\varepsilon(\omega, \cdot)$ of $(\mathcal{P}_\varepsilon(\omega))$ converges to $u(\omega, \cdot)$ in $C([0,T], L^2(\Omega))$, solution of the homogenized problem

$$(\mathcal{P}(\omega)) \begin{cases} \dfrac{du(\omega)}{dt}(t) - a\, \mathrm{div}(D^{hom}(\omega)\nabla u(\omega, t)) + b\, u(\omega, t) \\ \qquad - \displaystyle\int_0^t K(t-s)\mathrm{div}(D^{hom}(\omega)\nabla u(\omega, s))ds \\ \qquad\qquad = F(\omega, t, u(\omega, t)) \text{ for a.e. } t \in (0, T), \\ u(\omega, 0) = u^0(\omega) \in \mathrm{dom}(\partial\Phi^{hom}(\omega)) \end{cases}$$

with $D^{hom}(\omega) = (d_{i,j}^{hom}(\omega))_{i,j=1,\dots N}$,

$$\begin{cases} d_{i,j}^{hom}(\omega) = \dfrac{1}{2}(W^{hom}(\omega, e_i + e_j) + W^{hom}(\omega, e_i - e_j)), \\ \mathrm{dom}(\partial\Phi^{hom}(\omega)) = H_0^1(\Omega) \cap H^2(\Omega), \end{cases}$$

where $(e_i)_{i=1,\dots,N}$ is the canonical basis of $\mathbb{R}^N$. The homogenized reaction functional is given for every $u \in L^2(\Omega)$, $\mathbb{P}$-a.e. $\omega \in \Sigma$, and all $(t, x) \in [0, T] \times \mathbb{R}^N$ by

$$\begin{cases} F^{hom}(\omega, t, u)(x) = r^{hom}(\omega, t) \cdot g(u(x)) + q^{hom}(\omega, t), \\ r^{hom}(\omega, t) = \mathbb{E}^{\mathcal{I}}\left(\displaystyle\int_{(0,1)^N} r(\omega, t, y)\, dy \right), \\ q^{hom}(\omega, t) = \mathbb{E}^{\mathcal{I}}\left(\displaystyle\int_{(0,1)^N} q(\omega, t, y)\, dy \right). \end{cases}$$

Proof. Firstly, by using arguments from ergodic theory of additive processes, we obtain that for $\mathbb{P}$-a.e. $\omega \in \Sigma$,

$$r_\varepsilon(\omega, \cdot, \cdot) \rightharpoonup r^{hom}(\omega, \cdot)$$

for the $\sigma(L^\infty(0,T,L^2(\Omega,\mathbb{R}^l)),L^1(0,T,L^2(\Omega,\mathbb{R}^l)))$ topology as $\varepsilon\to 0$,

$$q_\varepsilon(\omega,\cdot,\cdot)\rightharpoonup q^{hom}(\omega,\cdot)$$

weakly in $L^2(0,T,L^2(\Omega))$, and $\sup_{\varepsilon>0}\|q(\omega,t,\frac{\cdot}{\varepsilon})\|_{L^2(\Omega)}<+\infty$ for all $t\in[0,T]$. For a proof refer to the proof of Theorem 7.1, ***Proof of*** (H$_5$), Lemma 7.2, and ***Proof of*** (H$_6$).

It remains to establish (STAB$_4$) and (STAB'$_5$) of Remark 5.3, i.e.: for $\mathbb{P}$-a.e. $\omega\in\Sigma$:

$$\Phi_\varepsilon(\omega)\overset{M}{\to}\Phi(\omega,\cdot)\text{ and }\Psi_\varepsilon(\omega,\cdot)_{\lfloor H_0^1(\Omega)}\overset{\Gamma_{w-H_0^1}}{\to}\Psi(\omega,\cdot)_{\lfloor H_0^1(\Omega)}$$

where

$$\Phi(\omega,u)=\begin{cases}a\displaystyle\int_\Omega D(\omega)\nabla u\cdot\nabla u\,dx+\frac{b}{2}\int_\Omega u^2dx & \text{if } u\in H_0^1(\Omega),\\ +\infty & \text{otherwise.}\end{cases}$$

$$\Psi(\omega,u)=\begin{cases}\displaystyle\int_\Omega D(\omega)\nabla u\cdot\nabla u\,dx & \text{if } u\in H_0^1(\Omega),\\ +\infty & \text{otherwise.}\end{cases}$$

Observe that the Γ-convergence of $\Psi_\varepsilon(\omega,\cdot)$ to $\Psi(\omega,\cdot)$ when $L^2(\Omega)$ is equipped with its strong topology, yields the Γ-convergence of $\Psi_\varepsilon(\omega,\cdot)_{\lfloor H_0^1(\Omega)}$ to $\Psi(\omega,\cdot)_{\lfloor H_0^1(\Omega)}$ when $H_0^1(\Omega)$ is equipped with its weak topology. This property is indeed a direct consequence of uniform coercivity:

$$\Psi_\varepsilon(\omega,u)\geq\alpha a\int_\Omega|\nabla u(x)|^2dx,\quad\text{for all } u\in H_0^1(\Omega)$$

(see Proposition B.5). Noticing that $\Phi_\varepsilon(\omega,\cdot)$ is a continuous perturbation of $a\Psi_\varepsilon(\omega,\cdot)$ by $\frac{b}{2}\|\cdot\|^2_{L^2(\Omega)}$, these two convergences are straightforward consequences of (Attouch *et al.*, 2014, Theorem 12.1.1, (ii), Theorem 12.4.7).

Finally, it is easily seen that the matrix $D^{hom}(\omega,\cdot)$ satisfies

$$\alpha|\xi|^2\leq\sum_{i,j=1}^N d_{i,j}^{hom}(\omega)\xi_i\xi_j\leq\beta|\xi|^2$$

for $\mathbb{P}$-a.e. $\omega\in\Sigma$, and all $\xi\in\mathbb{R}^N$. Hence $\mathrm{dom}(\partial\Phi(\omega))=H_0^1(\Omega)\cap H^2(\Omega)$. This completes the proof. □

10.2 Stochastic homogenization of nonlinear integrodifferential reaction-diffusion equations in one dimension space in the setting of a Poisson point process

We now consider the random environment described in Section 7.3.2 with $N=1$. Recall that the heterogeneities of size of order ε are intervals with radius $R>0$

whose centers are independently randomly distributed on a line with a given frequency λ, following a Poisson point process with intensity λ. This random environment is modeled by an ergodic dynamical system $(\Sigma, \mathcal{A}, \mathbb{P}_\lambda, (T_x)_{x\in\mathbb{R}})$ where $T_x\omega = \omega - x$ for all $(\omega, x) \in \Sigma \times \mathbb{R}$, and $\mathbb{P}_\lambda$ is characterized for every bounded Borel subset B of $\mathbb{R}$, every $k \in \mathbb{N}$, by

$$\mathbb{P}_\lambda(\#(\Sigma \cap B) = k) = \lambda^k \mathcal{L}_2(B)^k \frac{\exp(-\lambda \mathcal{L}_N(B))}{k!}.$$

Recall that $\mathbb{E}_\lambda[\#(\Sigma \cap B)] = \lambda \mathcal{L}_N(B)$.

Let Ω be an open bounded interval of $\mathbb{R}$. Let $\sigma^\pm$ be two scalar functions in $C^1(\mathbb{R})$ and $a^\pm$ two real numbers satisfying

$$0 < a^\pm \leq (\sigma^\pm)' \tag{10.4}$$

and set for all $\xi \in \mathbb{R}$

$$W^\pm(\xi) = \int_0^\xi \sigma^\pm(s)ds.$$

We assume that there exists $(\alpha, \beta) \in (\mathbb{R}_+^*)^2$ such that $\alpha\xi^2 \leq W^\pm(\xi) \leq \beta(1+\xi^2)$. Such a condition is fulfilled by assuming suitable conditions on $\sigma^\pm$, as for example growth conditions of order one. Given $R > 0$, for all $(\omega, x, \xi) \in \Sigma \times \mathbb{R} \times \mathbb{R}$, we define the random density W by

$$W(\omega, x, \xi) = \begin{cases} W^-(\xi) \text{ if } x \in \underset{i\in\mathbb{N}}{\cup} B_R(\omega_i), \\ W^+(\xi) \text{ otherwise,} \end{cases}$$

and the random integral functional $\Phi_\varepsilon : L^2(\Omega) \to [0, +\infty]$ by

$$\Phi_\varepsilon(\omega, u) = \begin{cases} \displaystyle\int_\Omega W\left(\omega, \frac{x}{\varepsilon}, \frac{du}{dx}(x)\right) dx & \text{if } u \in H_0^1(\Omega) \\ +\infty & \text{otherwise.} \end{cases}$$

It is easy to show that $\Phi_\varepsilon(\omega, \cdot)$ is a proper convex lsc functional with domain $H_0^1(\Omega)$, and that for all $\omega \in \Sigma$, its subdifferential (actually its distributional derivative) is given by

$$\begin{cases} \operatorname{dom}(\partial\Phi_\varepsilon(\omega,\cdot)) = \left\{ u \in H_0^1(\Omega) : \left(W'_\xi\left(\omega, \frac{\cdot}{\varepsilon}, \frac{du}{dx}\right)\right)' \in L^2(\Omega) \right\} \\ \partial\Phi_\varepsilon(\omega,\cdot) = -\left(W'_\xi\left(\omega, \frac{\cdot}{\varepsilon}, \frac{du}{dx}\right)\right)'. \end{cases}$$

On the other hand, we set for all $(\omega, x) \in \Sigma \times \mathbb{R}$,

$$a(\omega, x) = \begin{cases} a^-(x) \text{ if } x \in \underset{i\in\mathbb{N}}{\cup} B_R(\omega_i) \\ a^+(x) \text{ otherwise,} \end{cases}$$

and define the random quadratic integral functional $\Psi_\varepsilon : L^2(\Omega) \to [0, +\infty]$ by

$$\Psi_\varepsilon(\omega, u) = \begin{cases} \dfrac{1}{2}\displaystyle\int_\Omega a\left(\omega, \frac{x}{\varepsilon}\right)\left|\frac{du}{dx}(x)\right|^2 dx & \text{if } u \in H_0^1(\Omega) \\ +\infty & \text{otherwise.} \end{cases}$$

The subdifferential of $\Psi_\varepsilon(\omega,\cdot)$ (its distributional derivative) is given by

$$\begin{cases} \operatorname{dom}(\partial\Psi_\varepsilon(\omega,\cdot)) = \left\{u \in H_0^1(\Omega) : \left(a\left(\omega, \frac{\cdot}{\varepsilon}\right)\frac{du}{dx}\right)' \in L^2(\Omega)\right\} \\ \partial\Psi_\varepsilon(\omega,\cdot) = -\left(a\left(\omega,\frac{\cdot}{\varepsilon}\right)\frac{du}{dx}\right)'. \end{cases}$$

Condition (5.1) is uniformly fulfilled: take $\alpha_{\Psi_\varepsilon} = \frac{1}{2}\min(a^-, a^+)$. In the lemma below we state that (5.2) is uniformly satisfied for $\mathbb{P}_\lambda$ a.e. $\omega \in \Sigma$.

Lemma 10.1. *For $\mathbb{P}_\lambda$-a.e. $\omega \in \Sigma$, the subdifferentials $\partial\Phi_\varepsilon(\omega,\cdot)$ and $\partial\Psi_\varepsilon(\omega,\cdot)$ are connected as follows:*

$$\begin{cases} \operatorname{dom}(\partial\Phi_\varepsilon(\omega,\cdot)) \subset \operatorname{dom}(\partial\Psi_\varepsilon(\omega,\cdot)), \\ \langle\partial\Phi_\varepsilon(\omega,u), \partial\Psi_\varepsilon(\omega,u)\rangle \geq \|\partial\Psi_\varepsilon(\omega,u)\|^2_{L^2(\Omega)} \ \textit{for all } u \in \operatorname{dom}(\partial\Phi_\varepsilon(\omega,\cdot)). \end{cases}$$

Proof. For $\mathbb{P}_\lambda$-a.e. $\omega \in \Sigma$, set $\Omega_\varepsilon^-(\omega) := \Omega \cap [\frac{x}{\varepsilon} \in \cup_{i\in\mathbb{N}} B_R(\omega_i)]$ and $\Omega_\varepsilon^+(\omega) := \Omega \cap [\frac{x}{\varepsilon} \notin \cup_{i\in\mathbb{N}} B_R(\omega_i)]$. Let $u \in \operatorname{dom}(\partial\Phi_\varepsilon(\omega,\cdot))$, from (10.4) we have

$$\begin{aligned}
&\int_\Omega \left(a\left(\omega,\frac{\cdot}{\varepsilon}\right)\frac{du}{dx}\right)'^2 dx \\
&= \int_{\Omega_\varepsilon^-(\omega)} (a^-)^2\left(\frac{d^2u}{dx^2}\right)^2 dx + \int_{\Omega_\varepsilon^+(\omega)} (a^+)^2\left(\frac{d^2u}{dx^2}\right)^2 dx \\
&\leq \int_{\Omega_\varepsilon^-(\omega)} \left(\sigma^{-'}\left(\frac{du}{dx}\right)\frac{d^2u}{dx^2}\right)^2 dx + \int_{\Omega_\varepsilon^+(\omega)} \left(\sigma^{+'}\left(\frac{du}{dx}\right)\frac{d^2u}{dx^2}\right)^2 dx \\
&= \int_\Omega \left(W'_\xi\left(\omega,\frac{\cdot}{\varepsilon},\frac{du}{dx}\right)\right)'^2 dx < +\infty
\end{aligned}$$

so that $u \in \operatorname{dom}(\Psi_\varepsilon(\omega,\cdot))$.

Fix now $u \in \operatorname{dom}(\partial\Phi_\varepsilon(\omega,\cdot))$. We have from (10.4)

$$\begin{aligned}
&\langle\partial\Phi_\varepsilon(\omega,u), \partial\Psi_\varepsilon(\omega,u)\rangle \\
&= \int_\Omega \left(W'_\xi\left(\omega,\frac{x}{\varepsilon},\frac{du}{dx}\right)\right)'\left(a\left(\omega,\frac{x}{\varepsilon}\right)\frac{du}{dx}\right)' dx \\
&= \int_{\Omega_\varepsilon^-(\omega)} \left(\sigma^-\left(\frac{du}{dx}\right)\right)' a^-\frac{d^2u}{dx^2}dx + \int_{\Omega_\varepsilon^+(\omega)} \left(\sigma^+\left(\frac{du}{dx}\right)\right)' a^+\frac{d^2u}{dx^2}dx \\
&= \int_{\Omega_\varepsilon^-(\omega)} \sigma^{-'}a^-\left(\frac{d^2u}{dx^2}\right)^2 dx + \int_{\Omega_\varepsilon^+(\omega)} \sigma^{+'}a^+\left(\frac{d^2u}{dx^2}\right)^2 dx \\
&\geq \int_{\Omega_\varepsilon^-(\omega)} a^{-2}\left(\frac{d^2u}{dx^2}\right)^2 dx + \int_{\Omega_\varepsilon^+(\omega)} a^{+2}\left(\frac{d^2u}{dx^2}\right)^2 dx \\
&= \|\partial\Psi_\varepsilon(\omega,u)\|^2_{L^2(\Omega)}.
\end{aligned}$$

This completes the proof. □

Let K be a kernel as defined in Section 5.1, and a reaction functional defined as in the previous section with $N=1$, i.e. for all $(\omega,t)\in\Sigma\times[0,T]$, all $u\in L^2(\Omega)$ and all $x\in\Omega$,

$$F_\varepsilon(\omega,t,u)(x)=r\left(\omega,t,\frac{x}{\varepsilon}\right)\cdot g(u(x))+q\left(\omega,t,\frac{x}{\varepsilon}\right),$$

fulfilling the same conditions. Consider the random integrodifferential reaction-diffusion problem defined for $\mathbb{P}_\lambda$-a.e. $\omega\in\Sigma$ by

$$(\mathcal{P}_\varepsilon(\omega))\begin{cases}\dfrac{du_\varepsilon(\omega)}{dt}(t)+\partial\Phi_\varepsilon(\omega,u_\varepsilon(\omega,t))+\displaystyle\int_0^t K(t-s)\partial\Psi_\varepsilon(\omega,u_\varepsilon(\omega,s))\,ds\\ \qquad\qquad = F_\varepsilon(\omega,t,u_\varepsilon(\omega,t)) \text{ for a.e. } t\in(0,T),\\ u_\varepsilon(\omega,0)=u^0_\varepsilon(\omega)\in\operatorname{dom}(\partial\Phi_\varepsilon(\omega)).\end{cases}$$

From Proposition 5.2, $(\mathcal{P}_\varepsilon(\omega))$ admits a unique solution. A straightforward application of Theorem 5.5 yields

Corollary 10.2. *Assume that for $\mathbb{P}$ a.e. $\omega\in\Sigma$*

i) $\sup_{\varepsilon>0}\Phi_\varepsilon(u^0_\varepsilon(\omega))<+\infty$;

ii) $u^0_\varepsilon(\omega)\to u^0(\omega)$ *strongly in* $L^2(\Omega)$.

Then for $\mathbb{P}_\lambda$-a.e. $\omega\in\Sigma$, the solution $u_\varepsilon(\omega,\cdot)$ of $(\mathcal{P}_\varepsilon(\omega))$ converges in $C([0,T],L^2(\Omega))$ to the solution $u(\omega,\cdot)$ of the homogenized problem

$$(\mathcal{P}(\omega))\begin{cases}\dfrac{du}{dt}(\omega,t)-(\partial W^{hom}(u(\omega,t)))'-\displaystyle\int_0^t K(t-s)a^{hom}\frac{d^2u}{dx^2}(\omega,s)\,ds\\ \qquad\qquad \ni F(t,u(\omega,t)) \text{ for a.e. } t\in(0,T),\\ u(\omega,0)=u^0(\omega)\in\operatorname{dom}(\partial\Phi),\end{cases}$$

where W^{hom} is deterministic, given for all $\xi\in\mathbb{R}$ by

$$\begin{aligned}W^{hom}(\xi)&=\lim_{n\to+\infty}\inf\left\{\frac{1}{n^N}\int_{nY}W\left(\omega,y,\xi+\frac{du}{dy}(y)\right)dy:u\in H^1_0(nY)\right\}\\ &=\inf_{n\in\mathbb{N}^*}\mathbb{E}_\lambda\inf\left\{\frac{1}{n^N}\int_{nY}W\left(\cdot,y,\xi+\frac{du}{dy}(y)\right)dy:u\in H^1_0(nY)\right\}.\end{aligned}$$

The coefficient a^{hom} is given by

$$\begin{cases}a^{hom}=\dfrac{a^-a^+}{\theta a^-+(1-\theta)a^+},\\ \theta=1-\exp(2\lambda R)\end{cases}$$

and, $\partial\Phi^{hom}$, possibly multivalued,[1] is given by

$$\begin{cases}\operatorname{dom}(\partial\Phi^{hom})=\left\{v\in H^1_0(\Omega):\left(W^{hom'}\left(\dfrac{dv}{dx}\right)\right)'\in L^2(\Omega)\right\}\\ \partial\Phi^{hom}=-\left(W^{hom'}\left(\dfrac{dv}{dx}\right)\right)'.\end{cases}$$

[1] We use the convention of Remark 2.7.

The homogenized reaction functional is given for every $u \in L^2(\Omega)$, and all $(t,x) \in [0,T] \times \mathbb{R}$ by

$$F^{hom}(t,u)(x) = r^{hom}(t) \cdot g(u(x)) + q^{hom}(t),$$

$$r^{hom}(t) = \mathbb{E}_\lambda \left(\int_{(0,1)^N} r(\cdot,t,y)\, dy \right),$$

$$q^{hom}(t) = \mathbb{E}_\lambda \left(\int_{(0,1)^N} q(\cdot,t,y)\, dy \right).$$

Assume further that the Fenchel conjugate of $W^\pm$ satisfies condition (D_3^*). Then ∂W^{hom} is single valued and is the $\mathbb{P}_\lambda$ almost everywhere pointwise limit of $\partial W_n'(\omega,\cdot)$ where

$$W_n(\omega,\xi) = \inf \left\{ \frac{1}{n^N} \int_{nY} W\left(\omega, y, \xi + \frac{du}{dy}(y)\right) dy : u \in H_0^1(nY) \right\}.$$

Proof. The weak limit of the reaction term is obtained as in the proof of Corollary 10.1. In order to apply Theorem 5.5 it is enough to establish that for $\mathbb{P}_\lambda$-a.e. $\omega \in \Sigma$, the following variational convergences hold:

$$\Phi_\varepsilon(\omega) \overset{M}{\to} \Phi, \quad \Psi_\varepsilon(\omega)_{\lfloor H_0^1(\Omega)} \overset{\Gamma_{w-H_0^1}}{\to} \Psi$$

where

$$\Phi(u) = \begin{cases} \displaystyle\int_\Omega W^{hom}\left(\frac{du}{dx}\right) dx & \text{if } u \in H_0^1(\Omega) \\ +\infty & \text{otherwise,} \end{cases}$$

and

$$\Psi(u) = \frac{1}{2} \int_\Omega a^{hom} \left| \frac{du}{dx}(x) \right|^2 dx.$$

The first convergence is established in Chapter 7 in the proof of Theorem 7.1, ***Proof of*** (H_1). For the second convergence, note that for quadratic functionals in one dimension

$$F_\varepsilon(u) = \int_\Omega a_\varepsilon(x) \left| \frac{du}{dx}(x) \right|^2 dx,$$

with $0 \le \alpha \le a_\varepsilon \le \beta$, one has: $F_\varepsilon \overset{\Gamma_{w-H_0^1}}{\to} F$ if and only if $\frac{1}{a_\varepsilon}$ converges to $\frac{1}{a}$ for the $\sigma(L^\infty, L^1)$ topology, and F is given by

$$F(u) = \int_\Omega a(x) \left| \frac{du}{dx}(x) \right|^2 dx.$$

For a proof, see for instance (Attouch *et al.*, 2014, Theorem 12.3.1). Hence it remains to establish that for $\mathbb{P}_\lambda$-a.e. $\omega \in \Sigma$ the following convergence holds

$$\frac{1}{a\left(\omega, \frac{\cdot}{\varepsilon}\right)} \rightharpoonup \frac{\theta a^- + (1-\theta)a^+}{a^+ a^-} \quad \sigma(L^\infty, L^1).$$

This result is a direct consequence of the additive ergodic theorem (see (Attouch *et al.*, 2014, Theorem 12.4.2)) which states that for $\mathbb{P}_\lambda$-a.e. $\omega \in \Sigma$

$$\frac{1}{a\left(\omega, \frac{\cdot}{\varepsilon}\right)} \rightharpoonup \mathbf{E}_\lambda \int_{(0,1)} \frac{1}{a(\cdot, y)} dy.$$

An easy calculation gives

$$\mathbf{E}_\lambda \int_{(0,1)} \frac{1}{a(\cdot, y)} dy = \frac{\theta a^- + (1-\theta)a^+}{a^+ a^-}.$$

The last claim follows straightforwardly from Proposition 7.2, (iii). This completes the proof. □

Chapter 11

Stochastic homogenization of non diffusive reaction equations and memory effect

Let Ω be a bounded domain in $\mathbb{R}^N$, $N \in \mathbb{N}^*$, and $T > 0$. The main motivation of this chapter is the characterization of the weak limit in $H^1(0,T,L^2(\Omega))$ of the sequence $(u_\varepsilon(\omega,\cdot,\cdot))_{\varepsilon>0}$ as $\varepsilon \to 0$, where $u_\varepsilon(\omega,\cdot,\cdot)$ solves the non diffusive differential equation with a rapidly random oscillating reaction function

$$\begin{cases} -\dfrac{\partial u_\varepsilon}{\partial t}(\omega,t,x) = \dfrac{\partial \psi}{\partial \zeta}\left(\omega, \dfrac{x}{\varepsilon}, t, x, u_\varepsilon(\omega,t,x)\right), & \text{for a.e. } (t,x) \in (0,T)\times\Omega \\ u_\varepsilon(\omega,0,x) = u_0(x), \quad u_0 \in L^2(\Omega) \end{cases} \tag{11.1}$$

with $\omega \in \Sigma$. The reaction functional is deriving from a potential ψ which is a stationary random process, where $\zeta \mapsto \psi(\omega,y,t,x,\zeta)$ is strictly convex. As for all examples treated in Part 2, the stationarity expresses the fact that the evolution takes place in statistically homogeneous spatial environments (including periodic environments), and the scaling $\frac{x}{\varepsilon}$ accounts for small heterogeneities of size ε. We show that almost surely, $u_\varepsilon(\omega,\cdot,\cdot)$ weakly converges in $H^1(0,T,L^2(\Omega))$ towards the minimizer u of a functional which is non local in general, but which can be expressed as an inf-convolution of two integral functionals when the spatial environment presents two random phases. As a consequence, the limit u is a convex combination of the solutions of two non diffusive integro-differential equations. This illustrates in the scope of stochastic homogenization, the memory effect induced by the homogenization, phenomenon first pointed out in Mascharenas (1993), Tartar (1989, 1990), Toader (1999). This result is obtained as a corollary of a general theorem of homogenization specified below, which is itself a corollary of the principle of continuity, Theorem 6.1, established in Chapter 6.

Let us give some details on the content of this chapter. Let $(\Sigma,\mathcal{A},\mathbb{P},(T_z)_{z\in\mathbb{Z}^N})$ be a probability dynamical system that we assume to be ergodic, and $f : \Sigma\times\mathbb{R}^N\times[0,T]\times\mathbb{R}^N\times\mathbb{R}^2 \to \mathbb{R}$ a measurable map such that for all $\omega\in\Sigma$ and $y\in\mathbb{R}^N$, $f(\omega,y,\cdot,\cdot,\cdot,\cdot)$, belongs to $\mathcal{X}$, and which satisfies a covariance property with respect to $(T_z)_{z\in\mathbb{Z}^N}$ (for a relevant definition, see Section 11.1). In Proposition 11.1, we prove that the random Young measure μ_ε defined for every $\omega\in\Sigma$, by $\mu_\varepsilon(\omega) = dx\otimes\delta_{f(\omega,\frac{x}{\varepsilon})}$, almost surely narrow converges to the deterministic homogeneous Young measure $dx\otimes\mu^f$ where the constant disintegration probability measure $\mu_x := \mu^f$

on $\mathcal{X}$, is defined for every Borel set $\mathcal{B}(\mathcal{X})$ of $\mathcal{X}$ by

$$\mu^f(\mathcal{B}) = \int_\Sigma \left(\int_{(0,1)^N} \mathbb{1}_\mathcal{B}(f(\omega, y))dy \right) d\mathbb{P}(\omega).$$

Therefore, as a consequence of Theorem 6.1, Theorem 11.1 states the following stochastic homogenization result: for $\mathbb{P}$-a.e. ω in Σ, the random integral functional defined in $\mathcal{V}(0, T, L^2(\Omega))$ by

$$F_\varepsilon(\omega) : u \mapsto \int_{(0,T)\times\Omega} f\left(\omega, \frac{x}{\varepsilon}, t, x, u(t,x), \dot{u}(t,x)\right) dt \otimes dx + \int_\Omega \Theta(u((T,x)))\, dx$$

Γ-converges to the functional $F^{hom}(\omega)$ defined in $\mathcal{V}(0, T, L^2(\Omega))$ by

$$\begin{aligned} &F^{hom}(u) \\ &= \inf\left\{ \int_{(0,T)\times\Omega\times\Sigma\times Y} f\left(\omega, y, t, x, V(t,x,\omega,y), \dot{V}(t,x,\omega,y)\right) dt \otimes dx \otimes d\mathbb{P}(\omega) \otimes dy \right. \\ &\quad \left. + \int_{\Omega\times\Sigma\times Y} \Theta(V((T,x,\omega,y)))\, dx \otimes d\mathbb{P}(\omega) \otimes dy : V \in X_u^{hom} \right\} \end{aligned}$$

where

$$\begin{aligned} &X_u^{hom} \\ &= \left\{ V \in \mathcal{V}(0, T, L^2_{dx\otimes\mathbb{P}\otimes dy}(\Omega \times \Sigma \times Y)) : \int_{\Sigma\times Y} V(t,x,\omega,y) d\mathbb{P}(\omega) \otimes dy = u(t,x) \right\}. \end{aligned}$$

In the case when the ergodic dynamical system $(\Sigma, \mathcal{A}, \mathbb{P}, (T_z)_{z\in\mathbb{Z}^N})$ models a Poisson point process or a random checkerboard-like spatial environment, the homogenized functionals F^{hom} can be expressed as the inf-convolution of two integral functionals defined in $H^1((0,T), L^2(\Omega))$ (Corollary 11.1). When applying Corollary 11.1 to the specific sequence of functionals $(F_\varepsilon(\omega))_{\varepsilon>0}$ associated with the non diffusive reaction differential equation (11.1), we obtain that almost surely, the solution $u_\varepsilon(\omega, \cdot, \cdot)$ weakly converges to the unique minimizer of the corresponding functional F^{hom} (Corollary 11.2). In the context of a Poisson point process, or a random checkerboard-like spatial environment, this minimizer is expressed as a convex combination of the solutions of two non diffusive integro-differential equations (Corollary 11.3).

11.1 A general result of homogenization

In this section, $(\Sigma, \mathcal{A}, \mathbb{P}, (T_z)_{z\in\mathbb{Z}^N})$ is a probability dynamical system, assumed to be ergodic, and $f : \Sigma \times \mathbb{R}^N \times [0,T] \times \mathbb{R}^N \times \mathbb{R}^2 \to \mathbb{R}$ is a measurable function when $\Sigma \times \mathbb{R}^N \times [0,T] \times \mathbb{R}^N \times \mathbb{R}^2$ is equipped with the product σ-algebra (as usual, the topological spaces are equipped with their Borel σ-algebra). Let $\mathcal{X}$ be a Polish subspace of $(\mathbb{E}_{\alpha,\beta,\gamma}, \mathbf{d})$ (see Chapter 6); for all $\omega \in \Sigma$ and all y in $\mathbb{R}^N$, we assume that the function $f(\omega, y, \cdot, \cdot, \cdot, \cdot)$, denoted by $f(\omega, y)$, belongs to $\mathcal{X}$.

We assume that f satisfies the *covariance* property with respect to the dynamical system $(\Sigma, \mathcal{A}, \mathbb{P}, (T_z)_{z\in\mathbb{Z}^N})$: for all $z \in \mathbb{Z}^N$, for a.e. $x \in \mathbb{R}^N$, for all $\left(t,\zeta,\dot{\zeta}\right) \in [0,T]\times\mathbb{R}^2$ and for $\mathbb{P}$-a.e. $\omega \in \Sigma$,

$$f\left(T_z\omega, x, t, \zeta, \dot{\zeta}\right) = f\left(\omega, x+z, t, \zeta, \dot{\zeta}\right).$$

When the set $\mathcal{X}^{\mathbb{R}^N}$ is endowed with the product Borel σ-algebra denoted by $\widetilde{\mathcal{A}}$, it is easily seen that the map

$$\begin{aligned} \widetilde{f} : \Omega &\longrightarrow \mathcal{X}^{\mathbb{R}^N} \\ \omega &\mapsto (f(\omega,y))_{y\in\mathbb{R}^N} \end{aligned}$$

is $\left(\mathcal{A},\widetilde{\mathcal{A}}\right)$-measurable. For every $z \in \mathbb{Z}^N$, consider the shift map on $\mathcal{X}^{\mathbb{R}^N}$, namely the measurable map $\tau_z : \mathcal{X}^{\mathbb{R}^N} \to: \mathcal{X}^{\mathbb{R}^N}$ defined for every $\widetilde{\omega} \in \mathcal{X}^{\mathbb{R}^N}$ by

$$((\tau_z\widetilde{\omega})_x)_{x\in\mathbb{R}^N} = (\widetilde{\omega}_{x+z})_{x\in\mathbb{R}^N}.$$

Then it is easy to show that the *covariance* property implies that the law $\widetilde{\mathbb{P}} = \widetilde{f}\#\mathbb{P}$ of $\widetilde{f}$ is invariant under the group $(\tau_z)_{z\in\mathbb{Z}^N}$, that is $\tau_z\#\widetilde{\mathbb{P}} = \widetilde{\mathbb{P}}$ for all $z \in \mathbb{Z}^N$. We say that $\widetilde{f}$ is periodic in law. In what follows, to simplify the notation, we no longer distinguish $\widetilde{f}$ from f. Note that from ergodicity of $(\Sigma, \mathcal{A}, \mathbb{P}, (T_z)_{z\in\mathbb{Z}^N})$, we infer the ergodicity of the dynamical system $\left(\mathcal{X}^{\mathbb{R}^N}, \widetilde{\mathcal{A}}, f\#\mathbb{P}, (\tau_z)_{z\in\mathbb{Z}^N}\right)$.

For every $\omega \in \Sigma$ and every $\varepsilon > 0$, we consider the functional $F_\varepsilon(\omega) : \mathcal{V}(0,T,L^2(\Omega)) \to \mathbb{R}$ defined by

$$F_\varepsilon(\omega)(u) = \int_{(0,T)\times\Omega} f\left(\omega, \frac{x}{\varepsilon}, t, x, u(t,x), \dot{u}(t,x)\right) dxdt + \int_\Omega \Theta(u(T,x))dx.$$

Such integral functionals $F_\varepsilon(\omega)$ model potential energies of physical systems in evolution, with a rapidly random oscillating density with respect to the spatial variable (see for example (Anza Hafsa *et al.*, 2020c, Section 4.4)). The small parameter ε accounts for the size of spatial randomly distributed heterogeneities. The covariance property reflects the fact that these heterogeneities are statistically homogeneous.

For all $\omega \in \Sigma$, consider the Young measure $\mu_\varepsilon(\omega) \in \mathcal{Y}(\Omega,\mathcal{X})$ associated with $x \mapsto f\left(\omega, \frac{x}{\varepsilon}\right)$, i.e. $\mu_\varepsilon(\omega) = dx \otimes \delta_{f(\omega,\frac{x}{\varepsilon})}$. By using standard arguments, we can easily establish the measurability of the map

$$\begin{aligned} \Sigma &\longrightarrow \mathcal{Y}(\Omega,\mathcal{X}) \\ \omega &\mapsto \mu^\varepsilon(\omega) \end{aligned}$$

when $\mathcal{Y}(\Omega,\mathcal{X})$ is equipped with its Borel σ-algebras. Set $Y = (0,1)^N$, and denote by dy the restriction of the Lebesgue measure on Y. For any $\omega \in \Sigma$, we can consider the image on $\mathcal{X}$ of the measure dy by the function $f_Y(\omega) : y \mapsto f(\omega,y)$ from Y into $\mathcal{X}$, and the expectation of these probabilities, i.e. $\mathbb{E}(f_Y(\cdot)\#dy)$. Recall that this probability measure is defined for any $B \in \mathcal{B}(\mathcal{X})$, by

$$\begin{aligned} \mathbb{E}(f_Y(\cdot)\#dy)(B) &= \int_\Sigma \left(\int_Y \mathbb{1}_B(f_Y(\omega)(y))\, dy\right) d\mathbb{P}(\omega) \\ &= \int_{\Sigma\times Y} \mathbb{1}_B(f(\omega,y))d\mathbb{P}(\omega)\otimes dy, \end{aligned}$$

or, for any measurable function $\varphi : \mathcal{X} \to \mathbb{R}$, such that $\omega \mapsto \int_Y \varphi(f(\omega, y))dy$ belongs to $L^1_{\mathbb{P}}(\Sigma)$, by

$$\int_{\mathcal{X}} \varphi(\Lambda)\; d\mathbb{E}(f_Y(\cdot)\#dy)(\Lambda) = \mathbb{E}\int_Y \varphi(f(\cdot, y))dy. \tag{11.2}$$

Proposition 11.1. *For $\mathbb{P}$-a.e. ω in Σ, the sequence of Young measures $(\mu^\varepsilon(\omega))_{\varepsilon>0}$ narrow converges to the homogeneous random Young measure $\mu : \Omega \to \mathcal{Y}(\Omega, \mathcal{X})$ defined by $\mu = dx \otimes \mathbb{E}(f_Y(\cdot)\#dy)$.*

Proof. It is sufficient to establish the existence of a set Σ' of full probability of $\mathcal{A}$ such that, for all $\omega \in \Sigma'$,

$$\lim_{\varepsilon\to 0}\int_{\Omega\times\mathcal{X}} \psi(x,\Lambda)\; d\mu_\varepsilon(\omega)(x,\Lambda) = \int_{\Omega\times\mathcal{X}} \psi(x,\Lambda)\; dx \otimes d(\mathbb{E}(f_Y(\cdot)\#dy))(\Lambda)$$

when ψ has the form $\psi(x,\Lambda) = \mathbb{1}_A(x)\varphi(\Lambda)$, $A \in \mathcal{B}(\Omega)$ and φ belongs to a dense countable subset $\mathcal{D}$ of $\mathcal{C}_c(\mathcal{X})$ (see Valadier (1990, 1994)). According to the additive ergodic theorem, Theorem C.5, it is easy to establish that for $\mathbb{P}$-a.e. ω in Σ, $x \mapsto \varphi\left(f\left(\omega, \frac{x}{\varepsilon}\right)\right)$ almost surely weakly converges to $\mathbb{E}\left(\int_Y \varphi(f(\cdot,y))\; dy\right)(\omega)$ in $L^1(\Omega)$, hence, there exists N_φ in $\mathcal{A}$ with $\mathbb{P}(N_\varphi) = 0$, such that for all ω in $\Sigma \setminus N_\varphi$,

$$\begin{aligned}
\lim_{\varepsilon\to 0}\int_{\Omega\times\mathcal{X}} \psi(x,\Lambda)\; d\mu_\varepsilon(\omega) &= \lim_{\varepsilon\to 0}\int_\Omega \psi\left(x, f\left(\omega, \frac{x}{\varepsilon}\right)\right) dx \\
&= \lim_{\varepsilon\to 0}\int_\Omega \mathbb{1}_A(x)\varphi\left(f\left(\omega, \frac{x}{\varepsilon}\right)\right) dx \\
&= \int_\Omega \mathbb{1}_A(x)\mathbb{E}\left(\int_Y \varphi(f(\cdot,y))\; dy\right) dx \\
&= \int_\Omega\left(\int_{\mathcal{X}} \mathbb{1}_A(x)\varphi(\Lambda)\; d\mathbb{E}(f_Y(\cdot)\#dy)(\Lambda)\right) dx \\
&= \int_\Omega\left(\int_{\mathcal{X}} \psi(x,\Lambda)\; d\mathbb{E}(f_Y(\cdot)\#dy)(\Lambda)\right) dx \\
&= \int_{\Omega\times\mathcal{X}} \psi(x,\Lambda)\; dx \otimes d\mathbb{E}(f_Y(\cdot)\#dy)(\Lambda).
\end{aligned}$$

We have used (11.2) in the fourth equality. The thesis follows by considering the set of full probability $\Sigma' = \bigcup_{\varphi\in\mathcal{D}} (\Sigma \setminus N_\varphi)$. □

Theorem 11.1. *For $\mathbb{P}$-a.e. ω in Σ, the sequence of random integral functional $F_\varepsilon(\omega)$ Γ-converges as $\varepsilon \to 0$ to the non local functional $F^{hom}(\omega)$ defined in $\mathcal{V}(0,T,L^2(\Omega))$ by*

$$\begin{aligned}
&F^{hom}(u) \\
&= \inf\left\{\int_{(0,T)\times\Omega\times\Sigma\times Y} f(\omega,y,t,x,V(t,x,\omega,y),\dot{V}(t,x,\omega,y))\; dt \otimes dx \otimes d\mathbb{P}(\omega) \otimes dy\right. \\
&\qquad \left. + \int_{\Omega\times\Sigma\times Y} \Theta(V((T,x,\omega,y)))\; dx \otimes d\mathbb{P}(\omega) \otimes dy : V \in X_u^{hom}\right\}
\end{aligned} \tag{11.3}$$

where

$$X_u^{hom} = \Big\{V \in \mathcal{V}(0,T,L^2_{dx\otimes\mathbb{P}\otimes dy}(\Omega\times\Sigma\times Y)) : \int_{\Sigma\times Y} V(t,x,\omega,y)d\mathbb{P}(\omega)\otimes dy = u(t,x)\Big\} \tag{11.4}$$

and

$$\mathcal{V}(0,T,L^2_{dx\otimes\mathbb{P}\otimes dy}(\Omega\times\Sigma\times Y))$$

is the subspace of $H^1\left(0,T,L^2_{dx\otimes\mathbb{P}\otimes dy}(\Omega\times\Sigma\times Y)\right)$ *made up of functions* V *satisfying* $V(0,\cdot,\cdot,\cdot)=0$.

Proof. From Theorem 6.1, Proposition 11.1, and (11.2), for $\mathbb{P}$-a.e. ω in Σ, the sequence of functions $F_\varepsilon(\omega)$ Γ-converges to Φ_μ defined for every $u\in\mathcal{V}(0,T,L^2(\Omega))$, by

$$\begin{aligned}
&\Phi_\mu(u)\\
&= \inf_{U\in X_\mu(u)}\Big\{\int_{(0,T)\times\Omega}\mathbb{E}\left[\int_Y f(\cdot,y,t,U(t,x,f(\cdot,y)),\dot U(t,x,f(\cdot,y)))dy\right]dt\otimes dx\\
&\quad+\int_\Omega\mathbb{E}\left[\int_Y\Theta(U(T,x,f(\cdot,y)))\,dy\right]dx\Big\}\\
&= \inf_{U\in X_\mu(u)}\Big\{\int_{(0,T)\times\Omega\times\Sigma\times Y} f(\omega,t,y,U(t,x,f(\omega,y)),\dot U(t,x,f(\omega,y)))\,dt\otimes dxd\mathbb{P}dy\\
&\quad+\int_{\Omega\times\Sigma\times Y}\Theta(U(T,x,f(\omega,y)))\,dxd\mathbb{P}dy\Big\},
\end{aligned}\tag{11.5}$$

where, to shorten the notation $dxd\mathbb{P}dy$ stands for $dx\otimes d\mathbb{P}(\omega)\otimes dy$, and

$$X_\mu(u) = \Big\{U\in\mathcal{V}(0,T,L^2_{dx\otimes\mathbb{E}(f_Y(\cdot)\#dy)}(\Omega\times\mathcal{X})) : \int_{\Sigma\times Y}U(t,x,f(\omega,y))d\mathbb{P}(\omega)\otimes dy = u(t,x)\text{ a.e. }(t,x)\in[0,T]\times\Omega\Big\}.$$

The proof then consists in showing that $\Phi_\mu = F^{hom}$. For each $U\in X_\mu(u)$, set $V(t,x,\omega,y) = U(t,x,f(\omega,y))$ for all (t,x,ω,y). We claim that V belongs to X_u^{hom}. Indeed, since U belongs to $L^2\left(0,T,L^2_{dx\otimes\mathbb{E}(f_Y(\cdot)\#dy)}(\Omega\times\mathcal{X})\right)$, we have

$$\begin{aligned}
&\int_{(0,T)\times\Omega\times\Sigma\times Y}|V(t,x,\omega,y)|^2dt\otimes dx\otimes d\mathbb{P}(\omega)\otimes dy\\
&=\int_{(0,T)\times\Omega}\left(\int_{\Sigma\times Y}|U(t,x,f(\omega,y))|^2\,d\mathbb{P}(\omega)\otimes dy\right)dt\otimes dx\\
&=\int_{0,T}\left(\int_{\Omega\times\mathcal{X}}|U(t,x,\Lambda)|^2dx\otimes\mathbb{E}(f_Y(\cdot)\#dy)\right)dt<+\infty.
\end{aligned}$$

A similar calculation holds for $\dot{V}(t,x,\omega,y) = \dot{U}(t,x,f(\omega,y))$, and clearly $V(0,\cdot,\cdot,\cdot) = 0$. Thus v belongs to $\mathcal{V}\left(0,T,L^2_{dx\otimes\mathbb{P}\otimes dy}(\Omega\times\Sigma\times Y)\right)$. On the other hand

$$\int_{\Sigma\times Y} V(t,x,\omega,y)\,dy\otimes d\mathbb{P}(\omega) = \int_{\Sigma\times Y} U(t,x,f(\omega,y))\,dy\otimes d\mathbb{P}(\omega) = u(t,x)$$

for a.e. $(t,x)\in(0,T)\times\Omega$, which completes the claim.

Therefore, from (11.5), we deduce that

$$\Phi_\mu \geq F^{hom}.$$

Let us establish the converse inequality. Let $V\in X_u^{hom}$, we have to exhibit $U\in X_\mu(u)$ with $\mu = dx\otimes\mathbb{E}(f_Y(\cdot)\#dy)$, satisfying

$$\begin{aligned}
&\int_{(0,T)\times\Omega\times\mathcal{X}} \Lambda\left(t,x,U(t,x,\Lambda),\dot{U}(t,x,\Lambda)\right) dt\otimes d\mu(x,\Lambda)\\
&\quad+\int_{\Omega\times\mathcal{X}} \Theta(U(T,x,\Lambda))\,d\mu(x,\Lambda)\\
&\leq \int_{(0,T)\times\Omega\times\Sigma\times Y} f\left(\omega,y,t,x,V(t,x,\omega,y),\dot{V}(t,x,\omega,y)\right) dt\otimes dx\otimes d\mathbb{P}(\omega)\otimes dy\\
&\quad+\int_{\Omega\times\Sigma\times Y} \Theta(V((T,x,\omega,y)))\,dx\otimes d\mathbb{P}(\omega)\otimes dy.
\end{aligned}$$

Define the Borel measure η in $\mathcal{X}\times(\Sigma\times Y)$ by $\eta = G\#(\mathbb{P}\otimes dy)$ where G is the map

$$\begin{aligned}
&G:\Sigma\times Y\to\mathcal{X}\times(\Sigma\times Y)\\
&(\omega,y)\mapsto G(\omega,y) = (f(\omega,y,\cdot,\cdot,\cdot),\omega,y).
\end{aligned}$$

Then, the measure projection $\pi_{\mathcal{X}}\#\eta$ of η on $\mathcal{X}$ is the measure $\mathbb{E}(f_Y(\cdot)\#dy)$, that is μ_x. According to the disintegration theorem, Theorem E.11, there exists a family $(\eta_\Lambda)_{\Lambda\in\mathcal{X}}$ of probability measures on $\Sigma\times Y$ such that $\eta = \mathbb{E}(f_Y(\cdot)\#dy)\otimes\eta_\Lambda$, i.e. $\eta = \mu_x\otimes\eta_\Lambda$. To sum up, we have

$$\begin{cases}\eta = G\#(\mathbb{P}\otimes dy),\\ \eta = \mu_x\otimes\eta_\eta,\ \mu_x = \mathbb{E}(f_Y(\cdot)\#dy).\end{cases} \tag{11.6}$$

Set

$$U(t,x,\Lambda) := \int_{\Sigma\times Y} V(t,x,\omega,y)\,d\eta_\Lambda(\omega,y). \tag{11.7}$$

We leave it to the reader to justify that for every $(t,x,\Lambda)\in(0,T)\times\Omega\times\mathcal{X}$,

$$\begin{aligned}
\dot{U}(t,x,\Lambda) &:= \int_{\Sigma\times Y}\dot{V}(t,x,\omega,y)\,d\eta_\Lambda(\omega,y),\\
U(T,x,\Lambda) &:= \int_{\Sigma\times Y} V(T,x,\omega,y)\,d\eta_\Lambda(\omega,y).
\end{aligned}$$

First we have to establish that $U \in X_\mu(u)$. Clearly, $U(0,t,\Lambda) = 0$ for $dt \otimes \mu_x$-a.e. (t,Λ) in $(0,T) \times \mathcal{X}$. On the other hand, from (11.7), (11.6), and since $\mu = dx \otimes \mu_x$, we have by Jensen inequality,

$$
\begin{aligned}
&\int_{(0,T)} \left(\int_{\Omega\times\mathcal{X}} |U(t,x,\Lambda)|^2 d\mu(x,\Lambda) \right) dt \\
&= \int_{(0,T)} \left(\int_{\Omega\times\mathcal{X}} \left| \int_{\Sigma\times Y} V(t,x,\omega,y)\, d\eta_\Lambda(\omega,y) \right|^2 d\mu(x,\Lambda) \right) dt \\
&\overset{\text{Jensen}}{\leq} \int_{(0,T)} \left(\int_{\Omega\times\mathcal{X}\times\Sigma\times Y} |V(t,x,\omega,y)|^2\, d\mu(x,\Lambda) \otimes d\eta_\Lambda(\omega,y) \right) dt \\
&= \int_{(0,T)} \left(\int_{\Omega\times\mathcal{X}\times\Sigma\times Y} |V(t,x,\omega,y)|^2\, dx \otimes \mu_x \otimes d\eta_\Lambda(\omega,y) \right) dt \\
&= \int_{(0,T)} \left(\int_{\Omega\times\mathcal{X}\times\Sigma\times Y} |V(t,x,\omega,y)|^2\, dx \otimes d\eta(\Lambda,\omega,y) \right) dt \\
&= \int_{(0,T)} \left(\int_{\Omega\times\Sigma\times Y} |V(t,x,\omega,y)|^2\, dx \otimes \mathbb{P}(\omega) \otimes dy \right) dt
\end{aligned}
$$

which is finite because $V \in \mathcal{V}\left(0,T,L^2_{dx\otimes\mathbb{P}\otimes dy}(\Sigma\times\Omega\times Y)\right)$. A similar calculation leads to

$$
\int_{(0,T)} \left(\int_{\Omega\times\mathcal{X}} |\dot{U}(t,x,\Lambda)|^2 d\mu(x,\Lambda) \right) dt < +\infty.
$$

On the other hand we have for a.e. $(t,x) \in (0,T) \times \Omega$,

$$
\begin{aligned}
\int_{\mathcal{X}} U(t,x,\Lambda)\, d\mu_x(\Lambda) &= \int_{\mathcal{X}} \left(\int_{\Sigma\times Y} V(t,x,\omega,y)\, d\eta_\Lambda(\omega,y) \right) d\mu_x(\Lambda) \\
&= \int_{\mathcal{X}\times\Sigma\times Y} V(t,x,\omega,y)\, d\eta(\Lambda,\omega,y) \\
&= \int_{\Sigma\times Y} V(t,x,\omega,y)\, d\mathbb{P}(\omega) \otimes dy = u(t,x)
\end{aligned}
$$

because v belongs to X_u^{hom}, which completes the claim.

Finally, from (11.7 and (11.6) we have

$$
\begin{aligned}
&\int_{(0,T)\times\Omega\times\mathcal{X}} \Lambda\left(t,x,U(t,x,\Lambda),\dot U(t,x,\Lambda)\right) dt\otimes d\mu(x,\Lambda)\\
&\quad+\int_{\Omega\times\mathcal{X}} \Theta(U(T,x,\Lambda))d\mu(x,\Lambda)\\
&=\int_{(0,T)\times\Omega\times\mathcal{X}} \Lambda\left(t,x,\int_{\Sigma\times Y}\left(V(t,x,\omega,y),\dot V(t,x,\omega,y)\right)d\eta_\Lambda(\omega,y)\right) dt\otimes d\mu(x,\Lambda)\\
&\quad+\int_{\Omega\times\mathcal{X}} \Theta\left(\int_{\Sigma\times Y}(V(t,x,\omega,y)d\eta_\Lambda(\omega,y))\right) d\mu(x,\Lambda)\\
&\overset{\text{Jensen}}{\leq}\int_{(0,T)\times\Omega\times\mathcal{X}\times\Sigma\times Y} \Lambda\left(t,x,V(t,x,\omega,y),\dot V(t,x,\omega,y)\right) dt\otimes d\mu(x,\Lambda)\otimes d\eta_\Lambda(y)\\
&\quad+\int_{\Omega\times\mathcal{X}\times\Sigma\times Y} \Theta(V(T,x,\omega,y))d\mu(x,\Lambda)\otimes d\eta_\Lambda(y)\\
&=\int_{(0,T)\times\Omega\times\mathcal{X}\times\Sigma\times Y} \Lambda\left(t,x,V(t,x,\omega,y),\dot V(t,x,\omega,y)\right) dt\otimes dx\otimes d\eta(\Lambda,\omega,y)\\
&\quad+\int_{\Omega\times\mathcal{X}\times\Sigma\times Y} \Theta(V(T,x,\omega,y))dx\otimes d\eta(\Lambda,\omega,y)\\
&\leq\int_{(0,T)\times\Omega\times\Sigma\times Y} f\left(\omega,y,t,x,V(t,x,\omega,y),\dot V(t,x,\omega,y)\right) dt\otimes dx\otimes d\mathbb{P}(\omega)\otimes dy\\
&\quad+\int_{\Omega\times\Sigma\times Y} \Theta(V(T,x,\omega,y))dx\otimes d\mathbb{P}(\omega)\otimes dy
\end{aligned}
$$

which is the desired estimate and completes the proof. □

In the case when the dynamical system $\left(\mathcal{X}^{\mathbb{R}^N},\widetilde{\mathcal{A}},f\#\mathbb{P},(\tau_z)_{z\in\mathbb{Z}^N}\right)$ models a random checkerboard-like or a Poisson point process environment, we can specify the functional F^{hom}. We are going to illustrate this fact in the case of the latter which has already been described in details in Section 7.3.2. Let $\lambda>0$. Let Σ be the set of all locally finite sequences $(\omega_i)_{i\in\mathbb{N}}$ of $\mathbb{R}^N$, and $\mathcal{M}$ the set of all countable sums of locally finite families of Dirac measures, equipped with its standard σ-algebra. Recall that there exists a probability measure $\mathbb{P}_\lambda$ (the law of the Poisson point process) on Ω such that the random measure

$$\Sigma\to\mathcal{M},\ \omega\mapsto\mathcal{N}(\omega,\cdot)=\sum_{i\in\mathbb{N}}\delta_{\omega_i}$$

satisfies: for every bounded Borel set A in $\mathbb{R}^N$ and every $k\in\mathbb{N}$

$$\mathbb{P}_\lambda(\{\mathcal{N}(.,A)=k\})=[\lambda\hat{\mathcal{L}}(A)]^k\,\frac{\exp(-\lambda\mathcal{L}_N(A))}{k!},$$

so that $\mathbb{E}_\lambda(\mathcal{N}(.,A))=\lambda\mathcal{L}_N(A)$, and for every disjoint bounded Borel sets A, B, $\mathcal{N}(.,A)$ and $\mathcal{N}(.,B)$ are independent.

Let $r > 0$ and g, h be two elements of $\mathcal{X}$. Let $B(\omega_i, r)$ denote the open ball in $\mathbb{R}^N$ with center $\omega_i \in \mathbb{R}^N$ and radius r. Now consider the random function defined for all $(\omega, x) \in \Sigma \times \Omega$ by

$$f(\omega, x) = \begin{cases} g \text{ if } x \in \bigcup_{i \in \mathbb{N}} B(\omega_i, r), \\ h \text{ otherwise,} \end{cases}$$

or equivalently

$$f(\omega, x) = h + (g - h) \min(1, \mathcal{N}(\omega, B(x, r))).$$

Set

$$f_\varepsilon(\omega, x) = f\left(\omega, \frac{x}{\varepsilon}\right) = h + (g - h) \min\left(1, \mathcal{N}\left(\omega, B\left(\frac{x}{\varepsilon}, r\right)\right)\right)$$

for all $\varepsilon > 0$, all $x \in \mathbb{R}^N$ and all $\omega \in \Sigma$. Then, according to basic probabilistic arguments, one can show that the probability dynamical system $\left(\mathcal{X}^{\mathbb{R}^N}, \widetilde{\mathcal{A}}, f\#\mathbb{P}, (\tau_z)_{z \in \mathbb{Z}^N}\right)$ is ergodic. It models environments whose heterogeneities are independently distributed with a frequency λ.

Set $\theta = \mathbb{P}_\lambda([\mathcal{N}(\cdot, B(0, r)) \geq 1])$, i.e. $\theta = 1 - \exp(-\lambda \mathcal{L}_N(B(0, r)))$. As an application of Theorem 11.1, we have

Corollary 11.1. *For $\mathbb{P}_\lambda$-a.e. ω in Σ, the following assertions hold.*

i) The random integral functional $F_\varepsilon(\omega)$ Γ-converges as $\varepsilon \to 0$ to the inf-convolution (also called epi-sum) $(G + K_\theta)\Box(H + K_{1-\theta})$ of the functionals $G + K_\theta$ and $H + K_{1-\theta}$, defined for every u in $\mathcal{V}(0, T, L^2(\Omega))$ by

$$(G + K_\theta)\Box(H + K_{1-\theta})(u) = \inf_{u_1 + u_2 = u} \{(G + K_\theta)(u_1) + (H + K_{1-\theta})(u_2)\},$$

$$G(u) = \int_{(0,T)\times\Omega} \theta g\left(t, x, \frac{u}{\theta}, \frac{\dot{u}}{\theta}\right) dt \otimes dx,$$

$$H(u) = \int_{(0,T)\times\Omega} (1 - \theta) h\left(t, x, \frac{u}{1-\theta}, \frac{\dot{u}}{1-\theta}\right) dt \otimes dx,$$

$$K_\theta(u) = \int_\Omega \theta \Theta\left(\frac{u(T, x)}{\theta}\right) dx,$$

$$K_{1-\theta}(u) = \int_\Omega (1 - \theta)\Theta\left(\frac{u(T, x)}{1-\theta}\right) dx.$$

ii) Assume that for every $(t, x) \in \mathbb{R}_+ \times \mathbb{R}^N$, the functions $\left(\zeta, \dot{\zeta}\right) \mapsto g\left(t, x, \zeta, \dot{\zeta}\right)$, $\left(\zeta, \dot{\zeta}\right) \mapsto h\left(t, x, \zeta, \dot{\zeta}\right)$, and $\zeta \mapsto \Theta(\zeta)$ are strictly convex and belong to $C^1(\mathbb{R}^2)$, and $C^1(\mathbb{R})$ respectively. Then, the minimizer $u_\varepsilon(\omega)$ of $F_\varepsilon(\omega)$ weakly converges in $\mathcal{V}(0, T, L^2(\Omega))$ to the minimizer u of F^{hom}, given by $u = v + w$, where v and w satisfy the following system in $\mathcal{V}(0, T, L^2(\Omega))$,

$$(\mathcal{S}_{g,h}) \begin{cases} \dfrac{\partial g}{\partial \zeta}\left(t, x, \dfrac{v}{\theta}, \dfrac{\dot{v}}{\theta}\right) - \dfrac{\partial}{\partial t}\left(\dfrac{\partial g}{\partial \dot{\zeta}}\left(t, x, \dfrac{v}{\theta}, \dfrac{\dot{v}}{\theta}\right)\right) = 0 \\ \text{for a.e. } (t, x) \in (0, T) \times \Omega \\ \dfrac{\partial h}{\partial \zeta}\left(t, x, \dfrac{w}{1-\theta}, \dfrac{\dot{w}}{1-\theta}\right) - \dfrac{\partial}{\partial t}\left(\dfrac{\partial h}{\partial \dot{\zeta}}\left(t, x, \dfrac{\dot{w}}{1-\theta}, \dfrac{w}{1-\theta}\right)\right) = 0 \\ \text{for a.e. } (t, x) \in (0, T) \times \Omega, \end{cases}$$

subjected to the time-boundary conditions

$$(\partial \mathcal{S}_{g,h}) \begin{cases} \dfrac{d\Theta}{d\zeta}\left(\dfrac{v(T)}{\theta}\right) + \theta \dfrac{\partial g}{\partial \dot{\zeta}}\left(T, x, \dfrac{v(T)}{\theta}, \dfrac{\dot{v}(T)}{\theta}\right) = 0, \ v(0,x) = 0 \\ \textit{for a.e. } x \in \Omega, \\ \dfrac{d\Theta}{d\zeta}\left(\dfrac{w(T)}{1-\theta}\right) + (1-\theta)\dfrac{\partial h}{\partial \dot{\zeta}}\left(T, x, \dfrac{w(T)}{1-\theta}, \dfrac{\dot{w}(T)}{1-\theta}\right) = 0, \ w(0,x) = 0 \\ \textit{for a.e. } x \in \Omega. \end{cases}$$

Proof. Proof of i). For every $V \in X_u^{hom}$ we have

$$\begin{aligned} &\int_{(0,T)\times\Omega\times\Sigma\times Y} f\left(\omega, y, t, x, V(t,x,\omega,y), \dot{V}(t,x,\omega,y)\right) dt \otimes dx \otimes d\mathbb{P}(\omega) \otimes dy \\ &+ \int_{\Omega\times\Sigma\times Y} \Theta(V((T,x,\omega,y)))\, dx \otimes d\mathbb{P}(\omega) \otimes dy \\ &= \int_{(0,T)\times\Omega} \left(\int_{[\mathcal{N}(\cdot,B(\cdot,r))\geq 1]} g\left(t,x,V(t,x,\omega,y), \dot{V}(t,x,\omega,y)\right) d\mathbb{P}(\omega) \otimes dy \right) dt \otimes dx \\ &+ \int_{(0,T)\times\Omega} \left(\int_{[\mathcal{N}(\cdot,B(\cdot,r))= 0]} h\left(t,x,V(t,x,\omega,y), \dot{V}(t,x,\omega,y)\right) d\mathbb{P}(\omega) \otimes dy \right) dt \otimes dx \\ &+ \int_{\Omega} \left(\int_{[\mathcal{N}(\cdot,B(\cdot,r))\geq 1]} \Theta(V((T,x,\omega,y))) d\mathbb{P}(\omega) \otimes dy \right) dx \\ &+ \int_{\Omega} \left(\int_{[\mathcal{N}(\cdot,B(\cdot,r))= 0]} \Theta(V((T,x,\omega,y))) d\mathbb{P}(\omega) \otimes dy \right) dx \end{aligned} \tag{11.8}$$

Using Jensen's inequality, we infer that

$$\begin{aligned} &\int_{[\mathcal{N}(\cdot,B(\cdot,r))\geq 1]} g\left(t,x,V(t,x,\omega,y), \dot{V}(t,x,\omega,y)\right) d\mathbb{P}(\omega) \otimes dy \\ &\qquad \geq g\left(t, x, \frac{1}{\theta} \int_{[\mathcal{N}(\cdot,B(\cdot,r))\geq 1]} \left(V(t,x,\omega,y), \dot{V}(t,x,\omega,y)\right) d\mathbb{P}(\omega) \otimes dy\right) \end{aligned}$$

and

$$\begin{aligned} &\int_{[\mathcal{N}(\cdot,B(\cdot,r))= 0]} h\left(t,x,V(t,x,\omega,y), \dot{V}(t,x,\omega,y)\right) d\mathbb{P}(\omega) \otimes dy \\ &\qquad \geq h\left(t, x, \frac{1}{1-\theta} \int_{[\mathcal{N}(\cdot,B(\cdot,r))= 0]} \left(V(t,x,\omega,y), \dot{V}(t,x,\omega,y)\right) d\mathbb{P}(\omega) \otimes dy\right). \end{aligned}$$

Set

$$\begin{cases} u_1(x,t) = \displaystyle\int_{[\mathcal{N}(\cdot,B(\cdot,r))\geq 1]} V(t,x,\omega,y) d\mathbb{P}(\omega) \otimes dy, \ \text{for a.e. } (t,x) \in (0,T)\times\Omega \\ u_2(x,t) = \displaystyle\int_{[\mathcal{N}(\cdot,B(\cdot,r))= 0]} V(t,x,\omega,y) d\mathbb{P}(\omega) \otimes dy, \ \text{for a.e. } (t,x) \in (0,T)\times\Omega. \end{cases}$$

Then, $u_1 + u_2 = u$ in $\mathcal{V}(0,T,L^2(\Omega))$, and from above and (11.8) we obtain

$$F^{hom}(u) \geq (G + K_\theta)\square(H + K_{1-\theta})(u).$$

The converse inequality is obtained firstly by setting

$$V(t,x,\omega,y) = \begin{cases} \dfrac{u_1(t,x)}{\theta} \text{ if } \mathcal{N}(\omega, B(y,r)) \geq 1, \\ \\ \dfrac{u_2(t,x)}{1-\theta} \text{ if } \mathcal{N}(\omega, B(y,r)) = 0, \end{cases}$$

where u_1 and u_2 are any functions in $\mathcal{V}(0,T,L^2(\Omega))$ such that $u_1 + u_2 = u$, and secondly by taking V as admissible function in the formula expressing F^{hom}.

Proof of ii). The inf-convolution $(G + K_\theta)\square(H + K_{1-\theta})$ is a closed convex function as Γ-limit. Therefore, according to (Attouch *et al.*, 2014, Proposition 9.5.3), u minimizes $(G + K_\theta)\square(H + K_{1-\theta})$ iff $0 \in \partial((G + K_\theta)\square(H + K_{1-\theta}))(u)$. On the other hand, it is easy to show that $G + K_\theta$ and $H + K_{1-\theta}$ are Gâteaux-differentiable. More precisely, we have, for all $\xi \in \mathcal{V}(0,T,L^2(\Omega))$ and all $u \in \mathcal{V}(0,T,L^2(\Omega))$

$$\langle \partial G(v), \xi \rangle$$

$$= \int_{(0,T)\times\Omega} \theta \left(\frac{\partial g}{\partial \zeta}\left(t,x,\frac{v}{\theta},\frac{\dot{v}}{\theta}\right)\xi + \frac{\partial g}{\partial \dot{\zeta}}\left(t,x,\frac{v}{\theta},\frac{\dot{v}}{\theta}\right)\dot{\xi}\right) dt \otimes dx,$$

$$\langle \partial H(v), \xi \rangle$$

$$= \int_{(0,T)\times\Omega} (1-\theta) \left(\frac{\partial h}{\partial \zeta}\left(t,x,\frac{v}{1-\theta},\frac{\dot{v}}{1-\theta}\right)\xi + \frac{\partial h}{\partial \dot{\zeta}}\left(t,x,\frac{v}{1-\theta},\frac{\dot{v}}{1-\theta}\right)\dot{\xi}\right) dt \otimes dx,$$

$$\langle \partial K_\theta(v), \xi \rangle = \int_{(0,T)\times\Omega} \frac{d\Theta}{d\zeta}\left(\frac{v(T)}{\theta}\right)\dot{\xi} dt \otimes dx.$$

The two first equalities are obtained in a standard way. We are going to establish the formula expressing ∂K_θ. According to the convexity and the C^1-regularity of Θ, we have

$$\begin{aligned} K_\theta(v+\xi) &\geq K_\theta(v) + \int_\Omega \theta \frac{d\Theta}{d\zeta}\left(\frac{v(T)}{\theta}\right)\frac{\xi(T)}{\theta} dx \\ &= \int_\Omega \frac{d\Theta}{d\zeta}\left(\frac{v(T)}{\theta}\right)\left(\int_0^T \dot{\xi}(t)dt\right) dx \\ &= \int_{(0,T)\times\Omega} \frac{d\Theta}{d\zeta}\left(\frac{v(T)}{\theta}\right)\dot{\xi}\; dt \otimes dx, \end{aligned}$$

and the conclusion follows from the fact that $\xi \mapsto \int_{(0,T)\times\Omega} \frac{d\Theta}{d\zeta}\left(\frac{v(t)}{\theta}\right)\dot{\xi}\; dt \otimes dx$ is a linear continuous form in $\mathcal{V}(0,T,L^2(\Omega))$. By using the standard subdifferential calculus for convex functions, we know that $(G+K_\theta)\Box(H+K_{1-\theta})$ is Gâteaux-differentiable and that its subdifferential is given by the parallel sum (cf. Theorem F.12)

$$\partial((G+K_\theta)\Box(H+K_{1-\theta})) = \partial(G+K_\theta) \parallel \partial(H+K_{1-\theta})$$

(for the definition of $\partial(G+K_\theta) \parallel \partial(H+K_{1-\theta})$ see Definition F.14). Hence, from above $0 \in \partial((G+K_\theta)\Box(H+K_{1-\theta}))(u)$ iff there exists $(v,w) \in \mathcal{V}(0,T,L^2(\Omega))^2$ such that $v+w = u$ and $0 \in \partial(G+K_\theta)(v)\cap\partial(H+K_{1-\theta})(w)$, i.e. for all $\xi \in \mathcal{V}(0,T,L^2(\Omega))$,

$$\begin{cases} \displaystyle\int_{(0,T)\times\Omega} \theta\left(\frac{\partial g}{\partial \zeta}\left(t,x,\frac{v}{\theta},\frac{\dot v}{\theta}\right)\xi + \frac{\partial g}{\partial \dot\zeta}\left(t,x,\frac{v}{\theta},\frac{\dot v}{\theta}\right)\dot\xi\right) dt\otimes dx \\ \qquad\qquad\qquad\qquad \displaystyle + \int_{(0,T)\times\Omega} \frac{d\Theta}{d\zeta}\left(\frac{v(T)}{\theta}\right)\dot\xi\; dt\otimes dx = 0, \\ \displaystyle\int_{(0,T)\times\Omega} (1-\theta)\left(\frac{\partial h}{\partial \zeta}\left(t,x,\frac{v}{1-\theta},\frac{\dot v}{1-\theta}\right)\xi + \frac{\partial h}{\partial \dot\zeta}\left(t,x,\frac{v}{1-\theta},\frac{\dot v}{1-\theta}\right)\dot\xi\right) dt\otimes dx \\ \qquad\qquad\qquad\qquad \displaystyle + \int_{(0,T)\times\Omega} \frac{d\Theta}{d\zeta}\left(\frac{v(T)}{1-\theta}\right)\dot\xi\; dt\otimes dx = 0. \end{cases}$$

Taking $\xi(t,x) = a(t)b(x)$, where $a \in \mathcal{D}(0,T)$, $b \in \mathcal{D}(\Omega)$, and integrating by part with respect to the time variable, we obtain the system $(\mathcal{S}_{g,h})$. Taking $\xi(t,x) = a(t)b(x)$, where $a(t) = (\frac{t}{T})^n$, $b \in \mathcal{D}(\Omega)$, integrating by part with respect to the time variable, and letting $n \to +\infty$, we obtain the system $(\partial\mathcal{S}_{g,h})$. □

Remark 11.1. When $(\Sigma,\mathcal{A},\mathbb{P},(T_z)_{z\in\mathbb{Z}^N})$ models a random checkerboard-like environment, by using similar arguments, it is easy to show that the same result holds, where, this time, θ and $1-\theta$ denotes the probability presence of g and h respectively.

11.2 Stochastic homogenization of non diffusive reaction differential equations: emergence of memory effects

For $T > 0$, we consider the non diffusive reaction differential equation in $\mathcal{V}(0,T,L^2(\Omega))$ with a random reaction functional:

$$\begin{cases} \displaystyle -\frac{\partial u_\varepsilon}{\partial t}(\omega,t,x) = \frac{\partial \psi}{\partial \zeta}\left(\omega,\frac{x}{\varepsilon},t,x,u_\varepsilon(\omega,t,x)\right) \text{ for a.e. } (t,x)\in(0,T)\times\Omega \\ u_\varepsilon(\omega,0,x) = 0. \end{cases} \tag{11.9}$$

The potential $\psi : \Sigma\times\mathbb{R}^N\times[0,T]\times\mathbb{R}^N\times\mathbb{R}\to\mathbb{R}$, is a $\mathcal{A}\otimes\mathcal{B}(\mathbb{R}^N)\otimes\mathcal{B}([0,T])\otimes\mathcal{B}(\mathbb{R}^N)\otimes\mathcal{B}(\mathbb{R})$-measurable scalar function which satisfies the following properties:

- for $\mathbb{P}$-a.e. $\omega\in\Sigma$ and for all $(y,t,x,\zeta)\in\mathbb{R}^N\times[0,T]\times\mathbb{R}^N\times\mathbb{R}$, $\zeta\mapsto\psi(\omega,y,t,x,\zeta)$ is strictly convex of class C^1 and $\zeta\mapsto\frac{\partial\psi}{\partial\zeta}(\omega,y,t,x,\zeta)$ is a Lipschitz continuous function, uniformly with respect to (y,t,x);

• for $\mathbb{P}$-a.e. $\omega \in \Sigma$, $\psi(T_z\omega, y, t, x, \zeta) = \psi(\omega, y + z, t, x, \zeta)$ for all $(z, y, t, x, \zeta) \in \mathbb{Z}^N \times \mathbb{R}^N \times [0, T] \times \mathbb{R}^N \times \mathbb{R}$.

It is easy to show that for $\mathbb{P}$-a.e. $\omega \in \Sigma$, (11.9) admits a unique solution in $\mathcal{V}(0, T, L^2(\Omega))$ (apply for instance Theorem 2.3, see also Remark 6.2)).

From now on, we write $\psi(\omega, y)(t, x, \zeta)$ for $\psi(\omega, y, t, x, \zeta)$, and, for $\mathbb{P}$-a.e. $\omega \in \Sigma$, and for a.e. $y \in \mathbb{R}^N$, we make the following additional assumptions on $\psi(\omega, y)$: there exist $\alpha > 0$, $\beta > 0$, $\gamma : \mathbb{R}_+ \to \mathbb{R}_+$ with $\lim_{r\to 0} \gamma(r) = 0$ such that

$$\alpha(|\zeta|^2 - 1) \leq \psi(\omega, y)(t, x, \zeta) \leq \beta(1 + |\zeta|^2) \tag{11.10}$$

for all $(t, x, \zeta) \in [0, T] \times \mathbb{R}^N \times \mathbb{R}$, and

$$|\psi(\omega, y)(t_2, x_2, \zeta) - \psi(\omega, y)(t_1, x_1, \zeta)| \leq \gamma(|t_2 - t_1| + |x_2 - x_1|) \tag{11.11}$$

for all $\zeta \in \mathbb{R}$, all (t_1, t_2) in $[0, T]^2$, and all (x_1, x_2) in $\mathbb{R}^N \times \mathbb{R}^N$.

For every $u \in \mathcal{V}(0, T, L^2(\Omega))$, set

$$F^{hom}(u) = \inf\left\{\int_{(0,T)\times\Omega\times\Sigma\times Y} f(\omega, y, t, x, V(t, x, \omega, y), \dot{V}(t, x, \omega, y))\, dt \otimes dx \otimes d\mathbb{P}(\omega) \otimes dy + \frac{1}{2}\int_{\Omega\times\Sigma\times Y} |V(T, x, \omega, y)|^2\, dx \otimes d\mathbb{P}(\omega) \otimes dy : V \in X_u^{hom}\right\},$$

with $f\left(\omega, y, t, x, \zeta, \dot{\zeta}\right) = \psi(\omega, y, t, x, \zeta) + \psi^*\left(\omega, y, t, x, -\dot{\zeta}\right)$, where for all $(\omega, y, t, x) \in \Sigma \times \mathbb{R}^N \times (0, T) \times \mathbb{R}^N$, the function $\psi^*(\omega, y, t, x, \cdot)$ denotes the Legendre-Fenchel conjugate of $\psi(\omega, y, t, x, \cdot)$. From (11.10) and (11.11) we infer that $\psi(\omega, y)$ belongs to $\mathcal{X}$ for $\mathbb{P}$-a.e. $\omega \in \Sigma$ and all $y \in \mathbb{R}^N$. Observing that $\left(\zeta, \dot{\zeta}\right) \mapsto f\left(\omega, y, t, x, \zeta, \dot{\zeta}\right)$ is strictly convex, and that for $\mathbb{P}$-a.e. $\omega \in \Sigma$, $y \mapsto f(\omega, y, \cdot, \cdot, \cdot, \cdot)$, is Borel measurable, from Theorem 11.1, we deduce the following corollary.

Corollary 11.2. *For $\mathbb{P}$-a.e. $\omega \in \Sigma$, the solution $u_\varepsilon(\omega, \cdot, \cdot)$ of (11.9) weakly converges in $\mathcal{V}(0, T, L^2(\Omega))$ to the unique minimizer of F^{hom}.*

For $i = 1, 2$ let $(t, x, \zeta) \mapsto \psi_i(t, x, \zeta)$, be two Borel measurable functions from $(0, T) \times \mathbb{R}^N \times \mathbb{R}$ into $\mathbb{R}$, such that $\zeta \mapsto \psi_i(t, x, \zeta)$ is strictly convex of class C^1, and $\zeta \mapsto \frac{\partial \psi_i}{\partial \zeta}(t, x, \zeta)$ is Lipschitz continuous, uniformly with respect to all (t, x). Assume furthermore that each ψ_i satisfies (11.10), (11.11). Consider the ergodic dynamical system $\left(\mathcal{X}^{\mathbb{R}^N}, \widetilde{\mathcal{A}}, f\#\mathbb{P}, (\tau_z)_{z\in\mathbb{Z}^N}\right)$ modeling a random Poisson point process environment. Set $\theta = 1 - \exp(-\lambda\mathcal{L}_N(B(0, r)))$, and, for all $\omega \in \Sigma$ and every $y \in \mathbb{R}^N$,

$$\psi(\omega, y) = \begin{cases} \psi_1 & \text{if } y \in \bigcup_{i\in\mathbb{N}} B(\omega_i, r), \\ \psi_2 & \text{otherwise.} \end{cases}$$

Then from Corollary 11.1, $F^{hom} = (G + K_\theta)\Box(H + K_{1-\theta})$, where, in this specific case, G, H, K_θ and $K_{1-\theta}$ are the integral functionals defined in $\mathcal{V}(0,T,L^2(\Omega))$ by

$$\begin{aligned}
G(v) &= \int_{(0,T)\times\Omega} \theta\left(\psi_1\left(t,x,\frac{v}{\theta}\right) + \psi_1^*\left(t,\frac{-\dot v}{\theta}\right)\right) dt \otimes dx,\\
H(v) &= \int_{(0,T)\times\Omega} (1-\theta)\left(\psi_2\left(t,x,\frac{v}{1-\theta}\right) + \psi_2^*\left(t,\frac{-\dot v}{1-\theta}\right)\right) dt \otimes dx,\\
K_\theta(v) &= \frac{1}{2\theta}\int_\Omega |v(T,x)|^2 dx,\\
K_{1-\theta}(v) &= \frac{1}{2(1-\theta)}\int_\Omega |v(T,x)|^2 dx.
\end{aligned}$$

We deduce the following result which illustrates, in the scope of stochastic homogenization, the memory effect induced by the homogenization of some non-diffusive reaction equations.

Corollary 11.3. *For $\mathbb{P}$-a.e. $\omega \in \Sigma$, the solution $u_\varepsilon(\omega,\cdot)$ of (11.9) weakly converges in $\mathcal{V}(0,T,L^2(\Omega))$ as $\varepsilon \to 0$, to the unique minimizer u of F^{hom}, characterized by $u = v + w$ with*

$$(\mathcal{S}_\psi)\begin{cases}
\dfrac{\partial\psi_1}{\partial\zeta}\left(t,x,\dfrac{v}{\theta}\right) + \dfrac{\partial}{\partial t}\left(\dfrac{\partial\psi_1^*}{\partial\dot\zeta}\left(t,x,-\dfrac{\dot v}{\theta}\right)\right) = 0 & \text{for a.e. } (t,x)\in(0,T)\times\Omega\\
\dfrac{\partial\psi_2}{\partial\zeta}\left(t,x,\dfrac{w}{\theta}\right) + \dfrac{\partial}{\partial t}\left(\dfrac{\partial\psi_2^*}{\partial\dot\zeta}\left(t,x,-\dfrac{\dot w}{\theta}\right)\right) = 0 & \text{for a.e. } (t,x)\in(0,T)\times\Omega,
\end{cases}$$

and the time-boundary conditions

$$(\partial\mathcal{S}_\psi)\begin{cases}
\dfrac{v(T,x)}{\theta} - \theta\dfrac{\partial\psi_1^*}{\partial\dot\zeta}\left(T,x,-\dfrac{\dot v(T,x)}{\theta}\right) = 0,\ v(0,x) = 0 & \text{for a.e. } x\in\Omega,\\
\dfrac{w(T,x)}{1-\theta} - (1-\theta)\dfrac{\partial\psi_2^*}{\partial\dot\zeta}\left(T,x,-\dfrac{\dot w(T,x)}{1-\theta}\right) = 0,\ w(0,x) = 0 & \text{for a.e. } x\in\Omega.
\end{cases}$$

Equivalently, $u = \theta\widetilde{v} + (1-\theta)\widetilde{w}$ where $\widetilde{v}$ and $\widetilde{w}$ are the unique solutions of the integro-differential Cauchy problems

$$\begin{cases}
-\dfrac{\partial\widetilde v}{\partial t}(t,x) = \dfrac{\partial\psi_1}{\partial\zeta}\left(t,x,\dfrac{\widetilde v(T,x)}{\theta} + \displaystyle\int_t^T \dfrac{\partial\psi_1}{\partial\zeta}\left(\tau,x,\dfrac{\partial\psi_1}{\partial\zeta}(\tau,x,\widetilde v(\tau,x))\right)d\tau\right)\\
\text{for a.e. } (t,x)\in(0,T)\times\Omega\\
\widetilde v(x,0) = 0 \text{ for a.e. } x\in\Omega,
\end{cases}$$

$$\begin{cases}
-\dfrac{\partial\widetilde w}{\partial t}(t,x) = \dfrac{\partial\psi_2}{\partial\zeta}\left(t,x,\dfrac{\widetilde w(T,x)}{1-\theta} + \displaystyle\int_t^T \dfrac{\partial\psi_2}{\partial\zeta}\left(\tau,x,\dfrac{\partial\psi_2}{\partial\zeta}(\tau,x,\widetilde w(\tau,x))\right)d\tau\right)\\
\text{for a.e.} (t,x)\in(0,T)\times\Omega\\
\widetilde w(x,0) = 0 \text{ for a.e. } x\in\Omega.
\end{cases}$$

Proof. It remains to establish that (v, w) is solution of the system $((\mathcal{S}_\psi), (\partial\mathcal{S}_\psi))$ if and only if $\widetilde{v} = \frac{v}{\theta}$ and $\widetilde{w} := \frac{w}{1-\theta}$ satisfy each two integrodifferential equations above. By integrating over (t, T) the first equation of $(\mathcal{S}_\psi)$, we infer that for every (t, x)

$$\frac{\partial \psi_1^*}{\partial \dot{\zeta}}\left(t, x, -\frac{\dot{v}(x,t)}{\theta}\right) \\ = \frac{\partial \psi_1^*}{\partial \dot{\zeta}}\left(T, x, -\frac{\dot{v}(x,T)}{\theta}\right) - \int_T^t \frac{\partial \psi_1}{\partial \zeta}\left(\tau, x, \frac{v(x,\tau)}{\theta}\right) d\tau. \tag{11.12}$$

Recall (see (Attouch *et al.*, 2014, Theorem 9.5.1, Remark 9.5.2)), that

$$\frac{\partial \psi_1^*}{\partial \dot{\zeta}}(t, x, \cdot) = \left(\frac{\partial \psi_1}{\partial \zeta}\right)^{-1}(t, x, \cdot),$$

hence, from (11.12), we deduce

$$-\frac{\partial v}{\partial t}(t, x) \\ = \theta \frac{\partial \psi_1}{\partial \zeta}\left(t, x, \frac{\partial \psi_1^*}{\partial \dot{\zeta}}\left(T, x, -\frac{v(x,T)}{\theta}\right) - \int_T^t \frac{\partial \psi_1}{\partial \zeta}\left(\tau, x, \frac{\partial \psi_1}{\partial \zeta}\left(\tau, x, \frac{v(\tau,x)}{\theta}\right)\right) d\tau\right),$$

and, setting $\widetilde{v} = \frac{v}{\theta}$, the conclusion for $\widetilde{v}$ follows from the first equation of $(\partial\mathcal{S}_\psi)$ which yields

$$\frac{v(T,x)}{\theta^2} = \frac{\partial \psi_1^*}{\partial \dot{\zeta}}\left(T, x, -\frac{\dot{v}(T,x)}{\theta}\right)$$

for a.e. $x \in \Omega$. The conclusion for $\widetilde{w}$ is obtained by following the same arguments. The converse implication is obtained firstly by computing the derivative with respect to the time variable of each two integro-differential equations, giving the system $(\mathcal{S}_\psi)$; secondly by setting $t = T$ in these two integro-differential equations, giving $(\partial\mathcal{S}_\psi)$. □

In the case when the random function ψ is quadratic and does not depend on the time variable, the limit u can be expressed as seen in the example below.

Example 11.1. For every $T > 0$, consider the following non diffusive reaction differential equation in $\mathcal{V}(0, T, L^2(\Omega))$,

$$\begin{cases} -\dot{u}_\varepsilon(\omega, t) = \frac{\partial \psi}{\partial \zeta}(\omega, \frac{x}{\varepsilon}, x, u_\varepsilon(\omega, t)) \\ u_\varepsilon(\omega, 0) = 0, \end{cases} \tag{11.13}$$

with a potential function which does not explicitly depend on t, given for all $(\omega, y, x, \zeta) \in \Sigma \times \mathbb{R}^N \times \mathbb{R}^N \times \mathbb{R}$ by

$$\psi(\omega, y, x, \zeta) = \begin{cases} \dfrac{a_1(x)}{2}(\zeta - \zeta_1(x))^2 \text{ if } y \in \bigcup_{i \in \mathbb{N}} B(\omega_i, r), \\ \dfrac{a_2(x)}{2}(\zeta - \zeta_2(x))^2 \text{ otherwise,} \end{cases}$$

where for $i = 1, 2$, ζ_i is a bounded Borel function, and a_i is a Borel function satisfying $0 < \alpha' \leq a_i \leq \beta'$ for some constants α' and β' in $\mathbb{R}_+^*$. Then ψ_i^* is given by $\psi_i^*\left(x, \dot{\zeta}\right) = \frac{1}{2a_i(x)}\dot{\zeta}^2 + \zeta_i(x)\dot{\zeta}$ for all $\left(x, \dot{\zeta}\right) \in \mathbb{R}^N \times \mathbb{R}$. The systems $(\mathcal{S}_\psi)$ and $(\partial\mathcal{S}_\psi)$ become

$$\begin{cases} a_1(x)\dfrac{v}{\theta} - \dfrac{\ddot{v}}{a_1(x)\theta} = a_1(x)\zeta_1(x) & \text{for a.e. } (t,x) \in (0,T) \times \Omega, \\ a_2(x)\dfrac{w}{1-\theta} - \dfrac{\ddot{w}}{a_2(x)(1-\theta)} = a_2(x)\zeta_2(x) & \text{for a.e. } (t,x) \in (0,T) \times \Omega, \\ \dfrac{v(T)}{\theta} + \dfrac{\dot{v}(T)}{a_1(x)} = \theta\zeta_1(x),\ v(0,x) = 0 & \text{for a.e. } x \in \Omega, \\ \dfrac{w(T)}{1-\theta} + \dfrac{\dot{w}(T)}{a_2(x)} = (1-\theta)\zeta_2,\ w(0,x) = 0 & \text{for a.e. } x \in \Omega. \end{cases}$$

An easy calculation shows that v and w are given for all $t \in [0,T]$, and all $x \in \Omega$, by

$$\begin{cases} v(t,x) \\ = \theta\zeta_1(x)(1 - \exp(-a_1(x)t)) + c(x)(\exp(a_1(x)t) - \exp(-a_1(x)t)), \\ w(t,x) \\ = (1-\theta)\zeta_2(x)(1 - \exp(-a_2(x)t)) + d(x)(\exp(a_2(x)t) - \exp(-a_2(x)t)), \end{cases}$$

where $c(x)$ and $d(x)$ are constant with respect to t, depend on θ and T, and can be computed from the time-boundary conditions at $t = T$. Hence, the solution $u_\varepsilon(\omega,\cdot,\cdot)$ of (11.13) almost surely weakly converges in $\mathcal{V}(0,T,L^2(\Omega))$ to the function u, given by

$$\begin{aligned} u(t) &= \theta\zeta_1(x)(1 - \exp(-a_1(x)t)) + (1-\theta)\zeta_2(x)(1 - \exp(-a_2(x)t)) \\ &\quad + c(x)(\exp(a_1(x)t) - \exp(-a_1(x)t)) + d(x)(\exp(a_2(x)t) - \exp(-a_2(x)t)). \end{aligned}$$

It will be noticed that the barycenter of each solution of (11.13) at scale $\varepsilon = 1$ in $\bigcup_{i\in\mathbb{N}} B(\omega_i, r)$ and $\mathbb{R}^N \setminus \bigcup_{i\in\mathbb{N}} B(\omega_i, r)$ weighted by $\mathbb{P}([\mathcal{N}(\omega, B(0,r)) \geq 1])$ and $\mathbb{P}([\mathcal{N}(\omega, B(0,r)) = 0])$ respectively, is given for every $x \in \Omega$ by

$$t \mapsto \theta\zeta_1(x)(1 - \exp(-a_1(x)t)) + (1-\theta)\zeta_2(x)(1 - \exp(-a_2(x)t)).$$

Then this example shows that the solution $u_\varepsilon(\omega,\cdot,\cdot)$ of (11.13), almost surely weakly converges in $\mathcal{V}(0,T,L^2(\Omega))$ to this barycenter, perturbed by the function

$$t \mapsto +c(x)(\exp(a_1(x)t) - \exp(-a_1(x)t)) + d(x)(\exp(a_2(x)t) - \exp(-a_2(x)t)).$$

Appendix A

Grönwall type inequalities

The so called Grönwall's lemma or Grönwall-Bellman lemma, has many variants. We mainly use this version.

Lemma A.1. *Let $T > 0$, $\mathbf{m} \in L^1(0,T)$ such that $\mathbf{m} \geq 0$ a.e. in $(0,T)$, and $a \in \mathbb{R}_+$. Let $\phi : [0,T] \to \mathbb{R}$ be a continuous function satisfying*

$$\phi(t) \leq a + \int_0^t \phi(s)\mathbf{m}(s)ds \text{ for all } t \in [0,T].$$

Then

$$\phi(t) \leq a \exp\left(\int_0^t \mathbf{m}(s)ds\right) \text{ for all } t \in [0,T].$$

Proof. Set $\psi(t) = a + \int_0^t \phi(s)\mathbf{m}(s)ds$ so that $\phi(t) \leq \psi(t)$ for all $t \in [0,T]$. Since ψ is absolutely continuous, we have

$$\frac{d\psi}{dt}(t) = \phi(t)\mathbf{m}(t) \leq \psi(t)\mathbf{m}(t)$$

for a.e. $t \in (0,T)$. Hence

$$\frac{d}{dt}\left(\psi(t)\exp\left(-\int_0^t \mathbf{m}(s)ds\right)\right) \leq 0$$

for a.e. $t \in (0,T)$, so that the absolutely continuous function $t \mapsto \psi(t)\exp\left(-\int_0^t \mathbf{m}(s)ds\right)$ is nonincreasing in $[0,T]$. It follows that

$$\psi(t)\exp\left(-\int_0^t \mathbf{m}(s)ds\right) \leq \psi(0) = a$$

for all $t \in [0,T]$, which proves the claim. □

The following lemma generalizes the result stated in (Brezis, 1973, Lemma A.5).

Lemma A.2. *Let $T > 0$, $\mathbf{m} \in L^1(0,T)$ be such that $\mathbf{m} \geq 0$ a.e. in $(0,T)$. Let $p \in [1,+\infty)$ and $\phi : [0,T] \to [0,+\infty)$ be a continuous function satisfying*

$$\frac{1}{p}\phi^p(t) \leq \frac{1}{p}a^p + \int_0^t \phi^{p-1}(s)\mathbf{m}(s)ds \text{ for all } t \in [0,T]$$

where $a \in \mathbb{R}_+^*$, *and* $a \in \mathbb{R}_+$ *for* $p \in \mathbb{N}^*$. *Then*

$$\phi(t) \leq a + \int_0^t \mathbf{m}(s)ds \text{ for all } t \in [0,T].$$

Proof. We assume that $a > 0$ (for $p \in \mathbb{N}^*$ substitute $a + \varepsilon$ for a and make $\varepsilon \to 0$ in the last inequality of the proof). Set $\psi(t) = \frac{1}{p}a^p + \int_0^t \phi^{p-1}(s)\mathbf{m}(s)ds$. Then $\psi > 0$ and

$$\phi(s) \leq p^{\frac{1}{p}}\psi^{\frac{1}{p}}(s) \text{ for all } s \in [0,T]. \tag{A.1}$$

Hence, since ψ is absolutely continuous

$$\frac{d\psi}{dt}(s) = \mathbf{m}(s)\phi^{p-1}(s) \leq \mathbf{m}(s)p^{\frac{1}{q}}\psi^{\frac{1}{q}}(s) \text{ for a.e } s \in (0,T) \tag{A.2}$$

where q is the conjugate of p, i.e. $\frac{1}{p} + \frac{1}{q} = 1$. It follows from (A.2) that for a.e. $s \in (0,T)$

$$p^{-\frac{1}{q}}\frac{d\psi}{dt}(s)\psi^{-\frac{1}{q}}(s) \leq \mathbf{m}(s),$$

that is

$$p^{\frac{1}{p}}\frac{d}{dt}\left(\psi^{\frac{1}{p}}(s)\right) \leq \mathbf{m}(s).$$

Integrating over $(0,t)$, we infer that for all $t \in [0,T]$

$$p^{\frac{1}{p}}\psi^{\frac{1}{p}}(t) \leq p^{\frac{1}{p}}\psi^{\frac{1}{p}}(0) + \int_0^t \mathbf{m}(s)ds,$$

that is, according to (A.1), $\phi(t) \leq a + \int_0^t \mathbf{m}(s)ds$ for all $t \in [0,T]$. □

Appendix B

Basic notions on variational convergences

B.1 Γ-convergence

Let $(\mathbb{T}, \tau)$ be a topological space, $(F_n, F)_{n\in\mathbb{N}}$ a sequence of functionals mapping $\mathbb{T}$ into $\mathbb{R} \cup \{+\infty\}$. The following notion of convergence, equivalent to the convergence of the epigraph of F_n to the epigraph of F in the Kuratowski-Painlevé sense, is of central importance in Calculus of Variations and Homogenization theory.

Definition B.1. The sequence $(F_n)_{n\in\mathbb{N}}$ (sequentially) Γ-converges to F at x in $\mathbb{T}$ if and only if both following assertions hold:

(i) there exists a sequence $(x_n)_{n\in\mathbb{N}}$ of $\mathbb{T}$, converging to x, such that

$$F(x) \geq \limsup_{n\to+\infty} F_n(x_n),$$

(ii) for every sequence $(y_n)_{n\in\mathbb{N}}$, converging to x in $\mathbb{T}$,

$$F(x) \leq \liminf_{n\to+\infty} F_n(y_n).$$

When (i) and (ii) hold for every x in $\mathbb{T}$, we say that $(F_n)_n$ Γ-converges to F in $(\mathbb{T}, \tau)$ and we write $F = \Gamma\text{-}\lim F_n$.

The main interest of this concept is its variational nature made precise in item (i) below.

Proposition B.2. *Assume that $(F_n)_n$ Γ-converges to F.*

(i) Let $(x_n)_{n\in\mathbb{N}}$ be a sequence in $\mathbb{T}$ such that for every $n \in \mathbb{N}$ $F_n(x_n) \leq \inf\{ F_n(x) : x \in \mathbb{T} \} + \varepsilon_n$, where $\varepsilon_n > 0$, $\varepsilon_n \to 0$. Assume furthermore that $(x_n)_{n\in\mathbb{N}}$ is τ-relatively compact. Then any cluster point $\overline{x}$ is a minimizer of F and

$$\lim_{n\to+\infty} \inf_{x\in\mathbb{T}} F_n(x) = F(\overline{x}).$$

(ii) If $G : X \to \mathbb{R}$ is continuous, then $(F_n + G)_{n\in\mathbb{N}}$ Γ-converges to $F + G$.

For a proof and more about Γ-convergence, we refer the reader to Attouch (1984), Dal Maso (1993), Baides (2002).

B.2 Mosco-convergence

We now consider the case where $(\mathbb{T}, \tau)$ is a Banach space $(V, \|.\|)$. Being endowed with strong and weak topology, we have two notions of Γ-convergence. Given a sequence $(\Phi_n)_{n\in\mathbb{N}}$ of functionals $\Phi_n : V \to \mathbb{R} \cup \{+\infty\}$, according to Definition B.1, we denote by $\Gamma_w\text{-}\lim \Phi_n$ and $\Gamma_s\text{-}\lim \Phi_n$ the Γ-limits associated with the weak and the strong convergence in V respectively, when they exist.

Definition B.2 (Mosco-convergence). Let $(V, \|.\|)$ be a Banach space, and $(\Phi_n)_{n\in\mathbb{N}}$ a sequence of extended real-valued functions $\Phi_n : V \to \mathbb{R} \cup \{+\infty\}$. The sequence $(\Phi_n)_{n\in\mathbb{N}}$ Mosco converges to the extended real-valued function $\Phi : V \to \mathbb{R} \cup \{+\infty\}$ and we write $\Phi_n \overset{M}{\to} \Phi$ if

$$\Phi = \Gamma_w\text{-}\Phi_n = \Gamma_s\text{-}\Phi_n.$$

The argument, which naturally led us to introduce the Mosco convergence notion, is the bicontinuity of the Fenchel duality transformation in the context of convex functions. More precisely

Theorem B.2. *Let $(V, \|.\|)$ be a reflexive Banach space and $(\Phi_n, \Phi)_{n\in\mathbb{N}}$ a sequence of convex proper lower semicontinuous functions from V into $\mathbb{R} \cup \{+\infty\}$. The following statements are equivalent:*

(i) $\Phi_n \overset{M}{\to} \Phi$ *on* V;
(ii) $\Phi_n^* \overset{M}{\to} \Phi^*$ *on* V^*.

For a proof, we refer the reader to (Attouch *et al.*, 2014, Theorem 17.4.3).

The following Proposition whose proof is straightforward, states an equivalent formulation interesting from the practical point of view.

Proposition B.3. *Let $(V, \|.\|)$ be a reflexive Banach space, and $(\Phi_n, \Phi)_{n\in\mathbb{N}}$ a sequence of convex proper lower semicontinuous functions from V into $\mathbb{R} \cup \{+\infty\}$. The following statements are equivalent:*

(i) $\Phi_n \overset{M}{\to} \Phi$;
(ii) $\forall v \in V,\ \exists v_n \to v$ *such that* $\Phi_n(v_n) \to \Phi(v)$;
$\forall v \in V,\ \forall v_n \rightharpoonup v,\ \Phi(v) \le \liminf_{n\to+\infty} \Phi_n(v_n)$;
(iii) $\forall v \in V,\ \exists v_n \to v$ *such that* $\Phi_n(v_n) \to \Phi(v)$;
$\forall v^* \in V^*,\ \exists v_n^* \to v^*$ *such that* $\Phi_n^*(v_n) \to \Phi^*(v)$.

A sequence $(\Phi_n)_{n\in\mathbb{N}}$ of lower semicontinuous convex functions $\Phi_n : V \to \mathbb{R} \cup \{+\infty\}$ is said to be uniformly proper if there exists a sequence $(v_n^0)_{n\in\mathbb{N}}$ such that $\sup_{n\in\mathbb{N}} \Phi_n(v_n^0) < +\infty$. Uniformly proper sequences which Mosco-converge are uniformly bounded from below. More precisely

Lemma B.3. *Let $(V, \|\cdot\|)$ be a reflexive Banach space and $(\Phi_n)_{n\in\mathbb{N}}$ a sequence of uniformly proper lower semicontinuous convex functions $\Phi_n : V \to \mathbb{R}\cup\{+\infty\}$ which Mosco converges to some function $\Phi : V \to \mathbb{R}\cup\{+\infty\}$. Then there exists $r > 0$ such that*

$$\forall n \in \mathbb{N} \quad \forall v \in X \quad \|\Phi_n\| \geq -r(\|v\| + 1).$$

Proof. We provide a partial proof in which we assume that the sequence $(v_n^0)_{n\in\mathbb{N}}$ is bounded. For a complete proof, we refer the reader to (Attouch, 1984, Lemma 3.8). If we contradict the assertion, we obtain the existence of a subsequence $(n(k))_{k\in\mathbb{N}}$ with $n(k) \to +\infty$, and $(v_k)_{k\in\mathbb{N}}$ such that

$$\Phi_{n(k)}(v_k) < -k(\|v_k\| + 1).$$

Let v be a weak limit of a subsequence not relabeled of $(v_k)_{k\in\mathbb{N}}$. From Proposition B.3, assertion (ii), we have

$$\Phi(v) = \inf\left\{\liminf_{n\to+\infty} \Phi_n(w_n) : w_n \rightharpoonup v\right\} \leq \liminf_{k\to+\infty}(-k(\|v_k\| + 1)) = -\infty,$$

which contradict that $\Phi : V \to \mathbb{R}\cup\{+\infty\}$. $\square$

The next proposition states the connexion between the Mosco-convergence and the pointwise convergence of the Moreau-Yosida approximates. We first recall the definition of the Moreau-Yosida approximation of convex functions.

Definition B.3. Let $(V, \|.\|)$ be a reflexive Banach space and $\Phi : V \to \mathbb{R}\cup\{+\infty\}$ a convex lower semicontinuous and proper function. For $\lambda > 0$, the Moreau-Yosida approximation of index λ of Φ is the function $\Phi_\lambda : V \to \mathbb{R}$ defined for all $u \in V$ by

$$\Phi_\lambda(u) = \inf\left\{\Phi(v) + \frac{1}{2\lambda}\|u - v\|^2 : v \in V\right\}.$$

The subdifferential (cf. Definition B.5 below) of the Moreau-Yosida approximation of index λ of Φ is called Yosida approximation of index λ of $\partial\Phi$.

Proposition B.4. *Let $(V, \|\cdot\|_V)$ be a reflexive Banach space whose norm together with its dual is strictly convex, and such that weak convergence of sequences and convergence of their norms imply strong convergence. Let $(\Phi_n, \Phi)_{n\in\mathbb{N}}$ be a sequence of lower semicontinuous convex proper functions from V into $\mathbb{R}\cup\{+\infty\}$. The following assertions are equivalent:*

(i) $\Phi_n \overset{M}{\to} \Phi$.
(ii) $\forall \lambda > 0\ \forall u \in V\ \Phi_n^\lambda(u) \to \Phi^\lambda(u)$.

For a proof and other properties of Moreau-Yosida approximations, we refer the reader to (Attouch, 1984, Theorem 3.26) and to Dal Maso (1993).

B.3 Γ-convergence versus Mosco-convergence

This section is devoted to the following proposition.

Proposition B.5. *Let X and V be two Banach spaces with $V \hookrightarrow\hookrightarrow X$. Let $(\Psi_n, \Psi)_{n\in\mathbb{N}}$ be a sequence of convex proper lower semicontinuous functions from X into $\mathbb{R} \cup \{+\infty\}$ such that $\mathrm{dom}(\Psi_n) = \mathrm{dom}(\Psi) = V$ for all $n \in \mathbb{N}$. Assume that the sequence $(\Psi_{n\lfloor V})_{n\in\mathbb{N}}$ is equi-coercive in the following sense: for all $r \in \mathbb{R}$, there exists a weakly compact subset K_r of V such that for all $n \in \mathbb{N}$*

$$\left[\Psi_{n\lfloor V} \leq r\right] \subset K_r.$$

Then

$$\Psi_{n\lfloor V} \overset{\Gamma_{w\text{-}V}}{\to} \Psi_{\lfloor V} \Longrightarrow \Psi_n \overset{M}{\to} \Psi.$$

Proof. Assume that $\Psi_{n\lfloor V} \overset{\Gamma_{w\text{-}V}}{\to} \Psi_{\lfloor V}$. Let $u_n \in X$ and $u \in X$ such that $u_n \rightharpoonup u$ in X, and assume that $\liminf_{n\to+\infty} \Psi_n(u_n) < +\infty$. Then there exists a subsequence $(u_{\sigma(n)})_{n\in\mathbb{N}}$ of $(u_n)_{n\in\mathbb{N}}$ such that

$$\liminf_{n\to+\infty} \Psi_n(u_n) = \lim_{n\to+\infty} \Psi_{\sigma(n)}(u_{\sigma(n)}) = \lim_{n\to+\infty} \Psi_{\sigma(n)\lfloor V}(u_{\sigma(n)}). \tag{B.1}$$

According to the equi-coerciveness hypothesis and to the compact embedding $V \hookrightarrow\hookrightarrow X$, we can extract a subsequence (not relabeled) of $(u_{\sigma(n)})_{n\in\mathbb{N}}$ which weakly converges in V and strongly in X to some $v \in V$. Thus $v = u$ and $u_{\sigma(n)} \rightharpoonup u$ in V. Hence from (B.1) and $\Psi_{n\lfloor V} \overset{\Gamma_{w\text{-}V}}{\to} \Psi_{\lfloor V}$, we infer that

$$\Psi(u) = \Psi_{\lfloor V}(u) \leq \liminf_{n\to+\infty} \Psi_{\sigma(n)\lfloor V}(u_{\sigma(n)}) = \liminf_{n\to+\infty} \Psi_n(u_n). \tag{B.2}$$

If $\liminf_{n\to+\infty} \Psi_n(u_n) = +\infty$ there is nothing to prove.

Let $u \in X$ and assume that $\Psi(u) < +\infty$ so that $\Psi(u) = \Psi_{\lfloor V}(u)$. From $\Psi_{n\lfloor V} \overset{\Gamma_{w\text{-}V}}{\to} \Psi_{\lfloor V}$ and the compact embedding $V \hookrightarrow\hookrightarrow X$, we deduce that there exist a subsequence of $(\Psi_{n\lfloor V})_{n\in\mathbb{N}}$ and a subsequence of $(u_n)_{n\in\mathbb{N}}$ in V (non relabeled) such that $u_n \rightharpoonup u$ in V, $u_n \to u$ strongly in X, and satisfying

$$\lim_{n\to+\infty} \Psi_n(u_n) = \lim_{n\to+\infty} \Psi_{n\lfloor V}(u_n) = \Psi(u). \tag{B.3}$$

From (B.2) and (B.3) we deduce that there exists a subsequence of $(\Psi_n)_{n\in\mathbb{N}}$ such that $\Psi_n \overset{\mathrm{M}}{\to} \Psi$. This conclusion being valid for any subsequence of $(\Psi_n)_{n\in\mathbb{N}}$, we conclude that $\Psi_n \overset{\mathrm{M}}{\to} \Psi$, which completes the proof. □

B.4 Graph-convergence

Let us recall the classical notion of the Kuratowski-Painlevé convergence for sequence of sets: let $(A_n)_{n\in\mathbb{N}}$ be a sequence of subsets of a metric space (X, d), or more generally of a topological space. The lower limit of the sequence $(A_n)_{n\in\mathbb{N}}$ is the subset of X denoted by $\liminf A_n$ and defined by

$$\liminf A_n = \{x \in X : \exists x_n \to x, \forall n \in \mathbb{N}\ x_n \in A_n\}$$

The upper limit of the sequence $(A_n)_{n\in\mathbb{N}}$ is the subset of X denoted by $\limsup A_n$ and defined by

$$\limsup A_n = \{x \in X : \exists (n_k)_{k\in\mathbb{N}},\ \exists (x_k)_{k\in\mathbb{N}},\ \text{for all } k \in \mathbb{N}, x_k \in A_{n_k},\ x_k \to x\}.$$

The sets $\liminf A_n$ and $\limsup A_n$ are clearly two closed subsets of (X, d) satisfying

$$\liminf A_n \subset \limsup A_n.$$

The sequence $(A_n)_{n\in\mathbb{N}}$ is said to be convergent if the following equality holds:

$$\liminf A_n = \limsup A_n.$$

The common value A is called the limit of $(A_n)_{n\in\mathbb{N}}$ in the Kuratowski-Painlevé sense and denoted by $K\text{-}\lim A_n$. Therefore, by definition $A := K\text{-}\lim A_n$ if and only if

$$\limsup A_n \subset A \subset \liminf A_n,$$

so that $x \in A = K\text{-}\lim A_n$ if and only if the two following assertions hold:

$$\forall x \in A \quad \exists (x_n)_{n\in\mathbb{N}} \text{ such that } \forall n \in \mathbb{N} \quad x_n \in A_n \text{ and } x_n \to x;$$
$$\forall (n_k)_{k\in\mathbb{N}} \quad \forall (x_k)_{k\in\mathbb{N}} \text{ such that } \forall k \in \mathbb{N} \quad x_k \in A_{n_k} \quad x_k \to x \Longrightarrow x \in A.$$

One can show that there exists a topology which governs the Kuratowski-Painlevé set convergence if and only if the space (X, d) is locally compact. For a complete study of various types of set convergence and their associated topologies, namely Vietoris, Fell, Wijsmann, Attouch-Wets and Mosco convergence, we refer the reader to Beer (1993, 1994).

From now on $(V, \|.\|)$ is a Banach space and V^* is its topological dual space whose dual norm is denoted by $\|.\|_*$ and we recall that for $(u, u^*) \in V \times V^*$, we write $\langle u^*, u\rangle$ for $u^*(u)$. Given a multivalued operator $A : V \to 2^{V^*}$, for any $v \in V$ we write Av instead of $A(v)$. Let us recall some basic definitions

$\operatorname{dom} A = \{v \in X : Av \neq \emptyset\}$ denotes the domain of A;
$G(A) := \{(v, v^*) \in V \times V^* : v^* \in Av\}$ denotes the graph of A;
$R(A) := \{v^* \in V^* : \exists v \in V \quad v^* \in Av\}$ denotes the range of A.

We define the inverse operator $A^{-1} : V^* \to V$ of A by

$$A^{-1}(v^*) = \{v \in V : v^* \in Av\}.$$

Note that $\operatorname{dom}(A^{-1}) = R(A)$. Consider another multivalued operator $B : V \to 2^{V^*}$. The range of A with respect to B is the set

$$R_B(A) := \{v^* \in V^* : \exists v \in \operatorname{dom}(B) \quad v^* \in Av\}.$$

Definition B.4. An operator $A : V \to 2^{V^*}$ is said to be monotone, if $\langle u^* - v^*, u - v\rangle \geq 0$ whenever $(u, u^*) \in G(A)$ and $(v, v^*) \in G(A)$. It is maximal monotone, if it is monotone and if its graph is maximal among all the monotone operators mapping V to V^* when $V \times V^*$ is ordered by inclusion. An element (u, u^*) of $V \times V^*$ is said to be monotonically related to a monotone operator A provided

$$\langle u^* - v^*, u - v\rangle \geq 0 \text{ for all } (v, v^*) \in G(A).$$

A useful form of the definition of maximality for a monotone operator A is the following condition whose proof follows straightforwardly from the foregoing Definition B.4.

Proposition B.6. *Let $A : V \to 2^{V^*}$ be a monotone operator. Then A is maximal monotone if and only if whenever (u, u^*) is monotonically related to A then $u \in$ $\operatorname{dom} A$ and $u^* \in Au$.*

The most basic class of maximal monotone operators is the class of subdifferentials of convex functions (see (Attouch *et al.*, 2014, Theorem 17.4.1)).

Definition B.5. Let $(V, \|\cdot\|)$ be a normed space and $f : V \to \mathbb{R} \cup \{+\infty\}$ be a convex proper lower semicontinuous (lsc in short) function. We say that an element $u^* \in V^*$ belongs to the subdifferential of f at $u \in V$ if

$$\forall v \in V \quad f(v) \geq f(u) + \langle u^*, v - u\rangle_{(V^*,V)} \,.$$

We then write $u^* \in \partial f(u)$.

The terminology reflects the fact that when f is continuously differentiable and convex, the following inequality holds:

$$\forall v \in V \quad f(v) \geq f(u) + \langle \nabla f(u), v - u\rangle \,.$$

Moreover, this inequality characterizes $\nabla f(u)$. For this reason, when $u^* \in \partial f(u)$, we say either that u^* belongs to the subdifferential of f at u or that u^* is a subgradient of f at u. If $u^* \in \partial f(u)$, then necessarily $u \in \operatorname{dom} f$ (take $v_0 \in \operatorname{dom} f \neq \emptyset$, we have $f(v_0) - \langle u^*, v_0 - u\rangle \geq f(u)$ and $f(u) < +\infty$). It is important to notice that given $u \in \operatorname{dom} f$, the set $\partial f(u)$ may be empty. When the subdifferential is single valued, we say that f is Gâteaux-differentiable.

Given a sequence of operators, one can consider the lim inf and lim sup of the sequence of their graphs as subsets of $V \times V^*$. This leads to the following definition.

Definition B.6. A sequence $(A_n)_{n\in\mathbb{N}}$ of operators mapping V to V^* is said to be graph convergent to an operator $A : V \to 2^{V^*}$, if the sequence $(G(A_n))_{n\in\mathbb{N}}$ converges to the graph $G(A)$ of A in the sense of Kuratowski-Painlevé when $V \times V^*$ is endowed with the product norm.

From now on we identify the operators with their graphs so that we write A instead of $G(A)$ and $A = G\text{-}\lim_{n\to+\infty} A_n$ or $A_n \overset{G}{\to} A$ instead of $G(A) =$ $K\text{-}\lim_{n\to+\infty} G(A_n)$. When considering sequences of maximal monotone operators, thus subdifferentials, the definition of the graph convergence is reduced to

Proposition B.7. *Let $(A_n, A)_{n\in\mathbb{N}}$ be a sequence of maximal monotone operators mapping V to V^*. Then we have*

$$A = G\text{-}\lim_{n\to+\infty} A_n \iff A \subset \liminf_{n\to+\infty} A_n.$$

Proof. The only implication we have to establish is

$$A \subset \liminf_{n\to+\infty} A_n \Longrightarrow A = G\text{-}\lim_{n\to+\infty} A_n,$$

the converse being trivial. Thus, it remains to show that $\limsup A_n \subset A$ is automatically satisfied. Let $(u, u^*) \in \limsup A_n$, then there exists a subsequence $(n_k)_{k\in\mathbb{N}}$ of integers and $(u_k, u_k^*) \in A_{n_k}$ such that $(u_k, u_k^*) \to (u, u^*)$ in $V \times V^*$ as $k \to +\infty$.

In the other hand, since $A \subset \liminf A_n$, for all $(v, v^*) \in A$, there exists $(v_n, v_n^*) \in A_n$ such that $(v_n, v_n^*) \to (v, v^*)$ in $V \times V^*$. Passing to the limit in

$$\langle u_k^* - v_{n_k}^*, u_k - v_{n_k}\rangle \geq 0$$

(recall that A_{n_k} is monotone), we infer

$$\langle u^* - v^*, u - v\rangle \geq 0 \text{ for all } (v, v^*) \in A.$$

Therefore (u, u^*) is monotonically related to A and, according to Proposition B.6, $(u, u^*) \in A$, which completes the proof. □

Another equivalent formulation of the bicontinuity theorem, Theorem B.2, is the continuity of the "subdifferential map" when the class of maximal monotone operator is equipped with the graph-convergence.

Theorem B.3 (Attouch). *Let $(V, \|.\|)$ be a reflexive Banach space and $(\Phi_n, \Phi)_{n\in\mathbb{N}}$ a sequence of convex proper lower semicontinuous functions from V into $\mathbb{R}\cup\{+\infty\}$. The following statements are equivalent:*

(i) $\Phi_n \overset{M}{\to} \Phi$;

(ii) $\partial\Phi = G\text{-}\lim_{n\to+\infty} \partial\Phi_n$ *and the following normalisation condition (NC) holds: for every $n \in \mathbb{N}$ there exist $(u_n, u_n^*) \in \partial\Phi_n$ and $(\overline{u}, \overline{u}^*) \in \partial\Phi$ such that $u_n \to \overline{u}$ in V, $u_n^* \to \overline{u}^*$ in V^* and $\Phi_n(u_n) \to \Phi(\overline{u})$.*

For a proof, we refer the reader to (Attouch *et al.*, 2014, Section 17.4).

Denote by $A_n \overset{G_{s,s}}{\to} A$ the graph convergence in $V\times V^*$ of A_n to A when $V\times V^*$ is equipped with the strong product topology, by $A_n \overset{G_{w,s}}{\to} A$ the graph convergence in $V\times V^*$ of A_n to A when $V\times V^*$ is equipped with the weak-strong product topology, and by $A_n \overset{G_{s,w}}{\to} A$ the graph convergence in $V \times V^*$ of A_n to A when $V \times V^*$ is equipped with strong-weak product topology. On the other hand denote by $\Psi_n \overset{\Gamma_{s\text{-}V}}{\to} \Psi$ and $\Psi_n \overset{\Gamma_{w\text{-}V}}{\to} \Psi$ the sequential Γ-convergence of the functional $\Phi_n : V \to \mathbb{R}\cup\{+\infty\}$ toward the functional $\Phi : V \to \mathbb{R} \cup \{+\infty\}$ when V is equipped with its strong and weak topology respectively. The following theorem states the link between the variational convergence of convex functionals and the graph convergence of their subdifferentials. For a proof, refer to (Attouch, 1984, Theorems 3.66, 3.67, Proposition 3.68) or (Attouch *et al.*, 2014, Theorem 17.4.4).

Theorem B.4. *Let $(\Psi_n, \Psi)_{n\in\mathbb{N}}$ be a sequence of convex proper lower semicontinuous functions from V into $\mathbb{R}\cup\{+\infty\}$. Then the following implications hold:*

$$\Psi_n \overset{\Gamma_{s}-V}{\to} \Psi \Longrightarrow \partial\Psi_n \overset{G_{s,s}}{\to} \partial\Psi,$$

$$\Psi_n \overset{\Gamma_{w}-V}{\to} \Psi \Longrightarrow \partial\Psi_n \overset{G_{w,s}}{\to} \partial\Psi,$$

$$\Psi_n \overset{\Gamma_{w}-V}{\to} \Psi \text{ and } (\Psi_n)_{n\in\mathbb{N}} \text{ equi-coercive} \Longrightarrow \partial\Psi_n \overset{G_{s,w}}{\to} \partial\Psi.$$

Appendix C

Ergodic theory of subadditive processes

In what follows, a probability dynamical system $(\Sigma, \mathcal{A}, \mathbb{P}, (T_z)_{z\in\mathbb{Z}^N})$ is a probability space $(\Sigma, \mathcal{A}, \mathbb{P})$ endowed with a group $(T_z)_{z\in\mathbb{Z}^N}$ of $\mathbb{P}$-preserving transformations on $(\Sigma, \mathcal{A})$, i.e. a family of $(\mathcal{A}, \mathcal{A})$-measurable maps $T_z : \Sigma \to \Sigma$ satisfying:

$$T_z \circ T_{z'} = T_{z+z'},\ T_{-z} = T_z^{-1} \quad \text{for all } (z, z') \in \mathbb{Z}^N \times \mathbb{Z}^N;$$

$$T_z^{\#}\mathbb{P} = \mathbb{P} \quad \text{for all } z \in \mathbb{Z}^N.$$

We use the standard notation $T_z^{\#}\mathbb{P}$ to denote the image measure, or pushforward of $\mathbb{P}$ by T_z. The terminology "dynamical system" refers to the "evolution" of the elements (or alea) of Σ according to the group $(T_z)_{z\in\mathbb{Z}^N}$ as for standard differential dynamical systems, where for the latter, the semigroup $(S_t)_{t\geq 0}$ is generated by a gradient or a subdifferential of convex potential, which plays the role of the discrete group $(T_z)_{z\in\mathbb{Z}^N}$.

The probability dynamical system $(\Sigma, \mathcal{A}, \mathbb{P}, (T_z)_{z\in\mathbb{Z}^N})$ is said to be ergodic, if for all E of $\mathcal{A}$ we have

$$\forall z \in \mathbb{Z}^N \quad T_z E = E \Longrightarrow \mathbb{P}(E) = 0 \text{ or } \mathbb{P}(E) = 1.$$

A sufficient condition to ensure ergodicity is the so called mixing condition which expresses an asymptotic independence: for all sets E and F of $\mathcal{A}$

$$\lim_{|z|\to+\infty} \mathbb{P}(T_z E \cap F) = \mathbb{P}(E)\mathbb{P}(F). \tag{C.1}$$

Ergodicity is obtained from (C.1) by taking $E = F$. The defect of ergodicity is captured by the σ-algebra $\mathcal{I}$ of invariant sets of $\mathcal{A}$ by the group $(T_z)_{z\in\mathbb{Z}^N}$, i.e. $E \in \mathcal{I}$ iff $T_z E = E$ for all $z \in \mathbb{Z}^N$. We denote by $L^1_{\mathbb{P}}(\Sigma)$ the space of $\mathbb{P}$-integrable numerical functions. For $X \in L^1_{\mathbb{P}}(\Sigma)$, $\mathbb{E}^{\mathcal{I}}X$ denotes the conditional expectation of X given $\mathcal{I}$, that is to say, the unique $\mathcal{I}$-measurable function in $L^1_{\mathbb{P}}(\Sigma)$ satisfying for all $E \in \mathcal{I}$

$$\int_E \mathbb{E}^{\mathcal{I}}X(\omega)\, d\mathbb{P}(\omega) = \int_E X(\omega)\, d\mathbb{P}(\omega).$$

According to Proposition C.8 below, the function $\mathbb{E}^{\mathcal{I}}X$ is $(T_z)_{z\in\mathbb{Z}^N}$-invariant, i.e. for all $z \in \mathbb{Z}^N$

$$\mathbb{E}^{\mathcal{I}}X \circ T_z = \mathbb{E}^{\mathcal{I}}X.$$

Moreover, if the dynamical system $(\Sigma, \mathcal{A}, \mathbb{P}, (T_z)_{z\in\mathbb{Z}^N})$ is ergodic, then $\mathbb{E}^{\mathcal{I}}X$ is constant equal to the expectation value $\mathbb{E}X := \int_\Sigma X \, d\mathbb{P}$ of the function X.

Proposition C.8 (Invariance and $\mathcal{I}$-measurability). *A function $h : \Sigma \to \mathbb{R}$ is $\mathcal{I}$-measurable if and only if it is invariant under the group $(T_z)_{z\in\mathbb{Z}}$, i.e. $h \circ T_z = h$ for all $z \in \mathbb{Z}^N$.*

Proof. For the implication

$$(h \text{ is } \mathcal{I}\text{-measurable} \implies h \text{ is invariant}),$$

the claim is indeed the straightforward consequence of

$$T_z^{-1}h^{-1}(\{h(\omega)\}) = h^{-1}(\{h(\omega)\}) \iff h(T_z(\omega)) = h(\omega)$$

for all $z \in \mathbb{Z}^N$ and all $\omega \in \Sigma$. The other implication is immediate. □

We end this preamble by stating three basic convergence theorems concerning the conditional expectation with respect to any sub σ-algebra of $\mathcal{A}$.

Proposition C.9. *Let $\mathcal{F}$ be a sub σ-algebra of $\mathcal{A}$.*

i) Conditional monotone convergence theorem: if $(h_n)_{n\in\mathbb{N}}$ is a nonnegative and nondecreasing sequence of measurable functions $h_n : \Sigma \to [0, +\infty]$ with $h = \lim_{n\in\mathbb{N}} h_n$ $\mathbb{P}$-a.e. in Σ, then

$$\mathbb{E}^{\mathcal{F}} h_n \nearrow \mathbb{E}^{\mathcal{F}} h$$

$\mathbb{P}$-a.e. in Σ.

ii) Conditional Fatou's lemma: if $(h_n)_{n\in\mathbb{N}}$ is a nonnegative sequence of measurable functions $h_n : \Sigma \to [0, +\infty]$, then

$$\mathbb{E}^{\mathcal{F}}\left(\liminf_{n\to+\infty} h_n\right) \le \liminf_{n\to+\infty} \mathbb{E}^{\mathcal{F}} h_n$$

$\mathbb{P}$-a.e. in Σ.

iii) Conditional Lebesgue dominated convergence theorem: let $(h_n)_{n\in\mathbb{N}}$ be a sequence in $L^1_{\mathbb{P}}(\Sigma)$ such that $h_n \to h$, $\mathbb{P}$-a.e. in Σ, and assume that there exists $\widetilde{h} \in L^1_{\mathbb{P}}(\Sigma)$ such that $|h_n| \le \widetilde{h}$ for all $n \in \mathbb{N}$. Then $\mathbb{E}^{\mathcal{F}} h_n \to \mathbb{E}^{\mathcal{F}} h$, $\mathbb{P}$-a.e. in Σ.

Proof. Proof of i). Since $h_n \nearrow h$, we have that $(\mathbb{E}^{\mathcal{F}} h_n)_{n\in\mathbb{N}}$ is a nonnegative and nondecreasing sequence of functions $E^{\mathcal{F}} h_n : \Sigma \to [0, +\infty]$. Let $Y = \lim_{n\to+\infty} E^{\mathcal{F}} h_n$. We have to show that $Y = E^{\mathcal{F}} h$. Clearly Y is $\mathcal{F}$-measurable, as limit of $\mathcal{F}$-measurable functions. On the other hand, by the standard monotone convergence theorem, we have for all $A \in \mathcal{F}$

$$\int_\Sigma h \mathbb{1}_A \, d\mathbb{P} = \lim_{n\to+\infty} \int_\Sigma h_n \mathbb{1}_A \, d\mathbb{P} = \lim_{n\to+\infty} \int_\Sigma E^{\mathcal{F}} h_n \mathbb{1}_A \, d\mathbb{P} = \int_\Sigma Y \mathbb{1}_A \, d\mathbb{P}$$

which completes the proof of i).

Proof of ii). It suffices to apply the Conditional monotone convergence theorem to the sequence $(g_n)_{n\in\mathbb{N}}$ with $g_n = \inf_{k\geq n} g_k$.

Proof of iii). Applying conditional Fatou's lemma to the sequences $\left(h_n + \widetilde{h}\right)_{n\in\mathbb{N}}$ and $\left(\widetilde{h} - h_n\right)_{n\in\mathbb{N}}$, we obtain

$$\mathbb{E}^{\mathcal{F}}\left(h+\widetilde{h}\right) = \mathbb{E}^{\mathcal{F}}\left(\liminf_{n\to+\infty}\left(h_n+\widetilde{h}\right)\right) \leq \liminf_{n\to+\infty}\mathbb{E}^{\mathcal{F}}\left(h_n+\widetilde{h}\right),$$

$$\mathbb{E}^{\mathcal{F}}\left(\widetilde{h}-h\right) = \mathbb{E}^{\mathcal{F}}\left(\liminf_{n\to+\infty}\left(\widetilde{h}-h_n\right)\right) \leq \liminf_{n\to+\infty}\mathbb{E}^{\mathcal{F}}\left(\widetilde{h}-h_n\right).$$

Hence $\liminf_{n\to+\infty}\mathbb{E}^{\mathcal{F}}h_n \geq \mathbb{E}^{\mathcal{F}}h$ and $\limsup_{n\to+\infty}\mathbb{E}^{\mathcal{F}}h_n \leq \mathbb{E}^{\mathcal{F}}h$. □

C.1 Additive processes

We denote by $\mathcal{B}_b(\mathbb{R}^N)$ the family of bounded Borel subsets of $\mathbb{R}^N$.

Definition C.7 (Additive process). Let $(\Sigma, \mathcal{A}, \mathbb{P}, (T_z)_{z\in\mathbb{Z}^N})$ be a dynamical system. An additive process indexed by $\mathcal{B}_b(\mathbb{R}^N)$, and covariant with respect to the group $(T_z)_{z\in\mathbb{Z}^N}$, is a mapping $\mathbb{A} : \mathcal{B}_b(\mathbb{R}^N) \longrightarrow L^1_{\mathbb{P}}(\Sigma)$, $A \mapsto \mathbb{A}_A$, satisfying the three following conditions:

(i) for all $(A,B) \in \mathcal{B}_b(\mathbb{R}^N) \times \mathcal{B}_b(\mathbb{R}^N)$ such that $A\cap B = \emptyset$, $\mathbb{A}_{A\cup B} = \mathbb{A}_A + \mathbb{A}_B$;
(ii) for all $A \in \mathcal{B}_b(\mathbb{R}^N)$ and all $z \in \mathbb{Z}^N$, $\mathbb{A}_{z+A} = \mathbb{A}_A \circ T_z$;
(iii) there exists a nonnegative function h in $L^1_{\mathbb{P}}(\Sigma)$ such that $|\mathbb{A}_A| \leq h$ for all Borel sets A included in $[0,1[^N$.

Condition (ii) is referred to as the covariance property. It expresses the fact that, for the map $A \mapsto \mathbb{A}_A$, the spatial translations are transferred to the dynamic. Condition (iii) is referred to as the domination property.

Let denote by $\mathcal{I}_{int}$ the family of the half open intervals $[a,b[$ with a and b in $\mathbb{Z}^N$. A sequence $(B_n)_{n\in\mathbb{N}}$ of sets of $\mathcal{B}_b(\mathbb{R}^N)$ is said to be regular, if there exists a nondecreasing sequence $(I_n)_{n\in\mathbb{N}}$ of $\mathcal{I}_{int}$ and a constant $C_{reg} > 0$ such that $B_n \subset I_n$ and $\sup_{n\in\mathbb{N}} |I_n|/|B_n| \leq C_{reg}$.

For every $A \in \mathcal{B}_b(\mathbb{R}^N)$, we set $\rho(A) = \sup\{r \geq 0 \ : \ \exists x \in \mathbb{R}^N \ \ \overline{B}_r(x) \subset A\}$, where $\overline{B}_r(x)$ is the closed ball with radius $r > 0$ centered at $x \in \mathbb{R}^N$. The following theorem generalizes the pointwise Birkhoff ergodic theorem. For a proof we refer the reader to (Nguyen and Zessin, 1979, Corollary 4.20).

Theorem C.5. *Let $\mathbb{A}$ be an additive process covariant with respect to $(T_z)_{z\in\mathbb{Z}^N}$, and $(A_n)_{n\in\mathbb{N}}$ a regular sequence of convex sets of $\mathcal{B}_b(\mathbb{R}^N)$ satisfying $\lim_{n\to+\infty}\rho(A_n) = +\infty$. Then, for $\mathbb{P}$-a.e. $\omega \in \Sigma$,*

$$\lim_{n\to+\infty}\frac{\mathbb{A}_{A_n}}{|A_n|}(\omega) = \mathbb{E}^{\mathcal{I}}\mathbb{A}_{[0,1[^N}(\omega).$$

If moreover the dynamical system $(\Sigma, \mathcal{A}, \mathbb{P}, (T_z)_{z\in\mathbb{Z}^N})$ *is ergodic, then*

$$\lim_{n\to+\infty} \frac{\mathbb{A}_{A_n}}{|A_n|}(\omega) = \mathbb{E}\mathbb{A}_{[0,1[^N}.$$

C.2 Subadditive processes

Additive processes and the pointwise convergence result stated in Theorem C.5 can be generalized to subadditive processes.

Definition C.8. Let $(\Sigma, \mathcal{A}, \mathbb{P}, (T_z)_{z\in\mathbb{Z}^N})$ be a dynamical system. A subadditive process indexed by $\mathcal{B}_b(\mathbb{R}^N)$, and covariant with respect to $(T_z)_{z\in\mathbb{Z}^N}$, is a mapping $\mathbb{S} : \mathcal{B}_b(\mathbb{R}^N) \longrightarrow L^1_{\mathbb{P}}(\Sigma)$, $A \mapsto \mathbb{S}_A$, satisfying the four following conditions:

(i) for all $(A, B) \in \mathcal{B}_b(\mathbb{R}^N) \times \mathcal{B}_b(\mathbb{R}^N)$ such that $A \cap B = \emptyset$, $\mathbb{S}_{A\cup B} \le \mathbb{S}_A + \mathbb{S}_B$;
(ii) for all $A \in \mathcal{B}_b(\mathbb{R}^N)$ and all $z \in \mathbb{Z}^N$, $\mathbb{S}_{z+A} = \mathbb{S}_A \circ T_z$;
(iii) there exists a nonnegative function h in $L^1_{\mathbb{P}}(\Omega)$ such that $|\mathbb{S}_A| \le h$ for all Borel sets A included in $[0,1[^N$;
(iv) $\inf\left\{\int_\Sigma \frac{\mathbb{S}_I}{|I|}\, d\mathbb{P} : I \in \mathcal{I}_{int}\right\} > -\infty$.

The following pointwise convergence is obtained first for nonnegative subadditive processes, then by subtracting a suitable additive process to the general subadditive process considered, and applying Theorem C.5. For a proof, we refer the reader to Ackoglu and Krengel (1981), or to (Attouch *et al.*, 2014, Section 12.4.3) for a proof with complete description of the limit.

Theorem C.6. *Let* $\mathbb{S} : \mathcal{B}_b(\mathbb{R}^N) \longrightarrow L^1_{\mathbb{P}}(\Sigma)$ *be a subadditive process covariant with respect to* $(T_z)_{z\in\mathbb{Z}^N}$*, and* $(A_n)_{n\in\mathbb{N}}$ *a regular sequence of convex sets of* $\mathcal{B}_b(\mathbb{R}^N)$ *satisfying* $\lim_{n\to+\infty} \rho(A_n) = +\infty$*. Then, for* $\mathbb{P}$*-a.e.* $\omega \in \Sigma$,

$$\lim_{n\to+\infty} \frac{\mathbb{S}_{A_n}}{|A_n|}(\omega) = \inf_{m\in\mathbb{N}^*} \mathbb{E}^{\mathcal{I}} \frac{\mathbb{S}_{[0,m[^N}}{m^N}(\omega).$$

If moreover the dynamical system $(\Sigma, \mathcal{A}, \mathbb{P}, (T_z)_{z\in\mathbb{Z}^N})$ *is ergodic, then for* $\mathbb{P}$ *almost every* $\omega \in \Sigma$,

$$\lim_{n\to+\infty} \frac{\mathbb{S}_{A_n}}{|A_n|}(\omega) = \inf_{m\in\mathbb{N}^*} \mathbb{E} \frac{\mathbb{S}_{[0,m[^N}}{m^N} = \gamma(\mathbb{S}).$$

Appendix D

Large deviations principle

In the case of almost sure convergence of sequences of random functions, one may be interested in the rate of convergence towards 0 of the tail probabilities. Such a goal is achieved by trying to prove large deviations, which give an exact exponential rate for decay. The definition below of a Large Deviations Principle follows the presentation of Dembo and Zeitouni (1998):

Definition D.9. A family of probability measures $(\mu_n)_{n\geq 0}$ on a topological space $\mathcal{X}$ endowed with its Borel σ-algebra $\mathcal{B}$ satisfies a Large Deviations Principle (LDP) if there exists an increasing speed function $\alpha : \mathbb{N} \to \mathbb{N}$ such that $\lim_{n\to+\infty} \alpha(n) = +\infty$, and a rate function (also called Entropy function) $I : \mathcal{X} \to [0, +\infty]$, lower semicontinuous, such that for all measurable set $B \in \mathcal{B}$,

$$- \inf_{x\in \mathring{B}} I(x) \leq \liminf_{n\to+\infty} \frac{1}{\alpha(n)} \mathrm{Log}(\mu_n(B)) \leq \limsup_{n\to+\infty} \frac{1}{\alpha(n)} \mathrm{Log}(\mu_n(B)) \leq - \inf_{x\in\overline{B}} I(x).$$

The rate function I will be said to be *good* if it is inf-compact, i.e. its level sets $L_I(r) = \{x \in \mathcal{X} \; : \; I(x) \leq r\}$ are compact for all $r \in \mathbb{R}$.

Roughly speaking, the probabilities $\mu_n(B)$ behave like $\exp(-\alpha(n)\, I(B))$, where from now on $I(B)$ denotes $\inf_{x\in B} I(x)$.

We quote in the one-dimensional case ($\mathcal{X} = \mathbb{R}$), the two following important theorems for establishing LDPs. For a proof, we refer the reader to (Dembo and Zeitouni, 1998, Theorem 2.3.6) and (Dembo and Zeitouni, 1998, Theorem 4.4.2) respectively.

Theorem D.7 (Gärtner-Ellis). *Assume that* $\mathcal{X} = \mathbb{R}$. *Suppose that for all* $\lambda \in \mathbb{R}$, *the following limit exists*

$$\Lambda(\lambda) := \lim_{n\to\infty} \frac{1}{\alpha(n)} \mathrm{Log} \left(\int_{\mathbb{R}} e^{\alpha(n)\lambda x} \, \mu_n(dx) \right)$$

as an real extended, and that 0 *belongs to the interior of the domain* $\mathrm{dom}(\Lambda)$ *of* Λ. *Then, denoting by* I *the Legendre-Fenchel conjugate of* Λ,

$$I(x) = \sup_{\lambda\in\mathbb{R}} (\lambda x - \Lambda(\lambda)),$$

we have

(1) for every closed set $F \subset \mathbb{R}$,

$$\limsup_{n\to+\infty} \frac{1}{\alpha(n)} \mathrm{Log}(\mu_n(F)) \leq -I(F),$$

(2) for every open set $G \subset \mathbb{R}$,

$$\liminf_{n\to+\infty} \frac{1}{\alpha(n)} \mathrm{Log}(\mu_n(G)) \geq -I(G \cap E_{\mathrm{ex}}),$$

where E_{ex} is the set of exposed points of I whose exposing hyperplane belongs to the interior of $\mathrm{dom}(\Lambda)$.

We recall that an exposed point of I is a point $x \in \mathbb{R}$ such that there exists $\lambda \in \mathbb{R}$, called exposing hyperplane, such that

$$\forall y \in \mathbb{R} \setminus \{x\}\; I(y) > I(x) + \lambda(y - x).$$

Consequently, one will have a LDP if I is strictly convex on its domain. The LDP also holds when Λ is finite and differentiable everywhere (see (Dembo and Zeitouni, 1998, Exercice 2.3.20)). The following theorem has much stronger hypotheses, but its result is also stronger.

Theorem D.8 (Bryc). *Suppose that for all $f \in \mathcal{C}_b(\mathcal{X})$, the following limit exists,*

$$\Lambda_f = \lim_{n\to+\infty} \frac{1}{\alpha(n)} \mathrm{Log}\left(\int_{\mathbb{R}} e^{\alpha(n)f(x)}\, \mu_n(dx)\right),$$

then $(\mu_n)_{n\geq 0}$ satisfies a LDP with speed α and good rate function I defined by

$$I(x) = \sup_{f\in\mathcal{C}_b(\mathcal{X})} (f(x) - \Lambda_f).$$

Appendix E

Measure theory

E.1 Vector measures

We give a brief summary on the concept of vector measure used to define the time-delays operators in Chapter 3.

Definition E.10. Let $\mathbb{T}$ be a locally compact space and denote by $\mathcal{B}(\mathbb{T})$ its Borel σ-algebra. Given a Banach space $\mathbb{Y}$, we call $\mathbb{Y}$-valued Borel vector measure, any countably additive set function $\mathbf{m} : \mathcal{B}(\mathbb{T}) \to \mathbb{Y}$.

Let $\mathbb{X}$ be a Banach space. We denote by $\mathcal{E}(\mathbb{T}, \mathbb{X})$ the space of step functions, i.e. the space of functions $S : \mathbb{T} \to \mathbb{X}$ of the form $S = \sum_{i \in I} \mathbb{1}_{B_i} S_i$ where I is any finite set, $B_i \in \mathcal{B}(\mathbb{T})$, and $S_i \in \mathbb{X}$. Assume that there exists a continuous bilinear mapping $\mathbf{B} : \mathbb{Y} \times X \ni (u, v) \mapsto \mathbf{B}(u, v) \in \mathbb{X}$, it follows that we can integrate with respect to $\mathbf{m}$, step functions $S \in \mathcal{E}(\mathbb{T}, \mathbb{X})$, and define an element of $\mathbb{X}$, according to the formula

$$\int S d\mathbf{m} := \sum_{i \in I} \mathbf{B}(\mathbf{m}(B_i), S_i) \in \mathbb{X}.$$

The definition of $\int S d\mathbf{m}$ depends only on S and is independent of the particular way in which S is written (see (Dinculeanu, 1967, Chapter II, paragraph 7, 1, Proposition 1)). For every $B \in \mathcal{B}(\mathbb{T})$, set

$$\mu(B) = \sup \left\{ \sum_{i \in I} \|\mathbf{m}(B_i)\|_{\mathbb{Y}} : I \text{ finite}, \{B_i\}_{i \in I} \subset \mathcal{B}(\mathbb{T}) \text{ pairwise disjoint}, \ B = \bigcup_{i \in I} B_i \right\}.$$

Definition E.11. The set function $\mu : \mathcal{B}(\mathbb{T}) \to [0, +\infty]$ is called the variation of $\mathbf{m}$ and is denoted by $\|\mathbf{m}\|$. The vector measure $\mathbf{m}$ has *finite variation* if $\|\mathbf{m}\|(\mathbb{T}) < +\infty$.

The following proposition is an easy consequence of the definitions above.

Proposition E.10. *Let* $\mathbf{m}$ *be a vector measure with finite variation. Then* $\|\mathbf{m}\|$ *is a (scalar) nonnegative Borel measure and, for all* $S \in \mathcal{E}(\mathbb{T}, X)$*, we have*

$$\left\| \int S d\mathbf{m} \right\|_{\mathbb{X}} \leq C_{\mathbf{B}} \sup_{t \in \mathbb{T}} \|S(t)\|_{\mathbb{X}} \|\mathbf{m}\|(\mathbb{T}) \tag{E.1}$$

where $C_{\mathbf{B}}$ *is the norm of the bilinear mapping* $\mathbf{B}$*.*

Let denote by $\mathcal{E}^{\infty}(\mathbb{T}, \mathbb{X})$ the space of functions from $\mathbb{T}$ into $\mathbb{X}$ which are uniform limit of step functions. Then we have

Proposition E.11 (Extension of the integral)**.** *Let* $\mathbf{m}$ *be a* $\mathbb{Y}$*-valued Borel measure with finite variation. Then, the integral* $\int S d\mathbf{m}$*, defined for all* $S \in \mathcal{E}(\mathbb{T}, \mathbb{X})$*, can be extended in a standard way to the space* $\mathcal{E}^{\infty}(\mathbb{T}, \mathbb{X})$*, and for all* $u \in \mathcal{E}^{\infty}(\mathbb{T}, \mathbb{X})$*, we have*

$$\left\| \int u d\mathbf{m} \right\|_{\mathbb{X}} \leq \int \|u\|_{\mathbb{X}} d\|\mathbf{m}\|. \tag{E.2}$$

Sketch of the proof. First recall the standard extension process. Let $u \in \mathcal{E}^{\infty}(\mathbb{T}, \mathbb{X})$, $u = \lim_{n \to +\infty} S_n$ for the uniform norm, where $S_n \in \mathcal{E}(\mathbb{T}, \mathbb{X})$ for all $n \in \mathbb{N}$. Since $(S_n)_{n \in \mathbb{N}}$ is a Cauchy sequence for the uniform norm, from (E.1), $(\int S_n d\mathbf{m})_{n \in \mathbb{N}}$ is a Cauchy sequence in $\mathbb{X}$, thus converges to some element U in $\mathbb{X}$. Let us show that U does not depend on the choice of $(S_n)_{n \in \mathbb{N}}$. Let $(T_n)_{n \in \mathbb{N}}$ be any sequence in $\mathcal{E}(\mathbb{T}, \mathbb{X})$ uniformly converging to u. Then, from (E.1)

$$\left\| \int (T_n - S_n) d\mathbf{m} \right\|_{\mathbb{X}} \leq \sup_{t \in \mathbb{T}} \|(T_n - S_n)(t)\|_{\mathbb{X}} \|\mathbf{m}\|(\mathbb{T}),$$

so that $\lim_{n \to +\infty} \int T_n d\mathbf{m} = \lim_{n \to +\infty} \int S_n d\mathbf{m} = U$. Inequality (E.2) is easy to establish for functions in $\mathcal{E}^{\infty}(\mathbb{T}, \mathbb{X})$ (see (Dinculeanu, 1967, Chapter II, 7, 2, Proposition 2)). For the general case, argue by density. □

Proposition E.11 justifies the notation $\int u d\mathbf{m}$ for all u in $\mathcal{E}^{\infty}(\mathbb{T}, \mathbb{X})$. It is well known that $\mathcal{E}^{\infty}(\mathbb{T}, \mathbb{X})$ contains the space $C_c(\mathbb{T}, \mathbb{X})$ of continuous functions from $\mathbb{T}$ into $\mathbb{X}$ with compact support (see (Dinculeanu, 1967, Chapter III, 14, 5, Corollary 2)). Therefore, $\int u \, d\mathbf{m}$, which is sometimes written $\int_{\mathbb{T}} u(t) d\mathbf{m}(t)$, has a meaning for any $u \in C_c(\mathbb{T}, \mathbb{X})$, and defines an element of $\mathbb{X}$. Note also that for $B \in \mathcal{B}(\mathbb{T})$, and $u \in C_c(\mathbb{T}, \mathbb{X})$, $\mathbb{1}_B u$ is uniform limit of step functions so that $\int_{\mathbb{T}} \mathbb{1}_B u \, d\mathbf{m}$, written $\int_B u d\mathbf{m}$, has a meaning and defines an element of $\mathbb{X}$. We refer the reader to Dinculeanu (1967) for a comprehensive review on vector measures.

E.2 Young measures

For any topological space $\mathbb{T}$, we denote by $\mathcal{B}(\mathbb{T})$ its Borel σ-algebra. For any pair of measure spaces $(\mathcal{M}, \mathcal{A})$, $(\mathcal{M}', \mathcal{A}')$, any measurable map $f : \mathcal{M} \to \mathcal{M}'$ and any Borel measure ν in $(\mathcal{M}, \mathcal{A})$, we denote by $f_{\#}\nu$ its image measure on $(\mathcal{M}', \mathcal{A}')$ (or pushforward measure). Let Ω be an open bounded subset in $\mathbb{R}^N$ and $\mathcal{X}$ be a Polish space, namely, a topological space homeomorphic to a separable complete metric space. We call Young measure on $\Omega \times \mathcal{X}$, any positive measure μ on $\Omega \times \mathcal{X}$ such that its image by the projection π_Ω on Ω is the Lebesgue measure $\mathcal{L}_N \lfloor \Omega$ on Ω: $\pi_{\Omega\#}\mu = \mathcal{L}^N \lfloor \Omega$. We will denote by $\mathcal{Y}(\Omega, \mathcal{X})$ the space of all Young measures on $\Omega \times \mathcal{X}$.

Let denote by $\mathcal{F}c(\Omega, \mathcal{X})$ the space of all Carathéodory integrands, that is, the space of all functions $\psi : \Omega \times \mathcal{X} \to \mathbb{R}$, $\mathcal{B}(\Omega) \otimes \mathcal{B}(\mathcal{X})$ measurable such that $\psi(x, .)$ is bounded continuous on $\mathcal{X}$ for every $x \in \Omega$, and $x \mapsto \|\psi(x, .)\|$ is Lebesgue integrable. We equip $\mathcal{Y}(\Omega, \mathcal{X})$ with the narrow topology, that is the weakest topology which makes the maps

$$\mu \mapsto \int_{\Omega \times \mathcal{X}} \psi \, d\mu$$

continuous, where ψ runs through $\mathcal{F}c(\Omega, \mathcal{X})$.

Definition E.12. A sequence $(\mu_n)_{n \in \mathbb{N}}$ in $\mathcal{Y}(\Omega, \mathcal{X})$ is said to be tight if for all $\varepsilon > 0$ there exists K_ε, compact subset of $\mathcal{X}$, such that $\sup_{n \in \mathbb{N}} \mu_n(\Omega \times K_\varepsilon^c) < \varepsilon$.

Tight sequences are compact in the following sense.

Theorem E.9. *If $(\mu_n)_{n \in \mathbb{N}}$ is a tight sequence in $\mathcal{Y}(\Omega, \mathcal{X})$, then there exists a subsequence which narrow converges in $\mathcal{Y}(\Omega, \mathcal{X})$.*

The theorem below is fundamental. The second assertion is an extension of the continuity on $\mathcal{F}c(\Omega, \mathcal{X})$, to unbounded functions.

Theorem E.10. *(i) (Lower semicontinuity) For any function $\psi : \Omega \times \mathcal{X} \to [0, +\infty]$, $\mathcal{B}(\Omega) \otimes \mathcal{B}(\mathcal{X})$ measurable and such that $\psi(x, .)$ is lower semicontinuous for every x in Ω,*

$$\mu \mapsto \int_{\Omega \times \mathcal{X}} \psi \, d\mu$$

is lower semicontinuous on $\mathcal{Y}(\Omega, \mathcal{X})$,

(ii) (Extension of continuity) Let $\psi : \Omega \times \mathcal{X} \to \mathbb{R}$ be a $\mathcal{B}(\Omega) \otimes \mathcal{B}(\mathcal{X})$ measurable function such that $\psi(x, .)$ is continuous for every x in Ω and $(\mu_n)_{n \in \mathbb{N}}$ a sequence of Young measure narrow converging to μ in $\mathcal{Y}(\Omega, \mathcal{X})$. We assume that

$$\lim_{N \to +\infty} \left(\sup_n \int_{\{\psi \geq N\}} |\psi| \, d\mu_n \right) = 0.$$

Then

$$\int_{\Omega \times \mathcal{X}} \psi \, d\mu = \lim_{n \to +\infty} \int_{\Omega \times \mathcal{X}} \psi \, d\mu_n.$$

In Chapter 6, we systematically apply the disintegration theorem below together with Theorem E.10.

Theorem E.11. *Let $\mathbb{X}$ and $\mathbb{Y}$ be two Polish spaces (complete and separable metric spaces), μ be a Borel measure on the product $\mathbb{X} \times \mathbb{Y}$ and $\nu = \pi_{\mathbb{X}} \# \mu$ the image measure of μ by the projection $\pi_{\mathbb{X}}$ on $\mathbb{X}$. Then there exists a family of probability measure $(\mu_x)_{x \in \mathbb{X}}$ on $\mathbb{Y}$, unique up to equality ν-a.e. such that*

(i) $x \mapsto \int_{\mathbb{Y}} \psi(x,y)\, d\mu_x(y)$ *is ν-measurable,*

(ii) $\int_{\mathbb{X}\times\mathbb{Y}} \psi(x,y)\, d\mu(x,y) = \int_{\mathbb{X}} \left(\int_{\mathbb{Y}} \psi(x,y)\, d\mu_x(y) \right) d\nu(x)$

for each μ-integrable function ψ. The family $(\mu_x)_{x \in \mathbb{X}}$ is called a disintegration of the measure μ, and we write $\mu = \nu \otimes \mu_x$.

For the proof in the case where μ is a Young measure, i.e. $\nu = \mathcal{L}^N \lfloor \Omega$, we refer to (Attouch *et al.*, 2014, Theorem 4.2.4). For a proof in the general case, we refer to Tortrat (1977). For a general exposition of the theory of Young measures, and proofs of Theorems E.9, E.10, we refer the reader to Valadier (1990, 1994) and references therein. For a general study on Young measures generated by gradients, we refer the reader to Pedregal (1997).

Appendix F

Inf-convolution and parallel sum

Beyond the classical operations on extended real-valued functions, the following operation corresponds to the sum of the strict epigraph.

Definition F.13. Let V be a linear space and $f, g : V \to \mathbb{R} \cup \{+\infty\}$ two extended real-valued functions. The inf-convolution of f and g is the function $f \Box g : V \to \overline{\mathbb{R}}$ defined by

$$f \Box g(u) = \inf_{v+w=u} \{f(v) + g(w)\} .$$

For a proof of the following result we refer to (Attouch *et al.*, 2014, Propositions 9.2.1, 9.2.2). For other basic results on inf-convolution, consult Stromberg (1994) and references therein.

Proposition F.12. *i) Let $f, g : V \to \mathbb{R} \cup \{+\infty\}$ be two extended real-valued functions. Then*

$$\mathrm{epi}_S(f \Box g) = \mathrm{epi}_S f + \mathrm{epi}_S g,$$

where

$$\mathrm{epi}_S f := \{(v, \lambda) \in X \times \mathbb{R} : \lambda > f(v)\} .$$

ii) Let $f, g : V \to \mathbb{R} \cup \{+\infty\}$ be two convex functions, then their inf-convolution $f \Box g$ is still a convex function.

From now on, V is a Banach space and V^* is its topological dual.

Definition F.14. Let A and B be two operators of $V \times V^*$. The parallel sum of A and B is defined by

$$A \parallel B = (A^{-1} + B^{-1})^{-1}.$$

By using the definition of the inverse operator (see Appendix B.4), it is easily seen that,

$$A \parallel B = \{(u, u^*) \in X \times X^* : \exists (v, w) \in V \times V,\ v + w = u \text{ and } u^* \in A(v) \cap B(w)\} .$$

Theorem F.12. *Let V be a reflexive Banach space. Assume that $f : V \to \mathbb{R} \cup \{+\infty\}$ is a convex proper lower semicontinuous function and that $g : V \to \mathbb{R}$ is convex and Fréchet (respectively Gâteaux) differentiable. Then $f \Box g$ is Fréchet (respectively Gâteaux) differentiable and*

$$D(f \Box g) = \partial f \parallel Dg \text{ (respectively } \nabla(f \Box g) = \partial f \parallel \nabla g).$$

For a proof, we refer the reader to (Stromberg, 1994, Theorem 3.7).

Bibliography

Ackoglu, M. A. and Krengel, U. (1981). Ergodic theorem for superadditive processes. *J. Reine Angew. Math.* **323**, pp. 53–67.

Allaire, G., Mikelic, A. and Piatnitski, A. (2010). Homogenization approach to the dispersion theory for reactive transport through porous media. *SIAM J. Math. Anal.* 42, pp. 125–144.

Allaire, G., Pankratova, I. and Piatnitski, A. (2012). Homogenization of a nonstationary convection-diffusion equation in a thin rod and in a layer. *SeMA Journal* 58(1), pp. 53–95.

Allaire, G., Pankratova, I. and Piatnitski, A. (2012). Homogenization and concentration for a diffusion equation with large convection in a bounded domain. *J. Funct. Anal.* 262, pp. 300–330.

Amann, H. (2000). Compact embeddings of vector-valued Sobolev and Besov spaces. *Glasnik Mathematicki* 35, pp. 161–177.

Andreu-Vaillo, F., Mazón, J. M., Rossi and J. D., Toledo-Melero, J.J. (2010). Nonlocal Diffusion Problems. *Mathematical Surveys and Monographs* Vol. 165. American Mathematical Society and Real Sociedad Mathemática Espanola.

Anza Hafsa, O., Mandallena, J. P., and Michaille, G. (2019). Stability of a class of nonlinear reaction-diffusion equations and stochastic homogenization. *Asymptot. Anal.* vol. 115, no. 3-4, pp. 169–221.

Anza Hafsa, O., Mandallena, J. P., and Michaille, G. (2020). Convergence of a class of nonlinear distributed time delays reaction-diffusion equations and stochastic homogenization. *Nonlinear Differ. Equ. Appl.* 27, 20, 2020. https://doi.org/10.1007/s00030-020-0626-y.

Anza Hafsa, O., Mandallena, J. P., and Michaille, G. (2020). Continuity theorem for nonlocal functionals indexed by Young measures *J. Math. Pures Appl.* **136**, pp. 158–202.

Anza Hafsa, O., Mandallena, J. P., and Michaille, G. (2021). Convergence and stochastic homogenization of a class of two components nonlinear reaction-diffusion systems. *Asymptot. Anal.*, vol. 121, no. 3-4, pp. 259–305.

Anza Hafsa, O., Mandallena, J. P., and Michaille, G. (2021). Convergence and stochastic homogenization of nonlinear integrodifferential reaction-diffusion equations via Mosco×Γ-convergence. *To appear in Annales mathématiques Blaise Pascal* MIPA. Hal-02357210, 2019.

Anza Hafsa, O. and Mandallena, J. P. (2003). Interchange of infimum and integral. *Calc. Var. Partial Differential Equations* 18(4), pp. 433–449.

Armstrong, S., Kuusi, T., and Mourat, J. C. (2019). Quantitative Stochastic Homogenization and Large-Scale Regularity. *Grundlehren der mathematischen 352, A Series of Comprehensive Studies in Mathamatics, Springer.*

Armstrong, S. and Souganidis, P. E. (2012). Stochastic homogenization of Hamilton-Jacobi and degenerate Bellman equations in unbounded environments. *J. Math. Pures Appl.* (9), 97 (5), pp. 460–504.

Armstrong, S. and Souganidis, P. E. (2013). Concentration phenomena for neutronic multigroup diffusion in random environments. *Ann. Inst. H. Poincaré Anal. Non Linéaire* 30, pp. 419–439.

Attouch, H. (1977). Convergences de fonctionnelles convexes Journées d'Analyse Non Linéaire, Besançon. *Lecture Notes in Mathematics, Springer-Verlag* **665**, pp. 1–40.

Attouch, H. (1984). Variational convergence for functions and operators. *Applicable Mathematics Series. Pitman (Advanced Publishing Program)* Boston, MA.

Attouch, H., Buttazzo G., and Michaille, G. (2014). Variational analysis in Sobolev and BV spaces, Applications to PDEs and optimization. *MOS-SIAM Series on Optimization. Society for Industrial and Applied Mathematics (SIAM), Philadelphia, PA; Mathematical Optimization Society, Philadelphia, PA, second edition.*

Banks R. B. (1994). Growth and diffusion phenomena. *Texts in Applied Mathematics.* Springer-Verlag, Berlin Mathematical frameworks and applications.

Barbu, V. (1976). Nonlinear semigroups and differentail equations in Banach spaces. *Editura Academiei, Bucaresti Roumania, Noordhoff International Publishing, Leyden the Netherlands, 1976.*

Barbu, V., Malik, M. A. (1979). Semilinear Integro-Differential Equations in Hilbert Space. *Journal of Mathematics Analysis and Applications* bf 67, pp. 462–475.

Beer, G. (1993). Topologies on Closed and Closed Convex Sets. *Mathematics and Its Applications* Vol. 263, Kluwer.

Beer, G. (1994). Wijsman Convergence: A survey. *Mathematics and Its Applications, Spinger Set-Valued Analysis* **2**, issue 1-2, pp. 77–94.

Berestycki, H., Lions, P. L. (1980). Some applications of the method of super and subsolutions. *Bifurcation and nonlinear eigenvalue problems (Proc., Session, Univ. Paris XIII, Villetaneuse, 1978)*, vol. 782 of *Lecture Notes in Math.*, pp. 16–41. Springer, Berlin.

Berestycki, H. and Lions, P. L. (1980). Une méthode locale pour l'existence de solutions positives de problèmes semi-linéaires elliptiques dans $\mathbf{R}^N$. *J. Analyse Math.* 38, pp. 144–187.

Berezansky, L., and Braverman, E. (2006). Mackey-Glass Equation with Variable Coefficients. *Computers and Mathematics with Applications* **51**: 1–16.

Bouleau, N. (2000). Processus stochastiques et applications. *Collection Méthodes. Hermann.*

Braides, A. (2002). Γ-convergence for beginners. *Oxford Lecture Ser. Math. Appl.* **22**, Oxford University Press, Oxford, UK.

Brézis, H. (1973). Opérateurs maximaux monotones et semi-groupes de contractions dans les espaces de Hilbert. *North-Holland Publishing Co., Amsterdam-London; American Elsevier Publishing Co., Inc., New York.* North-Holland Mathematics Studies, No. 5. Notas de Matemática (50).

Brézis, H., and Ekeland, I. (1976). Un principe variationel associé à certaines équations paraboliques, Le cas dépendant du temps. *C.R. Acad. Sc. Paris* 282(A), pp. 1197–1198.

Brézis, H. and Pazy, A. (1972). Convergence and Approximation of Semigroups of Nonlinear Operators in Banach Spaces. *Journal (of Functional Analysis)* **9**, pp. 63–74.

Bryc, W. and Dembo, A. (1996). Large deviations and strong mixing. *Ann. Inst. Henri Poincaré, Probabiliés et Statistiques* 32(4), pp. 549–569.

Cardaliaguet, P., Lions, P. L., Souganidis, P. E. (2009). A discussion about the homogenization of moving interfaces. *J. Math. Pures Appl.* (9), 91 (4), pp. 339–363.

Cardaliaguet, P. and Souganidis, P. E. (2013). Homogenization and enhancement of the G-equation in random environments. *Comm. Pure Appl. Math.*, 66 (10), pp. 1582–1628.

Cazenave, T., and Haraux, A. (1998). An introduction to Semilinear Evolution Equation. *Oxford University Press*, Oxford Lecture Series in Mathematics and its Applications (13).

Chabi, E. and Michaille, G. (1994). Ergodic theory and application to nonconvex homogenization. *Set-Valued Anal.* Set convergence in nonlinear analysis and optimization, 2(1-2), pp. 117–134.

Chang, J. C. (2007). Local existence of retarded Volterra integrodifferential equations with Hille-Yosida operators. *Nonlinear Analysis* 66, pp. 2814–2832.

Chang, N., Chipot, M. (2003). Nonlinear nonlocal evolution problems. *Real Academia de Ciencias Exactas, Fisicas y Naturales. Revista. Serie A, Matematicas* 97(3), pp. 423–445.

Chang, N., Chipot, M. (2004). On some mixed boundary value problems with nonlocal diffusion. *Advances in Mathematical Sciences and Applications* 14 (1), pp. 1–24.

Chang, J. C., Lang, C. L., Shih, C-L., Yeh, C-N. (2010). Local existence of some nonlinear Volterra integrodifferential equations with infinite delay. *Taiwanese Journal of Mathematics* Vol. 14, No. 1, pp. 131–152.

Chipot, M., Lovat, M. (2001). Existence and uniqueness results for a class of nonlocal elliptic and parabolic problems. *Dynamics of Continuous, Discrete & Impulsive Systems. Series A. Mathematical Analysis* 8(1), pp. 35–51.

Chipot, M., Zhang, M. (2021). On some model problem for the propagation of interacting species in a special environment. *Discrete & Continuous Dynamical Systems* 41 (7), pp. 3141–3161. doi: 10.3934/dcds.2020401.

Cioranescu, D. and Piatnitski, A. (2006). Homogenization of porus medium with randomly pulsating microstructure. *Multiscale Model. Simul.* Vol. 5, **1**, pp. 170–183.

Ciomaga, A., Souganidis, P. E. and Tran H. V. (2015). Stochastic homogenization of interfaces moving with changing sign velocity. *J. Diff. Equations* 258, (4), pp. 1025–1057.

Collins, C. (1983). Length dependence of solutions of FitzHugh-Nagumo equations. *Trans. Amer. Math. Soc. Vol. 280*, **2**, pp. 809–832

Crandall, M. G., Londen, S. O. and Nohel, J. A. (1978). An Abstract Nonlinear Volterra Integrodifferential Equation. *J. Math. Anal. Appl.* **164**, pp. 701–735.

Dacorogna, B. (1989). Direct methods in the calculus of variations. *Appl. Math. Sci.* **78**, Springer-Verlag, Berlin.

Dal Maso, G. (1993). An introduction to Γ-convergence. *Progress in Nonlinear Differential Equations and their Applications* 8. Birkhäuser Boston Inc., Boston, MA.

Dal Maso, G. and Modica, L. (1986). Nonlinear stochastic homogenization and ergodic theory. *J. Reine Angew. Math.* **363**, pp. 27–43.

Dauvergne, F. (2006). Méthode de discrétisation pour la modélisation par éléments analytiques en hydrogéologie quantitative. Application aux écoulements en régimes permanents et transitoires. *Thèse de doctorat* Ecole Nationale Supérieure des Mines de Saint-Etienne. HAL Id: tel-00797147 https://tel.archives-ouvertes.fr/tel-00797147.

Dembo, A. and Zeitouni, O. (1998). Large Deviations Techniques and Applications. *Springer-Verlag New York Inc.* second edition.

Dikansky, A. (2005). FitzHugh-Nagumo equations in a nonhomogeneous medium. *Disc. Cont. Dyn. Sys. Supplement Volume (200)*, pp. 216–224.

Dinculeanu, N. (1967). Vector measures. *International Series of Monographs in Pure and Applied Mathematics* Vol. 95. Pergamon Press, Oxford-New York-Toronto, Ont.; VEB Deutscher Verlag der Wissenschaften, Berlin.

Duerinckx, M. and Gloria, A. (2016). Stochastic homogenization of nonconvex unbounded integral functionals with convex growth. *Arch. Rat. Mech. Anal.* 221 (3), pp. 1511–1584.

Duerinckx, M., Gloria, A., and Otto, F. (2020). The structure of fluctuations in stochastic homogenization. *Commun. Math. Phys.* 377, pp. 259–306.

Dupuit, J. (1863). Etudes théoriques et pratiques sur le mouvement des eaux dans les canaux découverts. *Paris, Dunod Editeur* BNF Gallica.

Friesecke, G. (1992). Exponentially growing solutions for a delay-diffusion equation with negative feedback. *J. Differential Equations* pp. 1–18.

Friesecke, G. (1993). Convergence to equilibrium for delay-diffusion equations with small delay. *Dynamic. Differential Equations* **5**, pp. 89–103.

Gloria, A., Neukamm, S., and Otto, F. (2015). Quantification of ergodicity in stochastic homogenization: optimal bounds via spectral gap on Glauber dynamics. *Invent. Math.* **199**(2), pp. 455–515.

Gloria, A., Neukamm, S., Otto, F. (2019). Quantitative estimates in stochastic homogenization for correlated coefficient fields. *arXiv:1910.05530v1.*

Gloria, A. and Otto, F. (2011). An optimal variance estimate in stochastic homogenization of discrete elliptic equations. *Ann. of Probab.* 39, No. 3, 779–856.

Gloria, A. and Otto, F. (2021). The corrector in stochastic homogenization: optimal rates, stochastic integrability, and fluctuations. *arXiv:1510.08290.*

Grasselli, M. and Lorenzi, A. (1991). Abstract nonlinear Volterra integrodifferential equations with nonsmooth kernels. *Atti della Accademia Nazionale dei Lincei. Classe di Scienze Fisiche, Matematiche e Naturali. Rendiconti Lincei. Matematica e Applicazioni* 2(1), pp. 43–53.

Grimmett, G. (1999). Percolation *Grundlehren der Mathematischen Wissenschaften [Fundamental Principles of Mathematical Sciences].* vol. 321, Springer-Verlag, Berlin, second edition.

Gudino Rojas, E. A. (2014). Recent developments in non-Fickian diffusion: a new look at viscoelastic materials. *These de Doutoramento em Matematica apresentada ao Departamento de Matematica da Faculdade de Ci ncias e Tecnologia da Universidade de Coimbra.*

Huang, W. (1998). Global Dynamics for a Reaction-Diffusion Equation with Time Delay. *journal of differential equations* **143**, pp. 293–326.

Iosifescu, O., Licht, C. and Michaille, G. (2001). Variational limit of a one dimensional discrete and statistically homogeneous system of material points. *Asymptot. Anal.* **28**, pp. 309–329.

Kosygina, E., Rezakhanlou, F. and Varadhan, S. R. S. (2006). Stochastic homogenization of Hamilton-Jacobi-Bellman equations. *Comm. Pure Appl. Math.* 59 (10), pp. 1489–1521.

Kozlov, S. M. (1979). The averaging of random operators. *Mat. Sb. (N.S.)* 109(151), pp. 188–202.

Londen, S. O. (1977). On an integral equation in a Hilbert space. *SIAM J. Math. Anal.* **8**, pp. 950–970.

Luckhaus, S. (1986). Global boundedness for a delay-differential equation. *Trans. Amer. Math. Soc.* **294**, pp. 767–774.

MacCamy, R. C. (1976). Stability theorems for a class of functional differential equations. *SIAM J. Appl. Math.* 30(3), pp. 557–576.

Mascharenas, M. L. (1993). Memory effect phenomena and Γconvergence. *Proc. Royal. Soc. Edinburgh Sec. A* **123**, pp. 311–322.

Messaoudi, K. and Michaille, G. (1994). Stochastic homogenization of nonconvex integral functionals. *Math. Modelling Numer. Anal.* **28**, no. 3, pp. 329–356.

Michaille, G., Michel, J. and Piccinini, L. (1998). Large deviations estimates for epigraphical superadditive processes in stochastic homogenization. *Prepublication ENS Lyon* 220.

Michaille, G. and Valadier, M. (2002). Young measures generated by a class of integrand: a narrow epicontinuity and applications to Homogenization. *J. Math. Pures Appl.* **81**, pp. 1277–1312.

Mielke, A., Reichelt, S. and Thomas, M. (2014). Two-scale homogenization of nonlinear reaction-diffusion systems with slow diffusion. *Netw. Heterog. Media* 9(2), pp. 353–382.

Moreau, J. J. (1967). Fonctionnelles convexes. *Cours Collège de France 1967, new edition CNR* Facoltà di Ingegneria di Roma, Roma, 2003.

Mosco, U. (1969). Convergence of convex sets and of solutions of variational inequalities. *Advances in Math.* 3, pp. 510–585.

Mosco, U. (1971). On the continuity of the Young-Fenchel transform. *J. Math. Anal. Appl.* 35, pp. 518–535.

Murray, J. D. (2002). Mathematical Biology: I. An Introduction, Third Edition. *Interdisciplinary Applied Mathematics* Vol. 17, Springer-Verlag New York Berlin Heidelberg.

Nandakumaran, A.K. and Visintin, A. (2015) Variational approach to homogenization of doubly-nonlinear flow in a periodic structure. *Nonlinear Analysis: TMA* **120**, pp. 14–29.

Neukamm, M., Schäffner and Schlömerkemper, A. (2017). Stochastic homogenization of nonconvex discrete energies with degenerate growth. *SIAM J. Math. Anal.* **49**, pp. 1761–1809.

Nguyen, X. X. and Zessin, H. (1979). Ergodic theorems for spatial processes. *Z. Wah. Vew. Gebiette* **48**, pp. 133–158.

Okubo, A. (1980). Diffusion and ecological problems: mathematical models. *Volume 10 of Biomathematics.* Springer-Verlag, Berlin-New York, 1980. An extended version of the Japanese edition, ıt Ecology and diffusion, Translated by G. N. Parker.

Pao, C. V. (1992). Nonlinear parabolic and elliptic equations. *Plenum Press* New York.

Papanicolaou, G. C. (1995). Diffusion in Random Media. *In: Keller, J.B., McLaughlin, D.W., Papanicolaou, G.C. (eds.)* Surveys in Applied Mathematics. Springer, Boston, MA.

Pedregal, P. (1997). Parametrized measures and variational principle. *Birkhäuser Boston.*

Perthame, B., Souganidis, P. E. (2011). A homogenization approach to flashing ratchets. *Nonlinear Differ. Equ. Appl.* 18, pp. 45–58.

Peter, M. A. (2009) Coupled reaction-diffusion processes inducing an evolution of the microstructure: analysis and homogenization. *Nonlinear Anal.* 70(2), pp. 806–821.

Piatnitski, A. and Zhizhina, E. (2017). Periodic homogenization of non-local operators with a convolution type kernel. *SIAM J. Math. Anal.* Vol. 49, No. 1, pp. 64–81.

Piatnitski, A. and Zhizhina, E. (2019). Homogenization of biased convolution type operator. *Asymptot. Anal.* vol. 115, no. 3-4, pp. 241–262.

Pohožaev, S. I. (1960). The Dirichlet problem for the equation $\Delta u = u^2$. *Soviet Math. Dokl.* 1, pp. 1143–1146.

Pujo-Menjouet, L. (2015). Blood Cell Dynamics: Half of a Century of Modelling. *Math. Model. Nat. Phenom.* Vol. 10, **6** pp. 182–205.

Reichelt, S. (2015). Two-Scale Homogenization of Systems of Nonlinear Parabolic Equations. *Dissertation, zur Erlangung des akademischen Grades doctor rerum naturalium (Dr. rer. nat.)* eingereicht an der Mathematisch-Naturwissenschaftlichen Fakultät der Humboldt-Universität zu Berlin.

Rennolet, C. (1979). *Existence and Boundedness of Solutions of Abstract Nonlinear lntegrodifferential Equations of Nonconvolution Type.* J. Math. Anal. Appl. **70**, pp. 42–60.

Rockefellar, R. T. (1966). Extension of Fenchel's duality theorem. *Duke Math. J.* **33**, pp. 81–89.

Ruan, S. (2006). Delay differential equations in single species dynamics. *Delay differential equations and applications*, vol. 205 of *NATO Sci. Ser. II Math. Phys. Chem.*, pp. 477–517. Springer, Dordrecht.

Ruan, S. and Wu, J. (1994). Reaction-Diffusion equations with infinite delay. *Canad. App. Math. Quart.* Vol. 2, **4** pp. 485–550.

Shonbek, M. E. (1978). Boundary Value Problems for the FitzHugh-Nagumo equations. *J. Differ. Equat.* **30**, pp. 119–147.

Souganidis, P. E. (1999). Stochastic homogenization of Hamilton-Jacobi equations and some applications. *Asymptot. Anal.*, 20 (1), pp. 1–11.

Strömberg, T. (1994) A Study of the Operation of Infimal Convolution. *Doctoral Thesis Department of Mathematics, Lule University of Technology S - 971 87* Lule, Sweden 1994.

Tartar, L. (1989). Nonlocal effects induced by homogenization. *Partial Differential Equations and the Calculus of Variation, Essay in Honor of Ennio De Giorgi II, Birkhäuser*, Boston, pp. 925–938.

Tartar, L. (1990). Memory effects and homogenization. *Arch. Rat. Mech. Anal.* 111 **2**, pp. 121–133.

Toader, A. M. (1999). Memory effect phenomena and Γconvergence: A time dependant case. *J. Convex Anal.* **6**, pp. 13–27.

Tortrat, A. (1977). Désintégration d'une probabilité. Statistiques exhaustives. *Séminaire de probabilités (Strasbourg)*, tome 11, pp. 539–565.

Tsoularis, A. (2001). Analysis of logistic growth models. *volume 2 of Research Letters in the Information and Mathematical Sciences*, Massey University, pp. 23–46.

Turchin, P. (2015). Quantitative Analysis of Movement: Measuring and Modeling Population Redistribution in Animals and Plants. *Beresta Books.*

Valadier, M. (1990). Young measures. *Methods of Nonconvex Analysis (A. Cellina ed.), Lecture Notes in Math.* **1446**, Springer-Verlag, Berlin pp. 152–188.

Valadier, M. (1994). A course on Young measures. *Workshop di Teoria della Misura e Analisi Reale, Grado, September 19–October 2, 1993, Rend. Istit. Mat. Univ. Trieste* **26** suppl., pp. 349–394.

Visintin, A. (2008). Extension of the Brezis-Ekeland-Nayroles principle to monotone operators. *Adv. Math. Sci. Appl.*, **18** pp. 633–650.

Wu, J. H. (1991). Semigroup and integral form of partial differential equations with infinite delay. *Differential Integral Equations* **4**, pp. 1325–1352.

Zuyev, S. A. and Quintanilla, J. (2003). Estimation of percolation thresholds via percolation in inhomogeneous media. *J. Math. Phys.* 44(12), pp. 6040–6046.

Zuyev, S. A. and Sidorenko, A. F. (1985). Continuous models of percolation theory. II *TMF*, Volume 62,Number 2, pp. 253–262.

Notation

$\#(\tau(\mathbb{R}^N))$ the cardinal of the set $\tau(\mathbb{R}^N)$. 78

$\mathbb{E}^{\mathcal{I}}\mathbf{h}$ the conditional expectation of $\mathbf{h}$ with respect to the σ-algebra $\mathcal{I}$. 194

$\mathcal{C}([0,T],X)$ the space of continuous functions from $[0,T]$ to X. 14

$\mathcal{C}_c((-\infty,T],L^2(\Omega))$ the space of continuous functions from $(-\infty,T]$ to $L^2(\Omega)$ with compact support. 76

$\mathcal{C}_\eta((-\infty,T],L^2(\Omega))$ the subset of $C_c((-\infty,T],L^2(\Omega))$ made up of functions u which are absolutely continuous on $[0,T]$, and whose restriction to $(-\infty,0]$ is equal to a given function $\eta \in C_c((-\infty,0],L^2(\Omega))$. 77

$\mathbf{Conv}_{\alpha,\beta}$ the class of functions satisfying conditions (D). 195

$\dot{u}$, $\dot{U}$ the distributional derivative with respect to t of $u \in H^1(0,T,L^2(\Omega))$ and $U \in H^1(0,T,L^2_\mu(\Omega \times \mathcal{X}))$. 174

$\mathbb{E}_\lambda$ the expectation operator with respect to the Poisson probability measure $\mathbb{P}_\lambda$. 215

$\xi \odot \xi'$ the Hadamard (or Schur) product of two elements ξ and ξ' in $\mathbb{R}^d$. 74

$\mathcal{H}_{N-1}$ the $N-1$-dimensional Hausdorff measure. 21

$T_z\#\mathbb{P}$ the image (or push-forward) measure of $\mathbb{P}$ by T_z. 194

$(G+K_\theta)\Box(H+K_{1-\theta})$ the inf-convolution (or epi-sum) of the functionals $G+K_\theta$ and $H+K_{1-\theta}$. 269

ψ^λ the Moreau-Yosida approximation of index λ of the extended real-valued function ψ. 50

$\mathbf{M}_1^+(\mathbb{R},L^\infty(\Omega))$ the set of positive vector Borel measures $\mathbf{m} : \mathcal{B}(\mathbb{R}) \to L^\infty(\Omega)$ satisfying $\|\mathbf{m}\|(\mathbb{R}) \leq 1$. 75

$\|\cdot\|_X$ the norm in the Hilbert space X. 13

$|\cdot|$ the norms of the euclidean spaces $\mathbb{R}^d$, $d \geq 1$. 13

$\partial(G+K_\theta) \parallel \partial(H+K_{1-\theta})$ the parallel sum of the operators $\partial(G+K_\theta)$ and $\partial(H+K_{1-\theta})$. 272

$\mathcal{P}_+(\mathbb{R})$ the set of Borel probability measures on $\mathbb{R}$ concentrated on $\mathbb{R}_+$. 80

$\xi \cdot \xi'$ the standard scalar product of two elements ξ, ξ' in $\mathbb{R}^d$. 13

$\langle \cdot,\cdot \rangle$ the scalar product in the Hilbert space X. 13

$L^1_{\mathbb{P}}(\Sigma)$ the space of $\mathbb{P}$-integrable real valued functions. 194

$L^2(0,T,X)$ the space of measurable functions on $(0,T)$ with value in X such that

$\|u\|_X^2$ is integrable. 13

$W^{1,1}(0,T,X)$ the space of absolutely continuous functions on $(0,T)$ with value in the Hilbert space X. 14

$H^1(0,T,L^2(\Omega))$ the space of function u in $L^2(0,T,L^2(\Omega))$ such that $\dot{u}$ belongs to $L^2(0,T,L^2(\Omega))$ in the distributional sense. 174

$\mathcal{V}(0,T,L^2(\Omega))$ the subspace of $H^1(0,T,L^2(\Omega))$ made up of the functions u satisfying $u(0,\cdot)=0$, equipped with the weak convergence of $H^1(0,T,L^2(\Omega))$. 174, 175

$H^1(0,T,L^2_\mu(\Omega\times\mathcal{X}))$ the space of function U in $L^2(0,T,L^2_\mu(\Omega\times\mathcal{X}))$ such that $\dot{U}$ belongs to $L^2(0,T,L^2_\mu(\Omega\times\mathcal{X}))$ in the distributional sense, where $\mu\in\mathcal{Y}(\Omega,\mathcal{X})$. 174

$\mathcal{V}(0,T,L^2_\mu(\Omega\times\mathcal{X}))$ the subspace of $H^1(0,T,L^2_\mu(\Omega\times\mathcal{X}))$ made up of the functions U satisfying $U(0,\cdot,\cdot)=0$. 174

$\mathcal{E}^\infty(\mathbb{R},L^2(\Omega))$ the space of functions $u:\mathbb{R}\to L^2(\Omega)$ which are uniform limit of step functions. 75

$\mathcal{E}([-M,T],L^2(\Omega))$ the space of step functions $v:[-M,T]\to L^2(\Omega)$. 225

$\partial\Phi$ the subdifferential of the scalar function Φ. 13

$\mathcal{Y}(\Omega,\mathcal{X})$ the topological space of Young measures on $\Omega\times\mathcal{X}$ equipped with the narrow topology. 174

$\mu=dx\otimes\mu_x$ writing of the Young measure $\mu\in\mathcal{Y}(\Omega,\mathcal{X})$ with its disintegration $(\mu_x)_{x\in\Omega}$. 174

Index